Multivariable Analysis

Satish Shirali · Harkrishan Lal Vasudeva

Multivariable Analysis

 Springer

Satish Shirali
Indian Institute of Science
Education and Research (Mohali)
Panchkula
India
satishshirali@usa.net

Harkrishan Lal Vasudeva
Indian Institute of Science
Education and Research (Mohali)
Chandigarh
India
harkrishanvasudeva@yahoo.co.in

ISBN 978-0-85729-191-2   e-ISBN 978-0-85729-192-9
DOI 10.1007/978-0-85729-192-9
Springer London Dordrecht Heidelberg New York

British Library Cataloguing in Publication Data
A catalogue record for this book is available from the British Library

Mathematics Subject Classification: 97I40, 97I50, 97I60

Printed on acid-free paper

Springer is part of Springer Science+Business Media (www.springer.com)

Preface

A thorough knowledge of multivariable analysis is an essential prerequisite for graduate studies in mathematics. The subject is presented in this book in a manner that would suit readers having a background of calculus in two and three variables, mathematical analysis in one variable, including compactness, and rudiments of matrices and determinants. The prerequisites with essential details are listed briefly in Chapter 1.

In Chapter 2, after a brief discussion of the basic algebraic theory of functions defined on subsets of $\mathbb{R}^n$ and having values in $\mathbb{R}^m$, the concepts of limit and continuity of these functions are defined. Also discussed is the invertibility of linear maps, which is a fundamental concern in the inverse and implicit function theorems at a subsequent stage. The chapter ends with a brief discussion of double sequences and series. Differentiation of functions from (subsets of) $\mathbb{R}^n$ into $\mathbb{R}^m$, their partial derivatives and equality of 'mixed' partial derivatives of second order are discussed in the next chapter. The approach to the inverse and implicit function theorems in Chapter 4 is via contraction mappings in $\mathbb{R}^n$. Use of compactness of a closed ball has been avoided, as it does not lend itself to the situation when $\mathbb{R}^n$ is replaced by an infinite-dimensional space. In the final section of the chapter, a second form of the implicit function theorem has been proved using the concept of connectedness, and also a two-dimensional version that is not a special case of the one in higher dimensions. The purpose of Chapter 5, on extrema, is to discuss from a theoretical perspective the methods of optimisation (determining points of extremum), constrained as well as unconstrained, of functions of several real variables. The reader is presumed to be familiar with the (pre-analysis) calculus techniques of solving optimisation problems in several variables, including the method of Lagrange multipliers. Instances are given when the Lagrange method appears to 'fail'. More important, a sufficient condition for a constrained extremum is proved, which few other books seem to cover. The next two chapters are devoted to Riemann integration and the transformation (change of variables) formula in $\mathbb{R}^n$. Fubini's theorem for continuous functions is also included. The treatment in Chapters 6 and 7 is more leisurely than elsewhere, and the details differ in some essential ways. Chapter 8 treats differential forms, chains and the general Stokes theorem in $\mathbb{R}^n$ without assuming the reader has a background in multilinear algebra. The formal introduction to the concepts involved is preceded by heuristic considerations in terms of vector analysis, which the reader is presumed to have encountered in calculus. The final section discusses the connections of differential forms with vector analysis in greater detail than is customary. The book closes with Chapter 9, in which solutions to most of the problems are presented, some in greater detail than others.

The book contains very general and complete versions of a number of important theorems and constructions. In order to mitigate the difficulties faced by the reader in assimilating the sophisticated versions of these, we have considered it expedient to include appropriate motivation. Complete definitions, explanations and proofs have been provided throughout. A large number of illustrative examples and problems for solution form an integral part of any book intended for self study or a course text. Accordingly, the book has a liberal sprinkling of both, with elaborate hints or solutions for most of the problems.

The reader with previous experience of the subject who wishes to find something different in this book is invited to browse the following items:

> Problem 2-3.P14, Problem 2-3.P15, Examples 3-2.3, Problem 3-3.P14, Remark 3-4.10, Problems 3-4.P22 and 3-4.P23, Problem 4-1.P10, Problem 4-1.P11, Remark 4-2.5, Examples 4-3.4, Theorem 4-4.1, Remark 4-4.3, Example 4-4.6, Problem 4-4.P3, Example 5-1.5(c), Theorem 5-2.9, Proposition 6-5.1, Problem 6-5.P3, Proposition 7-1.1, Proposition 7-1.3, Example 7-4.16(b), Examples 8-2.2(f) and (h), Problem 8-6.P7, parts of Section 8-7 and the problems therein.

Chapter 8 was written by Harkrishan L. Vasudeva with help from Satish Shirali.

Contents

1

Preliminaries

We shall find it convenient to use logical symbols such as $\forall$, $\exists$, $\ni$, $\Rightarrow$ and $\Leftrightarrow$. These are listed below with their meanings. A brief summary of set algebra, functions, elementary real analysis, matrices and determinants, which will be used throughout this book, is included in this chapter. Our purpose is descriptive and no attempt has been made to give proofs of the results stated. The reader is expected to be familiar with the material.

The words 'set', 'class', 'collection' and 'family' are regarded as synonymous and no attempt has been made to define these terms.

1-1 Sets and Functions

Throughout this book, the following commonly used symbols will be employed:

$\forall$ means 'for all' or 'for every'

$\exists$ means 'there exists'

$\ni$ means 'such that'

$\Rightarrow$ means 'implies that' or simply 'implies'

$\Leftrightarrow$ means 'if and only if'.

The concept of *set* plays an important role in every branch of modern mathematics. Although it is easy and natural to define a set as a collection of objects, it has been shown that this definition leads to a contradiction. The notion of set is therefore left undefined, and a set is described by simply listing its elements or by its properties. Thus $\{x_1, x_2, \dots, x_n\}$ is the set whose elements are $x_1, x_2, \dots, x_n$; and $\{x\}$ is the set whose only element is x. If X is the set of all elements x such that some property $P(x)$ is true, we shall write

$$X = \{x : P(x)\}.$$

The symbol $\varnothing$ denotes the empty set.

We write $x \in X$ if x is a member of the set X; otherwise $x \notin X$. If Y is a *subset* of X, that is, if $x \in Y$ implies $x \in X$, we write $Y \subseteq X$. If $Y \subseteq X$ and $X \subseteq Y$, then $X =$

S. Shirali, H.L. Vasudeva, *Multivariable Analysis*,
DOI 10.1007/978-0-85729-192-9_1, © Springer-Verlag London Limited 2011

Y. If $Y \subseteq X$ and $Y \neq X$, then Y is *proper subset* of X. Observe that $\varnothing \subseteq X$ for every set X.

We list below the standard notations for the most important sets of numbers:

$\mathbb{N}$ the set of all natural numbers

$\mathbb{Z}$ the set of all integers

$\mathbb{Q}$ the set of all rational numbers

$\mathbb{R}$ the set of all real numbers

$\mathbb{C}$ the set of all complex numbers.

Given two sets X and Y, we can form the following new sets from them:

$$X \cup Y = \{x : x \in X \text{ or } x \in Y\}$$
$$X \cap Y = \{x : x \in X \text{ and } x \in Y\}.$$

$X \cup Y$ and $X \cap Y$ are the *union* and *intersection* respectively of X and Y. If $\{X_\alpha\}$ is a collection of sets, where α runs through some indexing set Λ, we write

$$\bigcup_{\alpha \in \Lambda} X_\alpha \text{ and } \bigcap_{\alpha \in \Lambda} X_\alpha$$

for the union and intersection, respectively, of X_α :

$$\bigcup_{\alpha \in \Lambda} X_\alpha = \{x : x \in X_\alpha \text{ for at least one } \alpha \in \Lambda\}$$

$$\bigcap_{\alpha \in \Lambda} X_\alpha = \{x : x \in X_\alpha \text{ for every } \alpha \in \Lambda\}.$$

If $\Lambda = \mathbb{N}$, the set of all natural numbers, the customary notations are

$$\bigcup_{n=1}^{\infty} X_n \text{ and } \bigcap_{n=1}^{\infty} X_n .$$

If no two members of $\{X_\alpha\}$ have any element in common, then $\{X_\alpha\}$ is said to be a *pairwise disjoint collection of sets*.

If $Y \subseteq X$, the *complement* of Y in X is the set of elements that are in X but not in Y, that is

$$X \setminus Y = \{x : x \in X, x \notin Y\}.$$

The complement of Y is denoted by Y^c whenever it is clear from the context with respect to which larger set the complement is taken.

If $\{X_\alpha\}$ is a collection of subsets of X, then the following *De Morgan's laws* hold:

$$(\bigcup_{\alpha \in \Lambda} X_\alpha)^c = \bigcap_{\alpha \in \Lambda}(X_\alpha)^c \quad \text{and} \quad (\bigcap_{\alpha \in \Lambda} X_\alpha)^c = \bigcup_{\alpha \in \Lambda}(X_\alpha)^c .$$

The *Cartesian product* $X_1 \times X_2 \times \cdots \times X_n$ of the sets $X_1, X_2, \ldots, X_n$ is the set of all ordered n-tuples $(x_1, x_2, \ldots, x_n)$, where $x_i \in X_i$ for $i = 1, 2, \ldots, n$.

The symbol

$$f : X \to Y$$

means that f is a *function* (or *mapping* or *map*) from the set X into the set Y; that is, f assigns to each $x \in X$ an element $f(x) \in Y$. The elements assigned to members of X by f are often called values of f. If $A \subseteq X$ and $B \subseteq Y$, the *image* of A and *inverse image* of B are, respectively,

$$f(A) = \{ f(x) : x \in A \}$$

$$f^{-1}(B) = \{ x : f(x) \in B \}.$$

Note that $f^{-1}(B)$ may be empty even when $B \neq \varnothing$. The assertion that $C = f(A)$ is sometimes conveniently rephrased as 'f maps (the subset) A *onto* C'.

The *domain* of f is X and the *range* is $f(X)$; the *range space* is Y. If $f(X) = Y$, the function f is said to map X onto Y (or the function is said to be *surjective*). We write $f^{-1}(y)$ instead of $f^{-1}(\{y\})$ for every $y \in Y$. If $f^{-1}(y)$ consists of at most one element for each $y \in Y$, f is said to be *one-to-one* (or *injective*). If f is one-to-one, then f^{-1} is a function with domain $f(X)$ and range X. A function which is both injective and surjective is said to be *bijective*. One also speaks of a *bijection* or *one-to-one correspondence*. In the case when f is bijective, f^{-1} is a function with domain Y and range X, in which case, it is called the *inverse of f*. An inverse is unique if it exists and is referred to as *the* inverse of f. A map is said to be *invertible* if it has an inverse; thus being invertible is the same as being bijective.

It is sometimes necessary to consider a function f only on a subset S of its domain X. Technically, that makes it a different function and it is called the *restriction* of f to S. Introducing a new symbol to denote a restriction can clutter the notation and we shall avoid it as far as possible.

Let $g : U \to V$ and $f : X \to Y$ be maps, where X has a nonempty intersection with the range $g(U)$. Then the inverse image $Z = g^{-1}(X \cap g(U)) \subseteq U$ is nonempty and the function $f \circ g : Z \to Y$ such that $(f \circ g)(z) = f(g(z))$ is called the *composition* of f and g. For most theoretical purposes, it is sufficient to work with the case $X \supseteq g(U)$, because this ensures that $X \cap g(U) = g(U)$ and hence that $Z = U$.

If $\{X_\alpha : \alpha \in \Lambda\}$ is any family of subsets of X, then

$$f(\bigcup_{\alpha \in \Lambda} X_\alpha) = \bigcup_{\alpha \in \Lambda} f(X_\alpha)$$

and

$$f(\bigcap_{\alpha \in \Lambda} X_\alpha) \subseteq \bigcap_{\alpha \in \Lambda} f(X_\alpha).$$

Also, if $\{Y_\alpha : \alpha \in \Lambda\}$ is a family of subsets of Y, then

$$f^{-1}(\bigcup_{\alpha \in \Lambda} Y_\alpha) = \bigcup_{\alpha \in \Lambda} f^{-1}(Y_\alpha)$$

and

$$f^{-1}(\bigcap_{\alpha \in \Lambda} Y_\alpha) = \bigcap_{\alpha \in \Lambda} f^{-1}(Y_\alpha).$$

If Y_1 and Y_2 are subsets of Y, then

$$f^{-1}(Y_1 \setminus Y_2) = f^{-1}(Y_1) \setminus f^{-1}(Y_2).$$

Finally, if $f : X \to Y$ and $g : Y \to Z$, the *composite* function $g \circ f : X \to Z$ is defined by

$$(g \circ f)(x) = g(f(x)).$$

1-2 The Real Number System

In the present section, field axioms, linear ordering axioms and the least upper bound axiom of $\mathbb{R}$ are listed in detail. They fall naturally into three groups.

A.

For all real numbers x, y and z, we have

(i) $x + y = y + x$;

(ii) $(x + y) + z = x + (y + z)$;

(iii)   there exists $0 \in \mathbb{R}$ such that $x + 0 = x$;

(iv)   there exists a $w \in \mathbb{R}$ such that $x + w = 0$;

(v) $xy = yx$;

(vi) $(xy)z = x(yz)$;

(vii)   there exists $1 \in \mathbb{R}$ such that $1 \neq 0$ and $x \cdot 1 = x$;

(viii)   if x is different from 0; there exists a $w \in \mathbb{R}$ such that $xw = 1$;

(ix) $x(y + z) = xy + xz$.

The second group of properties possessed by the real numbers has to do with the fact that they are ordered. They can be phrased in terms of positivity of real numbers. When we do this, our second group of axioms takes the following form.

B.

The subset P of positive real numbers satisfies the following:

(i) P is closed with respect to addition and multiplication, that is, if $x, y \in P$, then so are $x + y$ and xy,

(ii) $x \in P$ implies $-x \notin P$,

(iii) $x \in \mathbb{R}$ implies $x = 0$ or $x \in P$ or $-x \in P$.

Any system satisfying the axioms of groups A and B is called an *ordered field*, for example, the ordered field of rational numbers.

In an ordered field, we define the notion $x < y$ to mean $y - x \in P$. We write $x \leq y$ to mean '$x < y$ or $x = y$'.

Absolute value is defined in any ordered field in the familiar manner:

$$|x| = \begin{cases} x & \text{if } x \geq 0 \\ -x & \text{if } x < 0. \end{cases}$$

It can be shown on the basis of this definition that the 'triangle inequality'

$$|x + y| \leq |x| + |y|$$

or equivalently,

$$|x - y| \leq |x - z| + |z - x|$$

holds.

From the two groups of axioms (A) and (B), it can be shown that $\mathbb{R} \supseteq \mathbb{Q} \supseteq \mathbb{N}$.

The third group contains only one axiom and it is this axiom that sets apart the real numbers from other ordered fields. Before stating this axiom, we need to define some terms. Let X be a nonempty subset of $\mathbb{R}$. If there exists M such that $x \leq M$ for all $x \in X$, then X is said to be *bounded above* and M is said to be *an upper bound of X*. If there exists m such that $x \geq m$ for all $x \in X$, then X is said to be *bounded below* and m is said to be *a lower bound of X*. If X is bounded above as well as below, then it is said to be *bounded*. A number M' is called the *least upper bound* (or *supremum*) of X if it is an upper bound and $M' \leq M$ for each upper bound M of X. The final axiom guarantees the existence of least upper bounds for nonempty subsets of $\mathbb{R}$ that are bounded above.

C.

Every nonempty subset of $\mathbb{R}$ that has an upper bound possesses a least upper bound.

We shall denote the least upper bound of X by sup X or by sup $\{x : x \in X\}$ or by $\sup_{x \in X} x$.

The *greatest lower bound* or *infimum* can be defined similarly. It follows from C above that every nonempty subset of $\mathbb{R}$ that has a lower bound possesses a greatest lower bound. The greatest lower bound of X is denoted by inf X or by inf $\{x : x \in X\}$ or by $\inf_{x \in X} x$. Note that $\inf_{x \in X} x = -\sup_{x \in X} -x$.

The following characterisation of supremum is used frequently.

1-2.1. Proposition. *Let X be a nonempty set of real numbers that is bounded above. Then $M = \sup X$ if and only if*

(i) $x \leq M$ *for all $x \in X$, and*

(ii) *given any $\varepsilon > 0$, there exists $x \in X$ such that $x > M - \varepsilon$.*

There is a similar characterisation of the infimum of a nonempty set of real numbers that is bounded below.

Certain kinds of subsets of $\mathbb{R}$ have a special role. If $a < b$, both real numbers, then the subset $\{x \in \mathbb{R} : a < x < b\}$ is called an *open interval* and is denoted by (a, b). Subsets of the form $\{x \in \mathbb{R} : x < b\}$ and $\{x \in \mathbb{R} : a < x \}$ are also called open intervals and denoted respectively by $(-\infty, b)$ and (a, ∞). The subsets $\{x \in \mathbb{R} : a \leq x \leq b\}$, $\{x \in \mathbb{R} : x \leq b\}, \{x \in \mathbb{R} : a \leq x\}$ are *closed intervals* and are denoted by $[a, b]$, $(-\infty, b]$, $[a, \infty)$, respectively. It is clear what $[a, b)$ and $(a, b]$ mean, and these intervals are neither open nor closed.

1-3 Sequences of Real Numbers

Functions that have the set $\mathbb{N}$ of natural numbers as domain play an important role in analysis. A function $f : \mathbb{N} \to S$, where S is any nonempty set, is called a *sequence in S* or a *sequence of elements of S*.

A sequence of real numbers is a map $x : \mathbb{N} \to \mathbb{R}$. Given such a map, we denote $x(n)$ by x_n, and this value is called the nth *term* of the sequence. The sequence itself is frequently denoted by $\{x_n\}_{n \geq 1}$. It is important to distinguish between the sequence $\{x_n\}_{n \geq 1}$ and its range $\{x_n : n \in \mathbb{N}\}$, which is a subset of $\mathbb{R}$.

A real number l is said to be a *limit of the sequence* $\{x_n\}_{n\geq 1}$ if for each $\varepsilon > 0$, there is a positive integer n_0 such that for all $n \geq n_0$, we have $|x_n - l| < \varepsilon$. It is easy to verify that a sequence has at most one limit. When $\{x_n\}_{n\geq 1}$ does have a limit, we denote it by $\lim x_n$. In symbols, $l = \lim x_n$ if

$$\forall\, \varepsilon > 0, \exists\, n_0 \ni n \geq n_0 \Rightarrow |x_n - l| < \varepsilon.$$

A sequence that has a limit is said to *converge* (or to be *convergent*).

If $\lim x_n$ and $\lim y_n$ both exist, then so do $\lim(x_n + y_n)$ and $\lim(x_n y_n)$; moreover, $\lim(x_n + y_n) = \lim x_n + \lim y_n$ and $\lim(x_n y_n) = (\lim x_n)(\lim y_n)$. If α is any real number, then $\lim(\alpha x_n) = \alpha(\lim x_n)$.

The sequence $x_n = (1 + \frac{1}{n})^n$ has a limit denoted by e; this number is irrational and lies between 2 and 3.

$\lim n^{1/n} = 1$.

A sequence $\{x_n\}_{n\geq 1}$ of real numbers is said to be *increasing* if it satisfies the inequalities $x_n \leq x_{n+1}$, $n = 1, 2, \ldots$, and *decreasing* if it satisfies the inequalities $x_n \geq x_{n+1}$, $n = 1, 2, \ldots$. We say that the sequence is *monotone* if it is either increasing or it is decreasing.

A sequence $\{x_n\}_{n\geq 1}$ of real numbers is said to be *bounded* if there exists a real number $M > 0$ such that $|x_n| \leq M$ for all $n \in \mathbb{N}$. The following simple criterion for the convergence of a monotone sequence is very useful.

1-3.1. Proposition. *A monotone sequence of real numbers is convergent if and only if it is bounded.*

Let $\{s_n\}_{n\geq 1}$ be a sequence in any set and let $n_1 < n_2 < \cdots < n_k < \cdots$ be a strictly increasing sequence of natural numbers. Then $\{s_{n_k}\}_{k\geq 1}$ is called a *subsequence* of $\{s_n\}_{n\geq 1}$

1-3.2. Bolzano–Weierstrass Theorem. *A bounded sequence of real numbers has a convergent subsequence.*

The convergence criterion described in Proposition 1-3.1 is restricted to monotone sequences. It is important to have a condition implying the convergence of a sequence of real numbers that is applicable to a larger class and preferably does not require knowledge of the value of the limit. The Cauchy criterion gives such a condition.

A sequence $\{x_n\}_{n\geq 1}$ of real numbers is said to be a *Cauchy sequence* if, for every $\varepsilon > 0$, there exists an integer n_0 such that $|x_n - x_m| < \varepsilon$ whenever $n \geq n_0$ and $m \geq n_0$. In symbols,

$$\forall\, \varepsilon > 0, \exists\, n_0 \text{ such that } (n \geq n_0, m \geq n_0) \Rightarrow |x_n - x_m| < \varepsilon.$$

1-3.3. Cauchy Convergence Criterion: *A sequence of real numbers converges if and only if it is a Cauchy sequence.*

When a sequence $\{s_n\}_{n \geq 1}$ is described in form $s_n = \sum_{k=1}^{n} a_k$, it is called a *series* and the number a_k is called its *k*th *term*, rather than s_k. The number s_n is then called the *n*th *partial sum* of the series. The limit $\lim s_n$, if it exists, is called the *sum* of the series. The symbol $\sum_{k=1}^{\infty} a_k$ denotes the series as well as the sum, if any. The context determines which of the two is intended. The series $\sum_{k=1}^{n} k^{-p}$ is convergent if and only if $p > 1$.

1-4 Limits of Functions and Continuous Functions

Mathematical analysis is primarily concerned with limit processes. We have already reviewed one of the basic limit processes, namely, convergence of a sequence of real numbers. In this section we shall recall the notion of the limit of a function, which is used in the study of continuity, differentiation and integration. The notion is parallel to that of the limit of a sequence. We shall also state the definition of continuity and its relation to limits.

A point $a \in \mathbb{R}$ is said to be a *limit point of a subset* $X \subseteq \mathbb{R}$ if every open interval $(a - \varepsilon, a + \varepsilon)$ in $\mathbb{R}$, where $\varepsilon > 0$, contains a point $x \neq a$ such that $x \in X$.

Let f be a real-valued function defined on a subset X of $\mathbb{R}$ and a be a limit point of X. We say that $f(x)$ *tends to l as x tends to a* if, for every $\varepsilon > 0$, there exists some $\delta > 0$ such that

$$|f(x) - l| < \varepsilon \quad \forall x \in X \text{ satisfying } 0 < |x - a| < \delta.$$

The number l is said to be *the limit of f(x) as x tends to a* and we write

$$\lim_{x \to a} f(x) = l \qquad \text{or} \qquad f(x) \to l \text{ as } x \to a.$$

Note that $f(a)$ need not be defined for the above definition to make sense. Moreover, the value l of the limit is uniquely determined when it exists.

If $\lim_{x \to a} f(x)$ and $\lim_{x \to a} g(x)$ both exist, then so do $\lim_{x \to a} (f(x) + g(x))$ and $\lim_{x \to a} (f(x)g(x))$; moreover,

$$\lim_{x \to a} (f(x) + g(x)) = \lim_{x \to a} f(x) + \lim_{x \to a} g(x)$$

and

$$\lim_{x \to a} (f(x)g(x)) = (\lim_{x \to a} f(x))(\lim_{x \to a} g(x)).$$

If α is any real number, then $\lim_{x \to a} (\alpha f(x)) = \alpha(\lim_{x \to a} f(x))$.

If $f(x) \leq h(x) \leq g(x)$ whenever $0 < |x - a| < \delta$ and if $\lim_{x \to a} f(x)$ and $\lim_{x \to a} g(x)$ both exist and are equal, then $\lim_{x \to a} h(x)$ also exists and $\lim_{x \to a} h(x) = \lim_{x \to a} f(x) = \lim_{x \to a} g(x)$.

The following important formulation of limit of a function is in terms of limits of sequences.

1-4.1. Proposition. *Let $f:X \to \mathbb{R}$ and let a be a limit point of X. Then $\lim_{x \to a} f(x) = l$ if and only if, for every sequence $\{x_n\}_{n \geq 1}$ in X that converges to a and $x_n \neq a$ for every n, the sequence $\{f(x_n)\}_{n \geq 1}$ converges to l.*

Let f be a real-valued function whose domain of definition is a set X of real numbers. We say that f is *continuous at the point* $x \in X$ if, given $\varepsilon > 0$, there exists a $\delta > 0$ such that for all $y \in X$ with $|y - x| < \delta$, we have $|f(y) - f(x)| < \varepsilon$. The function is said to be *continuous on X* if it is continuous at every point of X. If we merely say that a function is continuous, we mean that it is continuous on its domain.

It may checked that f is continuous at a limit point $a \in X$ if and only if $f(a)$ is defined and $\lim_{x \to a} f(x) = f(a)$. The following criterion of continuity of f at a point $a \in X$ follows immediately from the preceding criterion and Proposition 1-4.1.

1-4.2. Proposition. *Let f be a real-valued function defined on a subset X of $\mathbb{R}$ and $a \in X$ be a limit point of X. Then f is continuous at a if and only if, for every sequence $\{x_n\}_{n \geq 1}$ in X that converges to a and $x_n \neq a$ for every n, $\lim f(x_n) = f(\lim x_n) = f(a)$.*

This result shows that continuous functions are precisely those which send convergent sequences into convergent sequences; in other words, they 'preserve' convergence.

The next result, which is known as the *Bolzano intermediate value theorem*, guarantees that a continuous function on an interval assumes (at least once) every value that lies between any two of its values.

1-4.3. Intermediate Value Theorem: *Let I be an interval and $f:I \to \mathbb{R}$ be a continuous mapping on I. If $a, b \in I$ and $\alpha \in \mathbb{R}$ satisfy $f(a) < \alpha < f(b)$ or $f(a) > \alpha > f(b)$, then there exists a point $c \in I$ between a and b such that $f(c) = \alpha$.*

1-5 Compact Sets

The notion of compactness, which is of enormous significance in analysis, is an abstraction of an important property possessed by certain subsets of real numbers. The property in question asserts that every 'open cover' of a closed and bounded subset of $\mathbb{R}$ has a finite 'subcover'. This simple property of closed and bounded subsets has far reaching implications in analysis; for example, a real-valued continuous function defined on $[0,1]$, say, is bounded and uniformly continuous. In what follows, we shall define the notion of compactness in $\mathbb{R}$ and list some of its characterisations. To begin with, we recall the definitions of open and closed subsets of $\mathbb{R}$.

A subset G of $\mathbb{R}$ is said to be *open* if for each $x \in G$, there is an open interval $(x - \varepsilon, x + \varepsilon)$, $\varepsilon > 0$, which is contained in G. A subset of $\mathbb{R}$ is said to be *closed* if its complement is open.

Let X be a subset of $\mathbb{R}$. An *open cover* (*covering*) of X is a collection $\mathcal{C} = \{G_\alpha : \alpha \in \Lambda\}$ of open sets in $\mathbb{R}$ whose union contains X, that is,

$$X \subseteq \bigcup_\alpha G_\alpha .$$

If $\mathcal{C}'$ is a subcollection of $\mathcal{C}$ such that the union of sets in $\mathcal{C}'$ also contains X, then $\mathcal{C}'$ is called a *subcover* (or *subcovering*) from $\mathcal{C}$ of X. If $\mathcal{C}'$ consists of finitely many sets, then we say that $\mathcal{C}'$ is a *finite subcover* (or *finite subcovering*).

A subset X of $\mathbb{R}$ is said to be *compact* if every open cover of X contains a finite subcover. The following proposition characterises compact subsets of $\mathbb{R}$.

1-5.1. Heine–Borel Theorem: *Let X be a set of real numbers. Then the following statements are equivalent:*

(i) *X is closed and bounded.*

(ii) *X is compact.*

1-5.2. Proposition. *Let f be a real-valued continuous function defined on the closed bounded interval $I = [a,b]$. Then f is bounded on I and assumes its maximum and minimum values on I, that is, there are points x_1 and x_2 in I such that $f(x_1) \le f(x) \le f(x_2)$ for all $x \in X$.*

For our next proposition, we shall need the following definition. Let f be a real-valued continuous function defined on a set X. Then f is said to be *uniformly continuous on X* if, given $\varepsilon > 0$, there is a $\delta > 0$ such that for all $x, y \in X$ with $|x - y| < \delta$, we have $|f(x) - f(y)| < \varepsilon$.

1-5.3. Proposition. *If a real-valued function f is continuous on a closed and bounded interval I, then f is uniformly continuous on I.*

1-6 Derivatives and Riemann Integral

Let $S \subseteq \mathbb{R}$ and x be a limit point of S. A function $f: S \to \mathbb{R}$ is *differentiable at x* if

$$\lim_{h \to 0} \frac{f(x+h) - f(x)}{h}$$

exists, in which case, the limit is called the *derivative of f at x* and is denoted by $f'(x)$. It is often more convenient to write $\frac{d}{dx} f(x)$ for $f'(x)$. The *derivative function f'* is the one that maps each point of differentiability into the derivative at that point and is called simply *derivative of f*.

If $f'(x)$ and $g'(x)$ both exist, then so do $(f+g)'(x)$ and $(fg)'$; moreover,
$$(f+g)'(x) = f'(x) + g'(x) \text{ and } (fg)'(x) = f'(x)g(x) + f(x)g'(x).$$
If α is any real number, then $(\alpha f)'(x) = \alpha(f'(x))$. If x is a limit point of the set on which $g \neq 0$ and also belongs to the set, then

$$\left(\frac{f}{g}\right)'(x) = \frac{f'(x)g(x) - f(x)g'(x)}{g(x)^2} .$$

We assume that the reader is aware of trigonometric functions, the exponential and natural logarithm functions, and also of their limit and differentiation properties, such as

$$\frac{d}{dx} \sin x = \cos x, \quad \frac{d}{dx} \tan^{-1} x = \frac{1}{1+x^2}, \quad \frac{d}{dx} \ln x = \frac{1}{x} \quad \text{and so forth.}$$

The functions can be defined variously via limit processes and all their properties learned in calculus can be derived from there. The manner in which this is done will be of no consequence for the material in this book.

1-6.1. Proposition. Chain Rule: *Suppose $f: S \to \mathbb{R}$ is differentiable at $x \in S$ and g maps a set containing $f(S)$ into $\mathbb{R}$. If g is differentiable at $f(x) \in f(S)$, then the composition $g \circ f$ is differentiable at x and*

$$(g \circ f)'(x) = g'(f(x)) \cdot f'(x).$$

Let I denote an interval. A function $f: I \to \mathbb{R}$ is said to have a *local maximum* at $c \in I$ if there exists $\delta > 0$ such that $x \in I, |x - c| < \delta \Rightarrow f(x) \leq f(c)$. Similarly for a *local minimum*.

A function $f: I \to \mathbb{R}$ is said to be *increasing* if, for all $x_1, x_2 \in I$,
$$x_1 < x_2 \Rightarrow f(x_1) \leq f(x_2)$$

and *strictly increasing* if

$$x_1 < x_2 \Rightarrow f(x_1) < f(x_2).$$

Correspondingly for *decreasing* and *strictly decreasing*. A *monotone* function on an interval is one which is either increasing or decreasing.

1-6.2. Proposition. *If* $f:[a,b] \to \mathbb{R}$ *satisfies* $f'(x) \geq 0$ *for every* $x \in [a,b]$, *then* f *is increasing. Similarly, if* $f:[a,b] \to \mathbb{R}$ *satisfies* $f'(x) \leq 0$ *for every* $x \in [a,b]$, *then* f *is decreasing.*

1-6.3. Proposition. *If* $f:[a,b] \to \mathbb{R}$ *has a local maximum or a local minimum at* $c \in (a,b)$ *then* $f'(c) = 0$.

1-6.4. Proposition. *Suppose* $f:[a,b] \to \mathbb{R}$ *satisfies* $f'(c) = 0$, *where* $c \in (a,b)$. *If* $f''(c) < 0$ *then* f *has a local maximum at* c *and if* $f''(c) > 0$ *then* f *has a local minimum at* c.

1-6.5. Mean Value Theorem: *Suppose the continuous function* $f:[a,b] \to \mathbb{R}$ *is differentiable on* (a,b). *Then there exists some* $\xi \in (a,b)$ *such that*

$$f(b) - f(a) = f'(\xi)(b-a).$$

1-6.6. Taylor's Theorem: *Suppose* $n \in \mathbb{N}$ *and* $f:[a,b] \to \mathbb{R}$ *is a function such that* $f^{(n-1)}$ *is continuous on* $[a,b]$ *and* $f^{(n)}$ *exists on* (a,b). *Then there exists some* $c \in (a,b)$ *such that*

$$f(b) = \sum_{k=0}^{n-1} \frac{f^{(k)}(a)}{k!}(b-a)^k + \frac{f^{(n)}(c)}{n!}(b-a)^n.$$

1-6.7. Proposition. *Suppose* $f:[a,b] \to \mathbb{R}$ *has derivative zero at every point of its domain. Then the function is a constant.*

If the equation $f(x) = y$, where y is given and x is to be found, has a solution $x = r$, then the sequence $\{x_p\}$ generated by the scheme

$$x_{p+1} = x_p + f'(x_p)^{-1}(y - f(x_p))$$

converges to the solution r under appropriate but broad hypotheses. One such set of hypotheses is that on some interval containing r but not as an endpoint, $|f'|$ has a positive lower bound m, the second derivative $|f''|$ has an upper bound M and $|x_1 - r| < \frac{2m}{M}$. This way of approximating the solution is called *Newton's method*. Although we shall not make direct use of this, we shall be drawing a parallel between it and something else that we shall encounter.

By a partition P of an interval $[a,b]$ we mean a finite sequence of points x_k in the interval such that

$$P : a = x_0 < x_1 < \cdots < x_n = b.$$

For a bounded function $f:[a,b]\to\mathbb{R}$ and any partition, the nonempty set $\{f(x) : x_{k-1} \le x \le x_k\}$ is bounded above as well as below for each k. Consequently, it has supremum M_k and an infimum m_k. The *upper* and *lower sums of f over the partition P* are, respectively,

$$\sum_{k=1}^{n} M_k(x_k - x_{k-1}) \quad \text{and} \quad \sum_{k=1}^{n} m_k(x_k - x_{k-1}).$$

Their respective infimum and supremum are called the *upper* and *lower integrals* respectively of f and are denoted by $\overline{\int_a^b} f$ and $\underline{\int_a^b} f$.
Thus

$$\overline{\int_a^b} f = \inf\{\sum_{k=1}^{n} M_k(x_k - x_{k-1}) : \text{all partitions } P\}$$

and

$$\underline{\int_a^b} f = \sup\{\sum_{k=1}^{n} m_k(x_k - x_{k-1}) : \text{all partitions } P\}.$$

It turns out that $\overline{\int_a^b} f \ge \underline{\int_a^b} f$ for every bounded function f. If equality holds, then the function f is said to be *Riemann integrable*, or simply *integrable*, and the integral of f from a to b is the common value of the upper and lower integrals, denoted by

$$\int_a^b f.$$

Sometimes it is convenient to speak of f being *integrable on* $[a,b]$.

The integral exists, for instance, if f is continuous or monotone.

If the restriction of f to $[\alpha,\beta] \subseteq [a,b]$ is integrable, we say that f is integrable on $[\alpha,\beta]$.

If $f:[a,b]\to\mathbb{R}$ and $g:[a,b]\to\mathbb{R}$ are both integrable on $[a,b]$, then so are $f+g$, fg and αf (α a real number); moreover,

$$\int_a^b (f+g) = \int_a^b f + \int_a^b g \quad \text{and} \quad \int_a^b (\alpha f) = \alpha \int_a^b f.$$

Suppose $f:[a,b]\to\mathbb{R}$ is bounded and $a < c < b$. If f is integrable on $[a,b]$, then it is integrable on $[a,c]$ as well as $[c,b]$, and the equality

$$\int_a^b f = \int_a^c f + \int_c^b f$$

holds. Conversely, if f is integrable on $[a,c]$ as well as $[c,b]$, then it is integrable on $[a,b]$ and the foregoing equality holds. The equality holds without the proviso that $a < c < b$ if we agree that

$$\int_a^b f = -\int_b^a f.$$

If $f:[a,b]\to\mathbb{R}$ and $g:[a,b]\to\mathbb{R}$ are both integrable and $f(x) \le g(x)$ for each x $\in [a,b]$, then $\int_a^b f \le \int_a^b g$.

If f is integrable on $[a,b]$, and $[\alpha,\beta] \subseteq [a,b]$, then f is integrable on $[\alpha,\beta]$. If also $f \ge 0$ on $[a,b]$ then $\int_\alpha^\beta f \le \int_a^b f$.

1-6.8. Proposition. *Let* $f:[a,b]\to\mathbb{R}$ *be integrable. Then* $|f|:[a,b]\to\mathbb{R}$ *is also integrable and*

$$|\int_a^b f| \le \int_a^b |f|.$$

The following is known as the *fundamental theorem of integral calculus*.

1-6.9. Proposition. *Let* $f:[a,b]\to\mathbb{R}$ *be integrable and let*

$$F(x) = \int_a^x f, a < x < b.$$

Then F is continuous on $[a,b]$. *Moreover, if f is continuous at a point* $c \in [a,b]$, *then F is differentiable at c and*

$$F'(c) = f(c).$$

There are two versions of the *substitution rule* or *change of variables formula,* and it is the first version that we shall generalise to higher dimensions in Chapter 7 as *the transformation formula* for integrals.

1-6.10. Proposition. Version 1: *Suppose* $\varphi:[\alpha,\beta]\to[a,b]$ *is a bijection having a continuous derivative that vanishes nowhere. If* $(f\circ\varphi)|\varphi'|$ *is integrable on* $[\alpha,\beta]$ *and f is integrable on the image* $\varphi([\alpha,\beta]) = [a,b]$, *then*

$$\int_a^b f = \int_\alpha^\beta (f\circ\varphi)|\varphi'|.$$

The reason for the absolute value on the right side is that in case $\varphi' < 0$ everywhere, we have $\varphi(\alpha) = b$ and $\varphi(\beta) = a$. It is possible to deduce the integrability of $(f\circ\varphi)|\varphi'|$ from the remaining hypotheses. See Pugh [21, pp. 170–171].

1-6.10. Proposition. Version 2: *Let* $F:[a,b]\to\mathbb{R}$ *and* $\varphi:[\alpha,\beta]\to[a,b]$ *both be differentiable. If F' and* $(F'\circ\varphi)\varphi'$ *are both Riemann integrable, then*

$$\int_{\varphi(\alpha)}^{\varphi(\beta)} F' = \int_\alpha^\beta (F'\circ\varphi)\varphi'.$$

It is not presumed in either version of the above proposition that $\varphi(\alpha) < \varphi(\beta)$.

The next result is the *formula of integration by parts*.

1-6.11. Proposition. *Let f and g be differentiable functions on* $[a,b]$ *having integrable derivatives f' and g'. Then the products fg' and f'g are integrable, and*

$$\int_a^b fg' = f(b)g(b) - f(a)g(a) - \int_a^b f'g.$$

1-6.12. Proposition. *Suppose* $\{f_n\}_{n\geq 1}$ *is a sequence of Riemann integrable functions on* $[a,b]$ *with uniform limit* f. *Then* f *is Riemann integrable on* $[a,b]$ *and*
$$\lim\int_a^b f_n = \int_a^b f.$$

If a function $f:[a,\infty]\to\mathbb{R}$ is integrable on $[a,b]$ whenever $a < b$, the symbol $\int_a^\infty f$ means $\lim_{b\to\infty} \int_a^b f$, even if the limit does not exist. If it does not, we say that $\int_a^\infty f$ is *divergent*; otherwise *convergent*. In either case, $\int_a^\infty f$ is called the *improper integral of* f *over* $[a,\infty)$. If it is convergent, we speak of f being integrable over $[a,\infty)$.

1-6.13. Integral Test:.*Suppose* $f:[1,\infty)\to\mathbb{R}$ *is nonnegative-valued and decreasing. Then the series* $\sum_{n=1}^\infty f(k)$ *converges if and only if the improper integral* $\int_1^\infty f$ *converges.*

1-7 Matrices

We shall confine our attention to matrices whose entries are from the field $\mathbb{R}$ of real numbers, as these are the only matrices that will be used in subsequent discussions.

An array of mn real numbers with m rows and n columns,

$$\begin{bmatrix} a_{11} & a_{12} & \cdots & a_{1n} \\ a_{21} & a_{22} & \cdots & a_{2n} \\ \vdots & \vdots & \ddots & \vdots \\ a_{m1} & a_{m2} & \cdots & a_{mn} \end{bmatrix}$$

is called a real $m\times n$ *matrix*. When $m = n$, the array is called a *square matrix* of order n or simply a matrix of order $n\times n$. Its diagonal containing the entries $a_{11}, a_{22},\ldots, a_{nn}$ is called the *leading* or *main diagonal*.

In writing, a matrix is often denoted by a single letter A or X, or by any other symbol one cares to choose. For example, a common notation for the matrix of the definition is $A = [a_{ij}]$, where a_{ij} denotes the entry in the ith row and jth column. The square bracket is a conventional symbol and is indicative of the fact that we are not considering a determinant. A matrix with a single row

$$A = \begin{bmatrix} a_1 & a_2 & \cdots & a_n \end{bmatrix},$$

where $a_1, a_2, \ldots, a_n \in \mathbb{R}$, is called a *row matrix*, and a matrix with a single column

$$B = \begin{bmatrix} b_1 \\ b_2 \\ \vdots \\ b_m \end{bmatrix},$$

where $b_1, b_2, \ldots, b_m \in \mathbb{R}$, is called a *column matrix*. Row vectors can be converted into column vectors and vice versa by an operation that is called *transposition*. It is practical to define transposition for any matrix. The transpose of any $m \times n$ matrix $A = [a_{ij}]$ is the $n \times m$ matrix that has the first row of A as its first column, the second row of A as its second column, and so on. Thus the *transpose* of the matrix $A = [a_{ij}]$ is

$$A^T = \begin{bmatrix} a_{11} & a_{21} & \cdots & a_{m1} \\ a_{12} & a_{22} & \cdots & a_{m2} \\ \vdots & \vdots & \ddots & \vdots \\ a_{1n} & a_{2n} & \cdots & a_{mn} \end{bmatrix}.$$

The transpose of the row matrix $A = \begin{bmatrix} a_1 & a_2 & \cdots & a_n \end{bmatrix}$ is the column matrix

$$A^T = \begin{bmatrix} a_1 \\ a_2 \\ \vdots \\ a_n \end{bmatrix}$$

and the transpose of the column matrix

$$B = \begin{bmatrix} b_1 \\ b_2 \\ \vdots \\ b_m \end{bmatrix}$$

is the row matrix $B^T = \begin{bmatrix} b_1 & b_2 & \cdots & b_m \end{bmatrix}$.

The matrices $A = [a_{ij}]$ and $B = [b_{ij}]$ are equal if the number of rows (respectively, columns) in A equals the number of rows (respectively, columns) in B and $a_{ij} = b_{ij}$ for $1 \leq i \leq m$, $1 \leq j \leq n$.

Two matrices A and B are said to be *conformable for addition* if each has the same number of rows and the same number of columns as the other. The sum of two matrices $A = [a_{ij}]$ and $B = [b_{ij}]$ is defined only when they are conformable for addition. Their *sum* is then defined as the matrix having $a_{ij} + b_{ij}$ as the entry in the ith row and jth column. Thus,

$$A + B = [a_{ij} + b_{ij}].$$

The matrix $-A$, where $A = [a_{ij}]$, is that matrix whose entries are those of A multiplied by -1, that is,

$$-A = [-a_{ij}].$$

The matrix having every entry 0 is called a *null matrix* and is written O.

When α is a real number and $A = [a_{ij}]$ is a matrix, αA is defined to be the matrix each of whose entries is α times the corresponding entry of A, that is,

$$\alpha A = [\alpha a_{ij}].$$

By virtue of the definitions above, we are justified in writing

$$2A \text{ instead of } A + A$$
$$3A \text{ instead of } 5A - 2A.$$

Further, since the addition, subtraction and scalar multiplication of matrices is based on the addition, subtraction and scalar multiplication of corresponding entries, the laws that govern these operations also govern the analogous operations on matrices. More precisely, we have the following:

Let A, B, C be matrices that are conformable for addition and α, β be real numbers. Then

(i) $A + (B + C) = (A + B) + C$ (associative law)

(ii) $A + B = B + A$ (commutative law)

(iii) $A + O = A$

(iv) $A + (-A) = O$

(v) $\alpha(A + B) = \alpha A + \alpha B$

(vi) $\alpha(\beta A) = (\alpha\beta)A.$

Two matrices A and B (in that order) are *conformable for multiplication* if the number of columns in A is equal to the number of rows in B. The *product AB* is then defined to be the matrix whose entry in the ith row and jth column is

$$\sum_{k=1}^{n} a_{ik} b_{kj}.$$

Thus $AB = [\sum_{k=1}^{n} a_{ik} b_{kj}]$. A numerical example will perhaps be helpful. Take

$$A = \begin{bmatrix} 1 & -1 \\ 0 & 2 \end{bmatrix} \text{ and } B = \begin{bmatrix} 3 & 4 & 5 \\ 6 & 0 & 8 \end{bmatrix}.$$

The number of columns of A is equal to the number of rows of B. The entry in the first row and second column of AB equals

$$\sum_{k=1}^{2} a_{1k} b_{k2} = (1)(4) + (-1)(0) = 4.$$

The other entries of the product may be similarly computed. Upon doing so, we obtain

$$AB = \begin{bmatrix} -3 & 4 & -3 \\ 12 & 0 & 16 \end{bmatrix}.$$

The reader may note that A and B may be conformable for the product AB but not for the product BA, in which case the latter product is undefined.

In general, $AB \neq BA$ (even when both AB and BA are defined).

The other properties of matrix multiplication are similar to those for numbers, that is,

(i) $(\alpha A)B = \alpha(AB) = A(\alpha B)$ when α is real;

(ii) $A(BC) = (AB)C$ (associative law)

(iii) $(A+B)C = AC + BC$ (right distributive law)

(iv) $C(A+B) = CA + CB$ (left distributive law),

provided the matrices A, B and C are such that the expressions on the left are defined.

The square matrix of order n that has 1 in its leading diagonal places and 0 elsewhere is called the *identity matrix* of order n. It is denoted by I. Let A be a square matrix of order n; then $AI = IA = A$. Also, $I = I^2 = I^3 = \dots$.

For real numbers, $xy = 0$ implies that either x or y (or both, of course) must be zero. This law does not govern matrix products; that is, $AB = O$ does not necessarily imply that $A = O$ or $B = O$. Indeed, for the matrices

$$A = \begin{bmatrix} a & b \\ 0 & 0 \end{bmatrix} \quad \text{and} \quad B = \begin{bmatrix} b & 2b \\ -a & -2a \end{bmatrix},$$

the product AB is O. Again, AB may be O but not BA. For example, if

$$A = \begin{bmatrix} a & b \\ 0 & 0 \end{bmatrix} \quad \text{and} \quad B = \begin{bmatrix} b & 0 \\ -a & 0 \end{bmatrix},$$

then

$$AB = \begin{bmatrix} 0 & 0 \\ 0 & 0 \end{bmatrix} \quad \text{and} \quad BA = \begin{bmatrix} ab & b^2 \\ -a^2 & -ab \end{bmatrix}.$$

1-8 Determinants

Let $j_1, j_2, \dots, j_n$ be an ordering of the positive integers $1, 2, \dots, n$. An *inversion* occurs in this ordering whenever a greater integer precedes a smaller one. The *number of inversions* occurring in $j_1, j_2, \dots, j_n$ is the sum

$$k = \sum_{s=1}^{n-1} k_s,$$

where k_s is the number of integers greater than s that precede s in the given ordering $j_1, j_2, \ldots, j_n$.

Let $A = [a_{ij}]$ be an $n \times n$ square matrix of real numbers. The *determinant of A* is the number

$$|A| = \begin{vmatrix} a_{11} & a_{12} & \cdots & a_{1n} \\ a_{21} & a_{22} & \cdots & a_{2n} \\ \vdots & \vdots & \ddots & \vdots \\ a_{n1} & a_{n2} & \cdots & a_{nn} \end{vmatrix} = \sum (-1)^k a_{1j_1} a_{2j_2} \cdots a_{nj_n} ,$$

where in each term, the second (column) subscripts $j_1, j_2, \ldots, j_n$ are some ordering of $1, 2, \ldots, n$ and the sum is taken over all possible $j_1, j_2, \ldots, j_n$. For each term, the exponent k in $(-1)^k$ is the number of inversions occurring in $j_1, j_2, \ldots, j_n$. Besides the notation $|A|$ for the determinant of $A = [a_{ij}]$, we also write

$$D = \det A = \begin{vmatrix} a_{11} & a_{12} & \cdots & a_{1n} \\ a_{21} & a_{22} & \cdots & a_{2n} \\ \vdots & \vdots & \ddots & \vdots \\ a_{n1} & a_{n2} & \cdots & a_{nn} \end{vmatrix} .$$

It can be shown that

$$D = a_{j1}C_{j1} + a_{j2}C_{j2} + \cdots + a_{jn}C_{jn} , \quad j = 1, 2, \ldots, n \tag{3}$$

and

$$D = a_{1k}C_{1k} + a_{2k}C_{2k} + \cdots + a_{nk}C_{nk} , \quad k = 1, 2, \ldots, n , \tag{4}$$

where

$$C_{jk} = (-1)^{j+k} \det M_{jk} ,$$

and M_{jk} is a matrix of order $n-1$ obtained by deleting the jth row and kth column of A.

The following expansion of the third order determinant is instructive:

$$D = \begin{vmatrix} 1 & 3 & 0 \\ 2 & 6 & 4 \\ -1 & 0 & 2 \end{vmatrix} = (-1)^{1+1} \cdot 1 \begin{vmatrix} 6 & 4 \\ 0 & 2 \end{vmatrix} + (-1)^{1+2} \cdot 3 \begin{vmatrix} 2 & 4 \\ -1 & 2 \end{vmatrix} + (-1)^{1+3} \cdot 0 \begin{vmatrix} 2 & 6 \\ -1 & 0 \end{vmatrix}$$

$$= 12 - 3 \cdot 8 + 0 \cdot 6 = -12 .$$

This expansion has been implemented using the first row. The expansion using the third column gives

$$D = (-1)^{1+3} \cdot 0 \begin{vmatrix} 2 & 6 \\ -1 & 0 \end{vmatrix} + (-1)^{2+3} \cdot 4 \begin{vmatrix} 1 & 3 \\ -1 & 0 \end{vmatrix} + (-1)^{3+3} \cdot 2 \begin{vmatrix} 1 & 3 \\ 2 & 6 \end{vmatrix} = -12.$$

We list below some properties of determinants:

(i)   Interchange of two rows or columns multiplies the value of the determinant by -1.

(ii) Addition of a multiple of one row or column to another does not alter the value of the determinant.

(iii) Multiplying one row or column by k multiplies the value of the determinant by k.

(iv) Transposition leaves the value of the determinant unaltered.

(v)   A zero row or zero column renders the value of the determinant zero.

(vi) Proportional rows or columns (i.e., ones which are multiples of each other) render the value of the determinant zero; in particular, if two rows or columns of a (square) matrix are identical, then the determinant of the matrix is zero.

(vii) If A and B are square matrices of order n, then
$$\det (AB) = (\det A)(\det B).$$

We next discuss *elementary row* and *column operations* for matrices:

(i)   Interchanging two rows or two columns;

(ii) Multiplying a row or column by a nonzero real number;

(iii) Adding a multiple of a row or column to another;

(iv) Adding a row or column to another (special case of (iii)).

A square matrix of order n is called an *elementary matrix* if it can be obtained from the identity matrix of order n by a single elementary row or column operation of type (i), (ii) or (iii). Elementary operations can be represented by elementary matrices in the following manner. Let E be the elementary matrix obtained by performing an elementary row (respectively, column) operation on I. If the same elementary row (respectively, column) operation is performed on a square matrix A of order n, then the resulting matrix is the same as the product EA (respectively, AE).

For instance, suppose $A = [a_{ij}]$, $i,j = 1,2,3$, and

$$E_{12} = \begin{bmatrix} 0 & 1 & 0 \\ 1 & 0 & 0 \\ 0 & 0 & 1 \end{bmatrix}$$

is the elementary matrix obtained by interchanging the first and second rows in I. Then

$$E_{12}A = \begin{bmatrix} 0 & 1 & 0 \\ 1 & 0 & 0 \\ 0 & 0 & 1 \end{bmatrix} \begin{bmatrix} a_{11} & a_{12} & a_{13} \\ a_{21} & a_{22} & a_{23} \\ a_{31} & a_{32} & a_{33} \end{bmatrix} = \begin{bmatrix} a_{21} & a_{22} & a_{23} \\ a_{11} & a_{12} & a_{13} \\ a_{31} & a_{32} & a_{33} \end{bmatrix}.$$

Thus the resulting matrix is obtained from A by interchanging the first and second rows.

We record the following observation here: If in the elementary row (or column) operation of type (iii) above, the multiplying factor is 0, then the resulting elementary matrix is the identity matrix; otherwise it is a product of two elementary matrices of type (ii) with an elementary matrix of type (iv), the latter appearing in the middle. An illustration is shown when 5 times the third row is added to the first row:

$$\begin{bmatrix} 1 & 0 & 5 \\ 0 & 1 & 0 \\ 0 & 0 & 1 \end{bmatrix} = \begin{bmatrix} 1 & 0 & 0 \\ 0 & 1 & 0 \\ 0 & 0 & \frac{1}{5} \end{bmatrix} \begin{bmatrix} 1 & 0 & 1 \\ 0 & 1 & 0 \\ 0 & 0 & 1 \end{bmatrix} \begin{bmatrix} 1 & 0 & 0 \\ 0 & 1 & 0 \\ 0 & 0 & 5 \end{bmatrix}.$$

A square matrix is said to be *invertible* or *nonsingular* if there exists a square matrix B of the same order such that $AB = BA = I$.

Such a matrix B, which can be proved to be unique (if it exists), is called the *inverse of A* and is denoted by A^{-1}. The inverse of A can be obtained from what is called the *adjoint of A*, written as adj(A) whose (i,j)th entry is the cofactor of a_{ji}, that is, $(-1)^{j+i} \det(M_{ji})$, where M_{ji} is the submatrix of order $n-1$ obtained from A by deleting the jth row and the ith column. The relation between the inverse and the adjoint is that

$$A^{-1} = \frac{1}{\det(A)} \, \text{adj}(A).$$

The following statements for a square matrix A of order n are equivalent:

(a) A is invertible;

(b)   There exists a unique square matrix B of order n such that $AB = BA$ = I;

(c) A is a product of elementary matrices;

(d) $\det(A) \neq 0$.

See Artin [2, p. 16], Gopalkrishnan [12, p. 245], Singh [24, p. 40] or Hoffman and Kunze [14, p. 255].

We shall need the following simple consequence: In view of the observation recorded above, every invertible matrix is a product of elementary matrices of type (i), (ii) or (iv).

2

Functions Between Euclidean Spaces

2-1 Background

Solving equations of various sorts is one of the main concerns of mathematics. Equations in which there is more than one unknown or 'variable' naturally involve functions of more than one variable. The phrase 'several variables' is to be understood in the sense 'more than one variable but including the possibility of one variable as a special case'.

The kind of equations that are of concern here are limited (e.g., differential and difference equations are excluded). Some equations may have no solution:

$$x = x + 1 \qquad \text{(no solution, obviously)}$$
$$(x + 1)^2 - x^2 = 2(x + 6) + 18 \qquad \text{(no solution, almost as obviously)}.$$

Some equations may have many solutions, such as $\sin x = 0$. For others, the solution required may be a function:

$$x^2 + y^2 = 1, \text{ find } x \text{ in terms of } y.$$

A system of equations (sometimes called *simultaneous equations*) may ask for some variables to be expressed in terms of the remaining variables:

$$2x + 3y + 7z - 8w = 3$$
$$4x + 6y + 8z - 7w = 4,$$

where x and y are to be obtained in terms of z and w (can't be done!). For linear systems the subject of linear algebra provides complete answers in terms of matrices and determinants. In what follows, we shall be concerned more with nonlinear systems, in which the left hand sides have continuous partial derivatives.

The answers provided for questions about nonlinear equations are not as satisfactory as for linear systems in linear algebra. No general solution methods are available for nonlinear systems; the sufficient conditions for existence of a solution are not necessary conditions, and even the existence has a limitation that is rather too technical to describe at this stage. Because of the limitation, the solutions obtained are *local solutions* in mathematical parlance.

A system of two equations such as

$$f_1(x_1, x_2, y_1, y_2, y_3) = 0$$
$$f_2(x_1, x_2, y_1, y_2, y_3) = 0$$

S. Shirali, H.L. Vasudeva, *Multivariable Analysis*,
DOI 10.1007/978-0-85729-192-9_2, © Springer-Verlag London Limited 2011

can be regarded as a single equation for the single function f that maps the point (x_1,x_2,y_1,y_2,y_3) of $\mathbb{R}\times\mathbb{R}\times\mathbb{R}\times\mathbb{R}\times\mathbb{R} = \mathbb{R}^5$ into the point

$$(f_1(x_1,x_2,y_1,y_2,y_3),\, f_2(x_1,x_2,y_1,y_2,y_3)) \text{ of } \mathbb{R}\times\mathbb{R} = \mathbb{R}^2.$$

The equation for f can be written as simply $f(u) = 0$, where u is the element (x_1,x_2,y_1,y_2,y_3) in $\mathbb{R}^5$ and 0 on the right hand side denotes the element $(0,0)$ of $\mathbb{R}^2$. If the intention is to solve for (x_1,x_2) in terms of (y_1,y_2,y_3), then we naturally think of $\mathbb{R}^5$ as $\mathbb{R}^2\times\mathbb{R}^3$ and write the equation as

$$f(x,y) = 0 \in \mathbb{R}^2,$$

where $x = (x_1,x_2) \in \mathbb{R}^2$ and $y = (y_1,y_2,y_3) \in \mathbb{R}^3$. The domain of f is then understood to be a subset of $\mathbb{R}^2\times\mathbb{R}^3$. In order to carry over ideas of continuity and differentiability to such functions, we need to know more about $\mathbb{R}^n$ when n may be greater than 1. We discuss the relevant aspects of $\mathbb{R}^n$ in the next section.

2-2 Euclidean Spaces

We begin with a formal definition of what we mean by $\mathbb{R}^n$ and other relevant terminology.

2-2.1. Definition. *The Cartesian product $\mathbb{R}\times\mathbb{R}\times\cdots\times\mathbb{R}$ (n factors) consisting of all ordered n-tuples $x = (x_1,x_2,\ldots,x_n)$, where $x_k \in \mathbb{R}$ for $1 \leq k \leq n$, is denoted by $\mathbb{R}^n$. By the **kth** **coordinate** (or **component**) of x, we mean the number x_k. The **sum of** x, $y \in \mathbb{R}^n$ is the ordered n-tuple $x + y$ for which the kth component is given by $(x + y)_k = x_k + y_k$ for $1 \leq k \leq n$. For $\alpha \in \mathbb{R}$, the **product** αx is the ordered n-tuple for which the kth component is $(\alpha x)_k = \alpha\cdot x_k$ for $1 \leq k \leq n$. That is to say,*

$$(x_1,x_2,\ldots,x_n) + (y_1,y_2,\ldots,y_n) = (x_1 + y_1,\, x_2 + y_2,\, \ldots,\, x_n + y_n)$$

and $\qquad\qquad \alpha(x_1,x_2,\ldots,x_n) = (\alpha x_1,\, \alpha x_2,\, \ldots,\, \alpha x_n).$

*The set $\mathbb{R}^n$ with sum and product as defined above will be called **Euclidean n-space**.*

We shall generally speak of 'components' when $n > 3$ and 'coordinates' when $n \leq 3$, except in a context where established convention dictates otherwise.

Elements of Euclidean n-space are often referred to as **vectors** or as **points**, or sometimes also as **n-vectors** if necessary. In the context of Euclidean spaces, real numbers are often called **scalars**. The reader who has encountered 'plane vectors' in an elementary context, will recognise that when (x_1,x_2) is regarded as providing the coordinates of the 'terminal point' of a plane vector, then addition as described above corresponds to the parallelogram law. Similarly in three dimensions. Also, αx is the vector obtained from x by 'scaling' it by a factor α.

The vector in $\mathbb{R}^n$ with $x_k = 0$ for every k is called the **zero vector** of $\mathbb{R}^n$ and is often denoted simply as 0, because usually there is no danger of confusion. It is trivial to see that the zero vector satisfies $0 + x = x$ for any vector $x \in \mathbb{R}^n$. Also, given a vector $x \in \mathbb{R}^n$, the associated vector $-x$ such that $(-x)_k = -x_k$ for $1 \le k \le n$ has the property that $(-x) + x = 0$. In fact, the same laws of addition hold as for real numbers. It follows that cancellation and other properties of addition in $\mathbb{R}$ continue to be valid in $\mathbb{R}^n$ and that terms in a finite sum of vectors can be rearranged at will.

When the symbol x_p is employed for the pth term of a sequence $\{x_p\}$ (either finite or infinite) in $\mathbb{R}^n$, it does not represent the pth component of any single vector called x. In such a situation, we shall denote the kth component ($1 \le k \le n$) of the vector x_p by the symbol $x^{(p)}_k$. In the next paragraph we deal with a special finite sequence $\{e_j\}_{1 \le j \le n}$ of vectors, using subscripts to denote the order of the term in the sequence and not to indicate a component.

The vectors

$$e_1 = (1,0,0,\dots,0), \;\; e_2 = (0,1,\dots,0), \;\; \dots \;\; , \;\; e_n = (0,0,0,\dots,0,1)$$

constitute what is called the **standard basis** of the Euclidean space $\mathbb{R}^n$. By an easy computation based on Def. 2-2.1, $x_1 e_1 + x_2 e_2 + \cdots + x_n e_n = (x_1, x_2, \dots, x_n)$. Furthermore, the converse is also true, namely, that any $x = (x_1, x_2, \dots, x_n) \in \mathbb{R}^n$ can be expressed as

$$x = x_1 e_1 + x_2 e_2 + \cdots + x_n e_n.$$

Such an expression for x is unique in the sense that, whenever $\xi_1, \xi_2, \dots, \xi_n$ are real numbers for which the equality $x = \xi_1 e_1 + \xi_2 e_2 + \cdots + \xi_n e_n$ holds, the 'coefficients' $\xi_1, \xi_2, \dots, \xi_n$ must necessarily be the components of x; this is because

$$x = x_1 e_1 + x_2 e_2 + \cdots + x_n e_n = (x_1, x_2, \dots, x_n)$$

and also

$$x = \xi_1 e_1 + \xi_2 e_2 + \cdots + \xi_n e_n = (\xi_1, \xi_2, \dots, \xi_n),$$

so that

$$(\xi_1, \xi_2, \dots, \xi_n) = (x_1, x_2, \dots, x_n).$$

2-2.2. Definition. *The* **inner product of** $x, y \in \mathbb{R}^n$ *is the real number* $\sum_{k=1}^{n} x_k y_k$ *and is denoted by* $x \cdot y$. *It is also known as* **dot product** *or* **scalar product**. *Since* $x \cdot x$ *is a sum of squares, it is always nonnegative and therefore has a unique nonnegative square root. The nonnegative square root of* $x \cdot x$ *is called the* **Euclidean norm of** x *and is denoted by* $\|x\|_2$. *Thus*

$$\|x\|_2 = \left(\sum_{k=1}^{n} x_k^2 \right)^{1/2}.$$

The reason for the subscript 2 will become clear later. The function which maps each $x \in \mathbb{R}^n$ into $\|x\|_2$ is called the Euclidean norm on $\mathbb{R}^n$ and is denoted by $\| \ \|_2$.

The reader is cautioned that some authors define a Euclidean space not as simply $\mathbb{R}^n$ but instead as $\mathbb{R}^n$ with the Euclidean norm; see Rudin [22, p. 16]. However, Apostol [1, p. 47] and Spivak [26, p. 1] define it as we do, while Sohrab [25, pp. 28, 159] defines it both ways.

The following properties are easy to establish; they hold whenever x, y are any n-vectors and α, β are any real numbers. (In (3), the symbol '1' stands for the real number 1.)

(1) $\alpha(x + y) = \alpha x + \alpha y$;

(2) $(\alpha + \beta)x = \alpha x + \beta x$;

(3) $1x = x$;

(4) $\alpha(\beta x) = (\alpha\beta)x = \beta\,(\alpha x)$.

There are many more and they will be used as and when needed.

2-2.3. Proposition. *Suppose* $x, y, z \in \mathbb{R}^n$ *and* α *is any real number. Then*

(a) $\|-x\|_2 = \|x\|_2 \geq 0$; *also* $\|x\|_2 = 0$ *if and only if* $x = 0$.

(b) $\|\alpha x\|_2 = |\alpha|\,\|x\|_2$.

(c) $|x \cdot y| \leq \|x\|_2 \|y\|_2$. *If equality holds here and* $\|x\|_2 \neq 0$, *then there exists some real number* β *such that* $y = \beta x$. *Similarly if* $\|y\|_2 \neq 0$. *This is the Cauchy–Schwarz inequality.*

(d) $\|x + y\|_2 \leq \|x\|_2 + \|y\|_2$.   (triangle inequality)

(e) $\|x - z\|_2 \leq \|x - y\|_2 + \|y - z\|_2$.

(f) $\big|\,\|x\|_2 - \|y\|_2\,\big| \leq \|x - y\|_2$.

Proof. (a) Since $\|x\|_2$ is, by definition, the nonnegative square root of $\sum_{k=1}^{n} x_k^2$, we have $\|-x\|_2 = \|x\|_2 \geq 0$. Also, $\|x\|_2 = 0$ if and only if $\|x\|_2^2 = 0$, which means $\sum_{k=1}^{n} x_k^2 = 0$. But each term in the sum $\sum_{k=1}^{n} x_k^2$ is nonnegative. Therefore $\sum_{k=1}^{n} x_k^2 = 0$ if and only if each $x_k = 0$, or equivalently, $x = 0$.

(b) $\|\alpha x\|_2^2 = \sum_{k=1}^{n} \alpha^2 x_k^2 = \alpha^2 \big(\sum_{k=1}^{n} x_k^2\big) = \alpha^2 \|x\|_2^2 = (|\alpha|\,\|x\|_2)^2$.

(c) For $1 \leq k \leq n$, we have $(x_k \|y\|_2 - y_k \|x\|_2)^2 \geq 0$. Therefore

$$x_k^2 \|y\|_2^2 - 2x_k y_k \|x\|_2 \|y\|_2 + y_k^2 \|x\|_2^2 \geq 0. \tag{1}$$

By taking the sum over $k = 1, \ldots, n$, we obtain

$$\big(\textstyle\sum_{k=1}^{n} x_k^2\big)\|y\|_2^2 - 2\big(\textstyle\sum_{k=1}^{n} x_k y_k\big)\|x\|_2 \|y\|_2 + \big(\textstyle\sum_{k=1}^{n} y_k^2\big)\|x\|_2^2 \geq 0.$$

In view of the definition of norm, this inequality becomes

$$\|x\|_2^2\|y\|_2^2 - 2(\sum_{k=1}^{n} x_k y_k)\|x\|_2\|y\|_2 + \|x\|_2^2\|y\|_2^2 \geq 0,$$

from which it follows that

$$\|x\|_2^2\|y\|_2^2 - (\sum_{k=1}^{n} x_k y_k)\|x\|_2\|y\|_2 \geq 0. \tag{2}$$

If $\|x\|_2 = 0$, then $x = 0$ by (a), and hence $\sum_{k=1}^{n} x_k y_k$, i.e., $x{\cdot}y = 0$, which guarantees that $|x{\cdot}y| \leq \|x\|_2\|y\|_2$. Similarly if $\|y\|_2 = 0$. Suppose neither $\|x\|_2$ nor $\|y\|_2$ is 0. Then (2) shows that $\|x\|_2\|y\|_2 \geq \sum_{k=1}^{n} x_k y_k = x{\cdot}y$. By virtue of (a), it further follows that $\|x\|_2\|y\|_2 = \|x\|_2\|{-}y\|_2 \geq x{\cdot}(-y) = -(x{\cdot}y)$. Thus we see that the inequality in (c) is valid. For the other assertion in (c), consider the possibility that $|x{\cdot}y| = \|x\|_2\|y\|_2$. When this is the case and $x{\cdot}y \geq 0$, equality must hold in (2) and therefore also in (1), $1 \leq k \leq n$. Consequently, $x_k\|y\|_2 - y_k\|x\|_2 = 0$ for $1 \leq k \leq n$. Hence the number $\beta = \|y\|_2/\|x\|_2$ must satisfy $y_k = \beta x_k$ for $1 \leq k \leq n$. If instead $x{\cdot}y \leq 0$, then we obtain the same conclusion upon replacing y by $-y$.

(d) $\|x + y\|_2^2 = \sum_{k=1}^{n} (x_k + y_k)^2 = \sum_{k=1}^{n} x_k^2 + 2\sum_{k=1}^{n} x_k y_k + \sum_{k=1}^{n} y_k^2$

$$= \|x\|_2^2 + 2(x{\cdot}y) + \|y\|_2^2$$

$$\leq \|x\|_2^2 + 2\|x\|_2\|y\|_2 + \|y\|_2^2 \qquad \text{by part (c)}$$

$$= (\|x\|_2 + \|y\|_2)^2.$$

(e) $\|x - z\|_2 = \|(x - y) + (y - z)\|_2 \leq \|x - y\|_2 + \|y - z\|_2$ by part (d).

(f) $\|x\|_2 = \|(x - y) + y\|_2 \leq \|x - y\|_2 + \|y\|_2$ by part (d). Therefore,

$$\|x\|_2 - \|y\|_2 \leq \|x - y\|_2.$$

By an analogous argument, $\|y\|_2 - \|x\|_2 \leq \|y - x\|_2$. But what has been proved in part (a) shows that $\|y - x\|_2 = \|{-}(x - y)\|_2 = \|x - y\|_2$. Therefore,

$$\|y\|_2 - \|x\|_2 \leq \|x - y\|_2.$$

The two inequalities displayed above together yield $|\|x\|_2 - \|y\|_2| \leq \|x - y\|_2$. $\square$

It is worth noting here that the proofs of (e) and (f) depend exclusively on (a) and (d) while the proofs of (a)–(d) invoke the definition of the Euclidean norm.

2-2.4. Remark. Note that an element of $\mathbb{R}$ can be regarded as an ordered 1-tuple. The Euclidean 1-space $\mathbb{R}^1$ is thus the set of real 1-tuples with sum and product as in Def. 2-2.1. The mapping ϕ from $\mathbb{R}^1$ to $\mathbb{R}$ such that $x \to x_1$ is clearly a bijection satisfying

$$\phi(x + y) = (x + y)_1 = x_1 + y_1 = \phi(x) + \phi(y), \quad x{\cdot}y = x_1 y_1 = \phi(x)\phi(y),$$

and $\phi(\alpha x) = (\alpha x)_1 = \alpha x_1 = \alpha\phi(x) = \phi^{-1}(\alpha)\cdot x$ for any real number α.

This means that, for all intents and purposes,

(i)   the sum of x and y as elements of $\mathbb{R}^1$ has the same meaning as the sum of $\phi(x)$ and $\phi(y)$ in $\mathbb{R}$;

(ii) the scalar product of x and y as elements of $\mathbb{R}^1$ has the same meaning as the product of $\phi(x)$ and $\phi(y)$ in $\mathbb{R}$; and

(iii) the product of $\alpha \in \mathbb{R}$ and $x \in \mathbb{R}^1$ has the same meaning as the product of $\alpha \in \mathbb{R}$ and $\phi(x) \in \mathbb{R}$, which is essentially the same as the inner product of $\phi^{-1}(\alpha) \in \mathbb{R}^1$ and $x \in \mathbb{R}^1$. We may ignore the distinction between the real number $\phi(x) \in \mathbb{R}$ and the vector $x \in \mathbb{R}^1$, as long as we keep in mind that both the products of $\mathbb{R}^1$ are actually the same as the ordinary product in $\mathbb{R}$.

Since $\|x\| = [x_1{}^2]^{1/2} = |x_1| = |\phi(x)|$ and we identify x with $\phi(x)$, we may write $\|x\| = |x|$.

In working with the concepts of convergence, open set, continuity and so forth in $\mathbb{R}$, the fact that the norm, i.e., absolute value, possesses properties (a), (b) and (d) of Proposition 2-2.3 is needed at every turn. Availability of these properties for the norm in $\mathbb{R}^n$ enables us to extend the concepts to $\mathbb{R}^n$ by having the norm take over the role of absolute value; this will become evident as the chapter proceeds.

There are two other standard norms, $\|\ \|_1$ and $\|\ \|_\infty$ on $\mathbb{R}^n$ defined as:

$$\|x\|_1 = \sum_{j=1}^{n} |x_j|, \qquad \|x\|_\infty = \max\{|x_j| : 1 \le j \le n\}.$$

These two norms can be shown to satisfy the analogues of parts of (a), (b) and (d), but not (c) of Proposition 2-2.3. This is the reason for the common name 'norm'. Other norms are also possible, but we shall not need them. Since (e) and (f) of the proposition follow solely from (a) and (d), they are true of all norms. The proof that $\|\ \|_1$ is a norm is left as a problem [see 2-2.P1], but we shall prove it for $\|\ \|_\infty$ in Proposition 2-2.5 below.

Before proceeding, we point out that the inequalities $|x_j| \le \|x\|_1$, $|x_j| \le \|x\|_2$, $|x_j| \le \|x\|_\infty$ always hold for $1 \le j \le n$.

2-2.5. Proposition. *Suppose* $x, y, z \in \mathbb{R}^n$ *and* α *is any real number. Then*
(a) $\|-x\|_\infty = \|x\|_\infty \ge 0$; *also* $\|x\|_\infty = 0$ *if and only if* $x = 0$.
(b) $\|\alpha x\|_\infty = |\alpha|\,\|x\|_\infty$.
(c) $\|x+y\|_\infty \le \|x\|_\infty + \|y\|_\infty$.   (triangle inequality)

Proof. (a) This is obvious from the fact that $\|x\|_\infty = \max\{|x_j| : 1 \le j \le n\} = \max\{|-x_j| : 1 \le j \le n\}$.

(b) If $\alpha = 0$, then $\alpha x_j = 0$ for $1 \leq j \leq n$ and hence $\|\alpha x\|_\infty = \max\{|\alpha x_j| : 1 \leq j \leq n\}$ $= 0$, while at the same time, $|\alpha| \|x\|_\infty = 0 \cdot \|x\|_\infty = 0$. If $\alpha \neq 0$, then $|\alpha| > 0$ and $|\alpha x_j|$ $\leq |\alpha x_k|$ if and only if $|x_j| \leq |x_k|$. Hence

$$\max\{|\alpha x_j| : 1 \leq j \leq n\} = |\alpha| \max\{|x_j| : 1 \leq j \leq n\},$$

i.e., $\|\alpha x\|_\infty = |\alpha| \|x\|_\infty$.

(c) By definition of $\|\ \|_\infty$, we have $|x_j| \leq \|x\|_\infty$ and $|y_j| \leq \|y\|_\infty$ for $1 \leq j \leq n$. It follows that $|x_j + y_j| \leq |x_j| + |y_j| \leq \|x\|_\infty + \|y\|_\infty$ for $1 \leq j \leq n$. Appealing to the definition of $\|\ \|_\infty$ once again, we conclude that $\|x + y\|_\infty \leq \|x\|_\infty + \|y\|_\infty$. □

What makes these norms useful is that they are simpler to compute than the Euclidean norm (no root is involved) and also have the following relation to the latter:

2-2.6. Proposition. *For any $x \in \mathbb{R}^n$, we have*

(a) $\|x\|_\infty \leq \|x\|_2 \leq \|x\|_1 \leq n \cdot \|x\|_\infty$.

(b) $\|x\|_2 \leq (\sqrt{n}) \cdot \|x\|_\infty$.

Proof. (a) The first inequality follows from the fact that one of the terms in the sum $\sum_{j=1}^{n} |x_j|^2 = \|x\|_2^2$ equals $\|x\|_\infty^2$.

For the second inequality, we use induction on n: If $n = 1$, then $\|x\|_2 = \|x\|_1$ for any x. Suppose $\|x\|_2 \leq \|x\|_1$ for any n-vector x. Then for any $(n+1)$-vector x,

$$\|x\|_2^2 = \sum_{k=1}^{n+1} x_k^2 = \sum_{k=1}^{n} x_k^2 + x_{n+1}^2 \leq (\sum_{k=1}^{n} |x_k|)^2 + x_{n+1}^2 \leq (\sum_{k=1}^{n} |x_k|)^2 + x_{n+1}^2 +$$
$$2(\sum_{k=1}^{n} |x_k|)|x_{n+1}| = (\sum_{k=1}^{n+1} |x_k|)^2 = \|x\|_1^2.$$

Since $|x_k| \leq \|x\|_\infty$ for $1 \leq k \leq n$, it follows that $\|x\|_1 = \sum_{j=1}^{n} |x_j| \leq n \cdot \|x\|_\infty$, which proves the third inequality.

(b) Since $|x_k| \leq \|x\|_\infty$ for $1 \leq k \leq n$, we have $\|x\|_2^2 = \sum_{j=1}^{n} |x_j|^2 \leq n \cdot \|x\|_\infty^2$. □

The inequalities (a) of Proposition 2-2.6 render the three norms equivalent in the sense that whatever we have to say in connection with convergence, continuity or differentiability will usually be true with reference to one norm if and only if it is true with reference to the other two norms. Thus, any one of the three norms in $\mathbb{R}^n$ can take over the role of absolute value in $\mathbb{R}$. In view of the equivalence, we shall often not specify which norm is intended and denote the norm by $\|\ \|$, i.e., without any subscript. Whenever there is a need to work with a specific norm, we shall choose the one that seems convenient for the situation at hand.

Here is the first instance of our not specifying a norm on account of reasons just explained:

2-2.7. Definition. *A sequence $\{x_p\}_{p\geq 1}$ in $\mathbb{R}^n$ is said to* **converge** *to $x \in \mathbb{R}^n$ if for every $\varepsilon > 0$ there exists a natural number N such that*

$$\|x_p - x\| < \varepsilon \quad \text{whenever} \quad p \geq N.$$

The element x of $\mathbb{R}^n$ (which can easily be shown to be unique) is called the **limit** *of the sequence. In symbols, $x_p \to x$ or $\lim_{p\to\infty} x_p = x$. A sequence is said to be* **convergent** *it converges to some limit.*

An alternative formulation would be that $x_p \to x$ if and only if the associated real sequence $\|x_p - x\|$ converges to 0 in the usual sense of elementary analysis.

Proofs of the properties that

$$\lim_{p\to\infty} (x_p + y_p) = \lim_{p\to\infty} x_p + \lim_{p\to\infty} y_p \quad \text{and} \quad \lim_{p\to\infty} (\alpha_p x_p) = (\lim_{p\to\infty} \alpha_p)(\lim_{p\to\infty} x_p)$$

whenever the limits on the right sides exist are completely analogous to those for real sequences and will therefore not be taken up. A similar remark applies to the result that a convergent sequence $\{x_p\}_{p\geq 1}$ in $\mathbb{R}^n$ is bounded in the sense that there exists a real number M such that $\|x_p\| \leq M$ for all p.

2-2.8. Proposition. *A sequence $\{x_p\}_{p\geq 1}$ in $\mathbb{R}^n$ converges to $x \in \mathbb{R}^n$ if and only if the real sequence $\{x^{(p)}_j\}_{p\geq 1}$ converges to the real number x_j for each j. In other words, convergence in $\mathbb{R}^n$ is equivalent to componentwise convergence.*

Proof. Note that $(x_p - x)_j = x^{(p)}_j - x_j$ by Def. 2-2.1. Therefore $|x^{(p)}_j - x_j| = |(x_p - x)_j| \leq \|(x_p - x)\|$. This implies that if $x_p \to x$ then $x^{(p)}_j \to x_j$ for each j.

For the converse, suppose $x^{(p)}_j \to x_j$ for each j and consider any $\varepsilon > 0$. For each j there exists N_j such that $|(x_p - x)_j| < \varepsilon$ whenever $p \geq N_j$. Set $N = \max\{N_j : 1 \leq j \leq n\}$. Then, for every j, $1 \leq j \leq n$, $|(x_p - x)_j| < \varepsilon$ whenever $p \geq N$. It follows that $\max\{|(x_p - x)_j| : 1 \leq j \leq n\} < \varepsilon$ whenever $p \geq N$. But this means precisely that $\|x_p - x\|_\infty < \varepsilon$ whenever $p \geq N$. By Proposition 2-2.6, such an N exists for the other two norms as well. $\qquad\square$

2-2.9. Definition. *A sequence $\{x_n\}_{n\geq 1}$ in $\mathbb{R}^n$ is called a* **Cauchy sequence** *if for every $\varepsilon > 0$, there exists a natural number N such that*

$$\|x_p - x_q\| < \varepsilon \quad \text{whenever} \quad p \geq N \text{ and } q \geq N.$$

As in the case of $\mathbb{R}$, it is easy to prove that a convergent sequence in $\mathbb{R}^n$ is Cauchy. The least upper bound property of $\mathbb{R}$ has the important consequence that a Cauchy sequence in $\mathbb{R}$ is always convergent. This carries over to $\mathbb{R}^n$ with very little effort, as we shall now see. Thus $\mathbb{R}^n$ is '(Cauchy) complete'.

2-2.10. Theorem. *Any Cauchy sequence in $\mathbb{R}^n$ converges to some limit.*

Proof. Let $\{x_p\}_{p\geq 1}$ be a Cauchy sequence in $\mathbb{R}^n$. Now, $(x_p - x_q)_j = x^{(p)}_j - x^{(q)}_j$ by Def. 2-2.1, and we therefore have $|x^{(p)}_j - x^{(q)}_j| = |(x_p - x_q)_j| \leq \|x_p - x_q\|$. It follows that, for each j such that $1 \leq j \leq n$, the real sequence $\{x^{(p)}_j\}_{p\geq1}$ is Cauchy and hence converges to some limit in $\mathbb{R}$; denote its limit by x_j. Then the vector

$$x = (x_1, \ldots, x_n) \in \mathbb{R}^n$$

is seen to have the property that $x_p \to x$ in view of Proposition 2-2.8.   □

2-2.11. Proposition. *Suppose $\{x_p\}_{p\geq1}$ is a sequence in $\mathbb{R}^n$ and $x \in \mathbb{R}^n$. If for every $\varepsilon > 0$ and every $N \in \mathbb{N}$, there exists some integer $p \in \mathbb{N}$ satisfying $p \geq N$ as well as $\|x_p - x\| < \varepsilon$, then $\{x_p\}_{p\geq1}$ has a subsequence converging to x.*

Proof. Consider $\varepsilon = 1$ and $N = 1$. By hypothesis, some integer $p_1 \in \mathbb{N}$ satisfies $p_1 \geq 1$ as well as $\|x_{p_1} - x\| < 1$. Now consider $\varepsilon = \frac{1}{2}$ and $N = p_1 + 1$. Then by hypothesis, some integer $p_2 \in \mathbb{N}$ satisfies $p_2 > p_1$ as well as $\|x_{p_2} - x\| < \frac{1}{2}$. Next, we consider $\varepsilon = \frac{1}{3}$ and $N = p_2 + 1$, and apply the hypothesis once again. Proceeding in this manner, we obtain a subsequence $\{x_{p_q}\}_{q\geq1}$ such that $\|x_{p_q} - x\| < \frac{1}{q}$. This inequality implies that $\{x_{p_q}\}_{q\geq1}$ converges to x.   □

Problem Set 2-2

2-2.P1. Show that the analogues of Proposition 2-2.3 for all parts except (c) hold for $\| \ \|_1$.

2-2.P2. Show that $\|x\|_1 \leq n^{1/2}\|x\|_2$.

2-2.P3. Show that if $x \cdot y = 0$, then $\|x + y\|_2^2 = \|x\|_2^2 + \|y\|_2^2$.

2-2.P4. If a, b, c are positive real numbers, show that $abc(a + b + c) \leq a^3b + b^3c + c^3a$.

2-2.P5. If $x_p \to x$, show that $\|x_p\| \to \|x\|$.

2-2.P6. (a) Suppose $0 < p < q$ and $0 \leq a_j$ for $1 \leq j \leq n$. Then prove that

$$\left(\sum_{j=1}^{n} a_j^q\right)^{1/q} \leq \left(\sum_{j=1}^{n} a_j^p\right)^{1/p}.$$

(b) Show that $\lim_{p\to\infty} \|x\|_p = \|x\|_\infty$, where $x = (x_1, \ldots, x_n) \in \mathbb{R}^n$ and $\|x\|_p = \left(\sum_{j=1}^{n} |x_j|^p\right)^{1/p}$, $\|x\|_\infty = \max\{|x_i| : 1 \leq j \leq n\}$.

2-3 Simplest Functions Between Euclidean Spaces (Linear)

A function (or 'map') will often be referred to as a **transformation** when the domain is a subset of $\mathbb{R}^n$ with $n > 1$.

2-3.1. Definition. *A map (or mapping)* $A:\mathbb{R}^n \to \mathbb{R}^m$ *is called* **linear** *if*

$$A(x_1 + x_2) = A(x_1) + A(x_2) \text{ and } A(cx) = cA(x) \quad \forall\, x_1, x_2, x \in \mathbb{R}^n \text{ and } c \in \mathbb{R}.$$

When a map A is linear, we shall delete the parentheses '( )' in 'A(x)' whenever convenient. Thus the above conditions defining linearity can also be written as

$$A(x_1 + x_2) = Ax_1 + Ax_2 \text{ and } A(cx) = c(Ax) \quad \forall\, x_1, x_2, x \in \mathbb{R}^n \text{ and } c \in \mathbb{R}.$$

A linear map is sometimes called a **linear operator** *or a* **linear transformation**.

2-3.2. Examples. (a) $n = m = 1$. The map $A:\mathbb{R}^n \to \mathbb{R}^m$ defined by $Ax = 5x$ is easily seen to be linear: $5(x_1 + x_2) = 5x_1 + 5x_2$ and $5(cx) = c(5x)$. Instead of 5 any other number could have been taken of course; thus the map A such that $A(x) = ax$ is linear, whatever the number a may be.

In fact, these are the only linear maps when $n = m = 1$, because $A(x) = A(x \cdot 1) = x \cdot A(1) = ax$, where $a = A(1)$. This will also follow from more general considerations below.

(b) $n = 1$ but m is any positive integer. Let b be any vector in $\mathbb{R}^m$. Define the map $A:\mathbb{R}^n \to \mathbb{R}^m$ by $Ax = xb$ (product of the scalar x with vector b). Since

$$(x_1 + x_2)b = x_1 b + x_2 b \quad \text{and} \quad x(cb) = c(xb),$$

which is to say,

$$A(x_1 + x_2) = Ax_1 + Ax_2 \quad \text{and} \quad A(cb) = c(Ax),$$

the map A is linear. The special vector b that plays a role in describing A can be expressed as $b = 1b = A1$.

Conversely, any linear map $A:\mathbb{R}^n \to \mathbb{R}^m$ (where $n = 1$) is of this kind, because if we set $b = A1$, we have $Ax = A(x \cdot 1) = x \cdot A(1) = xb$.

Linear maps $A:\mathbb{R}^n \to \mathbb{R}^m$ with $n = 1$ will be referred to again and the reader would do well to keep this example in mind for ready retrieval when it is mentioned later on.

(c) $m = 1$ but n is any positive integer. Let $z \in \mathbb{R}^n$. Define a map $\mathbb{R}^n \to \mathbb{R}$ by $x \to z \cdot x$, the dot product of z and x in $\mathbb{R}^n$. Then elementary properties of the dot product lead to

$$z \cdot (x_1 + x_2) = z \cdot x_1 + z \cdot x_2, \quad z \cdot (cx) = c(z \cdot x),$$

which is the same as saying that the map $x \to z \cdot x$ from $\mathbb{R}^n$ to $\mathbb{R}$ is linear. This example will also be needed in the sequel.

(d) $n = m = 2$. Let a, b, c, d be real numbers. The map $A:\mathbb{R}^2 \to \mathbb{R}^2$ defined as $A(x,y) = (ax + by, cx + dy)$ is linear. A part of verifying the linearity is to check that, for any $(x, y) \in \mathbb{R}^2$ and $(x', y') \in \mathbb{R}^2$, the vector $(a(x + x') + b(y + y'), c(x + x') + d(y + y'))$ is the sum of $(ax + by, cx + dy)$ and $(ax' + by', cx' + dy')$. This is easily checked. The other part is to check that, for any scalar λ and any $(x, y) \in \mathbb{R}^2$, the vector

$$(a(\lambda x) + b(\lambda y), c(\lambda x) + d(\lambda y))$$

is the same as $\lambda(ax + by, cx + dy)$.

This too is easy to verify.

To solve the linear equations

$$ax + by = u$$
$$cx + dy = v,$$

where u and v are given, is to find (x, y) such that $A(x, y) = (u, v)$.

It is a consequence of the discussion below that the only linear maps of $\mathbb{R}^2$ into $\mathbb{R}^2$ are of the kind described in the foregoing example.

A linear map from a space $\mathbb{R}^n$ into itself is often called a **linear map** (or **linear operator**) *in* $\mathbb{R}^n$.

Any linear map A satisfies $A0 = 0$, because $A0 = A(0 + 0) = A0 + A0 = 2(A0)$. The fact that $A0 = 0$ for any linear map A will be used in future without explicit mention.

Let $e_1, e_2, \ldots, e_n$ be the standard basis

$$e_1 = (1,0,0,\ldots,0), \quad e_2 = (0,1,\ldots,0), \quad \ldots, \quad e_n = (0,0,0,\ldots,0,1)$$

of $\mathbb{R}^n$, and $A:\mathbb{R}^n \to \mathbb{R}^m$ be linear. If the (vector) values of $Ae_1, Ae_2, \ldots, Ae_n$ are given, then the value of Ax for any vector $x = x_1 e_1 + x_2 e_2 + \cdots + x_n e_n \in \mathbb{R}^n$ can be found easily, because

$$Ax = x_1(Ae_1) + x_2(Ae_2) + \cdots + x_n(Ae_n). \tag{1}$$

Some readers may prefer to express this informally as follows: If the vector x has scalar components $x_1, x_2, \ldots, x_n$ then Ax can be expressed in terms of $Ae_1, Ae_2, \ldots, Ae_n$ with the very same scalar coefficients, namely, $x_1, x_2, \ldots, x_n$. (Caution: It may be possible to express Ax this way with other coefficients as well.)

One very useful consequence of (1) is that if we know what A maps the n vectors $e_1, e_2, \ldots, e_n$ into, then we know what A maps all the infinitely many vectors of $\mathbb{R}^n$ into. Thus we can create a linear map by simply deciding what the vectors $Ae_1, Ae_2, \ldots, Ae_n$ should be and then leaving the rest to linearity via (1).

Now let $f_1, f_2, \ldots, f_m$ be the standard basis of $\mathbb{R}^m$. Consider any linear map $A:\mathbb{R}^n \to \mathbb{R}^m$. For each j from 1 to n, the vector $Ae_j \in \mathbb{R}^m$ has m components, which we shall name as a_{ij}:

Then
$$a_{ij} = (Ae_j)_i, \ 1 \leq i \leq m.$$

$$Ae_j = a_{1j} f_1 + a_{2j} f_2 + \cdots + a_{mj} f_m = \sum_{i=1}^{m} a_{ij} f_i \ , \ 1 \leq j \leq n.$$

The coefficients a_{ij} (which are mn in number) form an $m \times n$ matrix in the usual way. Observe that the summation takes place over the first index i, i.e., the row index, of a_{ij}. In view of (1), for any vector $x = x_1 e_1 + x_2 e_2 + \cdots + x_n e_n \in \mathbb{R}^n$, the image Ax is

$$Ax = x_1 \left(\sum_{i=1}^{m} a_{i1} f_i \right) + x_2 \left(\sum_{i=1}^{m} a_{i2} f_i \right) + \cdots + x_n \left(\sum_{i=1}^{m} a_{in} f_i \right)$$

$$= \left(\sum_{j=1}^{n} a_{1j} x_j \right) f_1 + \left(\sum_{j=1}^{n} a_{2j} x_j \right) f_2 + \cdots + \left(\sum_{j=1}^{n} a_{mj} x_j \right) f_m. \tag{2}$$

This shows that, for any linear map $A : \mathbb{R}^n \to \mathbb{R}^m$, there exist mn numbers a_{ij} ($1 \leq i \leq m$, $1 \leq j \leq n$) such that the image of any element $(x_1, x_2, \ldots, x_n) \in \mathbb{R}^n$ is $A(x_1, x_2, \ldots, x_n) = (y_1, y_2, \ldots, y_m) \in \mathbb{R}^m$, where

$$\left. \begin{aligned} y_1 &= a_{11} x_1 + a_{12} x_2 + \cdots + a_{1n} x_n \\ y_2 &= a_{21} x_1 + a_{22} x_2 + \cdots + a_{2n} x_n \\ &\vdots \\ y_m &= a_{m1} x_1 + a_{m2} x_2 + \cdots + a_{mn} x_n \end{aligned} \right\}. \tag{3}$$

In other words, if we represent x by the $n \times 1$ column matrix $[X]$ with entries $x_1, x_2, \ldots, x_n$, and represent $Ax = y$ by the $m \times 1$ column matrix $[Y]$ with entries $y_1, y_2, \ldots, y_m$, then $[Y]$ equals the matrix product $[A][X]$, where $[A]$ is the $m \times n$ matrix $[a_{ij}]$. We refer to the matrix $[A]$ with entries a_{ij} arising from the linear transformation A as the **matrix of** A. In particular, when $n = m = 1$, A is of the form $y = \alpha x$. Also, when $n = m = 2$, A is of the form described in Example 2-3.2(d).

In the reverse direction, any $m \times n$ matrix $[a_{ij}]$ gives rise to a linear transformation $A : \mathbb{R}^n \to \mathbb{R}^m$, namely the one defined by (2).

The correspondence described above between linear maps and their matrices is one-to-one and therefore a linear map can be completely specified through its matrix.

As with a map of any kind of a set X into itself, the inverse of A is a map $S : X \to X$ such that the compositions $A \circ S$ and $S \circ A$ are both equal to the 'identity map' $I : X \to X$ given by $I(x) = x$ for all $x \in X$. In this context, when $X = \mathbb{R}^n$, the identity map is clearly linear; moreover, the inverse, if any, is also linear. It will be denoted by A^{-1}. Considerable interest attaches to the question when a given linear map has an inverse. The elementary fact that a composition $A \circ B$ of invertible maps is invertible, with inverse $B^{-1} \circ A^{-1}$, will be used in Theorem 2-7.11.

If A and B are linear maps from $\mathbb{R}^n$ to $\mathbb{R}^m$, then the map $x \rightarrow (Ax + Bx)$ is easily seen to be a linear map. It is called the **sum** of A and B, and is denoted by $A + B$. Thus

$$(A + B)x = Ax + Bx \quad \text{whenever } x \in \mathbb{R}^n.$$

If $\lambda \in \mathbb{R}$, then the map $x \rightarrow \lambda(Ax)$ is also seen to be a linear map; it is called the **product** of λ and A, and is denoted by λA. The map $(-1)A$ will be denoted by the symbol $-A$. We shall have occasion to refer to the constant map $x \rightarrow 0$; this constant map will be denoted by O. Such properties as the following are easy to verify:

$$A + B = B + A, \quad A + (B + C) = (A + B) + C, \quad A + O = A,$$

$$A + (-A) = O, \quad \lambda(A + B) = \lambda A + \lambda B, \quad \lambda(\mu A) = (\lambda \mu)A,$$

and so on. Thus, linear maps behave rather like vectors in $\mathbb{R}^k$ for some k with regard to addition and to multiplication by scalars. One can even argue that k should be the product mn, but this is a matter we do not pursue here.

Linear maps, like other maps, can be composed whenever the range of one is a subset of the domain of the other. If $A:\mathbb{R}^n \rightarrow \mathbb{R}^m$ and $B:\mathbb{R}^m \rightarrow \mathbb{R}^p$ are linear, then the composition $B \circ A:\mathbb{R}^n \rightarrow \mathbb{R}^p$ is a linear map, as is easy to check. It is denoted simply by BA, without the symbol $\circ$ to indicate composition, and is called the **product** of the linear maps A and B. If $n = m = p$, then both the products AB and BA are defined, but they are not necessarily equal. In other words, multiplication of linear maps is not commutative (unless $n = 1$). Such properties as

$$A(B + C) = AB + AC, \quad (B + C)A = BA + CA, \quad (\lambda A)(\mu B) = (\lambda \mu)(AB),$$

and so on are easy to verify.

2-3.3. Remarks. (a) Let $A:\mathbb{R}^n \rightarrow \mathbb{R}^m$ and $B:\mathbb{R}^m \rightarrow \mathbb{R}^p$ be linear maps with matrices $[a_{ij}]$ and $[b_{ki}]$ respectively. Then the matrix of the composed map $BA:\mathbb{R}^n \rightarrow \mathbb{R}^p$ is the matrix product $[b_{ki}][a_{ij}]$.

Proof. Let $\{e_j : 1 \leq j \leq n\}$, $\{f_i : 1 \leq i \leq m\}$ and $\{g_k : 1 \leq k \leq p\}$ be the standard bases of $\mathbb{R}^n$, $\mathbb{R}^m$ and $\mathbb{R}^p$, respectively. Then

$$Ae_j = \sum_{i=1}^{m} a_{ij} f_i, \quad Bf_i = \sum_{k=1}^{p} b_{ki} g_k.$$

Hence, $\quad (BA)e_j = \sum_{i=1}^{m} a_{ij} Bf_i = \sum_{i=1}^{m} a_{ij} \left(\sum_{k=1}^{p} b_{ki} g_k \right) = \sum_{k=1}^{p} \left(\sum_{i=1}^{m} b_{ki} a_{ij} \right) g_k.$

But this says precisely that BA has matrix $[c_{kj}]$ with $c_{kj} = \sum_{i=1}^{m} b_{ki} a_{ij}$. By the definition of matrix product, the matrix of BA is therefore $[b_{ki}][a_{ij}]$.

We note a useful consequence of this result: The identity linear map I of $\mathbb{R}^n$ into itself has the matrix known by the familiar name of the identity matrix of order $n \times n$; combined with the above result, this shows that a linear map $A:\mathbb{R}^n \to \mathbb{R}^n$ is invertible if and only if its matrix is invertible.

(b) To solve the linear equations (3), where $y = (y_1, y_2, \dots, y_m)$ is given, is to find $x = (x_1, x_2, \dots, x_n)$ such that $Ax = y$. A unique solution x exists for every given y if and only if A is invertible; when this is the case, $x = A^{-1}y$. However, an inverse can exist only if $m = n$. We shall have no occasion to make use of this fact, but whenever we assume that some linear map $A:\mathbb{R}^n \to \mathbb{R}^m$ is invertible, we shall also assume that $m = n$.

If $A:\mathbb{R}^n \to \mathbb{R}^n$ is injective, then the only solution of (3) with each $y_i = 0$ is the one for which each $x_j = 0$. It is known from linear algebra that this implies that for every $(y_1, y_2, \dots, y_n) \in \mathbb{R}^n$, (3) has a solution $(x_1, x_2, \dots, x_n) \in \mathbb{R}^n$, which has the consequence that A is surjective. Thus an injective linear map $A:\mathbb{R}^n \to \mathbb{R}^n$ is surjective and hence invertible. We shall use this fact in the proof of the implicit function theorem (Theorem 4-3.2). It is also true that a surjective linear map $A:\mathbb{R}^n \to \mathbb{R}^n$ is injective and hence invertible.

(c) Consider (3) above with $m = n$ written in the form $[Y] = [A][X]$. If the inverse matrix $[A]^{-1}$ exists, then

$$[A]^{-1}[Y] = [A]^{-1}([A][X])$$
$$= ([A]^{-1}[A])[X] \quad \text{using associativity}$$
$$= [X].$$

Thus (3) has a unique solution $[X] = [A]^{-1}[Y]$ for every given $[Y]$. On the other hand, suppose we know that (3) has a unique solution $[X]$ for every given $[Y]$. This is the same as saying that it has a unique solution x for every given y. As in (b) above, this implies that A is invertible, and hence by (a), $[A]$ is an invertible matrix. Once again, the unique solution is given by $[X] = [A]^{-1}[Y]$.

2-3.4 Remarks. If a linear map A of $\mathbb{R}^n$ into itself merely multiplies the kth component by some nonzero a, i.e.,

$$A(x_1, \dots, x_n) = (y_1, \dots, y_n),$$

where

$$y_j = \begin{cases} x_j & \text{for } j \neq k \\ ax_j & \text{for } j = k, \end{cases}$$

then its matrix is the one obtained from the identity matrix by replacing the entry a_{kk} by a. Thus it is an elementary matrix; moreover, its determinant is a.

If A merely interchanges two components, i.e., $A(x_1, \dots, x_n) = (y_1, \dots, y_n)$, where

$$y_j = \begin{cases} x_l & \text{if } j = k \\ x_k & \text{if } j = l \\ x_j & \text{if } k \neq j \neq l \end{cases}$$

where $k \neq l$, then its matrix is the one obtained from the identity matrix by interchanging the kth and lth rows. Thus it is an elementary matrix; moreover, its determinant is -1.

If A merely adds one row to another row, i.e., there exist distinct indices k, l such that

$$A(x_1, \ldots, x_n) = (y_1, \ldots, y_n),$$

where

$$y_j = \begin{cases} x_j & \text{for } j \neq k \\ x_j + x_l & \text{for } j = k \ , \end{cases}$$

(so that $y_k = x_k + x_\ell$), then its matrix is the one obtained from the identity matrix by adding the lth row to the kth row. Thus it is an elementary matrix; moreover, its determinant is 1.

Since every invertible matrix is a product of elementary matrices of the above type (see last part of Chapter 1), it follows that every invertible linear map is a product of linear maps having elementary matrices of the type mentioned above. We shall make essential use of this fact in Proposition 7-4.1.

Problem Set 2-3

2-3.P1. Let $A:\mathbb{R} \to \mathbb{R}$ satisfy $A(cx) = cA(x)$ for any x and $c \in \mathbb{R}$. Set $a = A(1)$. Show that $A(x) = ax$ whenever $x \in \mathbb{R}$. Is it true that $A(x + y) = A(x) + A(y)$ whenever $x \in \mathbb{R}$ and $y \in \mathbb{R}$?

2-3.P2. Define $A:\mathbb{R}^2 \to \mathbb{R}^2$ by $A(x_1, x_2) = ((x_1^3 + x_2^3)^{1/3}, 0)$. Show that, for any $x = (x_1, x_2) \in \mathbb{R}^2$ and $c \in \mathbb{R}$, we have $A(cx) = cA(x)$. Is it true that

$$A(x + y) = A(x) + A(y) \quad \text{whenever } x \text{ and } y \in \mathbb{R}^2?$$

2-3.P3. Define $A:\mathbb{R}^2 \to \mathbb{R}^2$ by $A(x_1, x_2) = (3x_1 - 2x_2, 6x_1 + x_2)$. Show that

$$A(x + y) = A(x) + A(y) \text{ and } A(cx) = cA(x) \ \forall \ x, y \in \mathbb{R}^2 \text{ and } c \in \mathbb{R}.$$

2-3.P4. Find the range of the map $f:U \to \mathbb{R}^2$, where $U = \{(x,y) \in \mathbb{R}^2 : (x,y) \neq (0,0)\}$ and

$$f(x,y) = (f_1(x,y), f_2(x,y)) = \left(\frac{x^2 - y^2}{x^2 + y^2}, \frac{xy}{x^2 + y^2} \right) \in \mathbb{R}^2.$$

2-3.P5. Let $f:\mathbb{R} \to \mathbb{R}$ have a continuous derivative everywhere and let $\phi:\mathbb{R}^2 \to \mathbb{R}^2$ be the transformation

$$u = f(x), \qquad v = -y + xf(x).$$

If $f'(x_0) \neq 0$, show that the transformation is invertible on a subset of the form $I \times \mathbb{R}$, where I is a open set in $\mathbb{R}$ containing x_0 and that the inverse has the form

$$x = g(u), \qquad y = -v + ug(u).$$

2-3.P6. Let $A:\mathbb{R}^n \times \mathbb{R}^m \to \mathbb{R}^k$ be a linear map. Show that the maps $B:\mathbb{R}^n \to \mathbb{R}^k$ and $C:\mathbb{R}^m \to \mathbb{R}^k$ defined by $B(x) = A(x,0)$ and $C(y) = A(0,y)$ are also linear.

2-3.P7. Let $A:\mathbb{R}^n \to \mathbb{R}^k$ and $B:\mathbb{R}^m \to \mathbb{R}^k$ be linear maps. Show that the map $C:\mathbb{R}^n \times \mathbb{R}^m \to \mathbb{R}^k$ defined by $C(x,y) = A(x) + B(y)$ is linear.

2-3.P8. The Cartesian product $\mathbb{R}^n \times \mathbb{R}^m$ can be regarded as $\mathbb{R}^{n+m}$. Show that the map $A:\mathbb{R}^{n+m} \to \mathbb{R}^{n+m}$ defined by $A(a,b) = (0,b)$ is linear, and find its matrix when $m = 2, n = 3$.

2-3.P9. The equations $x_1 + x_2 + x_3 = 5, 2x_1 - x_2 + 4x_3 = 8$ can be expressed in terms of standard bases u_1, u_2, u_3 of $\mathbb{R}^3$ and v_1, v_2 of $\mathbb{R}^2$ as a single equation $f(x_1 u_1 + x_2 u_2 + x_3 u_3) = 5v_1 + 8v_2$, where $f:\mathbb{R}^3 \to \mathbb{R}^2$ is the function such that

$$f(x_1 u_1 + x_2 u_2 + x_3 u_3) = (x_1 + x_2 + x_3)v_1 + (2x_1 - x_2 + 4x_3)v_2.$$

One can also use coordinate language and avoid bringing in the standard basis explicitly by writing $f(x_1,x_2,x_3) = (5, 8)$, where $f(x_1,x_2,x_3)$ is defined as $(x_1 + x_2 + x_3, 2x_1 - x_2 + 4x_3)$. Now express the equations $p = e^x \cos y, q = e^x \sin y$ as a single equation using a suitable function; also rewrite the single equation in coordinate language.

2-3.P10. Eliminate the variable x_1 from the second and third equations in the system by using the first equation

$$x_1 + 5x_2 + 6x_3 = 9$$
$$2x_1 + 11x_2 + 13x_3 = 38$$
$$3x_1 + 12x_2 + 14x_3 = 2027.$$

(a) Now answer the following questions regarding the new system consisting of the two equations just obtained by eliminating x_1 together with the first of the three given equations: Must every solution of the new system be a solution of the given system (Yes or No)? And vice versa (Yes or No)?

(b) If a solution $x_2 = \beta$, $x_3 = \gamma$ of the two equations that have been obtained by elimination is known, what formula for x_1 can be derived from the first of the three given equations in terms of β and γ?

2-3.P11. In $\mathbb{R}^4$, find a common perpendicular (not 0) to the three given vectors

$$z_1 = (1,3,2,-1), \quad z_2 = (3,10,4,0), \quad z_3 = (4,13,7,4).$$

In other words find $x \in \mathbb{R}^4$ such that $x \cdot z_i = 0$ for $i = 1,2,3$.

2-3.P12. The range of the function f defined on the subset

$$D = \{(x_1,x_2,x_3) : x_1^2 + x_2^2 + (x_3 - \tfrac{1}{2})^2 = \tfrac{1}{2}\} \setminus \{(0,0,1)\}$$

of $\mathbb{R}^3$ by

$$y_1 = \frac{x_1}{1-x_3}, \; y_2 = \frac{x_2}{1-x_3}$$

is the whole of $\mathbb{R}^2$. The map is called the *stereographic projection*.

2-3.P13. Let $D = \{(x_1,x_2,x_3) \in \mathbb{R}^3 : x_1{}^2 + x_2{}^2 + x_3{}^2 \geq 1\}$ and $R = \{(y_1,y_2,y_3) \in \mathbb{R}^3 : 0 < y_1{}^2 + y_2{}^2 + y_3{}^2 \leq 1\}$. The function f defined by

$$y_1 = \frac{x_1}{r^2}, \; y_2 = \frac{x_2}{r^2}, \; y_3 = \frac{x_3}{r^2}$$

is called an inversion mapping of a part of $\mathbb{R}^3$ to another part. Show that the range of the mapping is R.

2-3.P14. Given any $\delta > 0$ and $a \in \mathbb{R}$, show that the function $F:\mathbb{R}^3 \to \mathbb{R}$ defined by $F(x,y,z) = xyz(x+y+z-1)$ takes positive as well as negative values in the δ-ball centred at $(a,0,0)$.

2-3.P15. Given any $\delta > 0$ and $b,c \in \mathbb{R}$ such that $b \neq 0 \neq c$ and $b+c = 1$, show that the function $F:\mathbb{R}^3 \to \mathbb{R}$ defined by $F(x,y,z) = xyz(x+y+z-1)$ takes positive as well as negative values in the δ-ball centred at $(0,b,c)$.

2-3.P16. Let $A:\mathbb{R}^2 \to \mathbb{R}^2$ be the linear map $A(x_1,x_2) = (x_1 + x_2, 0)$. The vectors $(1,0)$, $(0,1)$, $(3/5,4/5)$, $(12/13,5/13)$ and $(1/\sqrt{2}, 1/\sqrt{2})$ all have norm 1. Compute the norms of their images under A, i.e.,

$$A(1,0), \; A(0,1), \; A(3/5,4/5), \; A(12/13,5/13), \; A(1/\sqrt{2}, 1/\sqrt{2}).$$

Which is the largest?

Show that: $\|x\| \leq 1 \Rightarrow \|Ax\| \leq \sqrt{2}$, i.e., that $x_1{}^2 + x_2{}^2 \leq 1 \Rightarrow (x_1 + x_2)^2 + 0^2 \leq 2$.

2-3.P17. Let $A:\mathbb{R}^2 \to \mathbb{R}^2$ be the linear map $A(x_1,x_2) = (x_1 + x_2, 2x_1 - x_2)$. Compute

$$\|A(1,0)\|^2, \; \|A(0,1)\|^2, \; \|A(1/\sqrt{2}, 1/\sqrt{2})\|^2, \; \|A(12/13, 5/13)\|^2.$$

(a) Show that $\|A(x_1,x_2)\|^2 = 5x_1{}^2 + 2x_2{}^2 - 2x_1x_2$.
(b) Using (a), show that $\|x\| \leq 1 \Rightarrow \|Ax\| \leq \sqrt{6}$.
(c) Using (b), what can be said about $\sup \{\|Ax\| : \|x\| \leq 1\}$?

2-3.P18. Let $A:\mathbb{R}^2 \to \mathbb{R}^2$ be the linear map $A(x_1,x_2) = (x_1 + x_2, x_1 - x_2)$. Express $A(x_1,x_2)^2$ in terms of x_1 and x_2; hence find $\sup \{\|Ax\| : \|x\| \leq 1\}$.

2-3.P19. Let $A:\mathbb{R}^n \to \mathbb{R}^m$ be linear and let V a ball in $\mathbb{R}^n$ centred at 0 with some radius $a > 0$. If $Av = 0 \; \forall \; v \in V$, show that $Ax = 0 \; \forall \; x \in \mathbb{R}^n$.

2-4 Topology of Euclidean Spaces

In elementary analytic geometry, $\mathbb{R}^2$ and $\mathbb{R}^3$ are regarded as the plane and three-dimensional space, respectively, and the so called 'distance formula' says that the straight line distance between two points x and y is given by the Euclidean norm $\|x-y\|_2$. So, for a given $a \in \mathbb{R}^2$ or $\mathbb{R}^3$, and a given $r > 0$, the subset of $\mathbb{R}^2$ or $\mathbb{R}^3$ described as $\{x : \|x-a\|_2 < r\}$ is visualised as the disc or solid ball of diameter $2r$, centred at a and not including the periphery. The reader may check independently that $\{x \in \mathbb{R}^2 : \|x-a\|_\infty < r\}$ represents the inside of the square with vertical and horizontal sides of length $2r$, centred at a and not including the periphery. Similarly, $\{x \in \mathbb{R}^2 : \|x-a\|_1 < r\}$ represents the inside of the square with vertical and horizontal diagonals of length $2r$, centred at a and not including the periphery. See the figure. In general a subset of $\mathbb{R}^n$ of this kind is called a *ball*, regardless of what n is and what norm is used.

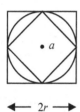

2-4.1. Definition. *The subset* $\{x \in \mathbb{R}^n : \|x-a\| < r\}$ *of* $\mathbb{R}^n$, *where* $a \in \mathbb{R}^n$ *and* $r > 0$, *is called the* **r-ball about a** *(or cen-tred at a). It is denoted by* **B(a,r)**. *The point (vector) a is called the* **centre of the ball** *and the positive number r is called its* **radius**.

The assertion $x_p \to x$ about a sequence $\{x_p\}_{p \geq 1}$ in $\mathbb{R}^n$ can now be reformulated as: For any $\varepsilon > 0$, the nth term x_n belongs to the ε-ball about x for sufficiently large n. More succinctly, all terms eventually lie in any given ball about x.

2-4.2. Definition. *A subset* $U \subseteq \mathbb{R}^n$ *is said to be* **open** *if every* $u \in U$ *has some ball about u that is contained in U*; in symbols:

$$\forall\, u \in U, \exists\, \delta > 0 \text{ such that } \xi \in \mathbb{R}^n, \|\xi - u\| < \delta \Rightarrow \xi \in U.$$

The entire space $\mathbb{R}^n$ is easily seen to be an open subset of itself. The empty subset is open 'by default' because there exists no element in it. The trivial observation that $0 < r < s \Rightarrow B(u,r) \subseteq B(u,s)$ leads to the conclusion that an intersection of two open subsets is again an open subset and hence so is the intersection of any finite number of open subsets. Indeed, if U_1 and U_2 are open and $u \in U_1 \cap U_2$, then there exist $\delta_1 > 0$ and $\delta_2 > 0$ such that $B(u,\delta_1) \subseteq U_1$ and $B(u,\delta_2) \subseteq U_2$, which implies that $B(u,\min\{\delta_1,\delta_2\}) \subseteq B(u,\delta_1) \cap B(u,\delta_2) \subseteq U_1 \cap U_2$. That the union of any number (even uncountable) of open subsets is again open hardly needs any argument.

An open subset of $\mathbb{R}^1 = \mathbb{R}$ in the sense defined above is the same as an open set of real numbers.

By Proposition 2-2.6, whether a subset is open or not does not depend on which norm is used: If a positive number that serves as the appropriate δ for one norm does not work with a different norm, then some other positive number will.

Example. The subset $(0,1)\times(2,7) \subseteq \mathbb{R}^2$ can be shown to be open. Any element u of it satisfies $0 < u_1 < 1, 2 < u_2 < 7$. Let $\delta_1 = \min\{u_1, 1-u_1\} > 0$. Then

and

$$|\xi_1 - u_1| < \delta_1 \Rightarrow \xi_1 - u_1 < 1 - u_1 \Rightarrow \xi_1 < 1$$

$$|\xi_1 - u_1| < \delta_1 \Rightarrow u_1 - \xi_1 < u_1 \Rightarrow 0 < \xi_1.$$

Thus,

$$|\xi_1 - u_1| < \delta_1 \Rightarrow 0 < \xi_1 < 1. \tag{1}$$

Similarly, $\delta_2 = \min\{u_2 - 2, 7 - u_2\} > 0$ has the property that

$$|\xi_2 - u_2| < \delta_2 \Rightarrow 2 < \xi_2 < 7. \tag{2}$$

It follows from (1) and (2) that, for $\delta = \min\{\delta_1, \delta_2\} > 0$, we have

$$\|\xi - u\|_\infty = \max\{|\xi_1 - u_1|, |\xi_2 - u_2|\} < \delta \Rightarrow \xi \in (0,1)\times(2,7).$$

The existence of such a positive number δ for every $u \in (0,1)\times(2,7)$ means by definition that $(0,1)\times(2,7)$ is open.

2-4.3. Proposition. *A ball is an open subset.*

Proof. Consider a ball $B(a,r) \subseteq \mathbb{R}^n$ and let $u \in B(a,r)$. Then $\|a - u\| < r$. Let $\delta = r - \|a - u\| > 0$. This positive number has the property that

$$\|x - u\| < \delta \Rightarrow \|x - u\| < r - \|a - u\| \Rightarrow \|x - u\| + \|a - u\| < r.$$

But $\|x - a\| \leq \|x - u\| + \|a - u\|$ by the triangle inequality. Therefore $\|x - u\| < \delta \Rightarrow \|x - a\| < r$. Thus $B(u, \delta) \subseteq B(a,r)$. The existence of such a positive δ for every $u \in B(a,r)$ means by definition that $B(a,r)$ is open. $\qquad \square$

2-4.4. Definition. *A subset $F \subseteq \mathbb{R}^n$ is said to be **closed** if its complement is open.*

2-4.5. Proposition. *A subset $F \subseteq \mathbb{R}^n$ is closed if and only if whenever all terms of a convergent sequence $\{x_p\}_{p\geq 1}$ belong to F, its limit also belongs to F.*

Proof. First suppose $F \subseteq \mathbb{R}^n$ is closed and $\{x_p\}_{p\geq 1}$ is a convergent sequence such that $x_p \in F$ for every $p \in \mathbb{N}$. We shall show that $\lim_{p\to\infty} x_p \in F$. If not, then the limit, which we shall denote by x, belongs to the complement of F. But the complement is given to be open and therefore some ball centred at x is contained in the complement. This means that no terms of the sequence can ever belong to the ball, which contradicts the fact that $x_p \to x$. Therefore the limit of the sequence has to belong to F.

For the converse, suppose $F \subseteq \mathbb{R}^n$ is not closed. We shall show that some convergent sequence with every term belonging to F has a limit that does not belong to F. Since F is not closed, the complement F^c is not open and therefore some $x \in F^c$ fails to have a ball centred at x and contained in F^c; this means every ball centred at x fails to be contained in F^c and thus has a nonempty intersection with F. In particular, for each $p \in \mathbb{N}$, the $\frac{1}{p}$-ball centred at x must contain some element $x_p \in F$. Therefore the sequence $\{x_p\}_{p \geq 1}$ not only has each term in F but also satisfies $\|x_p - x\| < \frac{1}{p}$ for each $p \in \mathbb{N}$, so that $x = \lim_{p \to \infty} x_p$. Thus $\{x_p\}_{p \geq 1}$ is a convergent sequence with every term belonging to F but having limit x that does not belong to F. $\qquad\qquad\qquad\qquad\qquad\qquad\qquad\qquad\qquad\qquad\qquad\qquad\qquad\qquad$ $\square$

2-4.6. Examples. (a) Since $\mathbb{R}^n$ is an open subset of itself, its complement, the empty subset is closed. Similarly, since the empty set is an open subset of $\mathbb{R}^n$, its complement $\mathbb{R}^n$ is closed. Thus each of the subsets $\varnothing$ and $\mathbb{R}^n$ is open as well as closed.

(b) Given any $a \in \mathbb{R}^n$ and any $r > 0$, the set $\{x \in \mathbb{R}^n : \|x - a\| \leq r\}$ is a closed sub-set of $\mathbb{R}^n$. To see why, consider any convergent sequence $\{x_p\}_{p \geq 1}$ with each x_p belonging to $\{x \in \mathbb{R}^n : \|x - a\| \leq r\}$. That is to say, $\|x_p - a\| \leq r$ for each $p \in \mathbb{N}$. In view of Proposition 2-4.5, we need only show that $\lim_{p \to \infty} x_p \in \{x \in \mathbb{R}^n : \|x - a\| \leq r\}$, i.e., $\|\lim_{p \to \infty} x_p - a\| \leq r$. Consider any $\varepsilon > 0$. By definition of limit of a se-quence, there exists $q \in \mathbb{N}$ such that $\|\lim_{p \to \infty} x_p - x_q\| < \varepsilon$. Now,

$$\|\lim_{p \to \infty} x_p - a\| \leq \|\lim_{p \to \infty} x_p - x_q\| + \|x_q - a\| < \varepsilon + r.$$

Since this holds for any arbitrary $\varepsilon > 0$, we have $\|\lim_{p \to \infty} x_p - a\| \leq r$.

The set $\{x \in \mathbb{R}^n : \|x - a\| \leq r\}$ is often called the *closed r-ball about a*; a ball will be understood to be open unless mentioned otherwise. A subset $X \subseteq \mathbb{R}^n$ is said to be *bounded* if there exists some $M > 0$ such that $\|x\| \leq M$ for every $x \in X$. Since $\|x - a\| < r \Rightarrow \|x - a\| \leq r \Rightarrow \|x\| \leq \|a\| + r$, we find that every ball is bounded.

2-4.7. Definition. *For a subset $X \subseteq \mathbb{R}^n$, a point $u \in \mathbb{R}^n$ is called an* **interior point** *of X if some ball about u is a subset of X. The set of all interior points of X is called the* **interior of X** *and is denoted by X°.*

2-4.8. Examples. (a) The interior of the closed ball $B_1 = \{x \in \mathbb{R}^n : \|x - a\| \leq r\}$ is the open ball $\{x \in \mathbb{R}^n : \|x - a\| < r\}$. Indeed, let $u \in \{x \in \mathbb{R}^n : \|x - a\| < r\}$. By Proposition 2-4.3, this ball is open and hence there exists $\alpha > 0$ such that $B(u, \alpha) \subseteq B_1$. So, u is an interior point of B_1. Also, no point u satisfying $\|u - a\| = r$ is an interior point of B_1, because for any $\alpha > 0$, the point $z = u + \frac{1}{2}\alpha \frac{u-a}{\|u-a\|}$ belongs to $B(u, \alpha)$ but not to B_1, as the following easy computation shows:

$$\|z-u\| = \|\frac{1}{2}\alpha\frac{u-a}{\|u-a\|}\| = \frac{\alpha}{2} < \alpha$$

but

$$\|z-a\| = \|u-a+\frac{1}{2}\alpha\frac{u-a}{\|u-a\|}\| = \|(1+\frac{1}{2}\frac{\alpha}{\|u-a\|})(u-a)\|$$

$$= (1+\frac{1}{2}\frac{\alpha}{\|u-a\|})\|u-a\| = \|u-a\| + \frac{\alpha}{2} > r.$$

(b) The interior of a set consisting of finitely many points is empty, because a ball always contains infinitely many points and cannot be a subset of any finite set.

It is clear from Def. 2-4.7 that, an interior point of a set belongs to that set; but not all points in a set are interior points. Also, an open set is one for which every point of it is an interior point; in other words, $X \subseteq \mathbb{R}^n$ is open if and only if $X = X^\circ$. This is immediate from Def. 2-4.2 and Def. 2-4.7.

2-4.9. Proposition. *The interior of any set is open.*

Proof. Let $X \subseteq \mathbb{R}^n$. Proposition 2-4.3 shows that, if $u \in X^\circ$, so that some ball B about u is contained in X, then for every $y \in B$, there is a ball about y that is contained in B and hence in X, thus making every $y \in B$ an interior point of X. This shows that, whenever $u \in X^\circ$, some ball B about u is contained in X°. Thus X° is an open set. $\square$

2-4.10. Definition. *For a subset $X \subseteq \mathbb{R}^n$, a point $u \in \mathbb{R}^n$ is called a* **closure point** *of X if every ball about u contains some point of X. The set of all closure points of X is called the* **closure of X** *and is denoted by $\overline{X}$.*

Clearly, a point of a set must be a closure point of it ($X \subseteq \overline{X}$) but not conversely: In $\mathbb{R}^2$, the point $u = (0,1)$ does not belong to the ball $X = \{x \in \mathbb{R}^2 : \|x\| < 1\}$ because $\|u\| = 1$. Now, any ball $\{x \in \mathbb{R}^2 : \|x-u\| < r\}$ about u contains the point $v = (0,1-s)$, where $0 < s < \min\{1,r\}$, because $\|v-u\| = \|(0,s)\| = |s| = s < r$. But this point v also belongs to the set X, because $\|v\| = |1-s| = 1-s < 1$. Thus any ball $\{x \in \mathbb{R}^2 : \|x-u\| < r\}$ about u contains a point of X, whereby u is seen to be a closure point of X, though it does not belong to X.

A closed set is one that contains each of its closure points. To see why, suppose first that $X \subseteq \mathbb{R}^n$ contains each of its closure points. We shall demonstrate that the complement X^c is open. With this in view, consider any $v \in X^c$. Then v is not a closure point and hence there exists a ball about v containing no point of X, i.e., is a subset of X^c. This means X^c is open. Suppose next that X^c is open and u is a closure point of X. We shall demonstrate that $u \in X$. If not, then some ball about u is contained in X^c and therefore contains no point of X; this contradicts the hypothesis that u is a closure point of X.

As noted before, it is always true that $X \subseteq \bar{X}$; we have just proved that X is closed if and only if $X \supseteq \bar{X}$. Therefore $X \subseteq \mathbb{R}^n$ is closed if and only if $X = \bar{X}$.

2-4.11. Proposition. *The closure of a set is closed.*

Proof. Let $X \subseteq \mathbb{R}^n$. We must show that the complement of the closure $\bar{X}$ is open. Accordingly, consider any $u \in (\bar{X})^c$. Then u is not a closure point and therefore some ball B about u contains no point of X, which means B is a subset of X^c. Proposition 2-4.3 shows that, for every $y \in B$, there is a ball about y that is contained in B and hence in X^c, thus ensuring that no $y \in B$ is a closure point of X, i.e., $y \in B \Rightarrow y \in (\bar{X})^c$. This shows that, whenever $u \in (\bar{X})^c$, some ball B about u is contained in $(\bar{X})^c$. Thus $(\bar{X})^c$ is an open set. $\qquad\square$

2-4.12. Definition. *The* **boundary** ∂X *of a subset* $X \subseteq \mathbb{R}^n$ *is the set* $\bar{X} \backslash X^\circ$ *of all points in the closure of X that do not belong to its interior.*

It is immediate from this definition that $x \in \partial X$ if and only if every ball about x contains a point of X as well as a point of the complement X^c.

2-4.13. Example. As noted in Example, 2-4.6(b), $X = \{x \in \mathbb{R}^2 : \|x\| \leq 1\}$ is closed and hence is its own closure $\bar{X}$. It has also been recorded in Example 2-4.8(a) that $X^\circ = \{x \in \mathbb{R}^2 : \|x\| < 1\}$. Therefore, the boundary is $\partial X = \{x \in \mathbb{R}^2 : \|x\| = 1\}$.

2-4.14. Proposition. *The boundary of any set is closed.*

Proof. For any $X \subseteq \mathbb{R}^n$, the closure $\bar{X}$ is closed [Proposition 2-4.11] and the interior X° is open [Proposition 2-4.9]. By 2-4.P4, the difference $\partial X = \bar{X} \backslash X^\circ$ is closed. $\qquad\square$

It follows from this Proposition and Example 2-4.13 that $\{x \in \mathbb{R}^2 : \|x\| = 1\}$ is closed. A direct proof using Proposition 2-4.5 is left to the reader in 2-4.P7.

Problem Set 2-4

2-4.P1. Show that a union of two closed sets is closed and that the intersection of any family of closed sets is closed.

2-4.P2. Show that a ball $\{x \in \mathbb{R}^n : \|x - a\| < r\}$ is not a closed subset of $\mathbb{R}^n$.

2-4.P3. Suppose $X \subseteq \mathbb{R}^n$ is a subset for which there exists some $u \in \mathbb{R}^n$ and some $M > 0$ satisfying $\|x - u\| \leq M$ for every $x \in X$. Show that X is bounded.

2-4.P4. Let $U \subseteq \mathbb{R}^n$ be open and $F \subseteq \mathbb{R}^n$ be closed. Show that the difference set $U \backslash F \subseteq \mathbb{R}^n$ is open and the difference set $F \backslash U$ is closed.

2-4.P5. Show that the interior of a set is the union of all open sets contained in that set and that the closure is the intersection of all closed sets containing that set.

2-4.P6. Show that $x \in \mathbb{R}^n$ is a closure point of a subset $X \subseteq \mathbb{R}^n$ if and only if some sequence in X converges to x.

2-4.P7. Using Proposition 2-4.5, but not concepts of interior, closure or boundary, show that $\{x \in \mathbb{R}^2 : \|x\| = 1\}$ is closed.

2-4.P8. Show that $(-1,0)$ is a boundary point of $E = \{(x_1, x_2) \in \mathbb{R}^2 : -1 \le x_1 \le 2\}$ and that $(1,0)$ is an interior point of $\{(x_1, x_2) \in \mathbb{R}^2 : 0 \le x_1 \le 2\}$.

2-4.P9. For any $F \subseteq \mathbb{R}^n$, show that $\overline{F} = F \cup \partial F = F^\circ \cup \partial F$.

2-5 Compact and Connected Subsets

The notion of compactness, which plays a significant role in analysis, was introduced into mathematics by M. Fréchet. However, he was working in a much more general framework of ideas than of Euclidean spaces.

One of the distinguishing characterisations of a bounded closed subset of $\mathbb{R}$ is that any sequence in it has a subsequence converging to a limit belonging to that subset. Another characterisation of such a subset is that any 'open cover' of it contains a finite subcover. In this section we shall prove the above characterisations for any bounded closed subset of $\mathbb{R}^n$. We begin with the following lemma.

As in Proposition 2-2.8, we shall denote the jth component of the pth term x_p of a sequence $\{x_p\}_{p \ge 1}$ by $x^{(p)}_j$.

2-5.1. Lemma. *Let $X \subseteq \mathbb{R}^n$ be bounded and suppose all terms of the sequence $\{x_p\}_{p \ge 1}$ belong to X. Then the sequence has a subsequence which is convergent, though its limit may not belong to X.*

Proof. For each j, $1 \le j \le n$, consider the real sequence formed by the jth components of the terms x_p, i.e., $\{x^{(p)}_j\}_{p \ge 1}$. Since $|x^{(p)}_j| \le \|x_p\|$, we know that each of them is a bounded sequence in $\mathbb{R}$. By the Bolzano–Weierstrass theorem (see Berberian [3; Proposition 3.5.9]), the first sequence, namely $\{x^{(p)}_1\}_{p \ge 1}$, has a subsequence $\{x^{(p_q)}_1\}_{q \ge 1}$ that converges to some limit in $\mathbb{R}$. Now, considering only those points of the original sequence $\{x_p\}$ which are numbered by p_q, we obtain a subsequence $\{x_{p_q}\}$ of the original sequence, in which $\{x^{(p_q)}_1\}_{q \ge 1}$ is convergent. Next, consider the sequence $\{x^{(p_q)}_2\}_{q \ge 1}$, that is, the sequence of second components of the latter sequence $\{x_{p_q}\}$. We can find a subsequence of the subsequence $\{x_{p_q}\}$ for which the sequence formed by its second components converges. Note that the sequence formed by its first components is a subsequence of $\{x^{(p_q)}{}_1\}_{q \geq 1}$ and therefore converges; thus the sequences formed by its first components and by its second components both converge. Repeating this procedure successively n times, we arrive at a subsequence of the given sequence $\{x_p\}$ for which all the sequences of the components of its points are convergent. Since convergence in $\mathbb{R}^n$ is equivalent to componentwise convergence [Proposition 2-2.8], the last mentioned subsequence of $\{x_p\}_{p \geq 1}$ has a limit in $\mathbb{R}^n$.   □

2-5.2. Theorem. *For a subset $K \subseteq \mathbb{R}^n$ to be bounded as well as closed, it is necessary and sufficient that every sequence with all its terms belonging to K have a convergent subsequence, the limit of which also belongs to K.*

Proof. First suppose $K \subseteq \mathbb{R}^n$ to be bounded as well as closed, and consider any sequence with all its terms belonging to K. Since K is bounded, the sequence has a convergent subsequence [Lemma 2-5.1], and since K is closed, the limit belongs to K [Proposition 2-4.5].

Conversely, suppose $K \subseteq \mathbb{R}^n$ is either not bounded or not closed. In the latter case, there exists a convergent sequence with all terms belonging to K but the limit is not in K [Proposition 2-4.5]; in particular, there is a sequence with all terms belonging to K but having no convergent subsequence whose limit belongs to K. In the former case, for each $p \in \mathbb{N}$, there exists some $x_p \in K$ such that $\|x_p\| \geq p$. The sequence $\{x_p\}_{p \geq 1}$ then has all terms belonging to K, but no subsequence $\{x_{p_q}\}_{q \geq 1}$ can be convergent because it is not bounded; in fact, it satisfies $\|x_{p_q}\| \geq p_q \geq q$.   □

2-5.3. Proposition. *Let $K \subseteq \mathbb{R}^n$ be bounded and ε be any positive number whatsoever. Then there is a finite number of ε-balls centred at points of K such that their union contains K.*

Proof. Suppose this is not so. Then the union of a finite number of ε-balls centred at points of K can never contain K. So, take any vector $x_1 \in K$. The ε-ball B_1 about x_1 cannot contain K and so there exists $x_2 \in K$ such that $x_2 \notin B_1$, i.e., $\|x_2 - x_1\| \geq \varepsilon$. Let B_2 be the ε-ball about x_2. Then the union $B_1 \cup B_2$ of finitely many (two, of course) ε-balls cannot contain K. So, there exists $x_3 \in K$ such that $x_3 \notin B_1 \cup B_2$, i.e., $\|x_3 - x_1\| \geq \varepsilon$, $\|x_3 - x_2\| \geq \varepsilon$. We can keep proceeding in this manner and obtain a sequence $\{x_p\}_{p \geq 1}$ of points in K having the property that

$$p > q \Rightarrow \|x_p - x_q\| \geq \varepsilon.$$

But this property guarantees that no subsequence can be a Cauchy sequence and hence that no subsequence can converge. Since all terms of $\{x_p\}_{p \geq 1}$ belong to the bounded set K, Lemma 2-5.1 is contradicted. Therefore we are led to the conclusion that our supposition is false, i.e., there exists a finite number of ε-balls centred at points of K such that their union contains K. $\square$

The phenomenon of a family of sets, finite or otherwise, whose union contains a given set needs to be studied further. Although the considerations we are about to enter into may seem outlandish at first, they are intimately connected with other ideas here and will be needed in Proposition 7-4.9. We introduce some terminology.

2-5.4. Definition. *For any subset $X \subseteq \mathbb{R}^n$, a family $\mathcal{U}$ of subsets of $\mathbb{R}^n$ whose union contains X is called a **cover of** (or **covering of**) X. If every set in the family is open, then the cover is said to be an **open cover** (or **open covering**). A subfamily of $\mathcal{U}$ whose union also contains X is called a **subcover of** $\mathcal{U}$ (or **subcovering of** $\mathcal{U}$).*

It is convenient to rephrase '$\mathcal{U}$ is a cover of X' by saying '$\mathcal{U}$ covers X'. Also, the phrase 'there is a subcover of $\mathcal{U}$' is usually recast as '$\mathcal{U}$ contains a subcover'.

2-5.5. Examples. (a) Let $X = \mathbb{R}^n$. Then the family $\mathcal{U}$ consisting of the sets

$$\{x \in \mathbb{R}^n : \|x\| \le p\}, p \in \mathbb{N},$$

is a cover of X. The subfamily $\{x \in \mathbb{R}^n : \|x\| \le 2p\}$, $p \in \mathbb{N}$, also covers X and is therefore a subcover of $\mathcal{U}$. The family $\{x \in \mathbb{R}^n : \|x\| < p\}$, $p \in \mathbb{N}$ is an open cover of X and its subfamilies

$$\{x \in \mathbb{R}^n : \|x\| < 7p\}, \quad p \in \mathbb{N}, \quad \text{and} \quad \{x \in \mathbb{R}^n : \|x\| < p^2\}, \quad p \in \mathbb{N},$$

are subcovers because they also cover X.

(b) Let $X = \{x \in \mathbb{R}^n : \|x\| \le 1\}$. Then the family $\mathcal{U}$ consisting of the sets

$$\{x \in \mathbb{R}^n : \|x\| < 1 - \frac{1}{p+1}\}, p \in \mathbb{N},$$

is not a cover of X because their union does not contain the points of X for which $\|x\| = 1$. If we enlarge the family by including the set

$$\{x \in \mathbb{R}^n : \frac{9}{10} < \|x\| < \frac{101}{100}\},$$

then the enlarged family is an open cover (union contains all elements of X and more). The open cover contains a finite subcover, for instance, the subfamily consisting of the two sets

$$\{x \in \mathbb{R}^n : \|x\| < 1 - \frac{1}{11}\} \quad \text{and} \quad \{x \in \mathbb{R}^n : \frac{9}{10} < \|x\| < \frac{101}{100}\}.$$

(c) Let $X = \{x \in \mathbb{R}^n : \|x\| < 1\}$. Then the family $\mathcal{U}$ consisting of the sets

$$\{x \in \mathbb{R}^n : \|x\| < 1 - \frac{1}{p+1}\}, \ p \in \mathbb{N},$$

is an open cover of X. This open cover contains no finite subcover. The open covers in part (a) also contain no finite subcovers.

2-5.6. Definition. *A set $K \subseteq \mathbb{R}^n$ is said to be* **compact** *if every open cover of it contains a finite subcover.*

2-5.7. Theorem. Heine–Borel: *A set $K \subseteq \mathbb{R}^n$ is compact if and only if it is bounded as well as closed.*

Proof. We prove the 'only if' part first. Let $K \subseteq \mathbb{R}^n$ be compact and $\{x_p\}_{p \geq 1}$ be a sequence with all its terms belonging to K. We shall show that it has a subsequence converging to a vector belonging to K.

Suppose this is not so. Then no vector $x \in K$ is the limit of a subsequence. By Proposition 2-2.11, given any $x \in K$, there exists $\varepsilon > 0$ and some $N \in \mathbb{N}$ (both depending on x) such that no integer p satisfies $p \geq N$ as well as $\|x_p - x\| < \varepsilon$. In other words, the ε-ball about x can contain x_p only if $p < N$; thus it contains x_p only for finitely many p (perhaps none). The family of all such balls about x with $x \in K$ is an open cover of K and, since K has been assumed compact, the cover must contain a finite subcover. This means the union of a finite family of the balls must contain K and hence contain x_p for all $p \in \mathbb{N}$. But this is a contradiction because each of the finitely many balls contains x_p only for finitely many p. Therefore our supposition that $\{x_p\}_{p \geq 1}$ has no subsequence converging to a limit in K must be false.

It follows by Theorem 2-5.2 that K is bounded as well as closed.

We now prove the 'if' part by contradiction. Assume K to be bounded as well as closed and suppose that it has an open cover $\mathcal{U}$ containing no finite subcover. By Proposition 2-5.3, there exists a finite family $\mathcal{B}_1$ of 2-balls centred at points of K that covers K. If the intersection with K of each of the finitely many 2-balls in $\mathcal{B}_1$ can be covered by a finite subfamily of $\mathcal{U}$, then all these finite subfamilies taken together make for a single finite subfamily of $\mathcal{U}$ that covers $K \cap (\cup \mathcal{B}_1) = K$, which is ruled out by what we have supposed. Therefore, the intersection with K of at least one of the 2-balls in $\mathcal{B}_1$, call it B_1, cannot be covered by a finite subfamily of $\mathcal{U}$. Being bounded, $K \cap B_1$ can be covered by a finite family $\mathcal{B}_2$ of $\frac{1}{2}$-balls centred at points of $K \cap B_1$, according to Proposition 2-5.3. By the same argument again, if the intersection with K of each of the finitely many $\frac{1}{2}$-balls in $\mathcal{B}_2$ can be covered by a finite subfamily of $\mathcal{U}$, then all these finite subfamilies taken together make for a single finite subfamily of $\mathcal{U}$ that covers $K \cap (\cup \mathscr{B}_2) \supseteq K \cap (K \cap B_1) = K \cap B_1$, contrary to our choice of B_1. Therefore the intersection with K of at least one of the $\frac{1}{2}$-balls in $\mathscr{B}_2$, call it B_2, cannot be covered by a finite subfamily of $\mathscr{U}$. It follows that B_2 cannot be covered by a finite subfamily of $\mathscr{U}$. Being bounded, $K \cap B_2$ can be covered by a finite family of $\frac{1}{2^2}$-balls centred at points of $K \cap B_2$, according to Proposition 2-5.3.

Continuing in this manner, we obtain a sequence of $\frac{1}{2^{p-1}}$-balls B_p such that, for each $p \in \mathbb{N}$,

$$B_p \text{ cannot be covered by a finite subfamily of } \mathscr{U} \tag{1}$$

and

$$B_{p+1} \text{ is centred at a point of } K \cap B_p. \tag{2}$$

Let $\{x_p\}_{p \geq 1}$ be the sequence of their centres. By (2), each x_p belongs to K and $\|x_{p+1} - x_p\| < \frac{1}{2^{p-1}}$. Therefore, for $p > q$, we have

$$\|x_p - x_q\| \leq \sum_{r=q}^{p-1} \|x_{r+1} - x_r\| < \sum_{r=q}^{p-1} \frac{1}{2^{r-1}} < \frac{1}{2^{q-2}},$$

which shows that $\{x_p\}_{p \geq 1}$ is a Cauchy sequence. By Theorem 2-2.10, it converges and its limit x belongs to K, as K is closed. But $\mathscr{U}$ is an open cover of K and therefore $x \in U$ for some $U \in \mathscr{U}$. Since U is open, for some $\varepsilon > 0$, the ε-ball centred at x is a subset of U. Now select some $p_0 \in \mathbb{N}$ such that

$$\|x_{p_0} - x\| < \frac{\varepsilon}{2} \quad \text{as well as} \quad \frac{1}{2^{p_0}} < \frac{\varepsilon}{2}.$$

This integer p_0 has the property that

$$\|y - x_{p_0}\| < \frac{1}{2^{p_0 - 1}} \Rightarrow \|y - x\| \leq \|y - x_{p_0}\| + \|x_{p_0} - x\| < \frac{\varepsilon}{2} + \frac{\varepsilon}{2} = \varepsilon \Rightarrow y \in U.$$

Consequently, the ball B_{p_0} is covered by the subfamily $\{U\}$ of $\mathscr{U}$ consisting of the single set U, which is not possible in view of (1). This contradiction proves the converse. $\square$

The following concept will be needed in Sections 3-3, 3-5 and 4-4.

2-5.8. Definition. *A subset E of $\mathbb{R}^n$ is called* **convex** *if, whenever a and b are in E and λ is any real number such that $0 \leq \lambda \leq 1$, the vector $b + \lambda(a - b)$, or what is the same thing, $\lambda a + (1 - \lambda)b$ is also in E.*

It is an elementary argument in $\mathbb{R}^n$ that any ball is a convex set.

The following proposition will be needed in Section 4-4.

2-5.9. Proposition. *If a convex set is a union of two disjoint open sets, then one of the open sets must be empty.*

Proof. Let $B \subseteq \mathbb{R}^n$ be a convex set. Suppose, if possible, that there exist open sets U and V such that

$$U \cup V = B, \quad U \cap V = \varnothing, \quad U \neq \varnothing, \quad V \neq \varnothing. \tag{1}$$

Since U and V are both nonempty, there exists some $a \in U$ and some $b \in V$. By convexity of B, we have $\lambda a + (1-\lambda)b \in B$ whenever $0 \leq \lambda \leq 1$. When $\lambda = 1$, we have $\lambda a + (1-\lambda)b = a \in U$. Therefore, $\{\lambda \in [0,1] : \lambda a + (1-\lambda)b \in U\}$ is nonempty and bounded below by 0. Let λ_0 be the infimum of this set. Then

$$0 \leq \lambda < \lambda_0 \Rightarrow \lambda a + (1-\lambda)b \notin U. \tag{2}$$

Also, $\lambda_0 \in [0,1]$, so that $c = \lambda_0 a + (1-\lambda_0)b \in B$.

Suppose $c \in U$. Since $b \in V$ and $U \cap V = \varnothing$, it follows that $c \neq b$, and hence $\lambda_0 > 0$. Since U is open, there exists some $\delta > 0$ such that

$$\|x - c\| < \delta \Rightarrow x \in U. \tag{3}$$

When $\lambda = \lambda_0 - \min\{\frac{\lambda_0}{2}, \frac{\delta}{2\|a-b\|}\}$, we have $0 \leq \lambda < \lambda_0 \leq 1$ and

$$\|\lambda a + (1-\lambda)b - c\| = \|(\lambda - \lambda_0)a - (\lambda - \lambda_0)b\|$$

$$= \|(\lambda - \lambda_0)(a - b)\| = |\lambda - \lambda_0| \cdot \|a - b\|$$

$$\leq \frac{\delta}{2} < \delta.$$

Therefore by (3), we have $\lambda a + (1-\lambda)b \in U$. But by (2), this is not possible, because $0 \leq \lambda < \lambda_0$. This contradiction shows that $c \notin U$.

Now suppose $c \in V$. Since $a \in U$ and $U \cap V = \varnothing$, it follows that $c \neq a$, and hence $\lambda_0 < 1$. Since V is open, there exists some $\delta' > 0$ such that

$$\|x - c\| < \delta' \Rightarrow x \in V. \tag{4}$$

By definition of infimum, there exists some λ such that

$$\lambda_0 \leq \lambda < \lambda_0 + \frac{\delta'}{2\|a-b\|}, \quad 0 \leq \lambda \leq 1 \quad \text{and} \quad \lambda a + (1-\lambda)b \in U. \tag{5}$$

Arguing as before, we can show $\|\lambda a + (1-\lambda)b - c\| < \delta'$. Therefore by (4), we have $\lambda a + (1-\lambda)b \in V$. In conjunction with (5), this implies $\lambda a + (1-\lambda)b \in U \cap V$, which is not possible, because $U \cap V = \varnothing$. This contradiction shows that $c \notin V$. Since it was already shown that $c \in B$ and $c \notin U$, the equality $U \cup V = B$ stands violated, thereby establishing that (1) can never hold. $\square$

The convex sets that come within the purview of the above proposition are necessarily open, because a union of open sets is open. However, the basic idea carries over to other kinds of sets. The formal definition is as follows:

2-5.10. Definition. *A subset E of $\mathbb{R}^n$ is called* **connected** *if, among any two intersections of open sets with E that are disjoint and have union equal to E, one must be empty.*

For an open set E, this is obviously equivalent to saying that among any two open sets that are disjoint with union equal to E, one must be empty.

Problem Set 2-5

2-5.P1. Show that the union of a finite number of compact subsets of $\mathbb{R}^n$ is compact.

2-5.P2. Show that any ball in $\mathbb{R}^n$ is a convex set.

2-5.P3. Let $K \subseteq \mathbb{R}^n$ be compact and $\{x_p\}_{p \geq 1}$ be a sequence in K. If $x \in \mathbb{R}^n$ has the property that any subsequence of $\{x_p\}_{p \geq 1}$ either converges to x or does not converge at all, show that $\{x_p\}_{p \geq 1}$ converges to x. Show also that the hypothesis that K is compact cannot be dropped.

2-5.P4. Show that the set of all points in $\mathbb{R}^n$ for which every component is an integer is not compact.

2-5.P5. Show that if a sequence $\{x_p\}_{p \geq 1}$ in $\mathbb{R}^n$ converges to x, then the set $\{x\} \cup \{x_p : p \in \mathbb{N}\}$ is compact.

2-5.P6. Show that $A = \{(x, y) \in \mathbb{R}^2 : x^2 - y^2 \geq 1\} \subseteq \mathbb{R}^2$ is disconnected.

2-6 Continuity

So far we have seen how the concepts of convergence, open set and compactness can be extended from $\mathbb{R}$ to $\mathbb{R}^n$ by replacing the absolute value by the norm. In this section we shall see that the same can be done with continuity and limits of functions defined on subsets of $\mathbb{R}^n$ with values in $\mathbb{R}^m$.

A precise description of continuity of a real-valued function f at a point x of its domain $S \subseteq \mathbb{R}$, with which the reader is undoubtedly familiar, is as follows:

$$\forall \, \varepsilon > 0, \exists \, \delta > 0 \text{ such that } \xi \in S, |\xi - x| < \delta \Rightarrow |f(\xi) - f(x)| < \varepsilon.$$

The definition of continuity when S is a subset of $\mathbb{R}^n$ and the range is a subset of $\mathbb{R}^m$ is the same except that absolute value is replaced by norm:

2-6.1. Definition. *For any subset $S \subseteq \mathbb{R}^n$, a map $f: S \to \mathbb{R}^m$ is* **continuous at** *a point $x \in S$ if:*

$$\forall\, \varepsilon > 0, \exists\, \delta > 0 \text{ such that } \xi \in S, \|\xi - x\| < \delta \Rightarrow \|f(\xi) - f(x)\| < \varepsilon.$$

If we denote $\xi - x$ by h, so that $\xi = x + h$, then continuity of f at x can be re-formulated as

$$\forall\, \varepsilon > 0, \exists\, \delta > 0 \text{ such that } x + h \in S, \|h\| < \delta \Rightarrow \|f(x + h) - f(x)\| < \varepsilon.$$

It is understood of course that the same symbol $\|\ \|$ is being used for the norms in $\mathbb{R}^n$ and $\mathbb{R}^m$. By Proposition 2-2.6, it makes no difference which norms we use. It is useful to reformulate continuity of f at x in terms of balls as below:

Given any ε-ball B_2 about $f(x)$, some δ-ball B_1 about x satisfies $f(S \cap B_1) \subseteq B_2$.

A map $f: S \to \mathbb{R}^m$ which is continuous at each point of its domain S is said to be simply **continuous**, or for emphasis, **continuous everywhere**.

Remark. If f is continuous at a point x, so are $\|f\|$ and αf, where α is any real number; if g is also continuous at x, then the sum $f + g$ and inner product $f \cdot g$ are continuous at x. Also, if the composition $g \circ f$ is defined, f is continuous at x and g is continuous at $f(x)$, then $g \circ f$ is continuous at x. These easily proven facts are called 'elementary properties' of continuous functions and will be used freely without reference.

2-6.2. Examples. (a) Perhaps the simplest example of a continuous map from $\mathbb{R}^n$ to $\mathbb{R}$, other than a constant, is $f(x) = x_j$, the jth component of $x \in \mathbb{R}^n$. To prove its continuity formally at a point $x \in \mathbb{R}^n$, consider any $\varepsilon > 0$. We must show that some $\delta > 0$ has the property that

$$\|h\| < \delta \Rightarrow |(x + h)_j - x_j| < \varepsilon,$$

i.e.,
$$\|h\| < \delta \Rightarrow |h_j| < \varepsilon.$$

One such δ is none other than $\delta = \varepsilon$, which is to say that

$$\|h\| < \varepsilon \Rightarrow |h_j| < \varepsilon.$$

This is true in view of the fact that

$$|h_j| \leq \|h\|.$$

(b) Consider the map from $\mathbb{R}^2$ to $\mathbb{R}$ defined by $f(x,y) = xy$. We shall use the norm $\|\ \|_2$ on $\mathbb{R}^2$, so that $\|(h,k)\| = \sqrt{(h^2 + k^2)}$. Since $|h|, |k| \leq \sqrt{(h^2 + k^2)}$, we have

$$|f(x + h, y + k) - f(x,y)| = |kx + hy + hk| \leq (|x| + |y| + |k|)\sqrt{(h^2 + k^2)}.$$

For $\|(h,k)\| = \sqrt{(h^2 + k^2)} \leq 1$, we have $|h| \leq 1$ as well as $|k| \leq 1$, and hence

$$|f(x + h, y + k) - f(x,y)| \leq (|x| + |y| + 1)\sqrt{(h^2 + k^2)}.$$

Therefore, the positive number

$$\delta = \min \left\{ 1, \frac{\varepsilon}{|x| + |y| + 1} \right\}$$

has the property that $\|(h,k)\| = \sqrt{(h^2 + k^2)} < \delta \Rightarrow |f(x + h, y + k) - f(x,y)| < \varepsilon.$
Note that δ depends upon (x,y) and ε.

(c) By (a), the functions from $\mathbb{R}^2$ to $\mathbb{R}$ that map (x,y) into x or into y are both continuous. It follows from 'elementary properties' of continuous functions that the function f discussed in (b) is continuous, as are functions described by polynomials in x and y or by expressions such as $e^x \sin(x + y^2)$.

(d) The function f on $\mathbb{R}^2$ defined by $f(x,y) = xy/(x^2 + y^2)$ for $(x,y) \neq (0,0)$ and $f(0,0) = a$ (any real number) is not continuous at $(0,0)$. In other words, there exists some $\eta > 0$ such that no matter what positive δ we take, some $(h,k) \in \mathbb{R}^2$ satisfies

$$\|(h,k) - (0,0)\| < \delta, \quad \text{but} \quad |f(h,k) - f(0,0)| \geq \eta.$$

We shall use the norm $\| \ \|_1$ on $\mathbb{R}^2$, so that $\|(h,k)\| = |h| + |k|$. To begin with, let b be a real number such that $b/(1 + b^2) \neq a$ and let

$$2\eta = \left| \frac{b}{1 + b^2} - a \right| > 0.$$

For $(h,k) = (h, bh)$, we have

$$\|(h,k) - (0,0)\| = \|(h,k)\| = \|(h, bh)\| = (1 + |b|)|h|$$

and

$$|f(h,k) - f(0,0)| = |f(h,k) - a| = \left| \frac{bh^2}{h^2 + b^2 h^2} - a \right| = \left| \frac{b}{1 + b^2} - a \right| = 2\eta.$$

This shows that, if $(h,k) \in \mathbb{R}^2$ is such that $k = bh$ and $0 < |h| < \delta/(1 + |b|)$, then it has the property claimed for it.

2-6.3. Definition. *A **limit point** of a nonempty subset $A \subseteq \mathbb{R}^n$ is an element $x \in \mathbb{R}^n$ (which may or may not belong to the subset) such that, for every $\delta > 0$, the ball $\{t \in \mathbb{R}^n : \|t - x\| < \delta\}$ contains at least one element that belongs to the set A but is different from x; in alternative formulation: for every $\delta > 0$, there exists $t \in A$ for which $0 < \|t - x\| < \delta$.*

*An **isolated point** of a nonempty subset $A \subseteq \mathbb{R}^n$ is an element of $\mathbb{R}^n$ which belongs to A but is not a limit point of it.*

Examples of limit points when $n = 1$ are given in Shirali and Vasudeva [23, 9-1.2]. Here we mention three examples when $n > 1$.

2-6.4. Examples. (a) Let A contain the set obtained from a ball of radius r by deleting the centre x, i.e., $A \supseteq \{t \in \mathbb{R}^n : 0 < \|t - x\| < r\}$. Then x may or may not belong to A but is a limit point of A. This is so because the ball $\{t \in \mathbb{R}^n : \|t - x\|$

$< \delta\}$ contains the element $t = x + sh$, where h is any element of $\mathbb{R}^n$ with $\|h\| = 1$ and s is any real number such that $0 < s < \min\{r, \delta\}$. Here $\|t - x\| = \|sh\| = s$.

(b) Let A be either the ball $\{t \in \mathbb{R}^n : \|t\| < r\}$ or the ball $\{t \in \mathbb{R}^n : \|t\| \leq r\}$. Any x such that $\|x\| = r$ is a limit point. To see why, consider any $\delta > 0$ and let $0 < s < \min\{1, \frac{\delta}{r}\}$. Then the element $t = (1 - s)x$ satisfies $\|t\| < r$ while $\|t - x\| = \|sx\| = sr$ so that $0 < \|t - x\| < \delta$.

(c) Let $A = \{(x,y) \in \mathbb{R}^2 : y \neq 0\}$. Then for any $a \in \mathbb{R}$, $(a, 0)$ is always a limit point. In fact, the ball $\{(x,y) \in \mathbb{R}^2 : \|(x,y) - (a, 0)\| < \delta\}$ contains the point $(a, \frac{\delta}{2})$, which is different from $(a, 0)$ and belongs to A.

2-6.5. Definition. *Let x be a limit point of the domain $A \subseteq \mathbb{R}^n$ of a function f with values in $\mathbb{R}^m$. An element $\lambda \in \mathbb{R}^m$ is called a **limit of f at x** if and only if for every $\varepsilon > 0$, there exists $\delta > 0$ such that*

$$\|f(x + h) - \lambda\| < \varepsilon \quad \text{whenever} \quad 0 < \|h\| < \delta \ \text{and} \ x + h \in A.$$

In symbols,

$$\forall \, \varepsilon > 0 \, \exists \, \delta > 0 \ \ni \ h \in \mathbb{R}^n, 0 < \|h\| < \delta \ and \ x + h \in A \ \Rightarrow \ \|f(x + h) - \lambda\| < \varepsilon.$$

The limit of f at x is also called the **limit of $f(t)$ as $t \to x$**.

As in the elementary case, there cannot be two distinct λ with this property, the reasons being analogous to why a sequence cannot have two different limits. Details are left to the reader, but we note here that the argument uses the fact that x is a limit point of A. Thus the limit, if it exists, is unique. Therefore we shall henceforth refer to it as *the* limit of f at x, and shall denote it by $\lim_{t \to x} f(t)$.

As is customary, we also write $f(t) \to \lambda$ **as** $t \to x$, if it is convenient to do so.

It is sometimes more convenient to express the above definition without explicit reference to h by introducing $t = x + h$:

$$\forall \, \varepsilon > 0 \ \exists \, \delta > 0 \ \ni \ \forall \, t \in A, 0 < \|t - x\| < \delta \ \Rightarrow \ \|f(t) - \lambda\| < \varepsilon.$$

2-6.6. Examples. (a) Let f be defined on the domain $\{(x,y) \in \mathbb{R}^2 : y \neq 0\}$ as $f(x,y) = x \sin(1/y)$. As seen in Example 2-6.4(c), $(0,0)$ is a limit point of the domain. We shall show that the limit of f at $(0,0)$ is 0. Indeed,

$$\|f(x,y) - 0\| = |x \sin(1/y)| \leq |x| \leq \|(x,y) - (0,0)\|.$$

Therefore, for any $\varepsilon > 0$, the positive number $\delta = \frac{\varepsilon}{2}$ has the property that $\|(x,y) - (0,0)\| < \delta \Rightarrow \|f(x,y) - 0\| \leq \frac{\varepsilon}{2} < \varepsilon$. So, the limit is 0, as claimed.

(b) The real function f defined by $f(x,y) = xy/(x^2 + y^2)$ for $(x,y) \neq (0,0)$ has no limit at $(0,0)$. In other words, for any real number a, there exists some $\eta > 0$ such that no matter what positive δ we take, some $(h, k) \in \mathbb{R}^2$ satisfies

$$\|(h,k) - (0,0)\| < \delta \text{ but } |f(h,k) - a| \geq \eta.$$

The argument is exactly as in Example 2-6.2(d).

Note however that for each $y \neq 0$, $\lim_{x \to 0} f(x,y) = 0$, so that the 'repeated limit'

$$\lim_{y \to 0} [\lim_{x \to 0} f(x,y)]$$

exists and is 0. The same is true of the other repeated limit in which the limit as $y \to 0$ is taken first. [Cf. 2-6.P2].

2-6.7. Proposition. *Let x be a limit point of the domain $A \subseteq \mathbb{R}^n$ of a function f with values in $\mathbb{R}^m$. Denote the component functions of f by f_j, $1 \leq j \leq m$, which is to say,*

$$f(t_1, t_2, \ldots, t_n) = (f_1(t_1, t_2, \ldots, t_n), f_2(t_1, t_2, \ldots, t_n), \ldots, f_m(t_1, t_2, \ldots, t_n)).$$

Then $\lim_{t \to x} f(t) = \lambda \in \mathbb{R}^m$ if and only if $\lim_{t \to x} f_j(t) = \lambda_j$ for each j.

Proof. For any choice of norm in $\mathbb{R}^m$, we have $|f_j(t) - \lambda_j| \leq \|f(t) - \lambda\|$. Therefore, if $\lim_{t \to x} f(t) = \lambda$, it follows that $\lim_{t \to x} f_j(t) = \lambda_j$ for each j.

For the converse, suppose $\lim_{t \to x} f_j(t) = \lambda_j$ for each j and consider any $\varepsilon > 0$. For each j, there exists δ_j such that

$$0 < \|t - x\| < \delta_j \Rightarrow |f_j(t) - \lim_{t \to x} f_j(t)| < \varepsilon.$$

Set $\delta = \min\{\delta_1, \ldots, \delta_m\} > 0$. Then

$$0 < \|t - x\| < \delta \Rightarrow |f_j(t) - \lim_{t \to x} f_j(t)| < \varepsilon \text{ for each } j$$

$$\Rightarrow \max\{|f_j(t) - \lambda_j| : 1 \leq j \leq m\} < \varepsilon$$

$$\Rightarrow \|f(t) - \lambda\|_\infty < \varepsilon.$$

By Proposition 2-2.6, such a δ exists for the other two norms as well. $\square$

2-6.8. Examples. (a) Let f map $\{(x,y) : y \neq 0\}$ into $\mathbb{R}^2$ as follows:

$$f(x,y) = (x \sin(1/y), 1 + y^{-3} \exp(-1/y^2)).$$

Since $x \sin(1/y)$ and $1 + y^{-3} \exp(-1/y^2)$ have limits 0 and 1, respectively, at $(0,0)$, it follows by the first part of Proposition 2-6.7 that f has limit $(0,1)$ at $(0,0)$. In other words, $f(x,y) \to (0,1)$ as $(x,y) \to (0,0)$.

(b) Let f map $\{(x,y) : (x,y) \neq (0,0)\}$ into $\mathbb{R}^2$ as follows:

$$f(x,y) = (xy/(x^2 + y^2), 1).$$

Here $xy/(x^2 + y^2)$ has no limit at $(0,0)$, as seen in Example 2-6.6(b). Therefore it follows by the second (converse) part of Proposition 2-6.7 that f has no limit at $(0,0)$.

2-6.9. Remark. Suppose f is a function with domain A having a limit point x, and $\lim_{t \to x} f(t) = \lambda$. Let f_B denote the restriction of f to a subset B of A, and suppose x is a limit point of the subset B as well. Then it is clear that $\lim_{t \to x} f_B(t) = \lambda$. Indeed, for every $\varepsilon > 0$, there exists $\delta > 0$ such that

$$0 < \|h\| < \delta \text{ and } x + h \in A \Rightarrow \|f(x + h) - \lambda\| < \varepsilon.$$

Since $B \subseteq A$, it is certainly true that

$$0 < \|h\| < \delta \text{ and } x + h \in B \Rightarrow \|f(x + h) - \lambda\| < \varepsilon$$

and hence that

$$0 < \|h\| < \delta \text{ and } x + h \in B \Rightarrow \|f_B(x + h) - \lambda\| < \varepsilon.$$

Since x is a limit point of B, this means that $\lim_{t \to x} f_B(t) = \lambda$. In practice, it is too cumbersome to introduce the notation for the subset B and the restriction f_B. These will be taken as understood and no explicit reference to the content of this remark will be made.

2-6.10. Proposition. *Suppose that x is a limit point of the domain $A \subseteq \mathbb{R}^n$ of an $\mathbb{R}^m$ valued function f and that $x \in A$. Then f is continuous at x if and only if $\lim_{t \to x} f(t)$ exists and equals $f(x)$. In case x is an isolated point of A, every $\mathbb{R}^m$ valued function f with domain A is continuous at x.*

Proof. The first part is an immediate consequence of the definitions of continuity and of limit. Suppose x is an isolated point of A. Then $x \in A$ and there exists $\delta > 0$ such that the ball $\{t \in \mathbb{R}^n : \|t - x\| < \delta\}$ contains no element of A that is different from x. Therefore $\|h\| < \delta$, $x + h \in A \Rightarrow x + h = x \Rightarrow |f(x + h) - f(x)| = 0$. Consequently, for any $\varepsilon > 0$,

$$\|f(x + h) - f(x)\| < \varepsilon \text{ whenever } \|h\| < \delta \text{ and } x + h \in A.$$

Thus, f is continuous at x. $\square$

While working with the definition of limit, we can usually omit writing '$x + h \in A$', because the subsequent reference to $f(x + h)$ makes it clear that $x + h$ is intended to be in the domain of f.

Remark. If $\lim_{t \to x} f(t)$ exists, then so do $\lim_{t \to x} \|f(t)\|$ and $\lim_{t \to x} (\alpha f)$, where α is any real number; moreover $\lim_{t \to x} \|f(t)\| = \|\lim_{t \to x} f(t)\|$ and $\lim_{t \to x} (\alpha f) = \alpha \lim_{t \to x} f(t)$. If $\lim_{t \to x} g(t)$ also exists, then $\lim_{t \to x} (f + g)$ and $\lim_{t \to x} (f \cdot g)$ exist as well; moreover,

$$\lim_{t \to x} (f + g) = \lim_{t \to x} f(t) + \lim_{t \to x} g(t) \quad \text{and} \quad \lim_{t \to x} (f \cdot g) = (\lim_{t \to x} f(t)) \cdot (\lim_{t \to x} g(t)).$$

These results about limits of sums, products, dot products and so on, are proved just as easily as in the case of $\mathbb{R}$. So also the results that (a) if $\lim_{t \to x} f(t)$ exists, then f is bounded near x, and (b) if the $\lim_{t \to x} f(t) \neq 0$, then $\|f\|$ is greater than some positive number near x except possibly at x. They will therefore be taken for granted and used without further ado. Proofs in the case of $\mathbb{R}$ may be found in Shirali and Vasudeva [23, Propositions 9-1.7 and 9-1.9].

A function f defined on a subset $S \subseteq \mathbb{R}^n$ with values in $\mathbb{R}^m$ is continuous at $x \in S$ if for every $\varepsilon > 0$, there exists a $\delta > 0$ such that $\|f(\xi) - f(x)\| < \varepsilon$ whenever $\|\xi - x\| < \delta$. In general, we cannot expect that for a fixed ε the same value of δ will serve equally well for every x in S. This might happen. If it does, the function is said to be uniformly continuous on S. More precisely, we have the following definition.

2-6.11. Definition. *For any subset* $S \subseteq \mathbb{R}^n$, *a map* $f: S \to \mathbb{R}^m$ *is* **uniformly continuous on** S *if:*

$$\forall\, \varepsilon > 0\ \exists\, \delta > 0\ \ni\ \xi \in S, x \in S, \|\xi - x\| < \delta\ \Rightarrow\ \|f(\xi) - f(x)\| < \varepsilon.$$

2-6.12. Examples. (a) Any constant function is trivially uniformly continuous. It is just as trivial to argue that the identity map I such that $I(x) = x$ for all $x \in \mathbb{R}^n$ is uniformly continuous (take $\delta = \varepsilon$).

(b) The function f on $S = \mathbb{R}^n$ given by $f(x) = \|x\|$ is uniformly continuous. This follows from the fact that $|\|x\| - \|y\|| \leq \|x - y\|$, whichever norm $\|\ \|$ we may use.

(c) Functions that are not uniformly continuous on $\mathbb{R}^n$ can be made up at will from the known instances in the single variable case by employing 2-6.P9.

2-6.13. Theorem. *A function continuous at each point of a compact subset of* $\mathbb{R}^n$ *is bounded. If real-valued, it has a maximum value and a minimum value.*

Proof. Let $K \subseteq \mathbb{R}^n$ be compact and $f: K \to \mathbb{R}^m$ be continuous at each point of K. If f is not bounded, then for each $p \in \mathbb{N}$, there exists $x_p \in K$ such that $\|f(x_p)\| > p$. The sequence $\{x_p\}_{p \geq 1}$ then satisfies

$$x_p \in K \quad \text{for each } p \in \mathbb{N} \tag{1}$$

and

$$\|f(x_p)\| > p \quad \text{for each } p \in \mathbb{N}. \tag{2}$$

By (1), Theorems 2-5.2 and 2-5.7, the sequence $\{x_p\}_{p \geq 1}$ has a convergent subsequence $\{x_{p_q}\}_{q \geq 1}$ with limit x belonging to K. Since f is continuous, it follows that the sequence $\{f(x_{p_q})\}_{q \geq 1}$ converges to $f(x)$ [see 2-6.P7] and is therefore bounded. But this contradicts (2). Therefore f must be bounded.

Now let $M = \sup \{f(x) : x \in K\}$. If $f(x) < M$ for all $x \in K$, then $1/(M - f(x))$ defines a continuous function on K that has no upper bound, in contradiction with what has just been proved. Therefore $f(x_0) = M$ for some $x_0 \in K$, so that f has a maximum value, namely, M. Similar considerations show that f also has a minimum value.   □

2-6.14. Theorem. *A function continuous at each point of a compact subset of $\mathbb{R}^n$ is uniformly continuous on it.*

Proof. Let $K \subseteq \mathbb{R}^n$ be compact and $f: K \rightarrow \mathbb{R}^m$ be continuous at each point of K. If f is not uniformly continuous, then

$$\exists \, \varepsilon > 0 \text{ such that } \forall \, \delta > 0, \text{ some } \xi \in K, x \in K \text{ satisfy}$$

$$\|\xi - x\| < \delta \quad \text{and} \quad \|f(\xi) - f(x)\| \geq \varepsilon.$$

Taking $\delta = \frac{1}{p}$, where $p \in \mathbb{N}$, we get sequences $\{x_p\}_{p \geq 1}$ and $\{\xi_p\}_{p \geq 1}$ such that

$$x_p \in K \quad \text{for each } p \in \mathbb{N}, \tag{1}$$

$$\xi_p \in K \quad \text{for each } p \in \mathbb{N}, \tag{2}$$

$$\|\xi_p - x_p\| < \frac{1}{p} \quad \text{for each } p \in \mathbb{N} \tag{3}$$

and

$$\|f(\xi_p) - f(x_p)\| \geq \varepsilon \quad \text{for each } p \in \mathbb{N}. \tag{4}$$

By (1) and Theorems 2-5.2, 2-5.7, the sequence $\{x_p\}_{p \geq 1}$ has a convergent subsequence $\{x_{p_q}\}_{q \geq 1}$ with limit x belonging to K. By (2) and Theorems 2-5.2, 2-5.7, the sequence $\{\xi_{p_q}\}_{q \geq 1}$ has a convergent subsequence $\{\xi_{p_{q_r}}\}_{r \geq 1}$ with limit ξ belonging to K. The corresponding subsequence $\{x_{p_{q_r}}\}_{r \geq 1}$ of $\{x_{p_q}\}_{q \geq 1}$ then also converges to x. It follows from (3) that

$$x = \xi \tag{5}$$

and from the continuity of f that the sequences $\{f(\xi_{p_{q_r}})\}_{r \geq 1}$ and $\{f(x_{p_{q_r}})\}_{r \geq 1}$ converge to $f(\xi)$ and $f(x)$, respectively [see 2-6.P7]. In view of (4), we have

$$\|f(\xi_{p_{q_r}}) - f(x_{p_{q_r}})\| \geq \varepsilon \quad \text{for each } r \in \mathbb{N}$$

and hence $\|f(\xi) - f(x)\| \geq \varepsilon$, contradicting (5). Therefore f must be uniformly continuous.   □

Problem Set 2-6

2-6.P1. Let x be a limit point of the domain $A \subseteq \mathbb{R}^n$ of a function f with values in $\mathbb{R}^m$. Suppose $y \in \mathbb{R}^m$ and δ, K are positive numbers such that $\|f(t) - y\| \leq K$ whenever $0 < \|t - x\| < \delta$ and $t \in A$. If $\lim_{t \to x} f(t)$ exists, show that $\|\lim_{t \to x} f(t) - y\| \leq K$.

2-6.P2(a). Suppose $x = (a, b)$ is a limit point of the domain $S \subseteq \mathbb{R}^n \times \mathbb{R}^m$ of an $\mathbb{R}^k$ valued function f and that $\lim_{t \to x} f(t)$ exists. Since $t \in \mathbb{R}^n \times \mathbb{R}^m$, it will be convenient to denote t by (u, v), where $u \in \mathbb{R}^n$ and $v \in \mathbb{R}^m$. Assume there exists a positive number μ such that $\lim_{v \to b} f(u, v)$ exists whenever $\|u - a\| < \mu$. Show that $\lim_{u \to a} [\lim_{v \to b} f(u, v)]$ exists and is equal to $\lim_{t \to x} f(t)$. Is it true that if there also exists a positive number ν such that $\lim_{u \to a} f(u, v)$ exists whenever $\|v - b\| < \nu$, then the 'repeated' limits $\lim_{u \to a} [\lim_{v \to b} f(u, v)]$ and $\lim_{v \to b} [\lim_{u \to a} f(u, v)]$ both exist and are equal to $\lim_{t \to x} f(t)$?

(b) Verify the result of part (a) for

$$f(x, y) = \begin{cases} \dfrac{1}{x} \sin(xy) & x \neq 0, y \neq 0 \\ 0 & x = 0 \text{ or } y = 0. \end{cases}$$

2-6.P3. For the function f of Example 2-6.6(a), show that $\lim_{y \to 0} [\lim_{x \to 0} f(x, y)]$ does not even make sense even though the other repeated limit exists and agrees with the limit as $(x, y) \to (0, 0)$.

2-6.P4. Show that any function $f: \mathbb{R}^2 \to \mathbb{R}$ such that

$$f(x, y) = \frac{x^6 + 9x^4 y^2 - 9x^2 y^4 - y^6}{(x^2 + y^2)^3} \quad \text{when } (x, y) \neq (0, 0)$$

cannot be continuous at $(0, 0)$, whatever the value of $f(0, 0)$ may be.

2-6.P5. For the function $f: \mathbb{R}^2 \to \mathbb{R}$ defined by $f(x, y) = x^2 y$, prove continuity at (x, y) by showing how to find $\delta > 0$ for a given $\varepsilon > 0$ so as to ensure that $\|(h, k)\| = \sqrt{(h^2 + k^2)} < \delta \Rightarrow |f(x + h, y + k) - f(x, y)| < \varepsilon$.

2-6.P6. Let $f: \mathbb{R}^2 \to \mathbb{R}$ be defined as follows:

$$f(x,y) = \begin{cases} \dfrac{x^3}{x^2 + y^2} & (x,y) \neq (0,0) \\ 0 & (x,y) = (0,0). \end{cases}$$

Show that f is continuous.

2-6.P7. Show that a map $f:S \to \mathbb{R}^m$, where $S \subseteq \mathbb{R}^n$, is continuous at a point $x \in S$ if and only if for any sequence $\{s_p\}_{p \geq 1}$ in S converging to x, the 'image' sequence $\{f(s_p)\}_{p \geq 1}$ converges to $f(\lim_{p \to \infty} s_p)$.

2-6.P8. (a) Show that the map $f:\mathbb{R}^2 \to \mathbb{R}^2$ given by $f(x_1,x_2) = (|x_1|,x_2)$ is continuous.
(b) For the (continuous) map $f:\mathbb{R}^2 \to \mathbb{R}^2$ given by $f(x_1,x_2) = (|x_1|,x_2)$, give an example of a boundary point u of $E = \{(x_1,x_2) \in \mathbb{R}^2 : -1 \leq x_1 \leq 2\}$ such that $f(u)$ is an interior point of $f(E)$.
(c) Let $E \subseteq U \subseteq \mathbb{R}^n$, where U is open, and let $f:U \to \mathbb{R}^m$ be continuous. If f is injective, show that $u \in U \cap \partial E \Rightarrow f(u) \in \partial(f(E))$, i.e., $f(U \cap \partial E) \subseteq \partial(f(E))$.

2-6.P9. Show that a map $f:S \to \mathbb{R}^m$, where $S \subseteq \mathbb{R}^n$, is uniformly continuous on S if and only if each component function f_j is uniformly continuous on S.

2-6.P10. Let $g:\mathbb{R} \to \mathbb{R}$ not be uniformly continuous. Show that the function $f:\mathbb{R}^n \to \mathbb{R}^m$ defined by $f(x_1,\ldots,x_n) = (g(x_1),0,\ldots,0)$ is also not uniformly continuous.

2-6.P11. Show that a map $f:S \to \mathbb{R}^m$, where $S \subseteq \mathbb{R}^n$, is continuous everywhere if and only if for any open set $V \subseteq \mathbb{R}^m$, the inverse image

$$f^{-1}(V) = \{x \in S : f(x) \in V\} \quad \text{(by definition)}$$

is the intersection of S with some open set $U \subseteq \mathbb{R}^n$.

2-6.P12. Let $K \subseteq \mathbb{R}^n$ be compact and $f:K \to \mathbb{R}^m$ be continuous at each point of K. Show that $f(K) \subseteq \mathbb{R}^m$ is compact. (This is usually paraphrased as: A continuous image of a compact set is compact.)

2-6.P13. Let $X \subseteq \mathbb{R}^n$ be bounded and $f:X \to \mathbb{R}^m$ be uniformly continuous on X. Show that f is bounded.

The result of the next problem can be obtained as a consequence of the following two results: (1) A continuous image of a connected set is connected (2) A connected subset of $\mathbb{R}$ is an interval. But here we ask for a direct proof. The result will be needed for 6-4.P7.

2-6.P14. Let $X \subseteq \mathbb{R}^n$ be connected and $f:X \to \mathbb{R}$ be continuous. Show that $f(X)$ is an interval.

2-6.P15. If the function $f:[a,b] \times [c,d] \to \mathbb{R}$ is continuous and $g:[a,b] \to \mathbb{R}$ is Riemann integrable, then

$$F(y) = \int_a^b g(x)f(x,y)\,dx$$

is defined for every $y \in [c,d]$ and the function F thus defined is continuous on $[c,d]$.

2-6.P16. Let x be a limit point of the domain $A \subseteq \mathbb{R}^n$ of a function f with values in $\mathbb{R}^m$. Show that f cannot have two distinct limits at x.

2-6.P17. [Needed in Proposition 7-4.13] The distance of a point $x \in \mathbb{R}^n$ from a nonempty subset $S \subseteq \mathbb{R}^n$ is defined as $d(x,S) = \inf\{\|x - s\| : s \in S\}$. Show that it is a continuous function of x.

2-6.P18. [Needed in 3-4.P23] Let $0 < \alpha < 1$ and $\Phi : [0,1] \times [0,1] \to \mathbb{R}^3$ be defined as

$$\Phi_1(u,v) = (1-\alpha(2u-1)\sin \pi v)\cos 2\pi v, \quad \Phi_2(u,v) = (1-\alpha(2u-1)\sin \pi v)\sin 2\pi v,$$
$$\Phi_3(u,v) = \alpha(2u-1)\cos \pi v.$$

The range of Φ is known as the 'Möbius band'. Show that

(a) $\Phi(u,1) = \Phi(1-u,0)$ for all $u \in [0,1]$;

(b) if $(u,v) \neq (u',v')$ but $\Phi(u,v) = \Phi(u',v')$, then $u = 1-u'$ and one among v,v' is 0 while the other is 1.

2-7 Norm and Invertibility of a Linear Map

So far, we have discussed limits and continuity in $\mathbb{R}^n$ on the basis of the norm, which played a role analogous to that of absolute value in $\mathbb{R}$. In this section we introduce a similar concept in the set $\mathfrak{L}(\mathbb{R}^n, \mathbb{R}^m)$ of all linear maps from $\mathbb{R}^n$ to $\mathbb{R}^m$. Although it is defined in a manner that bears little resemblance to the definition of norm in $\mathbb{R}^n$, it is denoted by the same symbol and plays a similar role. In particular, we can speak of convergence, open set, continuity and the like in the set of all linear maps. What is more, the norm in $\mathfrak{L}(\mathbb{R}^n, \mathbb{R}^m)$ has a relation to composition of linear maps, because of which its usefulness extends beyond being a mere analogue of the norm in $\mathbb{R}^n$.

We shall obtain a few facts about the set $\mathfrak{L}(\mathbb{R}^n, \mathbb{R}^m)$; for example, that each element of it is bounded on the set $\{x \in \mathbb{R}^n : \|x\| \leq 1\}$. As discussed earlier, an element of $\mathfrak{L}(\mathbb{R}^n, \mathbb{R}^m)$ can have an inverse in the same set. We shall show, among other things, that the subset consisting of elements having inverses is open. One of the results below is that, not only does the identity have an inverse (namely, itself) but also elements 'close to' the identity have inverses. We shall also prove that inversion is continuous.

2-7.1. Theorem. *For any linear map $A : \mathbb{R}^n \to \mathbb{R}^m$, $\sup\{\|Ax\| : \|x\| \leq 1\}$ is finite and does not exceed $\sum_{j=1}^{n} \|Ae_j\|$, where $e_1, e_2, \ldots, e_n$ is the standard basis of $\mathbb{R}^n$.*

Proof. For any $x = x_1e_1 + x_2e_2 + \cdots + x_ne_n \in \mathbb{R}^n$ (where $x_1, x_2, \ldots, x_n \in \mathbb{R}$, of course), we have $|x_j| \leq \|x\|$ for $1 \leq j \leq n$ regardless of which of the three equivalent norms is used. It follows that

$$\|Ax\| \leq \left\| \sum_{j=1}^{n} x_j (Ae_j) \right\| \leq \sum_{j=1}^{n} |x_j| \|Ae_j\| \leq \sum_{j=1}^{n} \|x\| \|Ae_j\| = \|x\| \left(\sum_{j=1}^{n} \|Ae_j\| \right).$$

Therefore, $\|x\| \leq 1 \Rightarrow \|Ax\| \leq \sum_{j=1}^{n} \|Ae_j\|$. It follows that

$$\sup \{\|Ax\| : \|x\| \leq 1\} \leq \sum_{j=1}^{n} \|Ae_j\|. \qquad \square$$

2-7.2. Definition. *The* **norm** *of a linear map* $A : \mathbb{R}^n \to \mathbb{R}^m$ *is*

$$\sup \{\|Ax\| : \|x\| \leq 1\}$$

and is denoted by $\|A\|$.

Since $x/\|x\|$ has norm 1 when $x \neq 0$, it follows from the above definition that $\left\| A\left(\frac{x}{\|x\|} \right) \right\| \leq \|A\|$ and hence that $\|Ax\| \leq \|A\| \|x\|$ for $x \neq 0$. However, this inequality holds trivially when $x = 0$. Therefore, we conclude that

$$\|Ax\| \leq \|A\| \|x\| \quad \text{whenever } x \in \mathbb{R}^n.$$

This inequality will be used without quoting any reference.

One can now reformulate Theorem 2-7.1 as saying that the norm of a linear map A is always a finite real number, which is nonnegative of course, and that it does not exceed $\sum_{j=1}^{n} \|Ae_j\|$, where $e_1, e_2, \ldots, e_n$ is the standard basis of $\mathbb{R}^n$.

Note that $\|A\| = 0$ if and only if $A = O$.

2-7.3. Examples. (a) Recall the example of a linear map $A : \mathbb{R}^n \to \mathbb{R}^m$ with $n = m = 1$ given in Example 2-3.2(a), which was $x \to ax$, where $a \in \mathbb{R}$ is fixed. Since $\|ax\| = |ax| = |a| |x| = |a| \|x\|$, the norm of this linear map, which by definition is

$$\sup \{\|ax\| : \|x\| \leq 1\},$$

works out to be $|a|$.

(b) For the linear map of $\mathbb{R}$ into $\mathbb{R}^m$ given by $x \to xb$, where $b \in \mathbb{R}^m$ is fixed (Example 2-3.2(b)), the norm is $\|b\|$. This can be seen from the equality $\|xb\| = |x| \|b\| = \|x\| \|b\|$, which has the immediate consequence that $\sup \{\|xb\| : \|x\| \leq 1\} = \|b\|$.

(c) For the linear map from $\mathbb{R}^n$ to $\mathbb{R}$ given in Example 2-3.2(c), which was $x \to z \cdot x$ (dot product), where $z \in \mathbb{R}^n$ is fixed, the norm is $\|z\|_2$ provided that we use the norm $\| \ \|_2$ in $\mathbb{R}^n$. To see why, note that $\|z \cdot x\| \leq \|z\|_2 \|x\|_2$, by the Cauchy–

Schwarz inequality, and that when $x = z/\|z\|_2$, assuming $z \neq 0$, we have $\|x\|_2 = 1$ while $\|z \cdot x\| = \|z\|_2$. In case $z = 0$, it is clear that the norm of the linear map $x \rightarrow z \cdot x$ is 0, while $\|z\|_2$ is also 0.

(d) Let $A:\mathbb{R}^2 \rightarrow \mathbb{R}^2$ be the linear map $A(x_1, x_2) = (x_1 + x_2, 0)$. Suppose that in $\mathbb{R}^2$ we use the norm $\|(x_1, x_2)\| = |x_1| + |x_2|$. Then $\|A(x_1, x_2)\| = |x_1 + x_2| + 0 \leq |x_1| + |x_2| = \|(x_1, x_2)\|$. Therefore $\|A\| = \sup \{\|Ax\| : \|x\| \leq 1\} \leq 1$. But since $\|A(1,0)\| = |1 + 0| + 0 = 1 = \|(1,0)\|$, we actually have $\|A\| = 1$. Next, suppose instead that we use the norm $\|(x_1, x_2)\| = \max \{|x_1|, |x_2|\}$. Then $\|A(x_1, x_2)\| = \max \{|x_1 + x_2|, 0\} = |x_1 + x_2| \leq 2\max \{|x_1|, |x_2|\} = 2\|(x_1, x_2)\|$. Therefore $\|A\| = \sup \{\|Ax\| : \|x\| \leq 1\} \leq 2$. But since $\|A(1,1)\| = \max \{|1 + 1|, 0\} = 2 = 2\|(1,1)\|$, we actually have $\|A\| = 2$. This illustrates how the norm of A can depend on which norm is being used in the domain and which in the range space.

(e) Let $A:\mathbb{R}^2 \rightarrow \mathbb{R}^2$ be the linear map $A(x_1, x_2) = (x_1 + x_2, x_1 - x_2)$. In the domain $\mathbb{R}^2$ we shall use the norm $\|(x_1, x_2)\| = \max \{|x_1|, |x_2|\}$, but in the range space $\mathbb{R}^2$, we shall take $\|(y_1, y_2)\|$ to be $\sqrt{(y_1^2 + y_2^2)}$. Then $\|A(x_1, x_2)\|^2 = (x_1 + x_2)^2 + (x_1 - x_2)^2 = 2(x_1^2 + x_2^2) \leq 4(\max \{|x_1|, |x_2|\})^2$. Therefore $\|A\| = \sup \{\|Ax\| : \|x\| \leq 1\} \leq 2$. But since $\|A(1,1)\| = \sqrt{[(1 + 1)^2 + 0]} = 2 = 2\|(1,1)\|$, we actually have $\|A\| = 2$.

2-7.4. Theorem. *A linear map $A:\mathbb{R}^n \rightarrow \mathbb{R}^m$ is uniformly continuous.*

Proof. If $\|A\| = 0$, then $A = O$ and is therefore uniformly continuous. So, suppose $\|A\| > 0$ and consider any two points ξ and x of $\mathbb{R}^n$. By linearity of A,

$$\|A(\xi) - A(x)\| = \|A(\xi - x)\| \leq \|A\| \|\xi - x\|.$$

For any given $\varepsilon > 0$, choose $\delta = \varepsilon/\|A\|$. Then

$$\|\xi - x\| < \delta \Rightarrow \|A(\xi) - A(x)\| < \|A\| \varepsilon/\|A\| = \varepsilon.$$

Thus A is uniformly continuous. $\square$

We have noted before that, with regard to addition and multiplication by scalars, linear transformations behave like vectors. The next theorem shows that they behave like vectors even with regard to the norm.

2-7.5. Theorem. *Suppose A and B are linear maps from $\mathbb{R}^n$ to $\mathbb{R}^m$ and that $\lambda \in \mathbb{R}$. Then*

$$\|A\| = 0 \Leftrightarrow A = O; \|A + B\| \leq \|A\| + \|B\|; \|\lambda A\| = |\lambda| \|A\|.$$

Proof. That $\|A\| = 0 \Leftrightarrow A = O$ has been noted before and is trivial to prove.

For any $x \in \mathbb{R}^n$ with $\|x\| \leq 1$,

$$\|(A + B)x\| = \|Ax + Bx\| \le \|Ax\| + \|Bx\| \le \|A\| + \|B\|.$$

Therefore $\|A\| + \|B\|$ is an upper bound of the set $\{\|(A+B)x\| : \|x\| \le 1\}$. Thus $\|A + B\| \le \|A\| + \|B\|$.

Since $\|(\lambda A)x\| = |\lambda|\,\|Ax\|$, then the set $\{\|(\lambda A)x\| : \|x\| \le 1\}$ is obtained from $\{\|Ax\| : \|x\| \le 1\}$ by multiplying each number in the latter by $|\lambda|$. It follows that the sup of the former set is $|\lambda|$ times that of the latter set (this is an elementary consequence of the definition of sup and should have been encountered by the reader while studying the concepts of sup and inf). But the two sups are, respectively, $\|\lambda A\|$ and $|\lambda|\,\|A\|$, by definition of the norm of a linear map. Therefore $\|\lambda A\| = |\lambda|\,\|A\|$. □

As mentioned in the opening paragraph of this section, the set of all linear maps from $\mathbb{R}^n$ to $\mathbb{R}^m$ will be denoted by the symbol $\mathcal{L}(\mathbb{R}^n, \mathbb{R}^m)$.

For a sequence $\{A_p\}_{p \ge 1}$ of linear maps, convergence to a limit and the Cauchy property are defined analogously to Def. 2-2.7 and Def. 2-2.9. The elementary properties mentioned for vectors just after Def. 2-2.7 carry over to linear maps exactly as they do for vectors. In particular, a convergent sequence of linear maps is 'bounded'. The analogue of Theorem 2-2.10 (Cauchy completeness) can be obtained without going through anything similar to Proposition 2-2.8. The following result can be phrased as '$\mathcal{L}(\mathbb{R}^n, \mathbb{R}^m)$ is Cauchy complete'.

2-7.6. Theorem. *Any Cauchy sequence of linear maps converges to some linear map.*

Proof. Let $\{A_p\}_{p \ge 1}$ be a Cauchy sequence of linear maps $A_p{:}\mathbb{R}^n{\to}\mathbb{R}^m$. We shall define a map $A{:}\mathbb{R}^n{\to}\mathbb{R}^m$ and argue that it is linear and that $\lim_{p \to \infty} A_p = A$, which is to say, for any $\varepsilon > 0$, there exists a natural number N such that

$$p \ge N \quad \Rightarrow \quad \|A_p - A\| < \varepsilon.$$

In order to define A, consider any $x \in \mathbb{R}^n$. The sequence $\{A_p(x)\}_{p \ge 1}$ in $\mathbb{R}^m$ is Cauchy because $\|A_p(x) - A_q(x)\| = \|(A_p - A_q)(x)\| \le \|A_p - A_q\|\,\|x\|$. But $\mathbb{R}^m$ is Cauchy complete by Theorem 2-2.10. Therefore $\{A_p(x)\}_{p \ge 1}$ converges to some limit in $\mathbb{R}^m$. We define $A(x)$ to be this limit. We have $A(x+y) = A(x) + A(y)$, because

$$A(x + y) = \lim_{p \to \infty} A_p(x+y) = \lim_{p \to \infty} (A_p(x) + A_p(y))$$

$$= \lim_{p \to \infty} A_p(x) + \lim_{p \to \infty} A_p(y) = A(x) + A(y).$$

A similar argument shows that $A(cx) = cA(x)$. Thus, A is linear. It remains to show that $\lim_{p \to \infty} A_p = A$.

For any $\varepsilon > 0$, there exists a natural number N such that

$$p \geq N, q \geq N \quad \Rightarrow \quad \|A_p - A_q\| < \tfrac{\varepsilon}{2} \quad \Rightarrow \quad \|(A_p - A_q)(x)\| < \tfrac{\varepsilon}{2} \text{ for } \|x\| \leq 1$$

$$\Rightarrow \quad \|A_p(x) - A_q(x)\| < \tfrac{\varepsilon}{2} \text{ for } \|x\| \leq 1,$$

whence, by taking the limit as $q \to \infty$, we find that

$$p \geq N \quad \Rightarrow \quad \|A_p(x) - A(x)\| \leq \tfrac{\varepsilon}{2} \quad \text{for } \|x\| \leq 1$$

$$\Rightarrow \quad \|A_p - A\| \leq \tfrac{\varepsilon}{2} \quad \text{by Def. 2-7.2.}$$

Thus, $\lim\limits_{p \to \infty} A_p = A$. $\qquad\qquad\qquad\qquad\qquad\qquad\qquad\qquad\qquad\square$

Since it follows by the usual argument from Theorem 2-7.5 that $\big|\|A_p\| - \|A\|\big| \leq \|A_p - A\|$, we can further obtain that $\lim\limits_{p \to \infty} A_p = A \Rightarrow \lim\limits_{p \to \infty} \|A_p\| = \|A\|$.

2-7.7. Theorem. *Suppose $A:\mathbb{R}^n \to \mathbb{R}^m$ and $B:\mathbb{R}^m \to \mathbb{R}^p$ are linear maps. Then*

$$\|BA\| \leq \|B\|\|A\|.$$

Proof. Consider any $x \in \mathbb{R}^n$ with $\|x\| \leq 1$. If $Ax = 0$, then $(BA)x = B(Ax) = B0 = 0$ and $\|(BA)x\| = 0 \leq \|B\|\|A\|$. If $Ax \neq 0$, then $\|Ax\| > 0$ and the vector $Ax/\|Ax\|$ has norm equal to 1. Therefore,

$$\|B(Ax/\|Ax\|)\| \leq \|B\|,$$

so that, from the linearity of B, it follows that $\|B(Ax)\| \leq \|B\|\|Ax\|$. But $(BA)x = B(Ax)$ and $\|Ax\| \leq \|A\|$ (because $\|x\| \leq 1$). Therefore,

$$\|(BA)x\| \leq \|B\|\|A\|.$$

This shows that $\|B\|\|A\|$ is an upper bound for $\{\|(BA)x\| : \|x\| \leq 1\}$. It follows from here that $\|BA\| \leq \|B\|\|A\|$. $\qquad\qquad\qquad\qquad\qquad\square$

The preceding theorem is used frequently and it is customary to use it without quoting it or giving any reference.

2-7.8. Theorem. *If $\{A_p\}_{p \geq 1}$ and $\{B_p\}_{p \geq 1}$ are convergent sequences of linear maps, then* $\lim\limits_{p \to \infty} (A_p B_p) = (\lim\limits_{p \to \infty} A_p)(\lim\limits_{p \to \infty} B_p)$.

Proof. This is proved exactly as in the real case by using the fact that a convergent sequence of linear maps is bounded, together with the following consequence of Theorem 2-7.7:

$$\|A_p B_p - AB\| \leq \|A_p\|\|B_p - B\| + \|B\|\|A_p - A\|. \qquad\qquad\square$$

2-7.9. Remarks. (a) Suppose the linear map $A \in \mathfrak{L}(\mathbb{R}^n, \mathbb{R}^m)$ has matrix $[a_{ij}]$. As noted after the definition of the norm of a linear map, $\|A\| \leq \Sigma_j \|Ae_j\|$, where

$e_1, e_2, \ldots, e_n$ is the standard basis of $\mathbb{R}^n$. In terms of the matrix $[a_{ij}]$ and standard basis $f_1, \ldots, f_m$ of $\mathbb{R}^m$, we therefore have

$$\|A\| \leq \Sigma_j \Sigma_i |a_{ij}| \|f_i\| = \Sigma_j \Sigma_i |a_{ij}|.$$

Moreover, by definition of matrix of a linear map, we have $a_{ij} = (Ae_j)_i$ for any i,j $(1 \leq i \leq m, 1 \leq j \leq n)$. Therefore,

$$|a_{ij}| = |(Ae_j)_i| \leq \|Ae_j\| \leq \|A\| \|e_j\| = \|A\|.$$

The two inequalities displayed above have the following consequence: Suppose that A is a mapping from a subset E of $\mathbb{R}^n$ into $\mathscr{L}(\mathbb{R}^n, \mathbb{R}^m)$, not necessarily linear, and that $A(x)$ has matrix $[a_{ij}(x)]$. Then A is continuous at a point of E if and only if all the mn $\mathbb{R}$-valued functions a_{ij} are continuous there. Indeed, the hypothesis that A maps into $\mathscr{L}(\mathbb{R}^n, \mathbb{R}^m)$ implies that $A(x) - A(y) \in \mathscr{L}(\mathbb{R}^n, \mathbb{R}^m)$ and has matrix $[a_{ij}(x) - a_{ij}(y)]$; therefore, for any $x, y \in E$,

$$\|A(x) - A(y)\| \leq \Sigma_j \Sigma_i |a_{ij}(x) - a_{ij}(y)|$$

and

$$|a_{ij}(x) - a_{ij}(y)| \leq \|A(x) - A(y)\|.$$

This is so, irrespective of which norm is used in $\mathbb{R}^n$ and $\mathbb{R}^m$. (The norm $\|A\|$ of a linear map A is defined in terms of the norms in $\mathbb{R}^n$ and $\mathbb{R}^m$ and its value therefore depends upon the norms chosen in the latter.)

(b) Now suppose $A: E \to \mathscr{L}(\mathbb{R}^n, \mathbb{R}^m)$ is continuous and $B \in \mathscr{L}(\mathbb{R}^m, \mathbb{R}^p)$. Then there is a map $C: E \to \mathscr{L}(\mathbb{R}^n, \mathbb{R}^p)$ given by $C(x) = B \circ (A(x))$. Since the linearity of B and $A(x)$ implies

$$\|C(x) - C(y)\| \leq \|B\| \|A(x) - A(y)\| \quad \text{for any } x, y \in E,$$

the map C is seen to be continuous. The reader is cautioned that C is not a composition of the continuous map A with B, because such a composition is not even possible, considering that the domain of B is $\mathbb{R}^m$ while the range of A is a subset of $\mathscr{L}(\mathbb{R}^n, \mathbb{R}^m)$. Therefore the reason for the continuity of C is not that a composition of continuous maps is continuous.

The following proposition shows that linear maps which are 'close to' the identity map have inverses.

2-7.10. Proposition. *Suppose $A \in \mathscr{L}(\mathbb{R}^n, \mathbb{R}^n)$ satisfies $\|A\| < 1$. Then the map $I - A$, where I denotes the identity map given by $I(x) = x$ for all $x \in \mathbb{R}^n$, has an inverse. Moreover, its norm satisfies $\|(I-A)^{-1}\| \leq 1/(1 - \|A\|)$.*

Proof. Consider the sequence $\{B_p\}_{p \geq 1}$ in $\mathscr{L}(\mathbb{R}^n, \mathbb{R}^n)$ given by

$$B_p = I + A + \cdots + A^{p-1}. \tag{1}$$

It satisfies $B_p(I-A) = (I-A)B_p = I-A^p$. By Theorem 2-7.7, $\|A^p\| \le \|A\|^p$ and we know from elementary analysis that $\|A\|^p \to 0$ as $p \to \infty$, because $\|A\| < 1$. It follows that $\lim_{p \to \infty} A^p = O$. Therefore, if $\{B_p\}_{p \ge 1}$ converges, then its limit B must be the inverse of $I - A$. In order to show that $\{B_p\}_{p \ge 1}$ converges, it is sufficient, in view of Theorem 2-7.6, to argue that it is a Cauchy sequence. But this is a consequence of the following computation for $p > q$, in which we use both (1) and Theorem 2-7.7:

$$\|B_p - B_q\| = \|A^q + \cdots + A^{p-1}\| \le \|A^q\| + \cdots + \|A^{p-1}\|$$

$$\le \|A\|^q + \cdots + \|A\|^{p-1} \le \|A\|^q/(1-\|A\|).$$

A similar computation yields $\|B_p\| \le 1/(1-\|A\|)$. Therefore

$$\|(I-A)^{-1}\| = \|\lim_{p \to \infty} B_p\| = \lim_{p \to \infty} \|B_p\| \le 1/(1-\|A\|). \qquad \square$$

The theorem below, which shows that the collection of invertible maps in $\mathfrak{L}(\mathbb{R}^n, \mathbb{R}^n)$ is open, will be required later for proving the inverse function theorem (Theorem 4-2.1).

2-7.11. Theorem. *Let Ω be the subset of $\mathfrak{L}(\mathbb{R}^n, \mathbb{R}^n)$ consisting of invertible maps. If $A \in \Omega$ and $\|B - A\| < 1/\|A^{-1}\|$, then $B \in \Omega$ and*

$$\|B^{-1}\| \le \frac{\|A^{-1}\|}{(1-\|A^{-1}\|\|A-B\|)}. \qquad (A)$$

Moreover, the map $A \to A^{-1}$ of Ω into itself is a continuous map.

Proof. First of all,

$$B = (B-A)+A = A(A^{-1}(B-A)+I). \qquad (1)$$

Moreover,

$$\|A^{-1}(B-A)\| \le \|A^{-1}\|\|B-A\| < 1, \quad \text{by hypothesis.}$$

Since $\|-A^{-1}(B-A)\| = \|A^{-1}(B-A)\|$, it follows by Proposition 2-7.10 and from the inequality above that $A^{-1}(B-A)+I$ is invertible, and hence by (1), B is also invertible.

Once again by Proposition 2-7.10, we have

$$\|(A^{-1}(B-A)+I)^{-1}\| \le 1/(1-\|A^{-1}(B-A)\|). \qquad (2)$$

But $1 - \|A^{-1}(B-A)\| \ge 1 - \|A^{-1}\|\|B-A\| > 0$ and hence by (1) and (2), we have

$$\|B^{-1}\| \le \|A^{-1}\|\|(A^{-1}(B-A)+I)^{-1}\| \le \|A^{-1}\|/(1-\|A^{-1}\|\|B-A\|),$$

thus establishing (A).

It remains to prove that the map $A \to A^{-1}$ of Ω into itself is continuous.

Consider any $A \in \Omega$ and let $\|B - A\| < 1/2\|A^{-1}\|$. Then on the one hand, we have $\|B - A\|\|A^{-1}\| < \frac{1}{2}$ and hence $1 - \|A^{-1}\|\|B - A\| > \frac{1}{2}$, so that

$$\frac{1}{(1 - \|A^{-1}\|\|A - B\|)} < 2. \tag{3}$$

On the other hand, we have $\|B - A\| < 1/\|A^{-1}\|$, so that (A) holds. Now,

$$\|B^{-1} - A^{-1}\| = \|B^{-1}(A - B)A^{-1}\| \leq \|B^{-1}\|\|B - A\|\|A^{-1}\|$$

$$\leq \frac{\|A^{-1}\|}{(1 - \|A^{-1}\|\|A - B\|)} \|B - A\|\|A^{-1}\| \quad \text{because (A) holds}$$

$$\leq 2\|B - A\|\|A^{-1}\|^2 \qquad \text{by (3)}.$$

This implies the desired continuity. In fact, for a given $\varepsilon > 0$, the number

$$\delta = \min\{1/2\|A^{-1}\|, \varepsilon/2\|A^{-1}\|^2\} > 0$$

is now seen to satisfy $\|B - A\| < \delta \implies \|B^{-1} - A^{-1}\| < \varepsilon$. $\square$

For certain purposes, it is necessary to know the relation between the norm $\|A\|_2$ when the norm of an element x of $\mathbb{R}^n$ or $\mathbb{R}^m$ is taken as $\|x\|_2$ and the norm $\|A\|_\infty$ when the norm of an element x of $\mathbb{R}^n$ or $\mathbb{R}^m$ is taken as $\|x\|_\infty$. In this direction, we have the following result:

2-7.12. Proposition. *Let* $A: \mathbb{R}^n \to \mathbb{R}^m$ *be a linear map and*

$$\|A\|_2 = \sup\{\|Ax\|_2 : \|x\|_2 \leq 1\}, \qquad \|A\|_\infty = \sup\{\|Ax\|_\infty : \|x\|_\infty \leq 1\}.$$

Then

$$\|A\|_2 \leq m^{1/2}\|A\|_\infty \qquad and \qquad \|A\|_\infty \leq n^{1/2}\|A\|_2.$$

Proof. From Proposition 2-2.6, it follows that

$$\|Ax\|_2 \leq m^{1/2}\|Ax\|_\infty \qquad \text{and} \qquad \|x\|_\infty \leq \|x\|_2. \tag{1}$$

Let $\|x\|_2 \leq 1$. Using both inequalities in (1), we get

$$\|Ax\|_2 \leq m^{1/2}\|Ax\|_\infty \leq m^{1/2}\|A\|_\infty\|x\|_\infty \leq m^{1/2}\|A\|_\infty\|x\|_2 \leq m^{1/2}\|A\|_\infty,$$

which implies that $\|A\|_2 \leq m^{1/2}\|A\|_\infty$. The other inequality follows by a similar argument. $\square$

Problem Set 2-7

In the first four problems here, take the norm to be $\|\ \|_2$.

2-7.P1. For the linear map $A:\mathbb{R}^2 \to \mathbb{R}^2$ defined by $A(x_1,x_2) = (x_1 + x_2, x_1 - x_2)$, find $\|A\|$.

2-7.P2. For the linear maps $A:\mathbb{R}^2 \to \mathbb{R}^2$ and $B:\mathbb{R}^2 \to \mathbb{R}^2$, defined by
$$A(x_1,x_2) = (x_1 + x_2, x_1 - x_2) \text{ and } B(x_1, x_2) = (x_1 + x_2, 0),$$
verify (by finding the norms involved) that $\|BA\| \leq \|B\|\|A\|$.

2-7.P3. For the linear maps $A:\mathbb{R}^2 \to \mathbb{R}^2$ and $B:\mathbb{R}^2 \to \mathbb{R}^2$, defined by
$$A(x_1, x_2) = (x_1, -x_1) \text{ and } B(x_1, x_2) = (x_1 + x_2, 0),$$
verify (by finding the norms involved) that $\|BA\| < \|B\|\|A\|$.

2-7.P4. Find the norm of the linear map of 2-3.P8.

2-7.P5. Let A be an invertible map in $\mathbb{R}^n$ and $\|A^{-1}(B - A)\| < 1$. Prove that B is invertible and that $\|B^{-1}\| \leq \|A^{-1}\|/(1 - \|A^{-1}(B - A)\|)$.

2-7.P6. If A and B are invertible linear operators in $\mathbb{R}^n$ such that $\|(B - A)A^{-1}\| < 1$, prove that $\|B^{-1}\| \leq \|BA^{-1}B^{-1}\|/(1 - \|(B - A)A^{-1}\|)$.

2-7.P7. Prove that, to every $A \in \mathfrak{L}(\mathbb{R}^n, \mathbb{R})$ there corresponds a unique $y \in \mathbb{R}^n$ such that $Ax = x \cdot y \ \forall \, x \in \mathbb{R}^n$. Note that, in view of Example 2-7.3(c), when we use the norm $\| \ \|_2$ in $\mathbb{R}^n$, we have $\|A\| = \|y\|$.

2-8 Double Sequences and Series

Given two series $\Sigma_m a_m$ and $\Sigma_n b_n$, one can set up the 'double series' $\Sigma_{m,n} f(m,n)$, where $f(m,n) = a_m b_n$. Since the ordered pairs (m,n) can be arranged in a sequence in a variety of ways, each of them provides a way of converting the double series into an ordinary series. The convergence and sum of the resulting series generally depend on the manner in which the ordered pairs have been arranged in a sequence.

In this section we propose considering double series $\Sigma_{m,n} f(m,n)$ that are not necessarily derived from separate series $\Sigma_m a_m$ and $\Sigma_n b_n$, and the question of their 'repeated' limits. The material will not be used in the sequel and can therefore be omitted without loss of continuity.

2-8.1. Definition. *By a **double sequence** we mean a function f defined on $\mathbb{N} \times \mathbb{N}$ to some set X. If $f(m,n) = a_{m,n}$, $(m,n) \in \mathbb{N} \times \mathbb{N}$, it is customary to denote the sequence f by $\{a_{m,n}\}_{m \geq 1, n \geq 1}$. The values of f, that is, the elements $a_{m,n}$ are called the **terms** of the sequence.*

We shall be interested only in real or complex-valued double sequences.

2-8.2. Definition. *If $l \in \mathbb{C}$, we write* $\lim_{m,n} f(m,n) = l$ *and say that the double sequence* **converges to** *l if the following condition holds: For every $\varepsilon > 0$, there exists an integer n_0 such that*

$$|f(m,n) - l| < \varepsilon \quad \text{whenever both } m \geq n_0 \text{ and } n \geq n_0.$$

It may be verified as usual that l is unique if it exists; it is called the **double limit** of $f(m,n)$.

In addition to the notion of double limit, there is the notion of repeated limit (or iterated limit) as described below:

For each fixed m, assume that $\lim_n f(m,n)$ exists; then the limit $\lim_m (\lim_n f(m,n))$, if its exists, is called a **repeated** (or **iterated**) **limit**. In like manner, one can consider the other repeated limit $\lim_n (\lim_m f(m,n))$.

2-8.3. Example. Let $f(m,n) = \frac{mn}{m^2 + n^2}$, $m,n \geq 1$. Then $\lim_m f(m,n) = 0$ and hence $\lim_n (\lim_m f(m,n)) = 0$. Also, $\lim_m (\lim_n f(m,n)) = 0$. But $f(m,n) = \frac{1}{2}$ if $m = n$ and $f(m,n) = \frac{2}{5}$ if $m = 2n$. So, $\lim_{m,n} f(m,n)$ does not exist.

2-8.4. Proposition. *If $\lim_{m,n} f(m,n) = l$ and the limit $\lim_n f(m,n)$ exists for each m, then the repeated limit $\lim_m (\lim_n f(m,n))$ also exists and has the value l.*
Proof. Let $l(m) = \lim_n f(m,n)$. For every $\varepsilon > 0$, there exists n_0 such that

$$|f(m,n) - l| < \tfrac{\varepsilon}{2} \quad \text{whenever both } m \geq n_0 \text{ and } n \geq n_0. \tag{1}$$

For each m, choose $n(m)$ such that

$$|l(m) - f(m,n)| < \tfrac{\varepsilon}{2} \quad \text{for } n \geq n(m). \tag{2}$$

Consider an $m \geq n_0$ and the corresponding $n(m)$ as in (2) and choose $n > \max \{n_0, n(m)\}$. Then the two inequalities (1) and (2) hold and hence

$$|l(m) - l| \leq |l(m) - f(m,n)| + |f(m,n) - l| < \varepsilon, \text{ provided } m \geq n_0.$$

Thus, $\lim_m l(m) = l$. □

2-8.5. Remarks. (a) A similar result holds if we interchange the roles of m and n.

(b) The existence of the double limit $\lim_{m,n} f(m,n)$ and of $\lim_n f(m,n)$ for every m implies the existence of the repeated limit $\lim_m (\lim_n f(m,n))$. Example 2-8.3 shows that the converse is not true.

We list below examples to illustrate various situations where the double limit and the two repeated limits may not all be equal or where some of the limits exist while others do not.

2-8.6. Examples. (a) Let $f(m,n) = \begin{cases} 1 & \text{if } n \geq m \\ 0 & \text{if } n < m. \end{cases}$

Here $\lim_m f(m,n) = 0$ and therefore $\lim_n(\lim_m f(m,n)) = 0$. But $\lim_n f(m,n) = 1$ and therefore $\lim_m(\lim_n f(m,n)) = 1$. So, the double limit cannot exist in view of Proposition 2-8.4.

(b) Let $f(m,n) = \dfrac{1-(-1)^n}{m} = \begin{cases} \dfrac{2}{m} & \text{if } n \text{ is odd} \\ 0 & \text{if } n \text{ is even.} \end{cases}$

Then $\lim_m f(m,n) = 0$ and therefore $\lim_n(\lim_m f(m,n)) = 0$. On the other hand, $\lim_n f(m,n)$ does not exist. So, the double limit cannot exist in view of Proposition 2-8.4.

(c) Let $f(m,n) = \begin{cases} 0 & \text{if } |m-n| \text{ is odd} \\ \dfrac{1}{\min\{m,n\}} & \text{if } |m-n| \text{ is even.} \end{cases}$

Then $\lim_{m,n} f(m,n) = 0$. But $\lim_n f(m,n)$ does not exist and hence $\lim_m(\lim_n f(m,n))$ cannot exist. Also, $\lim_m f(m,n)$ does not exist and hence $\lim_n(\lim_m f(m,n))$ cannot exist.

2-8.7. Definition. *Let f be a double sequence and let s be the double sequence defined by*

$$s(p,q) = \sum_{m=1}^{p}\sum_{n=1}^{q} f(m,n).$$

The double sequence $\{s(p,q)\}_{p\geq 1, q\geq 1}$ of **partial sums** *is denoted by $\Sigma_{m,n} f(m,n)$ and is called a* **double series**. *If it has limit l, that is, if $\lim_{p,q} s(p,q) = l$, then the double series $\Sigma_{m,n} f(m,n)$ is said to be* **convergent**; *l is called the* **sum** *of the series and we write $l = \Sigma_{m,n} f(m,n)$. Each number $f(m,n)$ is called a* **term** *of the series. The double series $\Sigma_{m,n} f(m,n)$ is said to* **converge absolutely** *if $\Sigma_{m,n} |f(m,n)|$ converges.*

It may be noted that the symbol $\Sigma_{m,n} f(m,n)$ may denote either the double series or its sum, depending on context.

2-8.8. Examples. (a) Consider the double series $\Sigma_{m,n} f(m,n)$, where $f(m,n)$ is described by the array, or 'infinite matrix'

$$
\begin{matrix}
1 & 1 & 1 & 1 & \cdots \\
1 & -1 & -1 & -1 & \cdots \\
1 & -1 & 0 & 0 & \cdots \\
1 & -1 & 0 & 0 & \cdots \\
 & & \vdots
\end{matrix}
$$

All the terms are zero except in the first two rows and first two columns. Here $s(p,q) = 2, p,q > 1$, and so the double series has sum 2.

(b) The partial sum $s(p,q)$ of the double series $\Sigma_{m,n}\, m^{-\alpha}n^{-\beta}$ can be represented as the product $(\sum_{m=1}^{p} m^{-\alpha})(\sum_{n=1}^{q} n^{-\beta})$. Since the series $\sum_{m=1}^{\infty} m^{-\alpha}$ and $\sum_{n=1}^{\infty} n^{-\beta}$ are convergent when $\alpha,\beta > 1$, the double series has sum equal to the product of the sums of the aforementioned series.

Since $f(p,q) = s(p,q) - s(p-1,q) - s(p,q-1) + s(p-1,q-1)$, it follows that when $\{s(p,q)\}_{p\geq1,q\geq1}$ converges, we can find μ so that $|f(p,q)| < \varepsilon$ for $p > \mu$, $q > \mu$. In fact, if μ is so chosen that $|s(p,q) - l| < \varepsilon/4$ when p and q are both larger than μ, then

$$|f(p,q)| \leq |s(p,q) - l| + |s(p-1,q) - l|$$
$$+ |s(p,q-1) - l| + |s(p-1,q-1) - l| < 4\left(\frac{\varepsilon}{4}\right) = \varepsilon.$$

This does not imply that $\lim_{p,q} f(p,q)$ will necessarily be zero when the repeated limits $\lim_q (\lim_p s(p,q))$ and $\lim_p (\lim_q s(p,q))$ both exist. For instance, consider the series $\Sigma_{m,n}\, f(m,n)$, where

$$f(m,n) = \begin{cases} 1 & \text{if } m = n+1, n = 1,2,\ldots \\ -1 & \text{if } m = n-1, n = 2,3,\ldots \\ 0 & \text{otherwise.} \end{cases}$$

Then $\lim_p (\lim_q s(p,q)) = -1$, $\lim_q (\lim_p s(p,q)) = 1$ and $f(m,n)$ does not have double limit zero. The reader may list the elements as an infinite matrix to see the validity of the above assertions.

2-8.9. Proposition. *A necessary and sufficient condition for the convergence of the double series $\Sigma_{m,n} f(m,n)$ is that for every $\varepsilon > 0$, there exists an integer μ such that*

$$|s(p,q) - s(m,n)| < \varepsilon \quad \text{for} \quad p \geq m \geq \mu \text{ and } q \geq n \geq \mu.$$

Proof. The condition is obviously necessary. We need only show that it is sufficient. Denote by σ_n the value of $s(m,n)$ when $m = n$. Then our condition yields

$$|\sigma_q - \sigma_m| < \varepsilon \text{ if } q \geq n \geq \mu.$$

Hence σ_n approaches a limit s, and so we can find μ_1 such that

$$|s - \sigma_n| < \tfrac{\varepsilon}{2} \text{ if } n \geq \mu_1.$$

Now the general condition also gives an integer μ_2 such that

$$|s(p,q) - \sigma_n| < \tfrac{\varepsilon}{2} \text{ if } p,q \geq n \geq \mu_2.$$

Let $\mu_3 = \max\{\mu_1,\mu_2\}$. Then

$$|s(p,q) - s| < \varepsilon \text{ if } p,q \geq \mu_3.$$ □

It is now easy to show that a double series is convergent if it is absolutely convergent.

We end this section with the following results on double series with non-negative terms.

2-8.10. Proposition. *Suppose* $\{s(p,q)\}\}_{p\geq1, q\geq1}$ *is the double sequence of partial sums of a double series with nonnegative terms. Then* $\lim_{p,q} s(p,q) = s$ *if and only if* $\lim_p s(p,p) = s$.

Proof. That $\lim_{p,q} s(p,q) = s$ implies $\lim_p s(p,p) = s$ is obvious. Assume that $\lim_p s(p,p) = s$ and consider an arbitrary $\varepsilon > 0$. There exists an integer μ such that

$$s \geq s(p,p) > s - \varepsilon \text{ whenever } p \geq \mu.$$

Since the terms of the double series are positive, it follows that

$$s \geq s(p+q,p+q) \geq s(p,q) \geq s - \varepsilon \text{ provided that } p,q \geq \mu.$$

Thus $\lim_{p,q} s(p,q) = s$.   □

2-8.11. Proposition. (a) *If the double series* $\Sigma_{m,n} f(m,n)$ *with nonnegative terms converges, then for each* m *(respectively,* n*) the series* $\Sigma_n f(m,n)$ *(respectively,* $\Sigma_m f(m,n)$*) converges and the following equality holds:*

$$\sum_{m=1}^{\infty}\sum_{n=1}^{\infty} f(m,n) = \Sigma_{m,n} f(m,n) = \sum_{n=1}^{\infty}\sum_{m=1}^{\infty} f(m,n).$$

(b) *Suppose that* $f(m,n) \geq 0$, $m,n = 1,2,\ldots$. *If either of repeated limits*

$$\lim_m (\lim_n s(m,n)), \ \lim_n(\lim_m s(m,n))$$

exists, so does the double limit and all three are equal.

Proof. (a) If $\Sigma_{m,n} f(m,n) = s$, then it is clear that $s(m,n) \leq s$; and consequently for any fixed value of m, the sum of any number of terms is less than or equal to s, which in turn, implies that the limit $\lim_n s(m,n)$, for any fixed m, is less than or equal to s. Now there exists an integer μ such that $s(m,n) > s - \varepsilon$ if $m,n > \mu$. Consequently,

$$s \geq \lim_n s(m,n) > s - \varepsilon \text{ if } m \geq \mu.$$

Hence

$$\lim_m (\lim_n s(m,n)) = s$$

or

$$\sum_{m=1}^{\infty}\sum_{n=1}^{\infty} f(m,n) = s.$$

In a similar way, we see that

$$\sum_{n=1}^{\infty} \sum_{m=1}^{\infty} f(m,n) = s.$$

(b) Suppose that $\lim_m (\lim_n s(m,n)) = s$. Then

$$s(m,m) \le s(m,n) \text{ if } n > m$$

and so,

$$s(m,m) \le \lim_n s(m,n) \le s.$$

Hence the sequence $\{s(m,m)\}$ converges to a limit σ, say. It now follows from Proposition 2-8.10 and (a) above that $s = \sigma$ and that $\lim_n (\lim_m s(m,n)) = \sigma$.   □

The restriction in the above proposition that the terms of the series be positive cannot be dropped. See 2-8.P2.

2-8.12. Example. $\sum_{m,n} (m+n)^{-\alpha}$ is convergent if $\alpha > 2$. One way to obtain this is to use the integral test (see Shirali and Vasudeva [23, Theorem 12-2.4]) twice. For each m, the function on $[1,\infty]$ given by $(m+x)^{-\alpha}$ is nonnegative and decreasing; moreover, its (improper) integral over $[1,\infty]$ is $\frac{(m+1)^{1-\alpha}}{\alpha-1}$. Therefore, $\sum_n (m+n)^{-\alpha} \le (m+1)^{-\alpha} + \frac{(m+1)^{1-\alpha}}{\alpha-1}$. Now, $(x+1)^{-\alpha} + \frac{(x+1)^{1-\alpha}}{\alpha-1}$ is nonnegative and decreasing on $[1,\infty]$, with integral equal to $\frac{2^{1-\alpha}}{\alpha-1} + \frac{2^{2-\alpha}}{(\alpha-1)(\alpha-2)}$. It follows that $\sum_m (\sum_n (m+n)^{-\alpha})$ is convergent. By Proposition 2-8.11, $\sum_{m,n} (m+n)^{-\alpha}$ is convergent.

A second way to arrive at the same conclusion is via Proposition 2-8.10. In the finite sum $s(p,p)$, we can rearrange terms to get

$$s(p,p) = \sum_{2 \le q \le 2p} \sum_{\substack{m+n=q \\ 1 \le m \le p \\ 1 \le n \le p}} (m+n)^{-\alpha}.$$

In the inner summation, there can be at most q terms because of the restriction that $m+n = q$. Therefore the sum is no greater than $q^{1-\alpha}$. Hence

$$s(p,p) \le \sum_{2 \le q \le 2p} q^{1-\alpha}.$$

Since $1-\alpha < -1$, the series $\sum_q q^{1-\alpha}$ is convergent and the above inequality implies that $s(p,p) \le \sum_q q^{1-\alpha}$. Thus the increasing sequence $s(p,p)$ is bounded above and hence converges. Now Proposition 2-8.10 shows that that the double series $\sum_{m,n} (m+n)^{-\alpha}$ is convergent.

Problem Set 2-8

2-8.P1. Consider the double series $\Sigma_{m,n} f(m,n)$, in which $f(m,n)$ is described by the array (or 'infinite matrix') below:

$\dfrac{1}{2}$	$-\dfrac{1}{4}$	$\dfrac{1}{4}$	$-\dfrac{1}{8}$	$\dfrac{1}{8}$	$-\dfrac{1}{16}$	$\dfrac{1}{16}$	$\cdots$
$\dfrac{1}{2^2}$	$-\dfrac{3}{4^2}$	$\dfrac{3}{4^2}$	$-\dfrac{7}{8^2}$	$\dfrac{7}{8^2}$	$-\dfrac{15}{16^2}$	$\dfrac{15}{16^2}$	$\cdots$
$\dfrac{1}{2^3}$	$-\dfrac{3^2}{4^3}$	$\dfrac{3^2}{4^3}$	$-\dfrac{7^2}{8^3}$	$\dfrac{7^2}{8^3}$	$-\dfrac{15^2}{16^3}$	$\dfrac{15^2}{16^3}$	$\cdots$
$\dfrac{1}{2^4}$	$-\dfrac{3^2}{4^4}$	$\dfrac{3^2}{4^4}$	$-\dfrac{7^3}{8^4}$	$\dfrac{7^3}{8^4}$	$-\dfrac{15^3}{16^4}$	$\dfrac{15^3}{16^4}$	$\cdots$
$\vdots$	$\vdots$	$\vdots$	$\vdots$	$\vdots$	$\vdots$	$\vdots$	$\vdots$

Show that the double series is not convergent, but $\sum_{m=1}^{\infty}(\sum_{n=1}^{\infty} f(m,n)) = 1$ while $\sum_{n=1}^{\infty}(\sum_{m=1}^{\infty} f(m,n))$ is undefined.

2-8.P2. For the convergent double series of Example 2-8.8(a), show that the 'repeated series' $\sum_{m=1}^{\infty}(\sum_{n=1}^{\infty} f(m,n))$ and $\sum_{n=1}^{\infty}(\sum_{m=1}^{\infty} f(m,n))$ are undefined.

2-8.P3. Show that the series $\Sigma_{m,n}(m^2 + n^2)^{-\alpha/2}$ converges if $\alpha > 2$.

3

Differentiation

3-1 Background

In the calculus of a function f of two real variables, i.e., of a two-dimensional vector variable (x, y), one usually works with the two partial derivatives $\partial f / \partial x$ and $\partial f / \partial y$ (to be formally defined in 3-4.1 below). The first of these is the limit of a certain quotient with numerator $f(x+t, y) - f(x, y)$. In the terminology of vectors, this numerator may be written as $f((x, y) + t(1, 0)) - f(x, y)$. If we now write simply x for $(x, y) \in \mathbb{R}^2$ and simply h for $(1, 0) \in \mathbb{R}^2$, then the numerator can be expressed quite compactly as $f(x+th) - f(x)$. With this notation, it becomes clearer that the partial derivative

$$\frac{\partial f}{\partial x} = \lim_{t \to 0} \frac{f(x+th) - f(x)}{t},$$

is the rate of change of f in the direction of h, where we have taken h as $(1, 0)$. If we change h to $(0, 1)$ instead, then the limit is the rate of change in the direction of $h = (0, 1)$, and this rate of change is usually denoted by $\partial f / \partial y$.

Now, what about the same limit with some other (nonzero) h, the so called *directional derivative* in the direction h? There are infinitely many other possibilities for h, but one works only with $h = (1, 0)$ and $h = (0, 1)$, which lead to the two familiar partial derivatives. This is because, in most situations (but by no means all!), the derivative in a general direction $h = (\alpha, \beta)$ works out to be

$$\alpha \frac{\partial f}{\partial x} + \beta \frac{\partial f}{\partial y}. \tag{1}$$

Thus, although each partial derivative is the rate of change in one coordinate direction only, the two partial derivatives *when taken in combination* reveal everything about the rate of change in all possible directions. This is what makes the vector with components equal to the respective partial derivatives, known as the *gradient* in Calculus, such a useful tool. Being a single object that tells all about the rate of change in any direction whatsoever, it is the natural two variable analogue of the elementary one variable derivative.

Since the gradient allows us to start with a direction $h = (\alpha, \beta)$ and, in most situations, compute the rate of change in that direction via (1), it maps (α, β) into a number. In other words, it is a map from $\mathbb{R}^2$ to $\mathbb{R}$. Besides, it is a *linear* map from $\mathbb{R}^2$ to $\mathbb{R}$.

S. Shirali, H.L. Vasudeva, *Multivariable Analysis*,
DOI 10.1007/978-0-85729-192-9_3, © Springer-Verlag London Limited 2011

In the n variable case, the analogue of the single variable derivative is a linear map from $\mathbb{R}^n$ to $\mathbb{R}$. One therefore expects that for a function f with values in $\mathbb{R}^m$, the analogue is a linear map from $\mathbb{R}^n$ to $\mathbb{R}^m$. We shall soon define the derivative of such a map f to be a linear map from $\mathbb{R}^n$ to $\mathbb{R}^m$ related to f in a certain way. However, the relation will not be that it maps each direction h into the derivative in that direction, because the existence of such a linear map does not imply continuity, as we shall see later in 3-2.P3. We shall instead define the derivative to be a linear map that generalises another aspect of the derivative in the one variable case, one that does imply continuity. We discuss this aspect and present the definition in the next section.

3-2 Derivatives of Functions Between Euclidean Spaces

Suppose f is a real-valued function of a single real variable x. The familiar definition of the derivative at some x in the interior of the domain of f is that it is the limit as $h \rightarrow 0$ of the quotient $[f(x+h)-f(x)]/h$. Since there is no division by vectors, such a quotient makes no sense when h is a vector. So we rephrase the definition without denominators in the following manner.

For $h \neq 0$, let $\omega(h) = [f(x+h)-f(x)]/h - a$, where a is some number, and let $\omega(0)$ be any number. Then a is the derivative if and only if $\omega(h) \rightarrow 0$ as $h \rightarrow 0$. Since it is immaterial what value $\omega(0)$ has, we shall say nothing about it. The description of ω can be rearranged without denominators as

$$f(x+h)-f(x) = ah + h\omega(h).$$

Thus the number a is the derivative $f'(x)$ if and only if there exists a function ω defined on a sufficiently small ball centred at 0 such that

$$f(x+h)-f(x) = ah + h\omega(h) \qquad \text{and} \qquad \omega(h) \rightarrow 0 \text{ as } h \rightarrow 0. \qquad (1)$$

This equivalent description of $f'(x)$ as being such a number a has no reference to any division by h; however, there is a reference to the product $h\omega(h)$. So we now set $u(h) = (h/|h|)\omega(h)$ for $h \neq 0$, so that (1) can be rephrased as

$$f(x+h)-f(x) = ah + |h|u(h) \qquad \text{and} \qquad u(h) \rightarrow 0 \text{ as } h \rightarrow 0.$$

Stated in this form, we can carry it over to the situation when f maps a subset of $\mathbb{R}^n$ to $\mathbb{R}^m$.

For reasons discussed in Section 3-1, the derivative in this general setting is a linear map A from $\mathbb{R}^n$ to $\mathbb{R}^m$, and the above term ah appears as the A-image of h, i.e., as Ah.

3-2.1. Definition. *Let x be an interior point of the domain $U \subseteq \mathbb{R}^n$ of a function $f: U \rightarrow \mathbb{R}^m$. If there exists a linear map $A: \mathbb{R}^n \rightarrow \mathbb{R}^m$ and an $\mathbb{R}^m$-valued function u on a ball H centred at $0 \in \mathbb{R}^n$ such that*

$$f(x + h) - f(x) = Ah + \|h\|u(h) \quad for\ h \in H \quad and \quad u(h) \to 0\ as\ h \to 0,$$

then A is the **linear** (or **Fréchet**) **derivative** of f at x; it is denoted by $f'(x)$. When $f'(x)$ exists, we say that f is (**Fréchet**) **differentiable** at x.

It should be noted that $f'(x) \in \mathfrak{L}(\mathbb{R}^n, \mathbb{R}^m)$. We shall often refer to the Fréchet derivative as simply the *derivative*. The name distinguishes it from the Gateaux derivative [see 3-3.P1], which will play only a minor role.

Observe that altering the value of $u(0)$ makes no difference. In a proof of differentiability, it is therefore sufficient to define u for nonzero h and establish that $u(h) \to 0$ as $h \to 0$.

3-2.2. Remarks. (a) There cannot be two distinct linear maps satisfying the conditions of Def. 3-2.1. To see why, suppose A and B are both linear maps of this kind with associated balls H and J and functions u and v. Then

$$Ah + \|h\|u(h) = Bh + \|h\|v(h)$$

for all h in $H \cap J$, which is again a ball centred at 0. Then

$$(A - B)h = \|h\|(v(h) - u(h)).$$

Let h be any nonzero vector in $H \cap J$. Then, for $t \in \mathbb{R}$, $0 < |t| \le 1$, we have

$$(A - B)(th) = |t|\,\|h\|(v(th) - u(th)), \text{ so that } \|(A - B)(h)\| = \|h\|\,\|(v(th) - u(th))\|.$$

Taking the limit as $t \to 0$, we find that $(A - B)(h) = 0$. But this holds for *every* h in the ball $H \cap J$ centred at 0. Therefore $A - B = O$ [2-3.P19], and hence $A = B$.

(b) If f is a constant, then the equality displayed in Def. 3-2.1 holds for any x with $A = O$ and $u(h) = 0$ for all $h \in H$. Therefore, f is differentiable with derivative O.

(c) When $n = 1$, the linear map $A = f'(x)$ must be of the form described in Example 2-3.2(b); that is, there must exist $b \in \mathbb{R}^m$ such that $Ah = hb$ (product of the scalar h with vector b). Upon dividing by the scalar h in the equality displayed in Def. 3-2.1 and then taking the limit as $h \to 0$, we find that

$$b = \lim_{h \to 0} \frac{f(x+h) - f(x)}{h}.$$

Thus $f'(x)$ can be described as the linear map $h \to hb$, where b is given by the above limit.

(d) When f is a linear map, the requirement of Def. 3-2.1 holds with $A = f$, $u(h) = 0$ and H any ball whatsoever. Therefore the derivative at any point x is none other than the map f itself, which is to say, $f'(x) = f$ for every x. In case f is of the form $f(x) = Ax + b$, where b is a fixed vector in $\mathbb{R}^m$, then again $f'(x) = A$. In particular, when $A = O$, which is to say, f is a constant map $f(x) = b$, the derivative is $f'(x) = O$ for every x.

(e) If f is differentiable at x, then it is also continuous at x. This follows from Def. 3-2.1 above, because

$$\|Ah + \|h\|u(h)\| \le \|h\|(\|A\| + \|u(h)\|),$$

and $\|A\| + \|u(h)\|$ remains bounded as $h\to 0$. The converse is not true. For example, the function $f:\mathbb{R}^2\to\mathbb{R}$ given by $f(x,y) = |x|$ is clearly continuous at $(0,0)$; however, it is not differentiable there, as will be shown in Example 3-2.3(c).

(f) If f and g are both differentiable at x, then $f+g$ and $f \cdot g$ are both differentiable at x and

$$(f+g)'(x) = f'(x) + g'(x),$$
$$(f \cdot g)'(x) = f'(x) \cdot g(x) + f(x) \cdot g'(x),$$

where the right hand side in the second equality is the real-valued map

$$h \to [f'(x)(h)] \cdot g(x) + f(x) \cdot [g'(x)(h)].$$

In conjunction with Remark 3-2.2(d), this shows that, if $g(x) = Ax + b - f(x)$, where A is a linear map and f is differentiable, then $g'(x) = A - f'(x)$.

We now present two illustrations of how to compute the derivative as a matrix directly from the definition (i.e., from 'first principles') and one example when the linear derivative does not exist. In Def. 3-2.1, the linear map A and the function u are specific to the x in question; therefore, they are functions of x as well, although our notation there may seem to suggest otherwise. The computations below will clarify this.

3-2.3. Examples. (a) Suppose we wish to find the derivative of the function f whose value at $(x, y) \in \mathbb{R}^2$ is $x^2 + y^2$. Denote the 'increment' vector by (h,k). Then

$$f(x + h, y + k) - f(x,y) = (x + h)^2 + (y + k)^2 - (x^2 + y^2)$$
$$= 2xh + 2yk + h^2 + k^2 = A(h,k) + \|(h,k)\|u(h,k),$$

where $A:\mathbb{R}^2\to\mathbb{R}$ is the map for which

$$A(h,k) = 2xh + 2yk \quad \text{and} \quad u(h,k) = \|(h,k)\| = \sqrt{(h^2 + k^2)}.$$

Since A is linear and $u(h,k)\to 0$ as $(h,k)\to 0$, the map A is the linear derivative of f at (x,y). It has the 1×2 matrix $[2x \quad 2y]$.

(b) Now suppose we wish to find the derivative of the function f whose value at $(x, y) \in \mathbb{R}^2$ is $x^2 y$. Denoting the increment vector by (h,k) we have

$$f(x + h, y + k) - f(x,y) = (x + h)^2(y + k) - x^2 y$$
$$= 2xyh + x^2 k + 2xhk + h^2 y + h^2 k$$
$$= A(h,k) + \|(h,k)\|u(h,k),$$

where $A:\mathbb{R}^2\to\mathbb{R}$ is the map for which

$$A(h,k) = 2xyh + x^2k \quad \text{and} \quad u(h,k) = \frac{2xhk + yh^2 + kh^2}{\sqrt{h^2 + k^2}} \quad \text{when } (h,k) \neq (0,0).$$

Since A is linear and $u(h,k) \to 0$ as $(h,k) \to 0$, the map A is the linear derivative of f at (x,y) despite the fact that we have not defined $u(h,k)$ for $(h,k) = (0,0)$. The linear derivative has the 1×2 matrix $[2xy \quad x^2]$.

(c) Consider the map $f: \mathbb{R}^2 \to \mathbb{R}$ defined as $f(x,y) = |x|$. Suppose f were differentiable at $(0,0)$ with derivative A, say. Taking $(h,0)$ as the increment vector (where $h \in \mathbb{R}$), we would then have

$$|h| = f(h,0) - f(0,0) = A(h,0) + |h| \cdot u(h,0)$$

for sufficiently small h, positive as well as negative, and $u(h,0) \to 0$ as $h \to 0$. Since $|-h| = |h|$, this implies

$$|h| = f(-h,0) - f(0,0) = A(-h,0) + |h| \cdot u(-h,0).$$

Adding the above equalities and using the linearity of A, we get

$$2|h| = |h| \cdot [u(h,0) + u(-h,0)].$$

In particular, for nonzero h, we would have $u(h,0) + u(-h,0) = 2$, contradicting the requirement that $u(h,0) \to 0$ as $h \to 0$.

(d) This example is relevant to Theorem 3-4.4; see Example 3-4.6(b). Let ϕ and ψ be real-valued functions defined on open intervals, which are differentiable at points α and β in their respective domains. Consider the real-valued map f defined on the Cartesian product of the intervals as $f(x,y) = \phi(x) + \psi(y)$. Differentiability of ϕ and ψ at the respective points α and β means

$$\phi(\alpha + h) - \phi(\alpha) = \phi'(\alpha)h + |h| \cdot u(h) \quad \text{where} \quad u(h) \to 0 \text{ as } h \to 0$$

and

$$\psi(\beta + k) - \psi(\beta) = \psi'(\beta)k + |k| \cdot v(k) \quad \text{where} \quad v(k) \to 0 \text{ as } k \to 0.$$

It follows that

$$f(\alpha + h, \beta + k) - f(\alpha, \beta) = \phi'(\alpha)h + \psi'(\beta)k + [|h| \cdot u(h) + |k| \cdot v(k)]$$

$$= \phi'(\alpha)h + \psi'(\beta)k + \sqrt{h^2 + k^2} \, w(h,k), \tag{1}$$

where

$$w(h,k) = \left[\frac{|h|}{\sqrt{h^2 + k^2}} u(h) + \frac{|k|}{\sqrt{h^2 + k^2}} v(k) \right].$$

Since $|w(h,k)| \leq |u(h)| + |v(k)|$, we see that $w(h,k) \to 0$ as $(h,k) \to 0$. Therefore by (1), f is differentiable at (α, β), the derivative being the linear map A given by $A(h,k) = \phi'(\alpha)h + \psi'(\beta)k$. Thus, it has the matrix $[\phi'(\alpha) \quad \psi'(\beta)]$.

We now formally define a concept that has been mentioned informally earlier in this chapter.

3-2.4. Definition. *Let x be an interior point of the domain $U \subseteq \mathbb{R}^n$ of a function $f: U \to \mathbb{R}^m$. When $0 \neq h \in \mathbb{R}^n$, the* **directional derivative at x in the direction h** *is defined to be*

$$D_h f(x) = \lim_{t \to 0} \frac{f(x+th) - f(x)}{t} \ .$$

3-2.5. Remarks. (a) The directional derivatives at x in the n 'coordinate directions' $h = e_j, 1 \leq j \leq n$, are known as the partial derivatives $(D_j f)(x)$. They will be studied in a later section with $m = 1$.

(b) If f is differentiable at x, then its directional derivative at x in any direction $h \neq 0$ exists and is given by $f'(x)h$. This is because, for sufficiently small nonzero $t \in \mathbb{R}$, we have

$$f(x + th) - f(x) = f'(x)(th) + \|th\| u(th) = t f'(x)(h) + |t| \cdot \|h\| u(th)$$

where $u(k) \to 0$ as $k \to 0$, which implies that

$$\frac{f(x+th) - f(x)}{t} = f'(x)(h) + |t| \cdot \|h\| u(th)/t,$$

where the second term on the right is easily seen to have limit 0 as $t \to 0$.

(c) The reader may note the consequence of part (b) above that, in the event of f being differentiable, the derivative in the direction h is linear in h.

3-2.6. Example. Part (c) of the foregoing remark can sometimes be used for establishing the nonexistence of a linear derivative. For example, let the function $f: \mathbb{R}^2 \to \mathbb{R}$ be defined as

$$f(0,0) = 0 \quad \text{and} \quad f(x,y) = x^3/(x^2 + y^2) \quad \text{when } (x, y) \neq (0,0).$$

Then $[f(th, tk) - f(0,0)]/t = h^3/(h^2 + k^2)$ and therefore the derivative in any direction (h, k) is $h^3/(h^2 + k^2)$. Since this is not linear in (h, k), the function cannot be differentiable at $(0,0)$ by Remark 3-2.5(c). It may be noted that, since $|x^3| \leq |x|(x^2 + y^2)$, we have $|f(x,y)| \leq |x|$, which shows that f is continuous at $(0,0)$. In this connection, see 3-2.P3.

3-2.7. Proposition. (a) *Suppose $f: E \to \mathbb{R}^m$ is differentiable at $x \in E \subseteq \mathbb{R}^n$ and $B: \mathbb{R}^m \to \mathbb{R}^p$ is linear. Then the composition $B \circ f: E \to \mathbb{R}^p$ is differentiable at x with derivative $B \circ (f'(x))$, i.e., the composition of the linear maps B and $f'(x)$.*

(b) *Suppose $B: \mathbb{R}^n \to \mathbb{R}^m$ is linear and $g: E \to \mathbb{R}^p$ is differentiable at $Bx \in E \subseteq \mathbb{R}^m$. Then the composition $g \circ B$ is differentiable at x with derivative $g'(Bx) \circ B$, i.e., the composition of the linear maps $g'(Bx)$ and B.*

Proof. (a) Since f is differentiable at x, there exists some ball H centred at $0 \in \mathbb{R}^n$ such that

$$f(x + h) - f(x) = Ah + \|h\|u(h) \quad \text{for } h \in H \quad \text{and} \quad u(h) \to 0 \text{ as } h \to 0,$$

where $A = f'(x)$ by definition. Applying B to both sides and using the linearity of B, we get

$$(B \circ f)(x + h) - (B \circ f)(x) = (B \circ A)h + \|h\|B(u(h)) \quad \text{for } h \in H.$$

Since $\|B(u(h))\| \le \|B\|\|u(h)\|$, the function $v(h) = B(u(h))$ has the property that $v(h) \to 0$ as $h \to 0$. Therefore, in view of the foregoing equality, $B \circ A = B \circ (f'(x))$ must be the derivative of $B \circ f$ at x.

(b) If $B = O$, there is nothing to prove. So, assume $\|B\| > 0$.

Since g is differentiable at Bx, there exists some ball K centred at $0 \in \mathbb{R}^m$ such that,

$$g(Bx + k) - g(Bx) = Ak + \|k\|u(k) \quad \text{for } k \in K \quad \text{and} \quad u(k) \to 0 \text{ as } k \to 0,$$

where $A = g'(Bx)$ by definition. Denote the radius of K by r. If H is the ball centred at $0 \in \mathbb{R}^n$ with radius $r/\|B\|$, then $h \in H \Rightarrow Bh \in K$. In particular, $h \in H$ implies that $Bx + Bh = B(x + h)$ belongs to the domain of g and thus $x + h$ belongs to the domain of $g \circ B$. Therefore, x is an interior point of the domain of $g \circ B$. Furthermore, we may substitute $k = Bh$ with $h \in H$ in the above equality. Using the linearity of B, the substitution leads to

$$(g \circ B)(x + h) - (g \circ B)(x) = A(Bh) + \|B(h)\|u(Bh) \quad \text{for } h \in H.$$

Setting $v(h) = \dfrac{\|Bh\|}{\|h\|} u(Bh)$ for $h \ne 0$, we have

$$(g \circ B)(x + h) - (g \circ B)(x) = A(Bh) + \|h\|v(h).$$

Also, v has the two properties

$$\|v(h)\| \le \|B\|\|u(Bh)\| \quad \text{and} \quad v(h) = 0 \text{ if } Bh = 0.$$

Together with the fact that $u(k) \to 0$ *as* $k \to 0$, the above two properties imply that $v(h) \to 0$ as $h \to 0$. Consequently, $A \circ B = g'(Bx) \circ B$ must be the derivative of $g \circ B$ at x. $\square$

3-2.8. Proposition. *Let x be an interior point of the domain $U \subseteq \mathbb{R}^n$ of a function $f : U \to \mathbb{R}^m$. Denote the standard basis of $\mathbb{R}^m$ by $e_1, e_2, \ldots, e_m$, so that*

$$f(u) = \sum_{j=1}^{m} f_j(u)e_j \qquad \text{for all } u \in U, \tag{A}$$

where the m real-valued functions f_j are the components of f. Then f is differentiable at x if and only if each f_j is differentiable at x.

Proof. First suppose that f is differentiable at x. For any j, $1 \le j \le m$, let $P_j : \mathbb{R}^m \to \mathbb{R}$ be the 'projection map' given by

$$P_j(a_1 e_1 + \cdots + a_m e_m) = a_j.$$

Then by (A), $f_j(u) = (P_j \circ f)(u)$ for all $u \in U$. Since P_j is linear, Proposition 3-2.7 implies that f_j is differentiable at x.

Now suppose that each f_j is differentiable at x. For any j, $1 \leq j \leq m$, let $Q_j : \mathbb{R} \to \mathbb{R}^m$ be the 'insertion map' given by

$$Q_j(\lambda) = \lambda e_j.$$

Then the composition $Q_j \circ f_j$ satisfies $(Q_j \circ f_j)(u) = Q_j(f_j(u)) = f_j(u)e_j$. Therefore by (A), $f = \sum_{j=1}^{m} (Q_j \circ f_j)$. But each $Q_j \circ f_j$ is differentiable at x by Proposition 3-2.7, and hence so is f, by Remark 3-2.2(f). $\square$

Problem Set 3-2

3-2.P1. Find the 1×2 matrix representation of the derivative of $x^3 + y$ directly from the definition (i.e., from first principles).

3-2.P2. Suppose x is an interior point of the domains of both the functions f and g. Let A and B be linear maps such that the directional derivative of f in any direction h is Ah while that of g is Bh. Show that $f + g$ has a directional derivative at x in any direction h and that it is given by $(A + B)h$.

3-2.P3. Let $\phi(x,y) = y^3/x$ when $x \neq 0$ and $\phi(0,y) = 0$. Show that ϕ has directional derivative 0 in every direction at $(0,0)$ but is not continuous there (and hence, by Remark 3-2.2(e), not differentiable either).

3-2.P4. Let $f : \mathbb{R}^n \to \mathbb{R}$, $n > 1$, be defined as $f(x) = \|x\|$. Prove that f is continuous but not differentiable at 0.

3-2.P5. Show directly from the definition, i.e., from first principles, that the map $f : \mathbb{R}^2 \to \mathbb{R}^3$ defined by $f(x, y) = (x^2, xy, y)$ is differentiable at every $(x, y) \in \mathbb{R}^2$ with derivative given by the matrix

$$\begin{bmatrix} 2x & 0 \\ y & x \\ 0 & 1 \end{bmatrix}.$$

3-2.P6. Find the 1×2 matrix representation of the derivative of $f(x,y) = (x^2 + y)^{10}$ directly from the definition (i.e., from first principles).

3-2.P7. Use the result of Example 3-2.3(b) and the fact that

$$(z + s)^{10} - z^{10} = 10z^9 s + |s| \cdot v(s), \text{ where } v(s) \to 0 \text{ as } s \to 0$$

to find the derivative of $x^{20}y^{10}$.

3-2.P8. Prove that there can be no real-valued function f with $D_h f(c) > 0$ for every direction h at a given point c, but there does exist one having $D_h f(c) > 0$ at every point c in a given direction h.

3-2.P9. Let $f_1, \ldots, f_n$ be real-valued functions on $\mathbb{R}$, each differentiable in an open interval (a, b) and let $E = \{(x_1, \ldots, x_n) \in \mathbb{R}^n : a < x_k < b, \ 1 \leq k \leq n\}$. Define $f: \mathbb{R}^n \to \mathbb{R}$ by $f(x_1, \ldots, x_n) = \sum_{k=1}^{n} f_k(x_k)$. Prove that f is differentiable at each $x \in E$ and that

$$f'(x)(h) = \sum_{k=1}^{n} f_k'(x_k) h_k \quad \text{whenever } h = (h_1, \ldots, h_n) \in \mathbb{R}^n.$$

3-2.P10. Let $f_1, \ldots, f_n$ be real-valued functions each differentiable in an open set $S \subseteq \mathbb{R}^n$. For each $x \in S$, define $f(x) = \sum_{k=1}^{n} f_k(x)$. Assume that for each k ($1 \leq k \leq n$), the limit

$$a_k(x) = \lim_{\substack{y \to x \\ y_k \neq x_k}} \frac{f_k(y) - f_k(x)}{y_k - x_k}$$

exists. Prove that f is differentiable at x and that $f'(x)(h) = \sum_{k=1}^{n} a_k(x) h_k$ for $h = (h_1, \ldots, h_n) \in \mathbb{R}^n$.

3-2.P11. Suppose f, g are mappings from $\mathbb{R}^n$ to $\mathbb{R}^m$ such that g is continuous at c, f is differentiable at c and $f(c) = 0$. Let $F(x) = g(x) \cdot f(x)$, the scalar product of $g(x)$ with $f(x)$. Prove that F is differentiable at c and that $F'(c)(h) = g(c) \cdot \{f'(c)(h)\}$.

3-2.P12. Show that the function $f: \mathbb{R}^2 \to \mathbb{R}$ defined by

$$f(x, y) = \begin{cases} x + y & \text{if } xy = 0 \\ 1 & \text{otherwise} \end{cases}$$

has partial derivatives at $(0,0)$. However, the directional derivative in any other direction does not exist. Conclude that the function is not differentiable at $(0,0)$. The fact that f is not differentiable at $(0,0)$ can also be concluded from the observation that it is not even continuous there. Indeed, $\lim_{(x,y) \to (0,0)} f(x,y)$ does not exist.

3-3 The Chain Rule and a Corollary

The reader probably remembers the chain rule from a course in calculus as a rule for computing the derivative of a composite function in terms of derivatives of its constituents. It is also employed for computing partial derivatives for real valued functions defined in $\mathbb{R}^n$. The version of it that we state below is also a rule for computing the derivative of a composite function in terms of derivatives of its constituents, and its connection with partial derivatives will be clarified in the next section after Theorem 3-4.2.

If the proof seems rather heavy going then it may be helpful to solve 3-2.P7 before proceeding.

3-3.1. Theorem. Chain Rule: *Suppose $E \subseteq \mathbb{R}^n$ and f maps E into $\mathbb{R}^m$. Let g map a subset of $\mathbb{R}^m$ into $\mathbb{R}^p$. If f is differentiable at $x \in E$ and g is differentiable at $f(x)$ $\in f(E)$, then the composition $g \circ f$ is differentiable at x and*

$$(g \circ f)'(x) = g'(f(x)) \circ f'(x).$$

Proof. By definition of differentiability, x is an interior point of E and $f(x)$ is an interior point of the domain of g. Therefore, continuity of f at x [Remark 3-2.2(e)] ensures that x is an interior point of the domain of $g \circ f$.

For h belonging to some ball H centred at $0 \in \mathbb{R}^n$,

$$f(x + h) - f(x) = f'(x)h + \|h\|u(h), \qquad \text{where } u(h) \to 0 \text{ as } h \to 0. \tag{1}$$

Again, for k belonging to some ball K centred at $0 \in \mathbb{R}^m$,

$$g(f(x) + k)) - g(f(x)) = g'(f(x))k + \|k\|v(k), \tag{2}$$

where $v(k) \to 0$ as $k \to 0$. Since f must be continuous at x, the ball H may be replaced by a smaller one if necessary so as to ensure that $h \in H \Rightarrow f(x + h) - f(x) \in K$. Then we may take $k = f(x + h) - f(x)$ in (2). But if we choose k in this manner, then we have $f(x) + k = f(x + h)$ and also $k = f'(x)h + \|h\|u(h)$ by (1). Therefore by (2),

$$g(f(x + h)) - g(f(x)) = g'(f(x))(f'(x)h + \|h\|u(h)) + \|k\|v(k)$$
$$= (g'(f(x)) \circ f'(x))h + \|h\|g'(f(x))u(h) + \|k\|v(k). \tag{3}$$

By (1), $\|k\| \leq \|h\|(\|f'(x)\| + \|u(h)\|)$, so that

$$\|k\|/\|h\| \text{ is bounded as } h \to 0. \tag{4}$$

Since $k \to 0$ as $h \to 0$ (by continuity of f at x), it follows from (4) that $(\|k\|/\|h\|)v(k) \to 0$ as $h \to 0$. Therefore, if for $h \in H$ we take

$$w(h) = [g'(f(x))u(h) + (\|k\|/\|h\|)v(k)] \text{ when } h \neq 0,$$

then $w(h) \to 0$ as $h \to 0$. Besides, it follows by using this definition of w in (3) that

$$g(f(x + h)) - g(f(x)) = (g'(f(x)) \circ f'(x))h + \|h\|w(h) \text{ whenever } h \in H. \qquad \square$$

Since it is customary to omit the composition symbol $\circ$ between linear maps, it is not necessary to write it between $g'(f(x))$ and $f'(x)$ on the right side of the equality in the chain rule. Accordingly, we shall often omit it in future.

3-3.2. Examples. (a) Let $f: \mathbb{R}^2 \to \mathbb{R}^2$ and $g: \mathbb{R}^2 \to \mathbb{R}$ be given respectively by $f(x, y) = (x^2, y^2)$ and $g(x, y) = x + y$. Then

$$f(x + h, y + k) - f(x, y) = ((x + h)^2, (y + h)^2) - (x^2, y^2)$$

$$= (2xh, 2yk) + (h^2, k^2)$$
$$= A(h, k) + \|(h, k)\| u(h, k),$$

where A is the linear map with matrix $\begin{bmatrix} 2x & 0 \\ 0 & 2y \end{bmatrix}$ and $u(h, k) = (h^2, k^2)/\|(h, k)\|$.

Recalling that $\|(h, k)\| = (h^2 + k^2)^{1/2}$, we have $u(h, k) \to 0$ as $\|(h, k)\| \to 0$. Therefore $f'(x, y) = A$. Thus,

$$[f'(x, y)](h, k) = (2xh, 2yk). \tag{1}$$

Also,

$$g(x + h, y + k) - g(x, y) = (x + h) + (y + k) - (x + y)$$
$$= h + k = B(h, k) + \|(h, k)\| v(h, k),$$

where B is the linear map with matrix $[1 \quad 1]$ and $v(h, k) = 0$. Therefore $g'(x, y) = B$. Hence

$$[g'(x, y)](h, k) = h + k. \tag{2}$$

According to the chain rule, the composition $g \circ f$, which is the map given by

$$(g \circ f)(x, y) = (x^2 + y^2),$$

must have derivative $(g \circ f)'(x, y) : \mathbb{R}^2 \to \mathbb{R}$ given by the composition

$$(g \circ f)'(x, y) = g'(f(x, y)) \circ f'(x, y),$$

which means

$$[(g \circ f)'(x, y)](h, k) = [g'(f(x, y))][f'(x, y)(h, k)].$$

In view of (1), the right side here can be written as

$$[g'(f(x, y))][f'(x, y)(h, k)] = [g'(f(x, y))](2xh, 2yk)$$
$$= 2xh + 2yk \quad \text{by (2)}.$$

Thus $(g \circ f)'(x, y)$ has the matrix $[2x \quad 2y]$. Alternatively, we can use Remark 2-3.3 to compute the matrix of $(g \circ f)'(x, y)$ as the product of matrices of $g'(f(x, y))$ and $f'(x, y)$, i.e.,

$$[1 \quad 1], \begin{bmatrix} 2x & 0 \\ 0 & 2y \end{bmatrix}$$

which works out to be $[2x \quad 2y]$.

(b) [Cf. 3-2.P6.] Let $f : \mathbb{R}^2 \to \mathbb{R}$ and $g : \mathbb{R} \to \mathbb{R}$ be given respectively by $f(x, y) = x^2 + y$ and $g(u) = u^{10}$. Then

$$f(x + h, y + k) - f(x, y) = 2xh + k + h^2 = A(h, k) + \|(h, k)\| h^2/\|(h, k)\|,$$

where $A(h, k) = 2xh + k$, i.e., A is the linear map from $\mathbb{R}^2$ to $\mathbb{R}$ with matrix $[2x \quad 1]$. Since $h^2/\|(h, k)\| \to 0$ as $\|(h, k)\| \to 0$, the derivative of f at (x, y) is the linear map A. Thus

$$[f'(x, y)](h, k) = 2xh + k. \tag{1}$$

Also, the linear map $g'(u) : \mathbb{R} \to \mathbb{R}$ is the one for which

$$[g'(u)](t) = 10u^9t. \tag{2}$$

According to the chain rule, the composition $g \circ f$, which is the map given by

$$(g \circ f)(x,y) = (x^2 + y)^{10}, \tag{3}$$

must have derivative $(g \circ f)'(x,y):\mathbb{R}^2 \to \mathbb{R}$ given by the composition

$$(g \circ f)'(x,y) = g'(f(x,y)) \circ f'(x,y),$$

which means

$$[(g \circ f)'(x,y)](h,k) = [g'(f(x,y))][f'(x,y)(h,k)].$$

In view of (1), the right side here can be written as

$$[g'(f(x,y))][f'(x,y)(h,k)] = [g'(f(x,y))](2xh + k)$$
$$= 10f(x,y)^9(2xh + k) \quad \text{by (2)}$$
$$= 10(x^2 + y)^9(2xh + k)$$
$$= 20x(x^2 + y)^9h + 10(x^2 + y)^9k.$$

Thus $(g \circ f)'(x,y)$ has the matrix $[20x(x^2 + y)^9 \quad 10(x^2 + y)^9]$. Alternatively, we can use Remark 2-3.3 to compute the matrix of $(g \circ f)'(x,y)$ as the product of matrices of $g'(f(x,y))$ and $f'(x,y)$, i.e.,

$$10(x^2 + y)^9[2x \quad 1],$$

which works out to be $[20x(x^2 + y)^9 \quad 10(x^2 + y)^9]$.

The reader may recall that this agrees with the solution of 3-2.P6, in which the matrix had to be computed directly without the chain rule for the function given by the right side of (3).

(c) [Cf. 3-2.P4.] Let $f:\mathbb{R}^n \to \mathbb{R}$ and $g:\mathbb{R} \to \mathbb{R}$ be given respectively by $f(x) = \|x\|^2$ and $g(u) = u^{1/2}$. Then, in terms of the dot product, we have

$$f(x + h) - f(x) = (x + h) \cdot (x + h) - x \cdot x = 2x \cdot h + h \cdot h = 2x \cdot h + \|h\| v(h),$$

where $v(h) = \|h\|$. Therefore

$$f'(x)(h) = 2x \cdot h. \tag{1}$$

Also, for $u \neq 0$, we have $g'(u) = \frac{1}{2}u^{-1/2}$, which means that, the linear map $g'(u):\mathbb{R} \to \mathbb{R}$ is the one for which

$$[g'(u)](t) = \frac{1}{2}u^{-1/2}t. \tag{2}$$

According to the chain rule, when $f(x) \neq 0$, the composition $g \circ f$, which is the map given by

$$(g \circ f)(x) = \|x\|,$$

must have derivative $(g \circ f)'(x):\mathbb{R}^2 \to \mathbb{R}$ given by the composition

$$(g \circ f)'(x) = g'(f(x)) \circ f'(x);$$

in other words,

$$[(g \circ f)'(x)](h) = [g'(f(x))][f'(x)(h)].$$

In view of (1), the right side here can be written as

$$[g'(f(x))][f'(x)(h)] = [g'(f(x))](2x \cdot h)$$
$$= \tfrac{1}{2} f(x)^{-1/2} (2x \cdot h) \quad \text{by (2), provided } f(x) \neq 0$$

$$= \frac{(x \cdot h)}{\|h\|}.$$

Thus $(g \circ f)'(x)$ is the linear map for which $[(g \circ f)'(x)](h) = (x \cdot h)/\|x\|$ when $f(x) \neq 0$, i.e., when $\|x\| \neq 0$. Alternatively, $f'(x)$ has matrix $[2x_1 \quad 2x_2 \quad \cdots \quad 2x_n]$ and $g'(f(x))$ has the 1×1 matrix with entry $\tfrac{1}{2} f(x)^{-1/2}$ and therefore $(g \circ f)'(x,y)$ has matrix given by the product

$$\tfrac{1}{2} f(x)^{-1/2} [2x_1 \quad 2x_2 \quad \cdots \quad 2x_n] = [x_1 \quad x_2 \quad \cdots \quad x_n]/\|x\|.$$

The reader will note that the next result was proved independently as Proposition 3-2.7. We now derive it from the chain rule.

3-3.3. Corollary. (a) *Suppose $E \subseteq \mathbb{R}^n$, f maps E into $\mathbb{R}^m$ and $g:\mathbb{R}^m \to \mathbb{R}^p$ is linear. If f is differentiable at $x \in E$, then the composition $g \circ f$ is differentiable at x and $(g \circ f)'(x) = g \circ (f'(x))$.*
(b) *Suppose $E \subseteq \mathbb{R}^m$, g maps E into $\mathbb{R}^p$ and $f:\mathbb{R}^n \to \mathbb{R}^m$ is linear. If g is differentiable at $f(x) \in E$, then the composition $g \circ f$ is differentiable at x and $(g \circ f)'(x) = g'(f(x)) \circ f$.*

Proof. (a) By the chain rule, $(g \circ f)'(x) = g'(f(x)) \circ f'(x)$. Since g is assumed linear, it follows by Remark 3-2.2(d) that $g'(f(x)) = g$ regardless of what $f(x)$ is. Consequently, $(g \circ f)'(x) = g \circ (f'(x))$, as claimed.

(b) This argument is similar to that of part (a). □

Before presenting the next result, we elaborate a step that will be used without detailed explanation later in the course of the forthcoming proof. It concerns the linear maps in Example 2-3.2(b) and 2-3.2(c).

In this paragraph, the norm is understood to be $\| \ \|_2$. Let $A:\mathbb{R} \to \mathbb{R}^n$ be the linear map given by $Ax = xb$, where b is a fixed vector in $\mathbb{R}^n$ and $C:\mathbb{R}^m \to \mathbb{R}$ be the linear map given in terms of the dot product as $Cx = z \cdot x$, where z is a fixed vector in $\mathbb{R}^m$. Then, as discussed in Examples 2-7.3(b), 2-7.3(c), the norms of these maps are $\|A\| = \|b\|$ and $\|C\| = \|z\|$. It follows by Theorem 2-7.7 that, when $n = m$, $CA:\mathbb{R} \to \mathbb{R}$ satisfies $\|CA\| \leq \|C\| \|A\| = \|b\| \|z\|$. In other words, the number $a \in \mathbb{R}$ such that $(CA)x = ax$ for every $x \in \mathbb{R}$ satisfies $|a| \leq \|b\| \|z\|$. Furthermore, if $B:\mathbb{R}^n \to \mathbb{R}^m$ is linear, then the number α for which

satisfies
$$(CBA)x = \alpha x \text{ for every } x \in \mathbb{R}$$

$$|\alpha| \leq \|b\| \|B\| \|z\|.$$

Recall from Def. 2-5.8 that a subset E of $\mathbb{R}^n$ is called *convex* if, whenever a and b are in E and t is any real number such that $0 \leq t \leq 1$, the vector $a + t(b-a)$, or what is the same thing, $tb + (1-t)a$, is also in E. Any ball is convex according to 2-5.P2.

The next few results are corollaries to the chain rule, all of which use the mean value theorem for functions of one variable. The exact form of the inequality asserted in any one of them depends on the norms used, but the essential substance is the same. However, we shall take cognisance of whether the norm used is $\| \ \|_2$ or $\| \ \|_\infty$.

3-3.4. Corollary. *Let E be a convex subset of $F \subseteq \mathbb{R}^n$ and $f: F \to \mathbb{R}^m$ be differentiable at each $x \in E$, with f' bounded above by $M > 0$; i.e., $\|f'(x)\|_2 \leq M$ for each $x \in E$. Then for any $a, b \in E$, we have*

$$\|f(b) - f(a)\|_2 \leq M \|b - a\|_2$$

and

$$\|f(b) - f(a)\|_\infty \leq n^{1/2} M \|b - a\|_\infty.$$

Proof. We need prove only the first inequality because the second follows from it upon using the inequalities $\|x\|_\infty \leq \|x\|_2$ and $\|x\|_2 \leq n^{1/2} \|x\|_\infty$ of Proposition 2-2.6. Since the argument for the first inequality involves working exclusively with the norm $\| \ \|_2$, the subscript will be omitted.

Denote $f(b) - f(a)$ by c, and let $\phi: [0, 1] \to \mathbb{R}$ be defined by

$$\phi(t) = c \cdot f(a + t(b-a)).$$

Since E is convex, therefore $a + t(b-a) \in E$ when $t \in [0, 1]$ and ϕ is defined on $[0, 1]$, as claimed. Also, ϕ is the composition of the maps

$$t \to a + t(b-a), \qquad x \to f(x) \quad \text{and} \quad x \to c \cdot x,$$

in that order. Consequently, ϕ is continuous. Moreover, the first map is differentiable on $(0, 1)$, with derivative

$$h \to h(b-a) \text{ for } h \in \mathbb{R}$$

by Remark 3-2.2(c), and the third has derivative

$$h \to c \cdot h \text{ for } h \in \mathbb{R}^m$$

by Remark 3-2.2(d). Their norms are, respectively, $\|b - a\|$ and $\|c\|$ [see Examples 2-7.3(b) and 2-7.3(c)]. By the chain Rule, the derivative $\phi'(t)$ exists for $0 < t < 1$ and equals the composition of the linear maps

$$h \to h(b-a) \qquad \text{for } h \in \mathbb{R},$$

and

$$h \to [f'(a + t(b-a))](h) \qquad \text{for } h \in \mathbb{R}^n$$

$$h \to c \cdot h \qquad \text{for } h \in \mathbb{R}^m,$$

in that order. Using the property that $\|ST\| \le \|S\| \|T\|$ for any linear maps S and T for which the product ST is defined, we find that

$$\|\phi'(t)\| \le \|b-a\| \|c\| M \quad \forall \, t \in (0,1).$$

However by the mean value theorem,

$$|\phi(1) - \phi(0)| = |\phi'(t)|(1-0) \quad \text{for some } t \in (0,1)$$

$$= |\phi'(t)| = \|\phi'(t)\|.$$

Moreover, $\phi(1) = c \cdot f(b)$ and $\phi(0) = c \cdot f(a)$. Hence

$$|c \cdot (f(b) - f(a))| \le \|b - a\| \|c\| M,$$

i.e., $\|c\|^2 \le \|b-a\| \|c\| M$ (because $c = f(b) - f(a)$).

So,

$$\|c\| \le \|b-a\| M.$$

Since $c = f(b) - f(a)$, this is the same as $\|f(b) - f(a)\| \le M \|b-a\|$. Therefore the first inequality is established and thus also the second. $\qquad\square$

3-3.5. Corollary. *Let E be a convex subset of $F \subseteq \mathbb{R}^n$ and $f : F \to \mathbb{R}^m$ be differentiable at each $x \in E$, with $f'(x) = O$ for each $x \in E$. Then f is constant on E.*

Proof. This follows immediately for Corollary 3-3.4 with $M = 0$. $\qquad\square$

3-3.6. Example. Let E be a convex subset of $F \subseteq \mathbb{R}^n$, $A : \mathbb{R}^n \to \mathbb{R}^m$ a linear map and $f : F \to \mathbb{R}^m$ be differentiable at each $x \in E$ with $f'(x) = A$. In other words, the derivative of f is constant. One can then show that there is a constant $b \in \mathbb{R}^m$ such that $f(x) = Ax + b$ for all $x \in E$. To see why, consider the map $g : F \to \mathbb{R}^m$ given by $g(x) = Ax$. It has the property that $g'(x) = A$ for all $x \in E$. Therefore $f - g$ has derivative O on E. Apply Corollary 3-3.5.

For the next Corollary (which we need only in Proposition 7-2.5), it is worth bearing in mind that, in view of Proposition 2-7.12, a subset of $\mathfrak{L}(\mathbb{R}^n, \mathbb{R}^m)$ is bounded in the sense of $\| \ \|_2$ if and only if it is bounded in the sense of $\| \ \|_\infty$.

3-3.7. Corollary. *Let E be a convex subset of $F \subseteq \mathbb{R}^n$ and $f : F \to \mathbb{R}^m$ be differentiable at each $x \in E$, with f' bounded above; let $a \in E$ and*

$$\|f'(x) - f'(a)\|_2 \le M \text{ for every } x \in E.$$

Then for any b ∈ E, we have

and

$$\|f(b)-f(a)-f'(a)(b-a)\|_2 \le M\|b-a\|_2$$

$$\|f(b)-f(a)-f'(a)(b-a)\|_\infty \le n^{1/2}M\|b-a\|_\infty.$$

Moreover, if

then

$$\|f'(x)-f'(a)\|_\infty \le M_\infty \quad \text{for every } x \in E,$$

$$\|f(b)-f(a)-f'(a)(b-a)\|_\infty \le n^{1/2}m^{1/2}M_\infty\|b-a\|_\infty.$$

Proof. We need prove only the first inequality because the second follows from it upon using the inequalities $\|x\|_\infty \le \|x\|_2$ and $\|x\|_2 \le n^{1/2}\|x\|_\infty$ of Proposition 2-2.6, and the third follows upon using Proposition 2-7.12.

Denote $f(b) - f(a) - f'(a)(b-a)$ by c, and let $\phi:[0,1] \to \mathbb{R}$ be defined by

$$\phi(t) = c\cdot[\,[f(a+t(b-a))-f(a)]-f'(a)(b-a)].$$

Since E is convex, then $a+t(b-a) \in E$ when $t \in [0,1]$ and ϕ is defined on $[0,1]$, as claimed. Also, ϕ is the composition of the maps

$$t \to a+t(b-a), \qquad x \to f(x)-f(a)-f'(a)(b-a) \quad \text{and} \quad x \to c\cdot x,$$

in that order. Consequently, ϕ is continuous. Moreover, the first map is differentiable on $(0,1)$ with derivative

$$h \to h(b-a) \text{ for } h \in \mathbb{R}$$

by Remark 3-2.2(c), and the third has derivative

$$h \to c\cdot h \text{ for } h \in \mathbb{R}^m$$

by Remark 3-2.2(d). Their norms are respectively $\|b-a\|$ and $\|c\|$ [see Examples 2-7.3(b) and 2-7.3(c)]. By the chain rule, the derivative $\phi'(t)$ exists for $0 < t < 1$ and equals the composition of the linear maps

$$h \to h(b-a) \qquad \text{for } h \in \mathbb{R},$$

and

$$h \to [f'(a+t(b-a))](h) \qquad \text{for } h \in \mathbb{R}^n$$

$$h \to c\cdot h \qquad \text{for } h \in \mathbb{R}^m,$$

in that order. Thus, for any $h \in \mathbb{R}$,

$$\phi'(t)(h) = c\cdot[f'(a+t(b-a))](h(b-a))$$
$$= hc\cdot[f'(a+t(b-a))](b-a),$$

which means that

$$\phi'(t) = c\cdot[f'(a+t(b-a))](b-a).$$

By the mean value theorem,

$$\phi(1) - \phi(0) = \phi'(t)\,(1 - 0) = \phi'(t) \quad \text{for some } t \in (0,1).$$

But $\phi(1) = c \cdot \big[[f(b) - f(a)] - f'(a)(b - a)\big]$ and $\phi(0) = c \cdot \big[-f'(a)(b - a)\big].$

Therefore

$$c \cdot [f(b) - f(a)] = c \cdot [f'(a + t(b - a))](b - a) \quad \text{for some } t \in (0,1).$$

Adding $c \cdot [-f'(a)(b - a)]$ to both sides and recalling that

$$c = f(b) - f(a) - f'(a)(b - a),$$

we get

$$\|c\|_2^2 = c \cdot [f'(a + t(b - a)) - f'(a)](b - a).$$

This leads to

$$\|c\|_2^2 \le \|c\|_2\, M \|b - a\|_2,$$

from where it follows that $\|c\|_2 \le M\|b - a\|_2$. In view of our definition of c, this is the same as $\|f(b) - f(a) - f'(a)(b - a)\|_2 \le M\|b - a\|_2$. Therefore, the first inequality is established and thus also the second and third.   □

3-3.8. Remark. The following consequence of the chain rule will be used in the sequel while proving the inverse function theorem (Theorem 4-2.1). Suppose a function ϕ is a composition of the form $\phi(x) = B(Ax + b - f(x))$, where A and B are linear maps and f is differentiable. By Remark 3-2.2(f) the function $g(x) = Ax + b - f(x)$ has derivative $g'(x) = A - f'(x)$, while by Remark 3-2.2(d), the function $x \to Bx$ has derivative B. Now it is a consequence of the chain rule that the composition ϕ has derivative $\phi'(x) = B(A - f'(x))$.

Problem Set 3-3

3-3.P1. By **Gateaux derivative** of a function at an interior point of its domain, we mean a linear map A such that the directional derivative at that point in every direction h exists and equals Ah. Thus, when the derivative exists, it is also the Gateaux derivative.

(a) Let $\phi(x,y) = y^3/x$ when $x \ne 0$ and $\phi(0,y) = 0$. Show that ϕ has a Gateaux derivative at $(0,0)$. (We have seen in 3-2.P3 that this function is not even continuous at $(0,0)$.)

(b) Suppose $E \subseteq \mathbb{R}^n$ and f maps E into $\mathbb{R}^m$. Let g map an open subset of $\mathbb{R}^m$ containing $f(E)$ into $\mathbb{R}^p$. If f is continuous and has Gateaux derivative A at $x \in E$ and if g is differentiable at $f(x)$, show that the composition $g \circ f$ has Gateaux derivative $g'(f(x)) \circ A$ at x.

3-3.P2. Suppose $E \subseteq \mathbb{R}^n$ and $f : E \to \mathbb{R}^m$ is of the form $f(x) = Ax + b$, where A is a linear map and $b \in \mathbb{R}^m$ is fixed. If x is an interior point of E, g maps an open set containing $f(x)$ into $\mathbb{R}^p$ and has Gateaux derivative G (defined in 3-3.P1) at $f(x)$, then show that $g \circ f$ has Gateaux derivative $G \circ A$ at x.

3-3.P3. Let E be a convex subset of $F \subseteq \mathbb{R}^n$ and $f: F \rightarrow \mathbb{R}^n$ have a Gateaux derivative $G(x)$, in the sense defined in 3-3.P1, at each $x \in E$. Let $G(x)$ be bounded above by $M > 0$; i.e., $\|G(x)\| \leq M$ whenever $x \in E$. Prove

$$\|f(b) - f(a)\| \leq M \|b - a\| \qquad \forall a, b \in E.$$

3-3.P4. Let $f: \mathbb{R}^n \rightarrow \mathbb{R}$ $(n > 1)$ have Gateaux derivative $G \neq 0$ at $c \in \mathbb{R}^n$ in the sense defined in 3-3.P1. Then G is a linear map from $\mathbb{R}^n$ to $\mathbb{R}$ and therefore has a norm $\|G\|$, which depends upon the norm used in $\mathbb{R}^n$.

(a) Suppose the norm used in $\mathbb{R}^n$ is $\|x\|_2 = \sqrt{\sum_{j=1}^{n} x_j^2}$. Show that there are precisely two elements h of norm 1 in $\mathbb{R}^n$ such that $\|G\| = |G(h)|$.

(b) Show that if the norm used in $\mathbb{R}^n$ is $\|x\|_1 = \sum_{j=1}^{n} |x_j|$, then there must be at least two elements h of norm 1 in $\mathbb{R}^n$ such that $\|G\| = |G(h)|$, but there can be more than two.

3-3.P5. If F is a continuous mapping of $[a, b]$ into $\mathbb{R}^k$ and is differentiable in (a, b), then show that there exists $c \in (a, b)$ such that $\|F(b) - F(a)\| \leq (b - a)\|F'(c)\|$. (Note: This is to be proved *not by using* Corollary 3-3.4, but by *modifying* its proof.)

3-3.P6. Let a_j, b_j, where $1 \leq j \leq n$, be $2n$ numbers with $a_j < b_j$. The set

$$\{x = \sum_{j=1}^{n} x_j e_j \in \mathbb{R}^n: a_j \leq x_j \leq b_j \text{ for } 1 \leq j \leq n\}$$

where $e_1, e_2, \ldots, e_n$ is the standard basis of $\mathbb{R}^n$, is called a **cuboid**.

(a) Show that a cuboid is a convex set.
(b) If x and $x + h$ both belong to a cuboid, where $h = h_1 e_1 + \ldots + h_n e_n$, show that

$$x + h_1 e_1 + \ldots + h_{j-1} e_{j-1} + t h_j e_j, \text{ where } 0 \leq t \leq 1 \text{ and } 1 < j \leq n,$$

also belongs to the cuboid.
(c) Does (b) hold if 'cuboid' is replaced by 'ball centred at x'?

3-3.P7. Let $x, e \in \mathbb{R}^n$, $\mu \in \mathbb{R}$, and f be a real-valued function defined on an open subset of $\mathbb{R}^n$ containing $\{x + t(\mu e) : 0 \leq t \leq 1\}$. Let $\phi:[0, 1] \rightarrow \mathbb{R}$ be the function $\phi(t) = f(x + t(\mu e))$. The derivative of ϕ at $t \in (0, 1)$ is the limit of a quotient and the derivative of f at $x + t(\mu e)$ in the direction of e is the limit of some other quotient. Find the two quotients and relate them with each other in order to

   (i) show that, when $\mu \neq 0$, one of the derivatives exists if and only if the other one does;

   (ii) find a relation between the derivatives that is valid even if $\mu = 0$ but assuming that the directional derivative exists;

(iii) express $f(x + \mu e) - f(x)$ in terms of the derivative of f in the direction of e at some $x + \theta(\mu e)$, $0 < \theta < 1$, by applying the mean value theorem to ϕ.

3-3.P8. Suppose that $f_1 : E \to \mathbb{R}$ is differentiable at $x_0 \in E \subseteq \mathbb{R}^n$, and $u_1, u_2, \ldots, u_m$ is the standard basis of $\mathbb{R}^m$. Show that the map $x \to f_1(x)u_1$ from E to $\mathbb{R}^m$ has derivative at x_0 given by $h \to [f_1'(x_0)(h)]u_1$. If $f_2 : E \to \mathbb{R}$ is also differentiable at $x_0 \in E$, then show that the map $\phi : E \to \mathbb{R}^m$ such that $\phi(x) = f_1(x)u_1 + f_2(x)u_2$ is differentiable at x_0, and find $\phi'(x_0)(h)$ in terms of $h, f_1'(x_0), f_2'(x_0), u_1$ and u_2.

3-3.P9. (a) Let $f : \mathbb{R} \to \mathbb{R}^2$ be defined by $f(t) = (\cos t, \sin t)$. Show that there is no $\theta \in (0, 2\pi)$ such that $f(2\pi) - f(0) = 2\pi f'(\theta)$.
(b) Prove the following mean value theorem: Let E be a convex open subset of $\mathbb{R}^n$ and $f : E \to \mathbb{R}^m$ have a derivative in every direction at each $x \in E$. Then for any $c \in \mathbb{R}^m$ and any $a, b \in E$, we have
$$(f(b) - f(a)) \cdot c = [(D_{b-a}f)(a + \theta(b - a)] \cdot c \quad \text{for some } \theta \in (0,1).$$

(c) For a general nonzero $c = (c_1, c_2) \in \mathbb{R}^2$ and $a = 0$, $b = 2\pi$, find the 'θ' of part (b) in terms of c for the function f of part (a).
(d) For the function $f : \mathbb{R} \to \mathbb{R}^2$ defined by $f(t) = (t - t^2, t - t^3)$, show that there is no $\theta \in (0, 1)$ such that $f(1) - f(0) = f'(\theta)$. With $a = 0$ and $b = 1$, determine the nonzero $c = (c_1, c_2) \in \mathbb{R}^2$ for which the 'θ' referred to in (b) fails to be unique.

3-3.P10. (a) Let B be an open ball in $\mathbb{R}^n$ and $f : B \to \mathbb{R}^m$ have the property that the directional derivative $D_u f(x)$ exists and is 0 for every $x \in B$ and every nonzero $u \in \mathbb{R}^n$. Show that f is constant on B.
(b) What can you conclude about f if $D_u f(x)$ exists and is 0 for every $x \in B$ but for a fixed (nonzero) $u \in \mathbb{R}^n$?

3-3.P11. Let S be an open connected subset of $\mathbb{R}^n$ and $f : S \to \mathbb{R}^m$ be differentiable at each point of S. If $f'(s) = O$ $\forall s \in S$, then prove that f is constant on S.

3-3.P12. Let $f(x, t)$ be a continuously differentiable function on $\mathbb{R}^2$ such that $\frac{\partial f}{\partial x} = \frac{\partial f}{\partial t}$. Suppose that $f(x, 0) > 0$ for all x. Prove that $f(x, t) > 0$ for all (x, t).

3-3.P13. Let $f : \mathbb{R}^n \to \mathbb{R}$ have the following properties:

(i) f is differentiable except perhaps at $0 \in \mathbb{R}^n$;

(ii) f is continuous at 0;

(iii) $\frac{\partial f}{\partial x_i}(p) \to 0$ as $p \to 0$, $1 \le i \le n$.

Prove that f is differentiable at 0.

3-3.P14. Show that each of the functions

$$\phi(x,y) = y\,\frac{x^4 + 4x^2y^2 - y^4}{(x^2 + y^2)^2} \quad \text{when } (x,y) \neq (0,0) \text{ and } \phi(0,0) = 0$$

and $$\psi(x,y) = x\,\frac{x^4 - 4x^2y^2 - y^4}{(x^2 + y^2)^2} \quad \text{when } (x,y) \neq (0,0) \text{ and } \psi(0,0) = 0$$

is continuous on $\mathbb{R}^2$, has a directional derivative in every direction at $(0,0)$, but is not differentiable there. Does it have a Gateaux derivative at $(0,0)$ in the sense of 3-3.P1?

3-3.P15. Show that the function $f:\mathbb{R}^2 \to \mathbb{R}$ defined by

$$f(x,y) = \begin{cases} \dfrac{xy^2}{x^2 + y^4} & \text{if } x \neq 0 \\ 0 & \text{if } x = 0 \end{cases}$$

has a finite derivative in every direction $h = (a_1, a_2) \neq (0,0)$ but is not continuous at $(0,0)$. Does f have a Gateaux derivative at $(0,0)$ in the sense of 3-3.P1?

3-4 Partial Derivatives

The reader who has solved 3-2.P1 and 3-2.P5 may have noticed that the entries in the matrix representing the derivative of a function are precisely the partial derivatives of the components of the function. So, one may ask whether a short-cut to finding the matrix of the derivative is to calculate partial derivatives. This turns out to be true if the function is independently known to be differentiable. In case the function is not differentiable, it can happen that the matrix of partial derivatives can nevertheless be formed, although the matrix formed in this manner obviously cannot represent a nonexistent derivative. Examples of such functions are given in 3-2.P3 and 3-4.P2.

Here we clarify the relationship between the matrix of partial derivatives and the matrix representing the derivative. The relationship between the chain rule and partial derivatives is also investigated.

Partial derivatives are directional derivatives of real-valued functions in the coordinate directions.

3-4.1. Definition. *The **partial derivative** $D_j f$, if it exists, of a real valued function f is its directional derivative in the jth coordinate direction:*

$$(D_j f)(x) = \lim_{t \to 0} \frac{f(x + te_j) - f(x)}{t}$$

where $e_1, e_2, \ldots, e_n$ is the standard basis of $\mathbb{R}^n$.

When $n = 2$ or 3, it is more convenient to denote an element of $\mathbb{R}^n$ without subscripts, for instance as (x,y) or (x,y,z); correspondingly, $x + te_1$ will be denoted by $(x + t,y)$ or $(x + t,y,z)$. When this notation is used, partial differentiation is often indicated by $\frac{\partial}{\partial x}$ or by a subscript, as in f_x. Thus $\frac{\partial p}{\partial z}$, p_z and $D_z p$ all mean the same thing. We shall use the alternative notation whenever convenient, especially for discussion of concrete examples.

Examples. (a) Let $f:\mathbb{R}^2 \to \mathbb{R}$ be given by $f(x,y) = |xy|^{1/2}$. Then

$$\frac{f(t,0) - f(0,0)}{t} = 0 = \frac{f(0,t) - f(0,0)}{t}$$

and consequently, $\frac{\partial f}{\partial x}(0,0) = \frac{\partial f}{\partial y}(0,0) = 0$.

(b) Let $f:\mathbb{R}^2 \to \mathbb{R}$ be given by

$$f(x,y) = \begin{cases} \dfrac{x^3 - y^3}{x^2 + y^2} & \text{if } (x,y) \neq (0,0) \\ 0 & \text{if } (x,y) = (0,0). \end{cases}$$

Then $\dfrac{f(t,0) - f(0,0)}{t} = \dfrac{1}{t}\dfrac{t^3 - 0}{t^2 + 0} = 1$ and $\dfrac{f(0,t) - f(0,0)}{t} = \dfrac{1}{t}\dfrac{0 - t^3}{0 + t^2} = -1$.

Therefore $\frac{\partial f}{\partial x}(0,0) = 1$ and $\frac{\partial f}{\partial y}(0,0) = -1$.

If f takes values in $\mathbb{R}^m$, then the m components of $f(x)$ provide real (i.e., scalar)-valued functions $f_1, f_2, \dots, f_m$. One can therefore speak of mn partial derivatives $D_j f_i(x)$ $(1 \leq i \leq m, 1 \leq j \leq n)$. Their relation to the derivative $f'(x)$ is the focus of the next result.

3-4.2. Theorem. *Let the subset $E \subseteq \mathbb{R}^n$ be open and $f:E \to \mathbb{R}^m$ be differentiable at $x \in E$. Then the mn partial derivatives $D_j f_i(x)$ $(1 \leq i \leq m, 1 \leq j \leq n)$ all exist. If $e_1, e_2, \dots, e_n$ is the standard basis of $\mathbb{R}^n$ and $u_1, u_2, \dots, u_m$ is the standard basis of $\mathbb{R}^m$, then*

$$f'(x)e_j = \sum_{i=1}^{m} (D_j f_i)(x) u_i, \quad 1 \leq j \leq n.$$

That is to say, the matrix of $f'(x)$ is the matrix whose (i,j)th entry is the partial derivative $(D_j f_i)(x)$.

Proof. Consider any j, $1 \leq j \leq n$. Since f is differentiable at x, therefore

$$f(x + te_j) - f(x) = f'(x)(te_j) + |t|u(te_j),$$

where $u(te_j) \to 0$ as $t \to 0$. It follows from this and the linearity of $f'(x)$ that

$$\lim_{t \to 0} \frac{f(x + te_j) - f(x)}{t} = f'(x)e_j.$$

Since

$$f(z) = \sum_{i=1}^{m} f_i(z) u_i \text{ for any } z \in E,$$

the above equality can be written as

$$\lim_{t \to 0} \sum_{i=1}^{m} \frac{(f_i(x + te_j) - f_i(x)) u_i}{t} = f'(x)e_j.$$

This implies that

$$\lim_{t \to 0} \frac{f_i(x + te_j) - f_i(x)}{t}$$

exists and equals the ith component of $f'(x)e_j$. Thus $(D_j f_i)(x)$ exists and

$$\sum_{i=1}^{m} (D_j f_i)(x) u_i = f'(x) e_j.$$

This has been shown to be true for any j from 1 to n.   □

The preceding theorem justifies computing partial derivatives and present-ing the matrix formed by them as the derivative, provided the existence of the derivative can be ascertained independently. It also justifies computing the ma-trix of the derivative directly and then presenting its entries as the partial derivatives. The matter of ascertaining the existence of the linear derivative in addition to computing partial derivatives will be taken up in Theorem 3-4.4.

The matrix of partial derivatives $[D_j f_i(x)]$ is called the **Jacobian matrix** of f at the point x. When it is a square matrix, its determinant is called the **Jacobian** of f at x. By Theorem 3-4.2, for a function known to be differentiable the Jaco-bian matrix represents the derivative.

For instance, the function $f : \mathbb{R}^2 \to \mathbb{R}^2$ given by $f(x, y) = (x^2, y^2)$ has been shown in Example 3-3.2(a) to have derivative with matrix $\begin{bmatrix} 2x & 0 \\ 0 & 2y \end{bmatrix}$. It follows that this is the Jacobian matrix, i.e.

$$\frac{\partial}{\partial x}(x^2) = 2x, \ \frac{\partial}{\partial y}(x^2) = 0, \ \frac{\partial}{\partial x}(y^2) = 0 \text{ and } \frac{\partial}{\partial y}(y^2) = 2y.$$

Of course, this does not mean that the partial derivatives are to be computed by following the procedure of Example 3-3.2!

Theorem 3-4.2 also enables us to clarify the relation between the chain rule (Theorem 3-3.1) and partial derivatives, of which there is no explicit mention in the latter. Suppose f and g are as in the chain rule. Then by Theorem 3-4.2,

$(g \circ f)'(x)$ has a $p \times n$ matrix whose (k, j)th entry is $(D_j (g \circ f)_k)(x)$;

$f'(x)$ has an $m \times n$ matrix whose (i,j)th entry is $(D_j f_i)(x)$;

$g'(f(x))$ has a $p \times m$ matrix whose (k, i)th entry is $(D_i g_k)(f(x))$.

The chain rule, when taken with Remark 2-3.3 as was done in Example 3-3.2(b), states that the first of the three matrices described above is the product of the latter two in the appropriate order. By the definition of matrix product, this amounts to

$$(D_j (g \circ f)_k)(x) = \sum_{i=1}^{m} (D_i g_k)(f(x))(D_j f_i)(x).$$

This equality is nothing but the usual version of the chain rule for computing partial derivatives, which the reader must have encountered in a course on the calculus of two or more variables. By Theorem 3-4.2 and the chain rule, the above procedure for computing partial derivatives is applicable when the functions concerned are differentiable.

The next theorem makes it possible to ascertain the existence of the derivative by examining partial derivatives. But it works only when the derivative exists in an open set and is continuous as well. On the other hand, this is most often so, hence the theorem turns out to be adequate in most situations. See Example 3-4.6(b) for a situation when it is not adequate.

In the notation of Theorem 3-4.2, $f'(x)$ is a linear map from $\mathbb{R}^n$ to $\mathbb{R}^m$. If $f'(x)$ exists for each $x \in E$, then f' is a map from E to the space of linear maps from $\mathbb{R}^n$ to $\mathbb{R}^m$. Since $E \subseteq \mathbb{R}^n$, it is clear what $\| \xi - x \|$ means when $\xi \in E, x \in E$. Moreover, since $f'(\xi) - f'(x)$ is a linear map from $\mathbb{R}^n$ to $\mathbb{R}^m$, $\| f'(\xi) - f'(x) \|$ denotes its norm in accordance with Def. 2-7.2. Thus it makes sense to speak of f' being continuous.

3-4.3. Definition. *A function f with values in $\mathbb{R}^m$ and domain $E \subseteq \mathbb{R}^n$ is said to be **continuously differentiable** (or **belong to class C^1**) at an interior point x of E if the derivative $f'(z)$ exists at every z in some open set containing x and the resulting map from that open set into the space $\mathfrak{L}(\mathbb{R}^n, \mathbb{R}^m)$ of linear maps from $\mathbb{R}^n$ to $\mathbb{R}^m$ is continuous at x. One also speaks of f being a C^1 **function** (or a C^1 **map**).*

3-4.4. Theorem. *Suppose $E \subseteq \mathbb{R}^n$ is open and f is a map from E to $\mathbb{R}^m$. Then f is continuously differentiable on E (i.e., at each point of E) if and only if each of the mn partial derivatives $D_j f_i$ ($1 \leq i \leq m$, $1 \leq j \leq n$) exists on E and is continuous there.*

Proof. Assume f' to be continuous on E. By Theorem 3-4.2, $D_j f_i$ exist on E and f' has matrix with $D_j f_i$ as its (i,j)th entry. Therefore for any x and $x + h$ in E, the matrix of $f'(x + h) - f'(x)$ has $(D_j f_i)(x + h) - (D_j f_i)(x)$ as its (i,j)th entry. By

Remark 2-7.9(a), $|(D_j f_i)(x+h) - (D_j f_i)(x)| \le \|f'(x+h) - f'(x)\|$. This immediately leads to the continuity of each partial derivative $D_j f_i$.

To prove the converse, assume that each $D_j f_i$ is continuous on E. Consider any i, $1 \le i \le m$ and any $x \in E$. For ease of notation, we denote f_i by g. We shall prove first that $g : E \to \mathbb{R}$ is continuously differentiable, that is to say, the map $g' : E \to \mathfrak{L}(\mathbb{R}^n, \mathbb{R})$ is continuous.

Let $e_1, e_2, \ldots, e_n$ be the standard basis of $\mathbb{R}^n$.

For $h = h_1 e_1 + \cdots + h_n e_n \in \mathbb{R}^n$ sufficiently small to ensure that $x + h \in E$, let

$$u(h) = [g(x+h) - g(x) - \sum_{j=1}^{n} (D_j g)(x) h_j]/\|h\| \quad \text{if } h \ne 0.$$

We shall show that $u(h) \to 0$ as $h \to 0$, so that g has derivative with matrix

$$[(D_1 g)(x) \quad (D_2 g)(x) \quad \cdots \quad (D_n g)(x)].$$

Consider any $\varepsilon > 0$. By the continuity of $D_j g$ for $1 \le j \le n$, there exists $\delta > 0$ such that the ball B of radius δ centred at x is contained in E and every $y \in B$ satisfies

$$|(D_j g)(x) - (D_j g)(y)| < \frac{\varepsilon}{\sqrt{n}} \quad \text{for} \quad 1 \le j \le n. \tag{1}$$

Let $h = h_1 e_1 + \cdots + h_n e_n$ satisfy $\|h\| < \delta$. Set $z_0 = 0$ and $z_j = h_1 e_1 + \cdots + h_j e_j$ for $1 \le j \le n$. Then $z_n = h$. Also, for any $t \in [0, 1]$,

$$\|z_0 + t h_1 e_1\| = |t h_1| \le |h_1| \le \|h\| < \delta$$

and, when $j > 1$,

$$\|z_{j-1} + t h_j e_j\| = (h_1^2 + \cdots + h_{j-1}^2 + t^2 h_j^2)^{1/2} \le \|h\| < \delta.$$

Therefore $x + z_{j-1} + t h_j e_j$ belongs to the ball B (including $j = 1$). It follows firstly that the map

$$t \to g(x + z_{j-1} + t h_j e_j)$$

is defined on $[0, 1]$, and secondly that (1) is applicable with $y = x + z_{j-1} + t h_j e_j$. Therefore, the mean value theorem applied to the function

$$G(t) = g(x + z_{j-1} + t h_j e_j), \quad t \in [0,1],$$

yields

$$g(x + z_j) - g(x + z_{j-1}) = G(1) - G(0) = G'(\theta_j), \quad \text{where } 0 < \theta_j < 1.$$

But by definition of $D_j g$, we have,

$$G'(\theta_j) = h_j (D_j g)(x + z_{j-1} + \theta_j h_j e_j).$$

Therefore

$$g(x + z_j) - g(x + z_{j-1}) = h_j (D_j g)(x + z_{j-1} + \theta_j h_j e_j), \quad \text{where } 0 < \theta_j < 1.$$

Now,

$$g(x+h) - g(x) = g(x+z_n) - g(x+z_0) = \sum_{j=1}^{n} (g(x+z_j) - g(x+z_{j-1}))$$

$$= \sum_{j=1}^{n} h_j (D_j g)(x + z_{j-1} + \theta_j h_j e_j).$$

Therefore

$$|\,\|h\| u(h)| = |g(x+h) - g(x) - \sum_{j=1}^{n} (D_j g)(x) h_j|$$

$$= |\sum_{j=1}^{n} [(D_j g)(x + z_{j-1} + \theta_j h_j e_j) - (D_j g)(x)] h_j|$$

$$\leq \sum_{j=1}^{n} \frac{\varepsilon}{\sqrt{n}} |h_j| \qquad \text{by (1)}$$

$$\leq \|h\| \left(n\left(\frac{\varepsilon^2}{n}\right) \right)^{1/2} \qquad \text{(Cauchy-Schwarz)}$$

$$\leq \|h\| \varepsilon.$$

This shows that $|u(h)| < \varepsilon$ when $0 < \|h\| < \delta$. Thus $u(h) \to 0$ as $h \to 0$, and g' has the matrix

$$[(D_1 g)(x) \quad (D_2 g)(x) \quad \cdots \quad (D_n g)(x)].$$

Since each $D_j g$ is continuous on E, so is the map $g': E \to \mathcal{L}(\mathbb{R}^n, \mathbb{R})$ [Remark 2-7.9(a)]. Since g can be any f_i, we conclude that every f_i is continuously differentiable.

Finally, we argue that f is itself continuously differentiable. Since each f_i has been shown to be differentiable, it follows by Proposition 3-2.8 that f is differentiable. It further follows by Theorem 3-4.2 that $f'(x)$ has matrix with (i, j)th entry given by the partial derivative $(D_j f_i)(x)$. Now by Remark 2-7.9(a),

$$\|f'(x+h) - f'(x)\| \leq \Sigma_j \Sigma_i |(D_j f_i)(x+h) - (D_j f_i)(x)|$$

when x and $x + h$ both belong to E. Since the partial derivatives have been assumed continuous, it is immediate from this inequality that f' is continuous. $\square$

3-4.5. Remarks. (a) The assumption of continuity of $D_1 g$ could have been avoided in proving the existence of g'. The existence result, without continuity of f_i', therefore remains valid if continuity is assumed for all the other partial derivatives. By rearranging the variables if necessary, we can work with the hypothesis that all partial derivatives with at most one exception are continuous.

(b) Since the equality

$$g(x+h) - g(x) = \sum_{j=1}^{n} h_j (D_j g)(x + z_{j-1} + \theta_j h_j e_j)$$

was derived in the proof of the theorem by using only the existence of partial derivatives, it can be used for establishing that (i) if the partial derivatives are bounded on E, then f is continuous on E [see 3-4.P7] and (ii) if the partial derivatives are continuous at x, then f is differentiable at x.

3-4.6. Examples. (a) We shall now show that the map of $\mathbb{R}^2$ to $\mathbb{R}^2$ given by

$$p = e^x \cos y, \qquad q = e^x \sin y$$

is differentiable by using Theorem 3-4.4 and find its Jacobian matrix.

(Although we shall avoid introducing the standard basis e_1, e_2 for $\mathbb{R}^2$, it should be understood that the given map f can be expressed in terms of the standard basis as

$$f(xe_1 + ye_2) = (e^x \cos y)e_1 + (e^x \sin y)e_2 \,.)$$

Here $f_1(x,y) = p = e^x \cos y$ and $f_2(x,y) = q = e^x \sin y$. Therefore, the partial derivatives are $D_1f_1 = e^x \cos y$, $D_2f_1 = -e^x \sin y$, $D_1f_2 = e^x \sin y$, $D_2f_2 = e^x \cos y$. These are all continuous and therefore the given map is differentiable by Theorem 3-4.4. The Jacobian matrix is

$$\begin{bmatrix} e^x \cos y & -e^x \sin y \\ e^x \sin y & e^x \cos y \end{bmatrix}.$$

We note for later purposes that the determinant of this matrix, the Jacobian, is never zero. It follows that the Jacobian matrix is invertible at every x and hence by Theorem 3-4.2, so is the derivative.

(b) In Example 3-2.3(d), take ϕ and ψ to be differentiable with derivatives discontinuous at α and/or β. For example, $\phi(t) = \psi(t) = t^2 \sin\frac{1}{t}$ for nonzero t and $\phi(0) = \psi(0) = 0$. As discussed there, f is then differentiable at $(0,0)$. However, the partial derivatives are $(D_1f)(x,y) = \phi'(x)$ and $(D_2f)(x,y) = \psi'(y)$, both of which are discontinuous at 0. So, the differentiability of f, which was proved in Example 3-2.3(d) cannot be deduced on the basis of Theorem 3-4.4.

We now turn our attention to an application of partial derivatives for testing whether a function, if differentiable, satisfies a certain condition that is purely 'algebraic' in the sense that the condition does not explicitly involve even limits, let alone differentiation.

Functions described by such expressions as $x^2 + xy - 2y^2$ or $\sqrt{(x^2y - y^3)}$, as opposed to $x + y^2$, are 'homogeneous' in an intuitively obvious sense. A precise definition would be as follows.

3-4.7. Definition. *A function $f:S \to \mathbb{R}$ with an open domain $S \subseteq \mathbb{R}^n$ is said to be* **homogeneous of degree p** *(where p is a real number) if*

$$f(\lambda x) = \lambda^p f(x) \quad \text{whenever} \quad x \in S, \lambda > 0, \lambda x \in S.$$

The restriction of a homogeneous function to any open subset of its domain is homogeneous of the same degree. Moreover, sums and constant multiples of homogeneous functions of common degree are again homogeneous of that degree. Therefore the examples below can be used to generate several others.

3-4.8. Examples. (a) The functions given by $x, y, x \pm y$, $\sqrt{(x^2 + y^2)}$, $|x|, |y|$ on $\mathbb{R}^2$ are homogeneous of degree 1. It may be noted that the last three among these would not be considered homogeneous if the definition were amended to include (as some authors prefer) all real λ.

(b) The function defined by $\sqrt{(xy)}$ on $\{(x, y) \in \mathbb{R}^2 : xy > 0\}$ is homogeneous of degree 1.

(c) The functions $x^2, xy, x|y|$ are homogeneous of degree 2 on $\mathbb{R}^2$.

(d) $\sqrt{(xyz)}$ describes a homogeneous function of degree $\frac{3}{2}$ on $\{(x, y, z) \in \mathbb{R}^3 : xyz > 0\}$

(e) The function defined on $\mathbb{R}^2$ by $x^2 + y$ is not homogeneous of any degree. Otherwise we would have $\lambda^2 x^2 + \lambda y = \lambda^p (x^2 + y)$ for all $(x, y) \in \mathbb{R}$ and all $\lambda > 0$. Choosing $(x, y) = (1, 0)$, we get $\lambda^2 = \lambda^p$ for all $\lambda > 0$; choosing $(x, y) = (0, 1)$, we get $\lambda = \lambda^p$ for all $\lambda > 0$. This is a contradiction.

(f) The relevance of this example to Theorem 3-4.9(b) below is mentioned in the remark that follows it. Let S be the union $S_1 \cup S_2$, where

and
$$S_1 = \{(x, y) \in \mathbb{R}^2 : x^2 + y^2 < 1\}$$
$$S_2 = \{(x, y) \in \mathbb{R}^2 : x^2 + y^2 > 1\}.$$

Define f to be $x + y$ on S_1 and $x - y$ on S_2. Note that, given $(x, y) \in S_2$, there exists $\lambda > 0$ such that $(\lambda x, \lambda y) \in S_1$ and therefore f is not homogeneous. However, its restrictions to S_1 and S_2 are.

The next proof uses the chain rule for computing partial derivatives as explained above after Theorem 3-4.2.

3-4.9. (a) **Euler's Theorem.** *If $S \subseteq \mathbb{R}^n$ is open and the differentiable function $f:S \rightarrow \mathbb{R}$ is homogeneous of degree p, then the following identity (called **Euler's relation**) holds*:

$$\sum_{j=1}^{n} x_j \cdot (D_j f)(x) = p \cdot f(x) \text{ for all } x \in S. \tag{A}$$

(b) *Let the open subset S of $\mathbb{R}^n$ satisfy the condition*

$$\lambda x \in S \quad \text{whenever} \quad x \in S \text{ and } \lambda > 0.$$

If the differentiable function $f:S \rightarrow \mathbb{R}$ satisfies (A), *then it is homogeneous of degree p.*

Proof. Consider any $x \in \mathbb{R}^n$ and $\lambda > 0$ such that $\lambda x \in S$. Since S is open, λ can 'vary within a small neighbourhood of itself'. To put it precisely, $\mu x \in S$ as long as $\mu \in (\lambda - \delta, \lambda + \delta)$; but we avoid bringing in μ and δ explicitly so as to keep the notation simple. Now let f be any differentiable function on S. Upon applying the chain rule to the composition $\phi(\lambda) = f(\lambda x)$, we get

$$\phi'(\lambda) = \sum_{j=1}^{n} x_j \cdot (D_j f)(\lambda x). \tag{1}$$

Note that this holds regardless of whether f is homogeneous (or $x \in S$ for that matter).

In order to prove (a), consider any $x \in S$. Then $\lambda x \in S$ for $\lambda = 1$. Therefore ϕ is defined on an open set containing 1. By the assumed homogeneity of f, we have $\phi(\lambda) = \lambda^p f(x)$. Hence it follows from (1) that

$$\sum_{j=1}^{n} x_j \cdot (D_j f)(\lambda x) = p \cdot \lambda^{p-1} f(x).$$

Upon substituting $\lambda = 1$ in this, we get (A).

In order to prove (b), consider any $x \in S$ and $\lambda > 0$. According to the hypothesis on the domain S, we must have $\lambda x \in S$. Thus, ϕ is defined on $(0, \infty)$ and satisfies (1). Since (A) is assumed to hold everywhere on S, we have

$$\sum_{j=1}^{n} \lambda x_j \cdot (D_j f)(\lambda x) = p \cdot f(\lambda x) \text{ for all } \lambda > 0.$$

But by (1), this states that $\lambda \phi'(\lambda) = p \cdot \phi(\lambda)$. It follows that

$$\frac{d}{d\lambda}\left(\frac{\phi(\lambda)}{\lambda^p}\right) = \phi'(\lambda)/\lambda^p - p \cdot \phi(\lambda)/\lambda^{p+1} = \left[p \cdot \phi(\lambda)/\lambda\right]/\lambda^p - p \cdot \phi(\lambda)/\lambda^{p+1} = 0.$$

Therefore, $\phi(\lambda)/\lambda^p$ is constant on $(0, \infty)$ and hence $\phi(\lambda)/\lambda^p = \phi(1) = f(x)$ for all $\lambda \in (0, \infty)$. This means f is homogeneous of degree p.   □

3-4.10. Remark. The condition on the domain in part (b) of the above theorem cannot be omitted. The function of Example 3-4.8(f) satisfies (A) on S_1 as well as S_2 and hence on its entire domain, but it is not homogeneous.

Problem Set 3-4

3-4.P1. Solve 3-2.P5, not from first principles this time but by using Theorems 3-4.2 and 3-4.4.

3-4.P2. Show that the function defined on $\mathbb{R}^2$ by $f(x,y) = xy/(x^2 + y^2)$ for $(x,y) \neq (0,0)$ and $f(0,0) = 0$ is not continuous at $(0,0)$ but both partial derivatives exist there. Also, show that no directional derivative exists in any direction (h, k) for which $h \neq 0 \neq k$.

3-4.P3. Given: $f(x, y) = xy/(x^2 + y^2)^{1/2}$ for $(x, y) \neq (0,0)$ and $f(0,0) = 0$. Show that f is continuous, possesses partial derivatives but is not differentiable at $(0,0)$.

3-4.P4. Show that the map of $\mathbb{R}^2$ into itself given by

$$x = p \cos q, \qquad y = p \sin q$$

is differentiable and that its Jacobian never vanishes except when $p = 0$.

3-4.P5. Find all points (p, q) where the Jacobian of the following map of $\mathbb{R}^2$ into itself vanishes:

$$x = \cos p \cosh q, \qquad y = \sin p \sinh q.$$

3-4.P6. Suppose $E \subseteq \mathbb{R}^n$ $(n \geq 2)$ is open and all partial derivatives of $g : E \to \mathbb{R}$ exist on E. If $D_2 g, \ldots, D_n g$ are continuous at some point $x \in E$, show that g is differentiable at x.

(Remark. Referenced just before Remark 4-3.1. The result remains true if we assume that, with any one exception, all the remaining partial derivatives are continuous at x. All we have to do is to rename the components $x_1, \ldots, x_n$ so as to have the exceptional component appear as the first one after renaming. The question is now reduced to the case considered in the problem above, and there is no need to work through the argument all over again. This practice of reducing a case to one that has already been handled has given rise to the following joke: A mathematician and a physicist are each given a lighted stove with a bowl of water to the stove's left and asked to heat the water. Both of them pick up the bowl and place it on the stove. Then they are each given a lighted stove with a bowl of water to the stove's *right* side and asked to heat the water. The physicist picks up the bowl and places it on the stove; the mathematician picks up the bowl, places it on the left of the stove and declares, 'I have reduced it to the previous case!')

3-4.P7. Suppose $E \subseteq \mathbb{R}^n$ is open and all partial derivatives of $g : E \to \mathbb{R}$ exist on E. If $D_1 g, \ldots, D_n g$ are bounded on E, then show that f is continuous on E.

3-4.P8. Let $f : \mathbb{R}^2 \to \mathbb{R}^3$ be defined by

$$f(x, y) = (\sin x \cos y, \; x + y, \; x^2 - y).$$

Find the Jacobian matrix.

3-4.P9. Show that the function $f : \mathbb{R}^2 \to \mathbb{R}$ of Example 3-2.6, which was defined as $f(0, 0) = 0$ and

$$f(x, y) = x^3/(x^2 + y^2) \quad \text{when } (x, y) \neq (0, 0),$$

has bounded partial derivatives everywhere but no Gateaux derivative at $(0, 0)$.

3-4.P10. Compute the Gateaux derivative where it exists for the function f, where $f(0, 0) = 0$ and

$$f(x, y) = x^2 y^2 \ln (x^2 + y^2) \quad \text{when } (x, y) \neq (0, 0).$$

3-4.P11. (a) Find a linear function of x and y which is a 'good' approximation for $F(x, y) = \arctan\left(\dfrac{x - y}{1 + xy}\right)$ when x and y are 'small'.

(b) Find a linear function of $x-3$ and $y-\frac{1}{2}$ which is a 'good' approximation for

$$F(x,y) = \arctan\left(\frac{x-y}{1+xy}\right)$$ when x and y are 'near' 3 and $\frac{1}{2}$, respectively.

3-4.P12. Let f be a real-valued function differentiable on an open ball centred at $(x_1,x_2) \in \mathbb{R}^2$ and let $y = (y_1,y_2)$ be in the ball. By considering the function

$$g(t) = f(ty_1 + (1-t)x_1, y_2) + f(x_1, ty_2 + (1-t)x_2),$$

prove that

$$f(y_1,y_2) - f(x_1,x_2) = (y_1 - x_1)D_1f(z_1,y_2) + (y_2 - x_2)D_2f(x_1,z_2),$$

where z_i is a point on the segment $\{ty_i + (1-t)x_i : 0 \le t \le 1\}$.

3-4.P13. Suppose that $F:S \to \mathbb{R}$ is differentiable on the open set S and $\sum_{j=1}^{n} x_j \cdot (D_j F)(x) = p \cdot F(x)$ for all $x \in S$. If x is a point of S for which there is an interval (t_0,t_1) such that $tx \in S \; \forall \; t \in (t_0,t_1)$ and also $1 \in (t_0,t_1)$, show that the relation $F(tx) = t^p F(x)$ holds for $t \in (t_0,t_1)$.

3-4.P14. Let x,y,z be differentiable functions of (u,v) on some open set in $\mathbb{R}^2$ and $J(x,y)$, $J(y,z)$, $J(z,x)$ be the Jacobians of (x,y), (y,z), (z,x), respectively. Prove that

$$x_u J(y,z) + y_u J(z,x) + z_u J(x,y) = 0 \quad \text{and} \quad x_v J(y,z) + y_v J(z,x) + z_v J(x,y) = 0.$$

3-4.P15. For any $n \times n$ matrix $A = [a_{ij}]$, denote the cofactor of the (i,j)th entry by A_{ij}. Then $\det A = \Sigma_j a_{ij}A_{ij}$ for each i. When B is also an $n \times n$ matrix, denote by A_B^i the matrix obtained by replacing the ith row of A by that of B. Then $A_{ij} = (A_B^i)_{ij}$.

(a) $\det A$ is a function of the n^2 variables a_{ij}; prove that $\dfrac{\partial}{\partial a_{ij}} A_{ik} = 0$ for any i,j,k and hence that $\dfrac{\partial}{\partial a_{ij}} \det A = A_{ij}$ for any i,j.

(b) Let each a_{ij} be a differentiable function on some common interval I in $\mathbb{R}$ and denote by B the matrix of derivatives $[a_{ij}'(x)]$. Prove by using the chain rule that $\dfrac{d}{dx} \det A = \Sigma_i \det A_B^i$.

(c) If furthermore A has an inverse $A^{-1} = [c_{ij}(x_0)]$ at some $x_0 \in I$, show that $\dfrac{d}{dx}$ det$A = \det A \, \Sigma_{ij} a_{ij}'c_{ji}$ at x_0. What does this say about $\dfrac{d}{dx}\big(\ln(\det A)\big)$?

3-4.P16. Find $x\dfrac{\partial z}{\partial x} + y\dfrac{\partial z}{\partial y}$ at points of differentiability of

$$z = \ln\left(\ln\left(\frac{x^5 + 7x^4 y + 86x^3 y^2 + 9y^5 + (x^2 + y^2)^{5/2}}{(x^2 + y^2)^{5/2}}\right)\right).$$

3-4.P17. [Needed in 3-4.P18–21] If the function $f:[a,b]\times[c,d]\to\mathbb{R}$ has a continuous partial derivative with respect to y at each point of $[a,b]\times[c,d]$, then the function F defined on $[c,d]$ by

$$F(y) = \int_a^b f(x,y)\,dx$$

has a continuous derivative given by

$$F'(y) = \int_a^b \frac{\partial f}{\partial y}(x,y)\,dx \quad (\textbf{Leibnitz's formula}).$$

Determine whether this extends to the case when $[c,d]$ is replaced by (c,∞) or $\mathbb{R}$.

3-4.P18. Use the Leibnitz Formula [see 3-4.P17] to show that

$$\int_0^{\pi/2} \ln(\alpha^2 - \sin^2\phi)\,d\phi = \pi\ln[\tfrac{1}{2}(\alpha + \sqrt{(\alpha^2 - 1)})] \quad \text{for } \alpha > 1.$$

3-4.P19. Show that the function $u(x) = \int_0^\pi \cos(n\phi - x\sin\phi)\,d\phi$, $x \in \mathbb{R}$, satisfies **Bessel's equation**, namely: $x^2 u'' + xu' + (x^2 - n^2)u = 0$.

3-4.P20. Suppose that f is a continuous function with continuous partial derivative $D_2 f$ on the rectangle $I = [a,b]\times[c,d]$. Further, suppose that the functions α and β on $[c,d]$ have values in $[a,b]$ and are continuously differentiable. Then the integral

$$F(y) = \int_{\alpha(y)}^{\beta(y)} f(x,y)\,dx$$

is defined for every $y \in [c,d]$ and F is continuously differentiable with derivative given by

$$F'(y) = f(\beta(y),y)\beta'(y) - f(\alpha(y),y)\alpha'(y) + \int_{\alpha(y)}^{\beta(y)} (D_2 f)(x,y)\,dx.$$

Here the intervals $[a,b]$, $[c,d]$ may be replaced by any other intervals containing more than one point.

3-4.P21. Let f be a continuous function on an interval containing 0. Consider the sequence of functions defined on that interval by

$$F_n(x) = \frac{1}{(n-1)!} \int_0^x (x-t)^{n-1} f(t)\,dt, \quad n \in \mathbb{N}.$$

Show that $F_n^{(n)}(x) = f(x)$ for each $n \in \mathbb{N}$.

3-4.P22. (The boundary of the unit disc in the $x_1 x_3$-plane of $\mathbb{R}^3$ is the entire disc and not just its circumference. Intuition suggests that a curve that reaches a point of the circumference orthogonally from within the disc and has a nonzero tangent at the point must exit the disc there. The next problem formulates the idea analytically.)

Let $D = \{(x_1, x_2, x_3) \in \mathbb{R}^3 : x_1^2 + x_2^2 \le 1, x_3 = 0\}$ and $\gamma:[0, 2\pi] \to \mathbb{R}^3$ be defined by

$$\gamma_1(\theta) = \cos\theta, \qquad \gamma_2(\theta) = \sin\theta, \qquad \gamma_3(\theta) = 0.$$

Suppose $\Gamma:[-1, 1] \to \mathbb{R}^3$ is differentiable at 0 and

(i) $\Gamma(t) \in D$ whenever $-1 \le t \le 0$;

(ii) $\Gamma_1(0) = \cos\theta_0$, $\Gamma_2(0) = \sin\theta_0$, $\Gamma_3(0) = 0$ where $0 \le \theta_0 \le 2\pi$;

(iii) $\Gamma_1{}'(0)^2 + \Gamma_2{}'(0)^2 > 0$;

(iv) $(\Gamma_1{}'(0), \Gamma_2{}'(0), \Gamma_3{}'(0)) \cdot (\gamma_1{}'(\theta_0), \gamma_2{}'(\theta_0), \gamma_3{}'(\theta_0)) = 0$.

Show that some $\delta > 0$ has the property that $0 < t < \delta \;\Rightarrow\; \Gamma(t) \notin D$.

3-4.P23.   Let $\Phi:[0,1] \times [0,1] \to \mathbb{R}^3$ be as in 2-6.P18. Fix any $u_0 \in [0,1]$. Consider the maps

$$\Gamma:[0,1] \to [0,1] \times [0,1]$$

and

$$\gamma:[-1,1] \to [0,1] \times [0,1]$$

given by

$$\Gamma(s) = (s, 0)$$

and

$$\gamma(t) = \begin{cases} (u_0, -t) & \text{if } -1 \le t \le 0 \\ (1 - u_0, 1 - t) & \text{if } 0 < t \le 1. \end{cases}$$

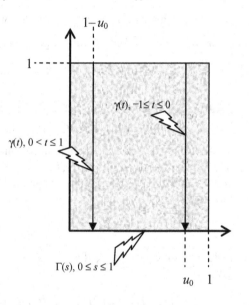

Then Γ has a continuous derivative everywhere but γ is manifestly discontinuous at 0. Show that

(a) $\Phi(\Gamma(u_0)) = \Phi(\gamma(0))$.

(b) $(\Phi \circ \Gamma)'(s) = (0, 0, 2\alpha)$ for all s $\in [0,1]$.

(c) $\Phi \circ \gamma$ is continuously differentiable on $[0,1]$.

(d)   Neither $(\Phi \circ \Gamma)'(u_0)$ nor $(\Phi \circ \gamma)'(0)$ vanishes but their inner product does.

In terms of the graph of the range M of Φ (the Möbius band) and the curve $\Phi \circ \Gamma$ lying in it, this can be interpreted as saying that, at every point $\Phi(\Gamma(u_0))$ of the curve other than $\Phi(\Gamma(\frac{1}{2}))$, it is crossed orthogonally by another curve, also lying in M, namely, $\Phi \circ \gamma$. How would you interpret this about the graph of $\Phi \circ \Gamma$ lying on the 'edge' of M?

3-5 Second Partial Derivatives

The partial derivatives $D_j f, j = 1, 2, \ldots, n$, of a real-valued function f defined on a subset of $\mathbb{R}^n$ are themselves real-valued functions defined on a subset of $\mathbb{R}^n$ and therefore can have partial derivatives. When they do, their partial derivatives $D_i (D_j f)$ are called 'second partial derivatives' of f. Example 3-5.2 below shows that $D_i (D_j f)$ need not always be equal to $D_j (D_i f)$. We shall prove theorems, one due to Schwarz and another due to Young, which assure the equality of $D_i (D_j f)$ and $D_j (D_i f)$ under different hypotheses.

We shall also prove a simple case of Taylor's theorem needed in the sequel.

We begin with a formal definition of second partial derivatives.

3-5.1. Definition. *If a partial derivative $D_j f$ of a function $f:S \to \mathbb{R}$, where $S \subseteq \mathbb{R}^n$, has a partial derivative $D_i (D_j f)$ at some point $x \in S$, then it is called a* **second partial derivative** *of f at x and is denoted by $(D_{ij} f)(x)$ or $D_{ij} f(x)$.*

In contrast, a partial derivative $D_j f$ is called a *first partial derivative*.

When $i \neq j$, the second partial derivatives $D_{ij} f$ and $D_{ji} f$ need not be equal everywhere even if the function f is differentiable everywhere. We present an instance of this before proving two sets of sufficient conditions for equality to hold.

3-5.2. Example. Define the function f on $\mathbb{R}^2$ as

$$f(x,y) = \begin{cases} \dfrac{xy(x^2 - y^2)}{x^2 + y^2} & \text{if } (x,y) \neq (0,0) \\ \quad 0 & \text{if } (x,y) = (0,0). \end{cases}$$

A routine computation shows that its partial derivatives are given by

$$(D_1 f)(x,y) = y\frac{x^4 + 4x^2 y^2 - y^4}{(x^2 + y^2)^2} \quad \text{when } (x,y) \neq (0,0) \text{ and } (D_1 f)(0,0) = 0$$

$$\text{and } (D_2 f)(x,y) = x\frac{x^4 - 4x^2 y^2 - y^4}{(x^2 + y^2)^2} \quad \text{when } (x,y) \neq (0,0) \text{ and } (D_2 f)(0,0) = 0.$$

Continuity of these partial derivatives follows from 3-3.P14. Thus f is differentiable everywhere. Furthermore, $(D_1 f)(0,y) = -y$ when $y \neq 0$ and $(D_2 f)(x,0) = x$ when $x \neq 0$. Hence $(D_{2\,1} f)(0,0) = -1$ while $(D_{1\,2} f)(0,0) = 1$.

For later purposes, we note that

$$(D_{2\,1}f)(x,y) = \frac{x^6 + 9x^4y^2 - 9x^2y^4 - y^6}{(x^2 + y^2)^3} \quad \text{when } (x,y) \neq (0,0).$$

The hypotheses of the next theorem do not imply that f is differentiable; so we cannot use the chain rule proved in Theorem 3-3.1.

3-5.3. Schwarz's Theorem. *Let x be a point in the domain $S \subseteq \mathbb{R}^n$ of a real-valued function f such that, in some ball centred at x, the derivatives $D_j f$, $D_i f$ and $D_{ij}f$ exist and $D_{ij}f$ is continuous at x. Then $D_{j\,i}f(x)$ exists and equals $D_{ij}f(x)$.*

Proof. For convenience of notation, we take $n = 2$, $i = 1$ and $j = 2$. Also, we denote x by (a,b), where $a,b \in \mathbb{R}$. Thus $D_{1\,2}f$ is assumed continuous at (a,b) and we must prove that $D_{2\,1}f(a,b)$ exists and equals $D_{1\,2}f(a,b)$.

Consider any $\varepsilon > 0$. Since $D_{1\,2}f$ has been assumed continuous at (a,b), there exists a positive δ (less than the radius of the ball mentioned in the hypothesis) such that for *any* $\theta,\theta' \in (0,1)$, we have

$$|h| < \delta, |k| < \delta \;\Rightarrow\; |D_{1\,2}f(a+\theta'h,\,b+\theta k) - D_{1\,2}f(a,b)| < \tfrac{\varepsilon}{2}. \qquad (1)$$

Let ϕ be the function with domain $[b, b+k]$, defined by

$$\phi(y) = f(a+h,y) - f(a,y).$$

Then by the mean value theorem, $\phi(b+k) - \phi(b) = k \cdot \phi'(b+\theta k)$, where $\theta \in (0,1)$. Now, $\phi'(y) = D_2 f(a+h,y) - D_2 f(a,y)$ by definition of partial derivative. Therefore,

$$\phi(b+k) - \phi(b) = k \cdot [D_2 f(a+h,b+\theta k) - D_2 f(a,b+\theta k)], \quad \text{where } \theta \in (0,1).$$

By a similar use of the mean value theorem again, we further have

$$\phi(b+k) - \phi(b) = k \cdot [h \cdot D_{1\,2}f(a+\theta'h,b+\theta k)], \quad \text{where } \theta' \in (0,1). \qquad (2)$$

However, $\phi(b+k) - \phi(b) = f(a+h,b+k) - f(a+h,b) - f(a,b+k) + f(a,b)$. It therefore follows from (2) that when $h \neq 0 \neq k$,

$$\frac{1}{k}\left[\frac{f(a+h,b+k) - f(a,b+k)}{h} - \frac{f(a+h,b) - f(a,b)}{h}\right] = D_{1\,2}f(a+\theta'h,b+\theta k).$$

In view of (1), this has the consequence that whenever $0 < |h| < \delta$, $0 < |k| < \delta$, we have

$$\left|\frac{1}{k}\left[\frac{f(a+h,b+k) - f(a,b+k)}{h} - \frac{f(a+h,b) - f(a,b)}{h}\right] - D_{1\,2}f(a,b)\right| < \tfrac{\varepsilon}{2}.$$

By hypothesis, the two quotients here have respective limits $D_1 f(a,b+k)$ and $D_1 f(a,b)$ as $h \to 0$. Therefore

$$0 < |k| < \delta \;\Rightarrow\; \left| \frac{D_1 f(a,b+k) - D_1 f(a,b)}{k} - D_{1\,2}f(a,b) \right| \le \tfrac{\varepsilon}{2} < \varepsilon.$$

Since such a positive δ has been shown to exist for an arbitrary positive ε, the second partial derivative $D_{2\,1}f(a,b)$ exists and equals $D_{1\,2}f(a,b)$. $\square$

Remark. The hypotheses of continuity in Schwarz's theorem cannot be dropped. For the function f of Example 3-5.2, the second partial derivative $(D_{2\,1}f)(x,y)$ was computed there and is seen to satisfy $\lim\limits_{x\to 0}(D_{2\,1}f)(x,0) = 1 = -\lim\limits_{y\to 0}(D_{2\,1}f)(0,y)$, which means it is not continuous at $(0,0)$; recall that its mixed partial derivatives $D_{2\,1}f(0,0)$ and $D_{1\,2}f(0,0)$ were shown to be unequal. The same example shows that the hypothesis of differentiability in the next theorem cannot be dropped: $D_1 f$ is not differentiable, as can be deduced from 3-3.P14, and the mixed partial derivatives $D_{2\,1}f(0,0)$ and $D_{1\,2}f(0,0)$ are not equal.

3-5.4. Young's Theorem. *Let x be a point in the domain $S \subseteq \mathbb{R}^n$ of a real-valued function f such that, in some ball centred at x, the derivatives $D_j f$, $D_i f$ exist and are differentiable at x. Then $D_{j\,i}f(x) = D_{i\,j}f(x)$.*

Proof. As in the preceding proof, we take $n = 2$, $i = 1$, $j = 2$ and denote x by (a,b), where $a, b \in \mathbb{R}$. Since the partial derivatives $D_1 f$ and $D_2 f$ are assumed differentiable at (a,b), there exist u_1 and u_2 such that, for $p = 1, 2$, we have

$$(D_p f)(a+h,b+k) - (D_p f)(a,b)$$
$$= h(D_{1p} f)(a,b) + k(D_{2p} f)(a,b) + (h^2 + k^2)^{\frac{1}{2}} u_p(h,k), \quad (1)$$

where

$$u_p(h,k) \to 0 \text{ as } (h,k) \to 0. \tag{2}$$

It is understood of course that (h,k) lies within a sufficiently small ball centred at $(0,0)$.

   In what follows, we work with $h = k \ne 0$. For each sufficiently small h, consider the functions

$$\psi(x) = f(x, b + h) - f(x, b) \text{ and } \phi(y) = f(a + h, y) - f(a, y).$$

By the mean value theorem and by definition of partial derivative (applied to f), we have

$$\psi(a + h) - \psi(a) = h \cdot \psi'(a + \theta h) \quad \text{for some } \theta \in (0,1)$$

$$= h \cdot [(D_1 f)(a + \theta h, b + h) - (D_1 f)(a + \theta h, b)]$$

$$= h \cdot [\theta h \cdot (D_{1\,1} f)(a,b) + h \cdot (D_{2\,1} f)(a,b)$$
$$+ |h| \cdot (1 + \theta^2)^{1/2} \cdot u_1(\theta h, h)$$
$$- \theta h \cdot (D_{1\,1} f)(a,b) - |h|\theta \cdot u_1(\theta h, 0)] \qquad \text{by (1)}$$

$$= h \cdot [h \cdot (D_{2\,1} f)(a,b) + |h| \cdot (1 + \theta^2)^{1/2} \cdot u_1(\theta h, h)$$
$$- |h|\theta \cdot u_1(\theta h, 0)]. \qquad (3)$$

A similar argument applied to ϕ leads to,

$$\phi(b+h) - \phi(b) = h \cdot [h \cdot (D_{1\,2} f)(a,b)$$
$$+ |h| \cdot (1+\theta'^2)^{1/2} \cdot u_2(h, \theta' h) - |h|\theta' \cdot u_2(0, \theta' h)]. \qquad (4)$$

But

$$\psi(a+h) - \psi(a) = f(a+h, b+h) - f(a+h, b) - f(a, b+h) + f(a,b) = \phi(b+h) - \phi(b).$$

Therefore, it follows from (3) and (4) that

$$(D_{2\,1} f)(a,b) + \frac{|h|}{h} (1 + \theta^2)^{1/2} \cdot u_1(\theta h, h) - \frac{|h|}{h} \theta \cdot u_1(\theta h, 0)$$

$$= (D_{1\,2} f)(a,b) + \frac{|h|}{h} (1 + \theta'^2)^{1/2} \cdot u_2(h, \theta' h) - \frac{|h|}{h} \theta' \cdot u_2(0, \theta' h).$$

Since this holds for all sufficiently small nonzero h, we may take the limit as $h \to 0$, which yields the desired equality in view of (2). $\qquad \square$

    A function f is said to be **twice continuously differentiable** on an open set if all the second partial derivatives $D_{i\,j} f_p$ of every component function f_p are continuous on the set. It is a consequence of either one of the preceding two theorems that $D_{j\,i} f_p = D_{i\,j} f_p$.

    We now prove a simple case of Taylor's theorem, which will be adequate for our purposes. It will be used in the proof of Theorem 5-2.1.

3-5.5. Proposition. *Let x be a point in the domain $S \subseteq \mathbb{R}^n$ of a real- valued function f such that the derivatives $D_j f$ ($1 \leq j \leq n$) exist and are differentiable everywhere in some open convex subset $B \subseteq S$ that contains x. Then for any h such that $x + h \in B$, there exists $\theta \in (0,1)$ such that*

$$f(x + h) = f(x) + \sum_{j=1}^n h_j \cdot D_j f(x) + \frac{1}{2} \sum_{j=1}^n h_j \left[\sum_{i=1}^n h_i D_{i\,j} f(x + \theta h) \right].$$

Proof. Since $D_j f$ are differentiable, they are continuous and hence f is also differentiable in B. Consider any $h \in \mathbb{R}^n$ with $x + h \in B$. By convexity of B, we have

$x + th \in B$ for all $t \in [0,1]$. Therefore, there is a function $\phi:[0,1] \to \mathbb{R}$ defined by $\phi(t) = f(x + th)$. It follows by the chain rule that it satisfies

$$\phi'(t) = \sum_{j=1}^{n} h_j \cdot D_j f(x + th) \quad \text{and} \quad \phi''(t) = \sum_{j=1}^{n} h_j [\sum_{i=1}^{n} h_i D_{ij} f(x + th)]. \tag{1}$$

By making use of Taylor's theorem for functions on an interval, we can assert that $\phi(1) = \phi(0) + \phi'(0) + \frac{1}{2}\phi''(\theta)$ for some $\theta \in (0,1)$. In view of (1), this is precisely the conclusion that was to be obtained. $\qquad\square$

We now derive the consequence that, if second partial derivatives of all component functions are bounded in a ball about a point x, then in Def. 3-2.1 of linear derivative, u satisfies the stronger condition that $\|u(h)\| \leq M\|h\|$, where M is some constant. We shall appeal to it only in Remark 4-2.5.

3-5.6. Corollary. *Let x be a point in the domain $S \subseteq \mathbb{R}^n$ of a function $f:S \to \mathbb{R}^m$ such that, in some open convex set $B \subseteq S$ that contains x, the mn derivatives $D_j f_k$ $(1 \leq j \leq n, 1 \leq k \leq m)$ exist and are differentiable everywhere in the entire ball. Suppose also that there exists $K > 0$ such that the n^2m second partial derivatives satisfy $|D_{ij}f_k| \leq K$ on B $(1 \leq i \leq n, 1 \leq j \leq n, 1 \leq k \leq m)$. Then for any h such that $x + h \in B$,*

$$\|f(x + h) - f(x) - f'(x)h\|_2 \leq \frac{1}{2} Knm^{\frac{1}{2}} \|h\|_2^2.$$

If either of the other two standard norms is used, then a similar inequality holds with some other constant on the right side.

Proof. By Proposition 3-5.5, there exist $\theta_k \in (0,1)$, $1 \leq k \leq m$, such that

$$f_k(x + h) = f_k(x) + \sum_{j=1}^{n} h_j \cdot D_j f_k(x) + \frac{1}{2} \sum_{j=1}^{n} h_j [\sum_{i=1}^{n} h_i D_{ij} f_k(x + \theta_k h)], \quad 1 \leq k \leq m.$$

The equality may be written as

$$|f_k(x + h) - f_k(x) - \sum_{j=1}^{n} h_j \cdot D_j f_k(x)| = |\frac{1}{2} \sum_{j=1}^{n} h_j [\sum_{i=1}^{n} h_i D_{ij} f_k(x + \theta_k h)]|.$$

From the inequality fulfilled by second partial derivatives and the Cauchy–Schwarz inequality, we now get

$$|f_k(x + h) - f_k(x) - \sum_{j=1}^{n} h_j \cdot D_j f_k(x)| \leq \frac{1}{2} K(\sum_{j=1}^{n} |h_j|)(\sum_{i=1}^{n} |h_i|) \leq \frac{1}{2} Kn\|h\|_2^2, \quad 1 \leq k \leq m.$$

In conjunction with Theorem 3-4.2, this leads to the desired inequality. The last part is an immediate consequence of Proposition 2-2.6. $\qquad\square$

Problem Set 3-5

3-5.P1. In the proof of Schwarz's theorem, can one infer the existence of the limit as $h \to 0$ of $D_{12}f(a + \theta'h, b + \theta k)$ from the continuity of $D_{12}f$?

3-5.P2. Show that it is possible to have $(D_{1\,2}f)(a,b) = (D_{2\,1}f)(a,b)$ even when $D_{1\,2}f$ and $D_{2\,1}f$ are not continuous at (a,b) by considering the function

$$f(x,y) = \begin{cases} x^2 y^2 / (x^2 + y^2) & \text{if } (x,y) \neq (0,0) \\ 0 & \text{if } (x,y) = (0,0). \end{cases}$$

3-5.P3. Show that it is possible to have $(D_{1\,2}f)(a,b) = (D_{2\,1}f)(a,b)$ even when $D_1 f$ is not differentiable at (a,b) by considering the function

$$f(x,y) = \begin{cases} x^2 y^2 / (x^2 + y^2) & \text{if } (x,y) \neq (0,0) \\ 0 & \text{if } (x,y) = (0,0). \end{cases}$$

3-5.P4. Let f and g be real-valued functions defined on $\mathbb{R}$, possessing continuous second derivatives f'' and g''. Define

$$F(x,y) = f(x + g(y)), \qquad (x,y) \in \mathbb{R}^2.$$

Determine $D_1 F$, $D_2 F$, $D_{1\,2} F$ and $D_{1\,1} F$ and show that $(D_1 F)(D_{1\,2} F) = (D_2 F)(D_{1\,1} F)$.

3-5.P5. If F is homogeneous of degree p and its partial derivatives are differentiable, then show that the equation

$$x^2 (D_{1\,1} F) + 2xy(D_{1\,2} F) + y^2 (D_{2\,2} F) = p(p-1)F$$

is valid.

3-5.P6. A weaker version of Schwarz's Theorem 3-5.3 is the following: Let x be a point in the domain $S \subseteq \mathbb{R}^n$ of a real-valued function f such that, in some ball centred at x, the derivatives $D_{i\,j}f$ and $D_{j\,i}f$ both exist and are continuous at x. Then $D_{j\,i}f(x) = D_{i\,j}f(x)$. Prove this weaker version directly without using Schwarz's Theorem.

3-5.P7. Prove that $D_{2\,1}f(0,0) \neq D_{1\,2}f(0,0)$ for the function on $\mathbb{R}^2$ such that $f(0,0) = 0$ and $f(x,y) = x^2 \arctan \frac{y}{x} - y^2 \arctan \frac{x}{y}$ for $(x,y) \neq (0,0)$. Here $u^2 \arctan \frac{v}{u}$ is understood to mean 0 when $u = 0 \neq v$.

3-5.P8. Let f be a homogeneous function of degree n with continuous second partial derivatives on an open subset of $\mathbb{R}^3$ and suppose (x,y,z) is a point where $z \neq 0$ and the determinants H and A of the respective matrices

$$\begin{bmatrix} D_{11}f & D_{12}f & D_{13}f \\ D_{21}f & D_{22}f & D_{23}f \\ D_{31}f & D_{32}f & D_{33}f \end{bmatrix} \quad \text{and} \quad \begin{bmatrix} D_{11}f & D_{12}f & D_1 f \\ D_{21}f & D_{22}f & D_2 f \\ D_1 f & D_2 f & \frac{n}{n-1} f \end{bmatrix}$$

are also nonzero. Show that $H = \dfrac{(n-1)^2}{z^2} A.$

3-5.P9. Let B be the determinant of the matrix

$$\begin{bmatrix} D_{11}f & D_{12}f & D_1 f \\ D_{21}f & D_{22}f & D_2 f \\ D_{31}f & D_{32}f & D_3 f \end{bmatrix},$$

where f is a homogeneous function with continuous second partial derivatives on an open subset of $\mathbb{R}^3$ and let H and A be the determinants as in 3-5.P8. Show that $B^2 = AH$ (although nothing is guaranteed to be nonzero!).

3-5.P10. Suppose the functions x, y, z of 3-4.P14 have continuous second partial derivatives. Then show that

$$J(x, J(y, z)) + J(y, J(z, x)) + J(z, J(x, y)) = 0.$$

3-5.P11. Let $y = y(x)$ be a twice continuously differentiable function satisfying $F(x, y) = 0$, where F has continuous second partial derivatives. Prove that, if $F_y \neq 0$, then

$$F_y^3 y'' = \det \begin{bmatrix} F_{xx} & F_{xy} & F_x \\ F_{xy} & F_{yy} & F_y \\ F_x & F_y & 0 \end{bmatrix}.$$

3-5.P12. Let f be a continuous function on the cuboid $[a,b] \times [c,d]$ in $\mathbb{R}^2$. For each interior point (x, y) of the cuboid, define

$$F(x, y) = \int_a^x \left(\int_c^y f(s, t)\, dt \right) ds.$$

Show that $D_{1\,2}\, F(x, y) = D_{2\,1}\, F(x, y) = f(x, y)$.

3-5.P13. Let x be a point in the domain $S \subseteq \mathbb{R}^n$ of a function $f : S \to \mathbb{R}^m$ such that, in some open convex subset $B \subseteq S$ that contains x, the mn derivatives $D_j f_k$ $(1 \leq j \leq n, 1 \leq k \leq m)$ exist and satisfy the following Lipschitz condition:

$$|D_j f_k(x + h) - D_j f_k(x)| \leq L \|h\|_2 \quad \text{whenever} \quad x + h \in B.$$

Then show that

$$\|f(x + h) - f(x) - f'(x)h\|_2 \leq L n^{1/2} m^{1/2} \|h\|_2^2 \quad \text{whenever} \quad x + h \in B.$$

4

Inverse and Implicit Function Theorems

4-1 Contraction Mapping Theorem

So far we have been concerned with maps from an open subset of $\mathbb{R}^n$ into $\mathbb{R}^m$. Soon we shall be considering maps from a set that is a subset of $\mathbb{R}^n$ into that very set, what are often called **self maps** of a set. For example, the map $T:[0, 1] \to [0, 1]$ given by $Tx = 1 - x$ is a self map. A trivial example would be the identity map T given by $Tx = x$ on *any* set X whatsoever. What we shall need is a property of a special kind of self maps called contractions or contraction maps of a closed subset of $\mathbb{R}^n$ (Theorem 4-1.6 below). Before proceeding to the theorem, we illustrate the ideas involved.

To begin with, we give some examples of self maps.

4-1.1. Examples. (a) $X = [0, 1]$, $Tx = 1 - x^p$, p some positive integer.

(b) $X = \{x \in \mathbb{R} : x \neq 0\}$, $Tx = x + (1/x)$.

(c) $X = \{x \in \mathbb{R} : x > 1\}$, $Tx = x + (1/x)$.

(d) $X = [3, 5]$, $Tx =$ integer part of x.

(e) $X = [1, 2]$, $Tx = x - (x^3 - 2)/16$.

It is a simple matter to verify that the maps described in (a)–(d) above are self maps. That (e) also describes a self map can be checked as follows: $T'(x) = 1 - \frac{3x^2}{16}$ is positive for $x \in [1, 2]$ and so $T(x)$ is an increasing function of x. Therefore, $\frac{17}{16} \leq T(x) \leq \frac{13}{8}$.

4-1.2. Definition. *A **fixed point** of a map $T:X \to Y$ is an element $x \in X$ such that $Tx = x$. (Obviously, any such x, if it exists, must also belong to Y.)*

4-1.3. Examples. For the identity map $Tx = x$, obviously every x in the domain is a fixed point. For the examples in 4-1.1, taken in reverse order, it is easily checked that fixed points are respectively $2^{1/3}$, $\{3, 4, 5\}$, none, none, and any root of the equation $x^p + x - 1 = 0$ that may belong to $[0, 1]$. Such a root exists; indeed, with $f(x) = x^p + x - 1$, we have $f(0) = -1$, $f(1) = 1$, so that an application of the intermediate value theorem yields the conclusion.

4-1.4. Definition. *When $X \subseteq \mathbb{R}^n$, a map $T:X \to X$ is called a **contraction** (or **contraction mapping, contraction map, shrinking map**) in X if there exists some $c \in [0, 1)$ such that*

S. Shirali, H.L. Vasudeva, *Multivariable Analysis*,
DOI 10.1007/978-0-85729-192-9_4, © Springer-Verlag London Limited 2011

$$\|Tx - Ty\| \le c\|x - y\| \quad x, y \in X.$$

Although the constant c cannot be unique, it is convenient to refer to it as *the contraction constant*. The same map can be a contraction in the sense of one norm in $\mathbb{R}^n$ but not another [see Example (c) below].

As the reader can easily see, every contraction is uniformly continuous.

Examples. (a) A constant self map is clearly a contraction, with any number c between 0 and 1 serving as the contraction constant.

(b) Let $b \in \mathbb{R}^n$ and $T:\mathbb{R}^n \to \mathbb{R}^n$ be defined by $Tx = \frac{1}{2}x + b$. Then T is a contraction with any number between $\frac{1}{2}$ and 1 serving as the contraction constant.

(c) A linear map $T:\mathbb{R}^n \to \mathbb{R}^n$ is a contraction if and only if for some $c \in [0,1)$ we have $\|Tx\| \le c\|x\|$ for all $x \in \mathbb{R}^n$, because $Tx - Ty = T(x-y)$. Take $n = 2$ and consider the map $T:\mathbb{R}^2 \to \mathbb{R}^2$ defined by

$$Tx = T(x_1, x_2) = (\frac{1}{2}x_1 + \frac{1}{4}x_2, \; \frac{2}{3}x_1 + \frac{1}{6}x_2).$$

We shall show that $\|Tx\|_\infty \le \frac{5}{6}\|x\|_\infty$ but there exists $x \in \mathbb{R}^2$ such that $\|Tx\|_1 > \|x\|_1$. The former inequality follows from the observation that

$$|\frac{1}{2}x_1 + \frac{1}{4}x_2| \le \frac{1}{2}|x_1| + \frac{1}{4}|x_2| \le \frac{1}{2}\|x\|_\infty + \frac{1}{4}\|x\|_\infty = \frac{3}{4}\|x\|_\infty$$

and

$$|\frac{2}{3}x_1 + \frac{1}{6}x_2| \le \frac{2}{3}|x_1| + \frac{1}{6}|x_2| \le \frac{2}{3}\|x\|_\infty + \frac{1}{6}\|x\|_\infty = \frac{5}{6}\|x\|_\infty.$$

For the other inequality, choose $x = (x_1, x_2) = (1,0)$; then $\|x\|_1 = 1$ and $\|Tx\|_1 = \frac{1}{2} + \frac{2}{3} = \frac{7}{6} > 1 = \|x\|_1$.

(d) Among the self maps illustrated in Examples 4-1.1, the last one is a contraction. This follows by applying the mean value theorem and noting that $0 < 1 - 3x^2/16 \le 13/16$ when $x \in [1,2]$; so we may take $c = 13/16$. In fact, the following general result holds.

4-1.5. Proposition. *Let I be an interval and $f:I \to I$ be differentiable. Assume that there exists a constant $K < 1$ such that $|f'(z)| \le K$ for all $z \in I$. Then f is a contraction.*

Proof. If $x, y \in I$, $x < y$, then $(f(x) - f(y))/(x - y) = f'(c)$, where $x < c < y$. Since $|f'(z)| \le K$ for all $z \in I$, it follows that $|f(x) - f(y)| \le K|x - y|$. $\square$

As another illustration of this result, we consider $Tx = \frac{1}{2}(x + \frac{2}{x})$, $x \in [1, \infty)$. Since $x + \frac{2}{x} \ge x + \frac{1}{x} \ge 2$, we see that T is a self map. Moreover, $|T'(x)| = |\frac{1}{2} - 1/x^2| \le \frac{1}{2}$. So, by the above proposition, T is a contraction.

The first four self maps listed in Examples 4-1.1 are not contractions: in (a), we have $|T0 - T1| = 1 = |0 - 1|$, while in (b), we have $|T(-1) - T(1)| = |(-2) - (2)| = 4$ whereas $|(-1) - (1)| = 2$. In (c),

$$x > y > 1 \Rightarrow |Tx - Ty| = |(x - y) + (1/x - 1/y)| = (x - y)(1 - 1/xy)$$

and therefore $|Tx - Ty|/(x - y)$ can be as close to 1 as desired. In (d), the map T is not even continuous.

We now come to the main theorem in $\mathbb{R}^n$ about contraction mappings. In a more general context than $\mathbb{R}^n$, the result is variously known as **contraction mapping theorem**, **shrinking lemma**, **Banach–Cacciopoli principle**, **contraction principle** and so forth. We shall refer to it as the *contraction principle*, although we restrict attention to $\mathbb{R}^n$.

4-1.6. Contraction Principle (in $\mathbb{R}^n$). *If T is a contraction map in a closed subset $X \subseteq \mathbb{R}^n$, then T has a unique fixed point.*

Proof. Since T is a contraction map, there exists $c \in [0, 1)$ such that

$$\|Tx - Ty\| \le c\|x - y\| \quad \text{whenever} \quad x, y \in X. \tag{1}$$

Now $|c| = c < 1$, and hence $c^p \to 0$ as $p \to \infty$.

Uniqueness is easily seen as follows. If $Tx_0 = x_0$ and also $Ty_0 = y_0$, then by (1), $\|x_0 - y_0\| \le c\|x_0 - y_0\|$. But $c < 1$. Hence $\|x_0 - y_0\| = 0$ and $x_0 = y_0$.

To prove existence, take any element $x_1 \in X$. Define a sequence in X inductively by setting $x_{p+1} = Tx_p$. For $p = 1$, it is trivial that

$$\|x_{p+1} - x_p\| \le c^{p-1}\|x_2 - x_1\|. \tag{2}$$

because the two sides are equal. Assume (2) holds for some $p \in \mathbb{N}$. Then

$$\|x_{p+2} - x_{p+1}\| = \|Tx_{p+1} - Tx_p\| \le c\|x_{p+1} - x_p\| \le c(c^{p-1}\|x_2 - x_1\|) = c^p\|x_2 - x_1\|.$$

Therefore (2) holds for *all* $p \in \mathbb{N}$.

Now, for $q > p$, we have

$$\|x_q - x_p\| \le \|x_q - x_{q-1}\| + \|x_{q-1} - x_{q-2}\| + \cdots + \|x_{p+1} - x_p\|$$

$$\le (c^{q-2} + c^{q-3} + \cdots + c^{p-1})\|x_2 - x_1\| \qquad \text{by (2)}$$

$$\le (c^{p-1} + c^p + c^{p+1} + \cdots)\|x_2 - x_1\|$$

$$= [c^{p-1}/(1-c)]\|x_2 - x_1\|.$$

Since $c^{p-1}/(1-c)$ tends to 0 as p tends to infinity, it follows that $\{x_p\}$ is a Cauchy sequence. By Theorem 2-2.10, the sequence must converge to some limit, say x_0, and by Proposition 2-4.5, $x_0 \in X$. Then for any $p \in \mathbb{N}$, we have

$$\|x_0 - Tx_0\| \le \|x_0 - x_p\| + \|x_p - Tx_p\| + \|Tx_p - Tx_0\|$$

$$\le \|x_0 - x_p\| + \|x_p - x_{p+1}\| + c\|x_p - x_0\|$$

$$\le \|x_0 - x_p\| + c^{p-1}\|x_2 - x_1\| + c\|x_p - x_0\| \quad \text{by (2)}.$$

Since $\|x_0 - x_p\| \to 0$ and $c^{p-1} \to 0$ as p tends to infinity, it follows that $\|x_0 - Tx_0\| = 0$. Therefore $Tx_0 = x_0$.   □

The proof of the theorem not only guarantees the existence of a fixed point but also provides a procedure for approximating it, because the sequence $\{x_p\}$ can be explicitly computed in terms of T and a chosen x_0. In the case of the contraction map T in $[1, 2]$ given by $Tx = x - (x^3 - 2)/16$, which was mentioned above, the sequence would be

$$1, 1 - (1^3 - 2)/16 = 17/16, (17/16) + [(17/16)^3 - 2]/16 = 66353/65536, \dots .$$

The first term could have been taken as any element of $[1, 2]$ instead of 1; the resulting sequence would still converge to the unique fixed point, which is $2^{1/3}$, as already observed.

Similarly, the self map of $[1, \infty)$ given by $Tx = \frac{1}{2}(x + \frac{2}{x})$, which has already been shown to be a contraction, is now seen to have a unique fixed point. It is easy to verify by an independent computation that the fixed point is $\sqrt{2}$.

The following corollary gives an estimate of the distance between x_p and x_0.

4-1.7. Corollary. *Let $T: X \to X$ be a contraction map in a closed subset $X \subseteq \mathbb{R}^n$ and*

$$\|Tx - Ty\| \le c\|x - y\| \qquad x, y \in X, \text{ where } c \in [0, 1).$$

If $x_1 \in X$ and $\{x_p\}_{p \ge 1}$ is the sequence defined inductively by $x_{p+1} = Tx_p$, then

$$\|x_0 - x_p\| \le [c^{p-1}/(1 - c)]\|x_2 - x_1\|, \text{ where } x_0 \text{ is the fixed point of } T.$$

Proof. In the proof of Contraction Principle 4-1.6, it was shown for $q > p$ that $\|x_q - x_p\| \le [c^{p-1}/(1 - c)]\|x_2 - x_1\|$. The inequality in question follows upon taking the limit as $q \to \infty$.   □

Examples. (a) In the case of the contraction map T in $[1, 2]$ given by $Tx = x - (x^3 - 2)/16$, we have $0 < T'(x) = 1 - 3x^2/16 \le 13/16$; so we may take $c = 13/16$. Then with $x_1 = 1$, we have

$$\|x_0 - x_p\| \le [c^{p-1}/(1-c)]\|x_2 - x_1\| \le [(\frac{13}{16})^{p-1}/(1 - \frac{13}{16})]\|x_2 - x_1\|$$

$$= \frac{16}{3}(\frac{13}{16})^{p-1}(\frac{17}{16} - 1) = \frac{1}{3}(\frac{13}{16})^{p-1}.$$

(b) For the contraction map T in $[1,\infty)$ given by $Tx = \frac{1}{2}(x + \frac{2}{x})$, we have $|T'(x)| = |\frac{1}{2} - 1/x^2| \le \frac{1}{2}$; so we may take $c = \frac{1}{2}$. Then with $x_1 = 1$, we have

$$\|x_0 - x_p\| \le \frac{c^{p-1}}{1-c}\|x_2 - x_1\| \le \frac{(\frac{1}{2})^{p-1}}{1 - \frac{1}{2}}\|x_2 - x_1\|$$

$$= 2\left(\frac{1}{2}\right)^{p-1}\left(\frac{3}{2} - 1\right) = \left(\frac{1}{2}\right)^{p-1}.$$

For the purpose of the inverse function theorem however, what matters is only the existence of a unique fixed point for every contraction map in a closed subset of $\mathbb{R}^n$.

We now present a generalisation of the contraction principle, which is useful in some situations; however, we shall have no occasion to use it later in this book and the reader may wish to omit it.

4-1.8. Corollary. *Let X be a closed subset $X \subseteq \mathbb{R}^n$ and $T:X \to X$ be a self map such that T^k is a contraction in X for some positive integer k. Then T has a unique fixed point.* (Note that T is not even assumed continuous!)

Proof. By the contraction principle, T^k has a unique fixed point; denote it by x_0. Then $T^k x_0 = x_0$ and hence $T^k(Tx_0) = T(T^k x_0) = Tx_0$, which means that Tx_0 is also a fixed point of T^k. But T^k has a *unique* fixed point and therefore $Tx_0 = x_0$. Thus x_0 is also a fixed point of T. Since any fixed point of T is also a fixed point of T^k, then T cannot have another fixed point. $\qquad\square$

Example. Under the hypotheses of the above corollary it can happen that T is not continuous. Indeed, let $T:\mathbb{R} \to \mathbb{R}$ be defined by $Tx = 0$ or 1 according as x is rational or irrational. Then $T^2(x) = 0$ for all $x \in \mathbb{R}$, which ensures that T^2 is a contraction. But T is discontinuous everywhere.

Problem Set 4-1

4-1.P1. Show that Contraction Principle 4-1.6 is not valid if the subset X of $\mathbb{R}^n$ is not closed. Determine the fixed points, if any, of the following maps:

(i) $T:\mathbb{R} \to \mathbb{R}$ defined by $T(x) = x^2$;
(ii) $T:\mathbb{R} \to \mathbb{R}$ defined by $T(x) = x + \alpha$;

(iii) $T:\mathbb{R}^2\to\mathbb{R}$ defined by $T(x,y) = x$.

4-1.P2. Show that if a self map $T:X\to X$ has the property that T^3 has a unique fixed point, then T has the same property.

4-1.P3. (a) Show that if a self map $T:X\to X\subseteq\mathbb{R}^n$ satisfies the condition that

$$\|Tx - Ty\| < \|x - y\| \text{ whenever } x \neq y,$$

then it has at most one fixed point.

(b) If the self map $T:X\to X\subseteq\mathbb{R}^n$ satisfies

$$\|Tx - Ty\| \leq \tfrac{1}{2}\|x - y\| \text{ for all } x,y\in X,$$

show that it has at most one fixed point.

4-1.P4. (a) Show that the map $T:[1,\infty)\to[1,\infty)$ such that $Tx = x + 1/x$ satisfies

$$\|Tx - Ty\| < \|x - y\| \text{ whenever } x \neq y,$$

but has no fixed point.

(b) Show that the map $f:\mathbb{R}\to\mathbb{R}$ such that $f(x) = x + (1 + e^x)^{-1}$ satisfies $0 < |f'(x)| < 1$ everywhere but has no fixed point.

4-1.P5. Let $T:X\to X$ be a self map of a set $X\subseteq\mathbb{R}^n$ and suppose

$$\|Tx - Ty\| < \|x - y\| \text{ whenever } x \neq y.$$

If X is compact, show that T has exactly one fixed point.

4-1.P6. Show that the map $T:[1,2]\to\mathbb{R}$ such that $Tx = x - (x^7 - 6)/500$ is a contraction map in $[1,2]$. What is the limit of the sequence

$$1, T(1), T(T(1)), T(T(T(1))), \dots ?$$

4-1.P7. Define $T:[0,3]\to[0,3]$ as $Tx = 1$ if $0 \leq x \leq 2$ and $Tx = 2$ if $2 < x \leq 3$. Find T^2. (This provides an example of a discontinuous self map with a unique fixed point.)

4-1.P8. Let $X\subseteq\mathbb{R}^n$ be closed and $T:X\to X$ be a self map. Assume that there is a real sequence $\{\alpha_n\}$ and a positive integer N such that

(i) $\|T^n x - T^n y\| \leq \alpha_n\|x - y\|$ for all $x,y\in X$ and all $n \geq N$; and

(ii) $\{\alpha_n\}$ has a subsequence converging to a limit < 1.

Show that T has a unique fixed point.

4-1.P9. For the self map $f:\mathbb{R}\to\mathbb{R}$ given by $f(x) = \tfrac{1}{3}(1 + x^3)$, show that

(a) there are three fixed points $u < v < w$;

(b) f maps the open interval (u,w) onto itself;

(c) for any $x\in(u,w)$, the sequence $f^n(x)$ converges to v;

(d) f is not a contraction map in (u,w).

4-1.P10. For the self map $f:\mathbb{R}\to\mathbb{R}$ given by $f(x) = 1/(x^4 + 1)$, it can be shown by direct computation that (i) $f^2(\frac{3}{4}) = f(f(\frac{3}{4})) > \frac{3}{4}$; (ii) $f(\frac{3}{4}) = \frac{256}{337} < \frac{16}{21}$; and (iii) $|f'(\frac{16}{21})| < 1$. Show that

(a) f has a unique fixed point $\alpha\in\mathbb{R}$ and that $\alpha\in[\frac{3}{4},1]$;

(b) $0 < x < \alpha < y \Rightarrow 0 < f(y) < \alpha < f(x)$;

(c) $\frac{3}{4} < f(\frac{3}{4})$ and f is a contraction in the interval $[\frac{3}{4}, f(\frac{3}{4})]$.

4-1.P11. Let S be a compact subset of $\mathbb{R}^n$ such that $0 < r < 1$, $x\in S \Rightarrow rx\in S$. If $T:S\to S$ satisfies $\|Tx - Ty\| \le \|x - y\|$ $\forall$ $x,y\in S$, show that T has a fixed point. Give an example to show that the fixed point need not be unique.

4-1.P12. Let T be a contraction map in a closed subset X of $\mathbb{R}^n$. For the case when X is also bounded, use Theorem 2-5.7 and Theorem 2-6.13 to show that T has a fixed point. Then extend to the case when X is unbounded. Finally, use the existence of a fixed point x_0 to show for any element $x_1\in X$ that the sequence in X defined inductively by setting $x_{p+1} = Tx_p$ converges to x_0. This proof due to Drager and Foote [9].

4-2 Inverse Function Theorem

Consider the question of expressing x in terms of y from $y = x^2 + 1$. This is asking for the inverse of the function $f:x\to x^2 + 1$. Since f is not injective, there is no inverse and one has to make do with either $x = g(y) = (y-1)^{\frac{1}{2}}$ or $x = g(y) = -(y-1)^{\frac{1}{2}}$, depending on whether one wants $x\ge 0$ or $x\le 0$. In other words, with x restricted to either one of two suitable subsets of the domain of f, one can get a continuous inverse for the restricted function. Unless a restriction is imposed on x, there can be no inverse, since f is not injective.

One thing that the inverse function theorem does is to provide a sufficient condition for such a restriction to be possible when the domain and range of f are both subsets of $\mathbb{R}^n$, which is to say $x,y\in\mathbb{R}^n$. Since we shall consider functions that are differentiable in the sense of Def. 3-2.1, the subset to which x is restricted must be open.

It may be noted in passing that discontinuous inverses are also possible: Take $g(y) = (y-1)^{\frac{1}{2}}$ for rational y and $g(y) = -(y-1)^{\frac{1}{2}}$ for irrational y, the restriction on x being that it should belong to the range of this discontinuous function, whatever it may be!

A familiar two-dimensional situation occurs when one wants to express the polar coordinates (r,θ) of a point in $\mathbb{R}^2$, other than the origin, in terms of its rectangular coordinates (x,y):

$$x = f_1(r,\theta) = r\cos\theta \,,\; y = f_2(r,\theta) = r\sin\theta \,,\; r \neq 0.$$

One cannot blithely take $\theta = \tan^{-1}(y/x)$, because this will restrict θ to lying between $-\pi/2$ and $\pi/2$. If one takes $r > 0$ and $-\pi < \theta \leq \pi$, then f becomes injective and an inverse g for the restricted f can be obtained, albeit a discontinuous one. Details are left to the reader in 4-2.P16. With a further restriction on (r,θ), such as $-\pi < \theta < \pi$, the inverse described therein can be shown to be continuous. Of course, any stronger restriction will also do, because if a function is injective on a subset of its domain, then it is injective on any nonempty subset of that subset.

For the function $f(x) = x^2 + 1$ discussed above, both instances of the restriction imposed on x in order to make the function injective were special for that function. If such a restriction were wanted for the function $f:\mathbb{R}^2 \to \mathbb{R}^2$ given by

$$y_1 = f_1(x_1,x_2) = x_1^2 + \cos x_2, \quad y_2 = f_2(x_1,x_2) = x_1 x_2,$$

it would not be easy to find one, or even to check whether it is possible in the first place.

Returning to the function $f(x) = x^2 + 1$, recall that there can be two different open subsets to which x can be restricted. When one restricts x to the open set $x > 0$, the inverse $g(y) = (y-1)^{\frac{1}{2}}$ 'accommodates' the value $x = 2$ and also any other value $x = a > 0$. Similarly, when one restricts x to the open set $x < 0$, the inverse accommodates any $a < 0$.

Thus the theorem is about a given a in the domain, which is to be accommodated in an open set U on which f is to be injective.

4-2.1. Inverse Function Theorem. *Let E be a subset of $\mathbb{R}^n$ and $f:E \to \mathbb{R}^n$ be continuously differentiable on E. Let $a \in E$ and $f'(a)$ be invertible. Then there exist subsets $U \subseteq E$, $V \subseteq \mathbb{R}^n$ such that*

(a) *U and V are open, $a \in U$, $f(U) = V$, f is injective on U and $f'(x)$ is invertible whenever $x \in U$; moreover,*

(b) *the inverse map $g:V \to U$ is continuously differentiable and satisfies*

$$g'(y) = f'(g(y))^{-1} \quad \text{whenever } y \in V$$

or equivalently,

$$g'(f(x)) = f'(x)^{-1} \quad \text{whenever } x \in U.$$

Proof. (a) Denote the linear map $f'(a)$ by A. Since f' is continuous at a, there exists an open ball $U \subseteq E$ centred at a such that

$$x \in U \Rightarrow \|f'(x) - A\| < \frac{1}{2\|A^{-1}\|} \Rightarrow \|A^{-1}\| \cdot \|f'(x) - A\| < \frac{1}{2}. \qquad (1)$$

Then $f'(x)$ is invertible whenever $x \in U$ (by Theorem 2-7.11).

For any $y \in \mathbb{R}^n$ and $x \in U$, define

$$\phi_y(x) = x + A^{-1}(y - f(x)) \qquad (2a)$$
$$= A^{-1}(Ax + y - f(x)). \qquad (2b)$$

Then by (2a),

$$x \text{ is a fixed point of } \phi_y \Leftrightarrow y = f(x). \qquad (3)$$

By (2b) and Remark 3-3.8, $\phi_y'(x) = A^{-1}(A - f'(x))$. Together with (1), this equality implies that $\|\phi_y'(x)\| < \frac{1}{2}$ for $x \in U$. It follows by Corollary 3-3.4 that

$$\|\phi_y(x_1) - \phi_y(x_2)\| \le \frac{1}{2}\|x_1 - x_2\| \qquad \forall\, x_1, x_2 \in U \ldots\ldots\ldots\ldots\ldots(4).$$

(Caution: This is not enough to guarantee that we have a contraction map, because we do not yet have a set that it maps into itself! We shall arrange for that later.) The inequality (4) shows that ϕ_y can have at most one fixed point, so that by (3), $f(x) = y$ for at most one $x \in U$. This further implies that f is injective on U. Let $V = f(U)$. Then f maps U injectively onto V, and $a \in U$. Therefore, there exists an inverse map $g: V \rightarrow U$, as illustrated in the accompanying figure. In order to complete the proof of part (a), it now remains only to show that V is open.

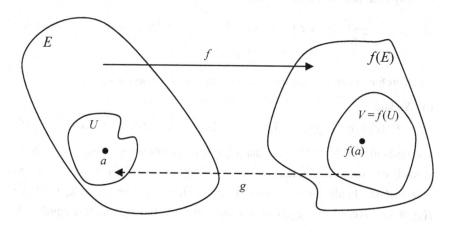

Consider an arbitrary $y_0 \in V$. Then $y_0 = f(x_0)$ for some $x_0 \in U$. Let $\varepsilon > 0$ be arbitrary but small enough to ensure that $\|x - x_0\| \le \varepsilon \Rightarrow x \in U$, so that the closed ball B of radius ε centred at x_0 is a subset of U; i.e.,

$$B = \{x \in \mathbb{R}^n : \|x - x_0\| \le \varepsilon\} \subseteq U.$$

We claim that the open ball of radius $\varepsilon/2\|A^{-1}\|$ centred at y_0 is contained in V; in other words,

$$y \in \mathbb{R}^n, \ \|y - y_0\| < \frac{\varepsilon}{2\|A^{-1}\|} \ \Rightarrow \ y \in f(U).$$

In order to prove this, let $\|y - y_0\| < \varepsilon/2\|A^{-1}\|$. Then for $x \in B$, we have

$$\|\phi_y(x) - x_0\| \leq \|\phi_y(x) - \phi_y(x_0)\| + \|\phi_y(x_0) - x_0\|$$

$$\leq \tfrac{1}{2}\|x - x_0\| + \|A^{-1}(y - f(x_0))\|, \text{ by (4) and (2a)}$$

$$\leq \frac{\varepsilon}{2} + \|A^{-1}\|\|y - y_0\|$$

$$\leq \frac{\varepsilon}{2} + \|A^{-1}\| \frac{\varepsilon}{2\|A^{-1}\|}$$

$$\leq \varepsilon,$$

so that $\phi_y(x) \in B$. Consequently, ϕ_y maps B into itself. Now by (4), ϕ_y is a contraction in B, which is a closed subset of $\mathbb{R}^n$. By the contraction principle in $\mathbb{R}^n$, ϕ_y has a fixed point in $B \subseteq U$. Therefore by (3), $y \in f(B) \subseteq f(U)$. This completes the argument that V is open, and (a) is established. Observe that a little more has actually been proved, namely, that for any $\varepsilon > 0$,

$$\|y - y_0\| < \frac{\varepsilon}{2\|A^{-1}\|}, \ y \in V \Rightarrow y \in f(B) \Rightarrow y = f(x) \text{ for some } x \in B \Rightarrow g(y) \in B$$

$$\Rightarrow \ \|g(y) - x_0\| \leq \varepsilon,$$

and therefore, continuity of the inverse map g has also been established.

(b) We know

$$f(g(y) + h) - f(g(y)) = f'(g(y))(h) + \|h\|u(h) \text{ for sufficiently small } h \qquad (5)$$

and $u(h) \to 0$ as $\|h\| \to 0$. Observe that $f'(g(y))$ is invertible because $g(y) \in U$. Now g has been proved continuous and we can therefore take k small enough to ensure that (5) holds for $h = g(y + k) - g(y)$. Then for every such k, we have $f(g(y + k)) - f(g(y)) = f'(g(y))(h) + \|h\|u(h)$. But the left side of this equality is just k; so $k = f'(g(y))(h) + \|h\|u(h)$. Applying $f'(g(y))^{-1}$ to both sides, we obtain

$$h = f'(g(y))^{-1}k - \|h\|f'(g(y))^{-1}(u(h)), \qquad (6)$$

which is the same as

$$g(y + k) - g(y) = f'(g(y))^{-1}k - \|h\|f'(g(y))^{-1}(u(h)).$$

Therefore, in order to arrive at the existence of $g'(y)$ and the equality $g'(y) = f'(g(y))^{-1}$, we need only show that $\frac{\|h\|}{\|k\|} f'(g(y))^{-1}(u(h)) \to 0$ as $\|k\| \to 0$. By continuity of g, we have $\|h\| \to 0$ as $\|k\| \to 0$. So it is sufficient to prove that $\frac{\|h\|}{\|k\|}$ remains bounded. Now (6) implies

$$\|h\| \leq \|f'(g(y))^{-1}\| \|k\| + \|h\| \|f'(g(y))^{-1}\| \|u(h)\|,$$

which further implies

$$\|h\|(1 - \|f'(g(y))^{-1}\| \|u(h)\|) \leq \|f'(g(y))^{-1}\| \|k\|.$$

Since $u(h) \to 0$ as $\|h\| \to 0$, it follows that $\frac{\|h\|}{\|k\|}$ remains bounded. We have thus proved, for all $y \in V$, the existence of $g'(y)$ and the equality $g'(y) = f'(g(y))^{-1}$.

Now, the equality shows that g' is the composition of g, f', and inversion of linear maps. All these are continuous maps (inversion is continuous by Theorem 2-7.11) and therefore g' is continuous. This establishes (b).   □

The conclusion of the above theorem is often summarised as 'f is locally invertible at a with a continuously differentiable local inverse' or 'f has a continuously differentiable local inverse at a'. The term 'local inverse' here refers to the function g.

It is possible to prove the inverse function theorem without using the contraction principle. Such proofs usually use the compactness of a closed ball in $\mathbb{R}^n$ (see Apostol [1]). However, proofs using compactness of a closed ball cannot be extended to the situation when $\mathbb{R}^n$ is replaced by an infinite-dimensional space, but the above proof can be (see Lang [17] or Brown and Page [4]). The treatment given above most closely resembles that of Rudin [22].

We shall now apply the above theorem to the examples discussed at the beginning of this section.

4-2.2. Examples. (a) For the function f defined on $\mathbb{R}$ by $f(x) = x^2 + 1$, the derivative is $f'(x) = 2x$, which is nonzero when $x \neq 0$. Therefore, by the inverse function theorem, every $a \neq 0$ lies in some open set U on which f is injective. Besides, f maps U onto an open set V and the inverse $g: V \to U$ is differentiable. Also, $g'(y) = (f'(g(y))^{-1} = 1/2g(y)$ $\forall$ $y \in V$. Thus f is locally invertible at every nonzero $a \in \mathbb{R}$ with a continuously differentiable local inverse.

(b) For the function f defined on $\mathscr{P} = \{(x_1, x_2) \in \mathbb{R}^2 : x_1 \neq 0\}$ by

$$y_1 = f_1(x_1, x_2) = x_1 \cos x_2, \quad y_2 = f_2(x_1, x_2) = x_1 \sin x_2, \quad x_1 \neq 0,$$

the Jacobian matrix is

$$\begin{bmatrix} \cos x_2 & -x_1 \sin x_2 \\ \sin x_2 & x_1 \cos x_2 \end{bmatrix}.$$

This is invertible when its determinant, which equals x_1, is nonzero. Therefore, by the inverse function theorem, every point of $\mathscr{P}$ lies in an open subset U of $\mathscr{P}$, on which f is injective. Besides, f maps U onto an open set V and the inverse map $g : V \to U$ is differentiable. Thus f has a continuously differentiable local inverse at every point of $\mathscr{P}$.

(c) For the function f defined on $\mathbb{R}^2$ by

$$f_1(x_1, x_2) = x_1^2 + \cos x_2, \quad f_2(x_1, x_2) = x_1 x_2 \quad \forall\, (x_1, x_2) \in \mathbb{R}^2,$$

the Jacobian matrix is

$$\begin{bmatrix} 2x_1 & -\sin x_2 \\ x_2 & x_1 \end{bmatrix}.$$

This is invertible when its determinant, which is equal to $2x_1^2 + x_2 \sin x_2$, is non-zero. Therefore, by the inverse function theorem, every point (a_1, a_2) of $\mathbb{R}^2$ for which $2a_1^2 + a_2 \sin a_2 \neq 0$ lies in an open subset U of $\mathbb{R}^2$ on which f is injective. Besides, f maps U onto V and the inverse function $g : V \to U$ is differentiable. Thus, f is locally invertible at every point (a_1, a_2) of $\mathbb{R}^2$ for which $2a_1^2 + a_2 \sin a_2 \neq 0$ and the local inverse is continuously differentiable.

(d) For the function f defined on $\mathbb{R}^2$ by

$$p = e^x \cos y, \quad q = e^x \sin y,$$

the Jacobian matrix is

$$\begin{bmatrix} e^x \cos y & -e^x \sin y \\ e^x \sin y & e^x \cos y \end{bmatrix}.$$

This is invertible when the determinant, which is equal to e^x, is nonzero. But this is so for all x. Therefore, by the inverse function theorem, every (a, b) in $\mathbb{R}^2$ lies in an open subset U of $\mathbb{R}^2$ on which f is injective. Besides, the inverse function $g : V \to U$ is differentiable. Thus, f has a continuously differentiable local inverse at every point of $\mathbb{R}^2$. Nonetheless, f is not invertible on $\mathbb{R}^2$ [see 4-2.P3].

4-2.3. Remark. In the proof of Theorem 4-2.1, we used the continuity of f' at a right at the outset. One may well ask whether the requirement of continuity at a could have been avoided in the simple case when the dimension n is 1. The answer is that it cannot be avoided even in this simple case: The function f defined as

$$f(x) = x + 2x^2 \sin(1/x) \text{ for } x \neq 0 \quad \text{and} \quad f(0) = 0$$

can be shown to have the properties that f' is bounded on $(-1, 1), f'(0) = 1$, but f is not injective on any open interval containing 0 [see 4-2.P8]. Of course, the trouble is that f' is not continuous at 0 and Theorem 4-2.1 is therefore not applicable.

4-2.4. Example. During the proof of Inverse Function Theorem 4-2.1, it was asserted in (3) that $x = g(y)$ is the fixed point of the contraction described in (2a) or (2b). As emphasised in Section 4-1, the contraction principle not only guarantees the existence of a fixed point but also provides an explicit sequence converging to it. Therefore, the proof also enables us to generate an explicit sequence converging to the required x, but valid in some sufficiently small ball about the 'accommodated' point (which has been designated as a in the statement of the theorem). We compute the first three terms of such a sequence for the simple example $f(x) = x^2 + 1$, $a = 2$, which was mentioned earlier and for which the inverse map can be written down explicitly as

$$x = g(y) = (y-1)^{\frac{1}{2}}.$$

Here $f'(a) = 4$ and hence $\phi_y(x) = x + \frac{1}{4}(y - f(x)) = x + \frac{1}{4}(y - (x^2 + 1))$. By applying the argument of the theorem with $a = 2$ and $f(a) = a^2 + 1 = 5$, we find that $g(y)$ is the fixed point of ϕ_y, and that ϕ_y is a contraction as long as y is in a suitable ball (interval) centred at $f(a) = 5$. For the approximating sequence that begins with the constant $x_1(y) = 2$ as the initial term, the next three terms are

$$x_2(y) = x_1(y) + \tfrac{1}{4}(y - (x_1(y)^2 + 1)) = 2 + \tfrac{1}{4}(y - 5),$$

$$x_3(y) = x_2(y) + \tfrac{1}{4}(y - (x_2(y)^2 + 1)) = 2 + \tfrac{1}{4}(y - 5) - \tfrac{1}{64}(y - 5)^2,$$

$$x_4(y) = x_3(y) + \tfrac{1}{4}(y - (x_3(y)^2 + 1))$$

$$= 2 + \tfrac{1}{4}(y - 5) - \tfrac{1}{64}(y - 5)^2 + \tfrac{1}{512}(y - 5)^3 - \tfrac{1}{16384}(y - 5)^4.$$

The reader may note that the fourth degree term in $x_4(y)$ does not agree with the Taylor expansion of g at $y = 5$, but the other terms do.

4-2.5. Remark. In the proof of Theorem 4-2.1, the value $f^{-1}(y)$ is obtained as the fixed point of the contraction ϕ_y given by $\phi_y(x) = x + A^{-1}(y - f(x))$, where $A = f'(a)$. In other words, as the limit of the sequence $\{x_p\}_{p \geq 1}$ generated by

$$x_{p+1}(y) = x_p(y) + f'(a)^{-1}(y - f(x_p(y))).$$

A modification of this way of generating the approximating sequence is to replace $f'(a)^{-1}$ by $f'(x_p(y))^{-1}$, which means we take

$$x_{p+1}(y) = x_p(y) + f'(x_p(y))^{-1}(y - f(x_p(y))).$$

Since we shall work with a fixed y, we shall keep the notation uncluttered by writing $x_p(y)$ as simply x_p from now on. Then the above equation becomes

$$x_{p+1} = x_p + f'(x_p)^{-1}(y - f(x_p)). \tag{1}$$

The one-dimensional version of this is easily recognised as Newton's method for solving $f(x) - y = 0$. In fact the multidimensional version is also known by the same name. We enter into an informal discussion of the following features it shares with the one-dimensional version:

(a) It converges if all the x_p lie in some open ball about $f^{-1}(y)$ on which all the second partial derivatives of the components of f as well as $\|f^{-1}\|$ are bounded.

(b) When it converges, it does so much faster than the sequence generated by the contraction. This explains why the convergence in 4-1.P10 is slower than that of Newton's method.

We work with the norm $\|\ \|_2$ in $\mathbb{R}^n$. Applying Corollary 3-5.6 to the function $f(x)-y$ with the ball mentioned in (a), we find that there is some constant $\beta > 0$ such that

$$\|(f(x)-y)-(f(x_p)-y)-f'(x_p)(x-x_p)\| \le \beta\|x-x_p\|^2$$

as long as x lies in the ball. Choose $x = f^{-1}(y)$, so that $f(x)-y = 0$. Then the above inequality becomes

$$\|-(f(x_p)-y)-f'(x_p)(x-x_p)\| \le \beta\|x-x_p\|^2.$$

By (a) above, there exists some $\alpha > 0$ such that $\|f'(x_p)^{-1}\| \le \alpha$ for all p. Therefore,

$$\|-f'(x_p)^{-1}(f(x_p)-y)-(x-x_p)\| = \|f'(x_p)^{-1}\left(-(f(x_p)-y)-f'((x_p)(x-x_p)\right)\|$$

$$\le \alpha\beta\|x-x_p\|^2.$$

But $-f'(x_p)^{-1}(f(x_p)-y) = x_{p+1}-x_p$ in view of (1). Therefore,

$$\|(x_{p+1}-x_p)-(x-x_p)\| \le \alpha\beta\|x-x_p\|^2,$$

which simplifies to

$$\|x-x_{p+1}\| \le \alpha\beta\|x-x_p\|^2.$$

This inequality shows that, if $\|x-x_p\|$ ever becomes less than $1/\alpha\beta$, then the sequence converges to $x = f^{-1}(y)$, because

$$\|x-x_{p+q}\| \le (\alpha\beta)^{-1}(\alpha\beta\|x-x_p\|)^{2^q}.$$

Contrast this with the estimate $\|x-x_{p+q}\| \le c^q\|x-x_p\|$ in the case of a contraction with contraction constant c.

The kind of ball we have assumed above can be shown to exist; the interested reader is referred to more advanced texts such as Chaudhary and Nanda [7], Kantorovich and Akilov [15] or Loomis and Sternberg [19].

Problem Set 4-2

4-2.P1. The point $(2, 4)$ in the plane lies on the graph of $y = f(x) = x^2$. Find an open set containing $y = 4$ such that the function $g(y) = y^{1/2}$ is defined on that open set, and show that

$$x = g(y) \Rightarrow y = x^2; \quad g(4) = 2.$$

The point $(-2, 4)$ also lies on the same graph. Find an open set containing $y = 4$ and a function g_1 defined on that open set such that

$$x = g_1(y) \Rightarrow y = x^2; \quad g_1(4) = -2.$$

The point $(0, 0)$ also lies on the same graph. Is there an open set containing $y = 0$ with a function g_2 defined on it such that $x = g_2(y) \Rightarrow y = x^2$ and $g_2(0) = 0$?

4-2.P2. The point $(1, e)$ lies on the graph of $y = xe^x$. Find an open set containing $y = e$ such that there is a continuous function $x = g(y)$ defined on it, for which $x = g(y) \Rightarrow y = xe^x$ and $g(e) = 1$. Formulate the corresponding question for the point $(-2, -2/e^2)$ and answer it.

4-2.P3. Let $f: \mathbb{R}^2 \rightarrow \mathbb{R}^2$ be defined as $f(x, y) = (e^x \cos y, e^x \sin y)$. Show that every point in $\mathbb{R}^2$ belongs to an open set on which f is one-to-one and that f is not injective on $\mathbb{R}^2$.

4-2.P4. Show that the Jacobian of the transformation

$$u = e^x \cos y, \quad v = e^x \sin y$$

from $\mathbb{R}^2$ to $\mathbb{R}^2$ is never 0. Does the inverse function theorem say that this transformation is invertible? Support your answer.

4-2.P5. [Needed in Proposition 7-2.4] If f is a continuously differentiable mapping of an open set $E \subseteq \mathbb{R}^n$ into $\mathbb{R}^n$ and if $f'(x)$ is invertible for every $x \in E$, then prove that f is an open mapping of E into $\mathbb{R}^n$. (Note: The phrase 'f is an open mapping of E into $\mathbb{R}^n$' means that f maps every open subset of E into an open subset of $\mathbb{R}^n$.)

4-2.P6. Let U and V be open subsets of $\mathbb{R}^n$ and let $f: U \rightarrow V$ be continuously differentiable and bijective (i.e., injective as well as surjective), so that the inverse map $g: V \rightarrow U$ exists. Suppose $f'(x)$ is invertible for every $x \in U$. Show that g' exists on the entire given set V.

4-2.P7. Let U be an open subset of $\mathbb{R}^n$ and $f: U \rightarrow \mathbb{R}^n$ be a continuously differentiable map such that $f'(x)$ is invertible for every $x \in U$. Suppose V is an open subset of U such that its closure $\overline{V}$ is bounded (hence compact) and contained in U, and f is injective on the closure. Show that the image $f(\overline{V})$ is the closure of an

open set. The result of this problem is useful in the study of integration of differential forms; see Rudin [22, p. 270].

4-2.P8. Show that the function f defined on $(-1, 1)$ by
$$f(x) = x + 2x^2 \sin(1/x) \text{ for } x \neq 0 \quad \text{and} \quad f(0) = 0$$
has the property that $f'(0) = 1$ and that f is not injective on any open interval
containing 0.

4-2.P9. Let X be any nonempty set and $\phi:\mathbb{R}^n \to \mathbb{R}^n$, $\psi:X \to \mathbb{R}^n$ be any maps. Define the map $\Phi:\mathbb{R}^n \times X \to \mathbb{R}^n \times X$ by $\Phi(s,x) = (\phi(s) + \psi(x), x)$. Prove (a) if ϕ is
injective, then so is Φ; and (b) if ϕ is surjective, then so is Φ.

4-2.P10. Let $f:\mathbb{R}^3 \to \mathbb{R}^3$ be defined by
$$y_1 = 2x_1^4 + x_3 \cos x_2 - x_1 x_3,$$
$$y_2 = (x_1 + x_3)^2 - 4\sin x_2,$$
$$y_3 = \ln(x_2 + 1) + 5x_1 + \cos x_3 - 1.$$

Show that the Jacobian matrix has rows
$$[8x_1^3 - x_3 \quad -x_3 \sin x_2 \quad \cos x_2 - x_1], \quad [2(x_1 + x_3) \quad -4\cos x_2 \quad 2(x_1 + x_3)]$$
$$\text{and} \quad [5 \quad \tfrac{1}{x_2 + 1} \quad -\sin x_3],$$
and that f has a continuous local inverse at $(0,0,0)$.

4-2.P11. State conditions on f and g under which the equations $x = f(u,v)$, $y = g(u,v)$ can be solved for u, v in an open set containing (x_0, y_0). If the solution is $u = F(x,y)$, $v = G(x,y)$ and if J is the determinant of the Jacobian matrix of the
map $(u,v) \to (f(u,v), g(u,v))$, show that
$$\frac{\partial F}{\partial x} = \frac{1}{J}\frac{\partial g}{\partial v}, \qquad \frac{\partial F}{\partial y} = -\frac{1}{J}\frac{\partial f}{\partial v}, \qquad \frac{\partial G}{\partial x} = -\frac{1}{J}\frac{\partial g}{\partial u}, \qquad \frac{\partial G}{\partial y} = \frac{1}{J}\frac{\partial f}{\partial u}.$$

4-2.P12. Show that the mapping $f:\mathbb{R}^2 \to \mathbb{R}^2$ defined by $f(x,y) = (x^2 - y^2, 2xy)$
maps the open set $U = \{(x,y) \in \mathbb{R}^2 : (x,y) \neq (0,0)\}$ 'two-to-one' onto itself and is
hence not invertible. Verify that every point of U belongs to an open set on
which the mapping has a differentiable (local) inverse.

4-2.P13. Let $f:U \to \mathbb{R}^2$ be the map described in 2-3.P4. Using the range of the
map derived there and the fact that $x^2 + y^2$, $x^2 - y^2$ and xy describe continuously
differentiable functions, but without computing the linear derivative f', show
that f' is not invertible at any point of U.

4-2.P14. Suppose $f_1, f_1, \ldots, f_n$ and h are continuously differentiable real-valued
homogeneous functions of the same degree on an open set $U \subseteq \mathbb{R}^n$. Suppose also

that h vanishes nowhere on U. Show that the Jacobian of $\frac{f_1(x)}{h(x)},\ldots,\frac{f_n(x)}{h(x)}$ with respect to $(x_1,\ldots,x_n)$ is zero everywhere. Deduce the result of 4-2.P13.

4-2.P15. For $x > 0$, $y > 0$, $z > 0$, let

$$u = \frac{x+y+z}{(x^2+y^2+z^2)^{1/2}}, \quad v = \frac{x^2-yz+z^2}{x^2+2y^2+65z^2} \quad \text{and} \quad z = \frac{x^3+6x^2y-7y^3}{(x^2+y^2)^{3/2}}.$$

Show that the Jacobian $\frac{\partial(u,v,w)}{\partial(x,y,z)}$ is zero everywhere. Formulate a result that in-

cludes this problem as well as 4-2.P14 special cases.

4-2.P16. Consider the map g of $\{(x,\,y) \in \mathbb{R}^2 : (x,\,y) \neq (0,0)\}$ into itself given by $g(x,y) = (r,\theta)$, where

$$r = (x^2+y^2)^{1/2}, \quad \theta = \cos^{-1}(x/(x^2+y^2)^{1/2}) \qquad \text{if } y \geq 0$$
$$\text{and} \qquad \theta = -\cos^{-1}(x/(x^2+y^2)^{1/2}) \qquad \text{if } y < 0.$$

Show that
(a) $(r,\theta) \in (0,\infty) \times (-\pi,\pi]$;
(b) $x = r\cos\theta$, $y = r\sin\theta$;
(c) if $(r,\theta) \in (0,\infty) \times (-\pi,\pi]$ and $x = r\cos\theta$, $y = r\sin\theta$, then $g(x,y) = (r,\theta)$;
(d) g is not continuous at $(-1,0)$.

4-2.P17. The map $f(x) = xe^x$ of $E = (-1,\infty)$ into $\mathbb{R}$ has a positive derivative eve-rywhere on its domain and therefore has an inverse. Choosing $a = 1$ in Theorem 4-2.1, describe the map ϕ_y. For the approximating sequence for $f^{-1}(y)$ starting with $x_1(y) = a$ as the first term, compute the next two terms. Which of them, if any, are partial sums of the Taylor series of f^{-1} at $y = e$? According to the theo-rem, the sequence converges to $f^{-1}(y)$ for all y in a suitable ball centered at some point; what is that point?

4-2.P18. Show that the map $(u,v) = f(x,y) = (x+y^2, x^3+y)$ of $\mathbb{R}^2$ into itself has a local inverse at $(0,0)$ and find the second and third terms of an approximating sequence for the local inverse, valid in some ball centred at $(0,0)$.

4-3 Implicit Function Theorem

We now return to the type of question discussed in Section 2-1, which is to ex-press (i.e., solve for) some variables in terms of the remaining from a system of equations, e.g.,

$$p^2+q^2-rp+\sin(s+p)=0, \quad p^3q+\cos(p+2q+r-s)-q-1=0.$$

134 Inverse and Implicit Function Theorems

In this example, we have two equations in four variables. From past experience, the reader will surely recognise that one can at best expect to solve for *two variables* in terms of the rest, because there are *two equations*. In general, if there are *n* equations in $n + m$ variables, one hopes to solve for *n* variables in terms of the other *m*. Of course, this is not always possible (see the linear example in Section 2-1). To ignore such exceptions and proceed until one is forced to take them into account may be entirely appropriate for some other intellectual pursuit, but in mathematics the tradition is that one makes as sure as possible before proceeding.

In the same spirit as the inverse function theorem, one can address the question of the existence and uniqueness of a local solution, and its differentiability and try to obtain an answer with a degree of certainty and precision.

For an equation $f(x,y) = 0$, where $x \in \mathbb{R}^p$, $y \in \mathbb{R}^q$, a **solution for x in terms of y** is a function g on some domain in $\mathbb{R}^q$ (usually consisting of more than one point!) such that $f(g(y),y) = 0$ for every y in that domain. Often one seeks a solution satisfying some additional requirement such as continuity and/or $g(b) = a$, where a and b are given. The latter kind of requirement is often expressed as $x(b) = a$ when one wishes to avoid introducing a letter to denote the solution function.

The distinction between a solution in the genuine mathematical sense of a function that fulfills a prestated requirement as opposed to an 'expression' obtained by skillful use of established computational procedures is illustrated by the following example: Solve the equation $x = xt$ for x as a continuous function of t such that $x(0) = 0$ and state the largest possible interval on which a solution is possible. According to received wisdom, one reacts to the given equation $x = xt$ by saying that $t = 1$ or $x = 0$. Therefore the demand for a solution for x in terms of t may seem perverse at first sight; nevertheless, the demand is perfectly legitimate and it takes one step from the observation that $t = 1$ or $x = 0$ to arrive at the solution $x(t) = 0$, $t \in \mathbb{R}$. This final step requires some imagination, not computational skill. For a generalisation of this example, see 4-3.P8.

The implicit function theorem(Theorem 4-3.2) below will provide a sufficient condition in order that a continuous solution g of $f(x,y) = 0$ for x in terms of y satisfying the requirement that $g(b) = a$ should exist and be unique. However, our formulation of the theorem does not explicitly mention the word 'solution'.

With the notation introduced in the opening paragraph of this section, when $n = 2$ and $m = 1$, we have a system of two equations in three variables:

$$f_1(x_1,x_2,y) = 0, \quad f_2(x_1,x_2,y) = 0.$$

Expressing x_1, x_2 in terms of y from here is the same as expressing x_1, x_2, y in terms of z_1, z_2, z_3 from

$$f_1(x_1,x_2,y) = z_1, \quad f_2(x_1,x_2,y) = z_2, \quad y = z_3$$

and then substituting $z_1 = 0$ and $z_2 = 0$. It may seem as though this makes the question more complicated, but this seemingly more complicated question calls for the inverse of the mapping $(x_1, x_2, y) \rightarrow (z_1, z_2, z_3)$ just described. Since the vectors (x_1, x_2, y) and (z_1, z_2, z_3) are of the same dimension, the question of the inverse of such a mapping has been tackled already in the inverse function theorem, and one can therefore reasonably expect to answer the present question in terms of that theorem.

For general n and m, we can think of the system of equations in vector form as the single equation $f(x, y) = 0 \in \mathbb{R}^n$, where $x \in \mathbb{R}^n$ and $y \in \mathbb{R}^m$. For a given y in an open subset of $\mathbb{R}^m$, we seek to find some x such that $f(x, y) = 0$. Now,

$$f(x, y) = 0 \Leftrightarrow (f(x, y), y) = (0, y).$$

This can be expressed in terms of the map $F : (x, y) \rightarrow (f(x, y), y)$ as

$$f(x, y) = 0 \Leftrightarrow F(x, y) = (0, y)$$

or as $f(x, y) = 0 \Leftrightarrow (x, y) = F^{-1}(0, y),$ but provided that F^{-1} exists.

If this inverse exists (perhaps only a local inverse), then the x that we seek to find is the first component (n-dimensional) of $(x, y) = F^{-1}(0, y)$. Since (x, y) and $(f(x, y), y)$ both lie in $\mathbb{R}^n \times \mathbb{R}^m = \mathbb{R}^{n+m}$, the existence of F^{-1} can be handled through the inverse function theorem. Another instance of 'reducing a case to one that has been already handled' (see the Remark after 3-4.P6).

4-3.1. Remark. Let the map ϕ from an open subset E of $\mathbb{R}^n$ to $\mathbb{R}^m$ be differentiable at some point $x_0 \in E$, and let A be a linear map from $\mathbb{R}^n$ to $\mathbb{R}^k$. Then the map $\Phi : E \rightarrow \mathbb{R}^m \times \mathbb{R}^k$ defined by $\Phi(x) = (\phi(x), Ax)$ can be shown to be differentiable at x_0 with derivative given by

$$\Phi'(x_0)(h) = (\phi'(x_0)(h), Ah), \qquad h \in \mathbb{R}^n.$$

This is usually regarded as obvious, but a proof is given here:

Φ is the sum of the two mappings $x \rightarrow (\phi(x), 0)$ and $x \rightarrow (0, Ax)$. The second of these is a linear map while the first one is the composition of ϕ with the linear map $y \rightarrow (y, 0)$. Using the chain rule and the fact that the derivative of a linear map is itself [Remark 3-2.2(d)], we find that the two mappings have the respective derivatives

$$h \rightarrow (\phi'(x_0)(h), 0) \quad \text{and} \quad h \rightarrow (0, Ah).$$

Since the derivative of a sum of two functions is the sum of their derivatives [Remark 3-2.2(f)], therefore

$$\Phi'(x_0)(h) = (\phi'(x_0)(h), 0) + (0, Ah) = (\phi'(x_0)(h), Ah).$$

The proof given above involves only a straightforward use of the chain rule and what are called 'elementary properties of the derivative', i.e., Remarks 3-2.2(b)–(d),(f). Details are therefore not normally expected to be given, and it suffices to say instead 'by elementary properties of the derivative and the chain rule'.

4-3.2. Implicit Function Theorem. *Let* $f: E \to \mathbb{R}^n$ *be a continuously differentiable map from an open subset* $E \subseteq \mathbb{R}^n \times \mathbb{R}^m$ *into* $\mathbb{R}^n$ *such that* $f(a, b) = 0$ *for some* $(a, b) \in E$. *Let* A_1 *and* A_2 *be linear maps of* $\mathbb{R}^n$ *and* $\mathbb{R}^m$ *respectively into* $\mathbb{R}^n$ *defined by* $A_1 h = f'(a, b)(h, 0)$ *and* $A_2 k = f'(a, b)(0, k)$, *so that* $f'(a, b)(h, k) = A_1 h + A_2 k \; \forall \; h \in \mathbb{R}^n, k \in \mathbb{R}^m$. *Suppose* A_1 *is invertible. Then*
(a) *there exist open sets* $U \subseteq E$, $W \subseteq \mathbb{R}^m$ *with* $(a, b) \in U$, $b \in W$ *and a unique map* $g: W \to \mathbb{R}^n$ *such that*

$$(g(y), y) \in U, \; f(g(y), y) = 0 \; \forall \; y \in W;$$

(b) *for every* $(x, y) \in U$ *such that* $f(x, y) = 0$, *we have* $y \in W$ *and* $x = g(y)$;
(c) *moreover, g is continuously differentiable and*

$$g(b) = a, \quad g'(b) = -A_1^{-1} A_2.$$

Proof. (a) Define $F: E \to \mathbb{R}^n \times \mathbb{R}^m$ by setting $F(x, y) = (f(x, y), y)$. Then by elementary properties of the derivative and the chain rule, F is differentiable on E and $F'(x, y)$ at any (x, y) is given by

$$F'(x, y)(h, k) = (f'(x, y)(h, k), k) \; \forall \; (h, k) \in \mathbb{R}^n \times \mathbb{R}^m.$$

It follows from this that F is continuously differentiable. (In fact, $\|F'(x_1, y_1) - F'(x_2, y_2)\| \leq \|f'(x_1, y_1) - f'(x_2, y_2)\|$.) It also follows that

$$F'(a, b)(h, k) = (A_1 h + A_2 k, k) \; \forall \; (h, k) \in \mathbb{R}^n \times \mathbb{R}^m.$$

Hence $F'(a, b)(h, k) = 0 \Rightarrow (A_1 h + A_2 k, k) = 0 \Rightarrow A_1 h + A_2 k = 0, k = 0$
$$\Rightarrow A_1 h = 0, k = 0$$
$$\Rightarrow h = 0, k = 0 \quad \text{because } A_1 \text{ is invertible.}$$

Therefore, $F'(a, b)$ is injective and thus invertible (by Remark 2-3.3(b); surjectivity can also be proved directly, i.e., without using Remark 2-3.3(b) but instead using the simple idea in 4-2.P9.) By the inverse function theorem, $\exists$ open sets $U \subseteq E$, $V \subseteq \mathbb{R}^n \times \mathbb{R}^m$ such that $(a, b) \in U$, F maps U injectively onto V and the inverse map $F^{-1}: V \to U$ is continuously differentiable. Now $F(a, b) = (f(a, b), b) = (0, b)$. So $(0, b) \in V$. Let $W = \{y \in \mathbb{R}^m : (0, y) \in V\}$. Then $b \in W$ and W is open. For any $y \in W$, we have $(0, y) \in V$ and hence $\exists \; (x, z) \in U$ such that $F(x, z) = (0, y)$. But $F(x, z) = (f(x, z), z)$. Therefore $(f(x, z), z) = (0, y)$, so that $y = z$ and $f(x, y) = 0$. If $f(x', y) = 0$ with $(x', y) \in U$, then $(f(x', y), y) = (0, y)$, i.e., $F(x', y) = (0, y) = F(x, y)$. But F is injective on U. So $x' = x$. Hence, there exists a unique x for which $f(x, y) = 0$ and $(x, y) \in U$. Call this x as $g(y)$. Then (a) is established.

(b) Let $(x, y) \in U$ and $f(x, y) = 0$. Then $F(x, y) \in V$. But $F(x, y) = (f(x, y), y)$ and $f(x, y) = 0$. This means $(0, y) \in V$, so that $y \in W$. By definition of g above, $g(y)$ is the unique ξ such that $f(\xi, y) = 0$ and $(\xi, y) \in U$. Therefore $x = \xi = g(y)$.

(c) Since $f(a, b) = 0$, $(a, b) \in U$ and $b \in W$, therefore $g(b) = a$. Now, for any $y \in W$, we have $f(g(y), y) = 0$ by (a), so that $F(g(y), y) = (0, y)$, from which it follows that $(g(y), y) = F^{-1}(0, y)$. Thus g is the composition of the maps

$$y \rightarrow (0, y), \quad (x, y) \rightarrow F^{-1}(x, y), \quad (x, y) \rightarrow x.$$

The first and third are linear while the second is continuously differentiable. It follows that g is continuously differentiable. Since $f(g(y), y) = 0 \; \forall \; y \in W$, the mapping $y \rightarrow f(g(y), y)$ must have derivative 0 everywhere. On the other hand, using elementary properties of the derivative and the chain rule once again, we find that the derivative of the mapping $y \rightarrow f(g(y), y)$ at b maps $k \in \mathbb{R}^m$ into

$$f'(g(b), b)(g'(b)k, k) = f'(a, b)(g'(b)k, k) \qquad \text{(because } g(b) = a\text{)}$$
$$= A_1 g'(b)k + A_2 k \qquad \text{(by hypothesis)}.$$

Since this must be equal to 0 for all $k \in \mathbb{R}^m$, then $g'(b) k = -A_1^{-1} A_2 k$ for all $k \in \mathbb{R}^m$. This completes the proof of (c). $\qquad\qquad\qquad\qquad\qquad\qquad\qquad\qquad\qquad\qquad\quad$ □

The conclusion of the above theorem is often summarised as 'the equation $f(x, y) = 0$ is locally solvable uniquely at (a, b) with a continuously differentiable local solution' or 'f has a continuously differentiable unique local solution at a'. The term 'local solution' here refers to the function g.

4-3.3. Remark. For application to concrete cases it is necessary to know how to compute the maps A_1 and A_2 from f. By Theorem 3-4.2, the linear map $f'(a, b)$ from $\mathbb{R}^n \times \mathbb{R}^m$ into $\mathbb{R}^n$ is represented by the $n \times (m + n)$ matrix of partial derivatives $D_j f_i(a, b)$, $1 \leq i \leq n$, $1 \leq j \leq m + n$. Consider any linear map A from $\mathbb{R}^n \times \mathbb{R}^m$ into $\mathbb{R}^n$ with matrix $[\alpha_{ij}]$, and its associated linear maps A_1 and A_2 as described in the above theorem. The matrix of A_1 consists of the n^2 entries α_{ij} with $1 \leq i \leq n$, $1 \leq j \leq n$, while the matrix of A_2 consists of the remaining entries α_{ij}. Therefore, when $A = f'(a, b)$, the matrix of A_1 has entries $D_j f_i(a, b)$, $1 \leq i \leq n$, $1 \leq j \leq n$. The invertibility of A_1 is equivalent to its determinant being nonzero. The matrix is called the *Jacobian matrix* of f *with respect to* x (or with respect to $x_1, \ldots, x_n$), and its determinant is known as the *Jacobian* of f (or of its component functions $f_1, \ldots, f_n$) *with respect to* x (or with respect to $x_1, \ldots, x_n$). When component notation is being used, it is standard practice to denote it by

$$\partial(f_1, \ldots, f_n)/\partial(x_1, \ldots, x_n).$$

Thus

$$\frac{\partial(f_1, \ldots, f_n)}{\partial(x_1, \ldots, x_n)} = \begin{vmatrix} \partial f_1/\partial x_1 & \partial f_1/\partial x_2 & \cdots & \partial f_1/\partial x_n \\ & & \ddots & \\ \partial f_n/\partial x_1 & \partial f_n/\partial x_2 & \cdots & \partial f_n/\partial x_n \end{vmatrix}.$$

Some authors express the hypothesis of invertibility of A_1 by saying that this determinant should be nonzero, e.g., Apostol [1]. (The theorem stated in [1, p. 374] differs from Theorem 4-3.2 and is a variant of Theorem 4-4.1 below.) Although there is no standard name or notation for the linear map A_1, a symbol such as $\partial f/\partial x$ or f_x could be used. The latter is found in Graves [13], where Theorem 2 on p.138 is essentially the same as Theorem 4-3.2 above. Since f is a function of two *vector variables*, the symbol $D_1 f$ can also be used, but preferably with explanation. This is the notation used in Burkill and Burkill [5], where the theorem stated on p. 216 is a variant of Theorem 4-3.2 above.]

In particular, if $n = m = 1$, so that $f'(a, b)$ has the 1×2 matrix

$$[(\partial f/\partial x)(a, b) \quad (\partial f/\partial y)(a, b)],$$

then A_1 has the 1×1 matrix with entry $(\partial f/\partial x)(a, b)$ and A_2 has the 1×1 matrix with entry $(\partial f/\partial y)(a, b)$. Therefore A_1 is invertible if and only if $(\partial f/\partial x)(a, b) \neq 0$. Besides,

$$A_1^{-1}A_2 = \frac{(\partial f / \partial y)(a,b)}{(\partial f / \partial x)(a,b)}.$$

This makes it possible to state the implicit function theorem in this case without explicit reference to linear maps. Also, it is possible to give a simple proof without using the inverse function theorem or the contraction principle. See Theorem 4-4.4 below. The reader will also find the article by Kumaresan [16] very useful.

These facts regarding A_1 and A_2 may be taken for granted in presenting any discussion of concrete examples, and no explanation is required as to why A_1 and A_2 are represented by matrices formed by partial derivatives in the manner described above.

4-3.4. Examples. In order to gain a better understanding of the implicit function theorem and appreciate the rather complicated nature of its hypotheses and conclusion, it is useful for one to consider applications to a few simple concrete examples. However, it should not be inferred that applications to these kinds of examples constitute the only *raison d'être* of the theorem. The result is needed for validating a computational procedure, known as 'method of Lagrange multipliers' to be discussed in another chapter.

Consider the problem of solving for one variable in terms of the other in each of the following cases:

(a) $x^2 + y^2 + 1 = 0$
(b) $x^2 + y^2 - 1 = 0$
(c) $x^2 + y^2 - 1 = 0$, $x(1) = 0$
(d) $x^2 - y^2 = 0$, $x(0) = 0$
(e) $x^2 + y^2 - 1 = 0$, $x(0) = 1$
(f) $x^3 - y^3 - 3xy - y = 0$, $y = g(x)$ near $(2, 1)$
(g) $\sin(x + y) - e^{xy} + 1 = 0$, $x(0) = 0$
(h) $x^3 - y^3 = 0$, $x(0) = 0$.

These examples serve to highlight various aspects of the implicit function theorem and we shall take them up one by one.

(a) Since no pair of real values of x and y can satisfy this equation, the need to solve it should not arise in practice. If it ever does, one would conclude that something went wrong before one arrived at it! This illustrates why the requirement that the equation be satisfied at some point is needed.

(b) No pair of values of x and y satisfying the equation has been specified. Here are some solutions for x in terms of y:

$$g(y) = (1 - y^2)^{1/2} \text{ for } 0 \le y \le 1; g(y) = -(1 - y^2)^{1/2} \text{ for } 0 \le y \le 1;$$
$$g(y) = (1 - y^2)^{1/2} \text{ for } 0 \le y \le 1/2 \text{ and } -(1 - y^2)^{1/2} \text{ for } 1/2 < y \le 1;$$
$$g(y) = (1 - y^2)^{1/2} \text{ for } 0 \le y < 1/47, -(1 - y^2)^{1/2} \text{ for } 1/47 \le y \le 1/21$$
$$\text{and } (1 - y^2)^{1/2} \text{ for } 1/21 < y \le 1;$$
$$g(y) = (1 - y^2)^{1/2} \text{ for rational } y \text{ and } -(1 - y^2)^{1/2} \text{ for irrational } y.$$

The first two are continuous and the rest are discontinuous. The reader is invited to add to this mélange of solutions. If no specific point (a, b) satisfying the equation is required to be 'accommodated', then the solution may not be unique.

(c) The given values $(a = 0, b = 1)$ do satisfy the given equation. Since they are stated in the form $x(1) = 0$, the solution required is for x as a function of y. The first two solutions listed above under (b) both satisfy the requirement that $x(0) = 1$, and both are continuous. Neither is differentiable when $y = 1$ and neither is defined on an open set containing 1. Thus the conclusion of the implicit function theorem that there exists a unique solution in an open set containing b does not hold. But the theorem is not contradicted, because the hypothesis about the linear map A_1 represented (as discussed above in Remark 4-3.3) by the partial derivative of $x^2 + y^2 - 1$ with respect to x when $x = 0$, $y = 1$ being invertible is not fulfilled, as the value of this partial derivative when $x = 0$, $y = 1$ is found to be 0.

(d) The given values $(a = 0, b = 0)$ satisfy the given equation. Here are two continuous solutions:

$$g(y) = |y| \text{ and } g(y) = -|y| \text{ both for all real } y.$$

Neither is differentiable when $y = 0$, but — in contrast to (c) — both are defined on an open set containing 0. Thus the conclusion of the implicit function theorem that there exists a unique solution in an open set containing b does not hold. But the theorem is not contradicted, because the hypothesis about the linear map A_1 represented (as discussed above in Remark 4-3.3) by the partial derivative of $x^2 - y^2$ with respect to x when $x = 0$, $y = 0$ being invertible is not fulfilled, as the value of this partial derivative when $x = 0$, $y = 0$ is found to be 0.

(e) The implicit function theorem is applicable. The function of (x, y) given by the left hand side of the equation, i.e., $f(x, y) = x^2 + y^2 - 1$, is differentiable and its partial derivative with respect to x (the variable for which we have to solve) is found to be $2x$, which is nonzero for the given values $x = 1$, $y = 0$. Also, the given values $x = 1$, $y = 0$ satisfy the equation. Therefore, by the implicit function theorem, there is an open subset of $\mathbb{R}^2$ containing $(1,0)$ and an open interval containing 0, on which there is a unique function g such that $(g(y), y)$ lies in the aforementioned open subset of $\mathbb{R}^2$ and $g(y)^2 + y^2 - 1 = 0$; moreover, this g is continuously differentiable and satisfies $g(0) = 1$. Its derivative when $y = 0$ is (see Remark 4-3.3 for explanation of the quotient)

$$\frac{(\partial f / \partial y)(1,0)}{(\partial f / \partial x)(1,0)} = 0.$$

In this simple case, g can be explicitly computed as $g(y) = (1 - y^2)^{1/2}$.

(f) The implicit function theorem is applicable. The function of (x, y) given by the left hand side of the equation, i.e., $f(x, y) = x^3 - y^3 - 3xy - y$, is differentiable and its partial derivative with respect to y (the variable for which we have to solve) is found to be

$$-3y^2 - 3x - 1,$$

which is nonzero for the given values $x = 2$, $y = 1$. (As explained in Remark 4-3.3, this means that the linear map denoted by A_1 in the implicit function theorem is invertible.) Also, the given values $x = 2$, $y = 1$ satisfy the equation. Therefore by the implicit function theorem, there is an open subset of $\mathbb{R}^2$ containing $(2,1)$ and an open interval containing 2, on which there is a unique function g such that $(x, g(x))$ lies in the aforementioned open subset of $\mathbb{R}^2$ and $x^3 - g(x)^3 - 3xg(x) - g(x) = 0$; moreover, this g is continuously differentiable and satisfies $g(2) = 1$. Its derivative when $x = 2$ is (see Remark 4-3.3 for explanation of the quotient)

$$\frac{(\partial f / \partial x)(2,1)}{(\partial f / \partial y)(2,1)} = -\frac{9}{10}.$$

(g) In this case, computing x in terms of y explicitly, as one tries to do in elementary calculus, is no joke. In fact, it is a hopeless undertaking. It is not even clear from computational procedures whether the required function for x in terms of y exists, let alone calculating an expression for it in the elementary sense. However, the given values $x = 0$, $y = 0$ do satisfy the equation and the x-partial derivative of $f(x, y) = \sin(x + y) - e^{xy} + 1$ when $x = 0$ and $y = 0$ has the nonzero value 1. (As explained in Remark 4-3.3, this means that the linear map denoted by A_1 in the implicit function theorem is invertible.) Therefore, by the implicit function theorem, there exists an open subset of $\mathbb{R}^2$ containing $(0,0)$ and an open interval containing 0, on which there is a unique function g such that $(g(y), y)$ lies in the aforementioned open subset of $\mathbb{R}^2$ and $x = g(y)$ is a solution of the given equation; moreover, this g is continuously differentiable and satisfies $g(0) = 0$. Its derivative when $y = 0$ is (see 4-3.3 for explanation of the quotient)

$$\frac{(\partial f/\partial y)(0,0)}{(\partial f/\partial x)(0,0)} = 1.$$

(h) When $f(x, y) = x^3 - y^3$, the partial derivative $\partial f/\partial x$ vanishes at $(0,0)$ and the implicit function theorem does not apply. However, there is a unique solution given by $g(y) = y$, and it is differentiable. [Cf. 4-4.P3.]

For another discussion of concrete examples that help gain a better understanding of the implicit function theorem see the article by Kumaresan [16] quoted above.

4-3.5. Remark. In (e), (f) and (g), the value of $dx/dy = g'(y)$ that has been computed is, of course, the same as what one would have obtained by 'implicit differentiation' in elementary calculus. However, implicit differentiation simply assumes that a differentiable solution for x in terms of y exists; the new element introduced by the implicit function theorem is that the existence of a differentiable solution is assured before one rushes in to compute. As already mentioned in the first two paragraphs of Section 4-3, such caution is the stuff that mathematics is made of.

4-3.6. Examples. (a) We now take up the question stated at the beginning of this section:

$$p^2 + q^2 - rp + \sin(s + p) = 0, \quad p^3q + \cos(p + 2q + r - s) - q - 1 = 0.$$

These equations hold when each of the four variables is 0. Can we solve for p and q in terms of r and s 'near' $(0, 0, 0, 0)$? Setting

$$f_1(p, q, r, s) = p^2 + q^2 - rp + \sin(s + p),$$
$$f_2(p, q, r, s) = p^3q + \cos(p + 2q + r - s) - q - 1,$$

we find that $\partial(f_1, f_2)/\partial(p, q)$ has the value -1 when $(p,q,r,s) = (0,0,0,0)$. Since this value is nonzero, it follows by the implicit function theorem that there exists a local solution for p and q in terms of r and s. We can also find the matrix form of the linear derivative at $(0,0)$ for the solution. In the notation of the theorem, the required linear derivative is $-A_1^{-1}A_2$, where

$$A_1 = \begin{bmatrix} \partial f_1/\partial p & \partial f_1/\partial q \\ \partial f_2/\partial p & \partial f_2/\partial q \end{bmatrix} \quad \text{and} \quad A_2 = \begin{bmatrix} \partial f_1/\partial r & \partial f_1/\partial s \\ \partial f_2/\partial r & \partial f_2/\partial s \end{bmatrix}.$$

Upon computing the partial derivatives and substituting $(p,q,r,s) = (0,0,0,0)$, we obtain

$$A_1^{-1} = A_1 = \begin{bmatrix} 1 & 0 \\ 0 & -1 \end{bmatrix} \text{ and } A_2 = \begin{bmatrix} 0 & 1 \\ 0 & 0 \end{bmatrix},$$

so that the required value of the linear derivative at $(0,0,0,0)$ is

$$-A_1^{-1}A_2 = \begin{bmatrix} 0 & -1 \\ 0 & 0 \end{bmatrix}.$$

(b) The following equations hold when $(x,y,t) = (0,0,0)$:

$$x\cos(x+y) - t = 0, \qquad xe^x + ye^y - \sin t = 0.$$

The question is whether there exists a (local) differentiable solution for x,y in terms of t near $t = 0$ such that $x = y = 0$ when $t = 0$ and, if so, what the values of $x'(0)$ and $y'(0)$ are.

To find A_1, we compute

$$\frac{\partial}{\partial x}(x\cos(x+y) - t) = \cos(x+y) - x\sin(x+y) = 1 \text{ when } x = y = t = 0$$

$$\frac{\partial}{\partial y}(x\cos(x+y) - t) = -x\sin(x+y) = 0 \text{ when } x = y = t = 0$$

$$\frac{\partial}{\partial x}(xe^x + ye^y - \sin t) = (1+x)e^x = 1 \text{ when } x = y = t = 0$$

$$\frac{\partial}{\partial y}(xe^x + ye^y - \sin t) = (1+y)e^y = 1 \text{ when } x = y = t = 0.$$

Consequently, $A_1 = \begin{bmatrix} 1 & 0 \\ 1 & 1 \end{bmatrix}$ and hence $A_1^{-1} = \begin{bmatrix} 1 & 0 \\ -1 & 1 \end{bmatrix}$. Therefore the local solution in question exists and is unique. Also,

$$\frac{\partial}{\partial t}(x\cos(x+y) - t) = -1 \text{ and } \frac{\partial}{\partial t}(xe^x + ye^y - \sin t) = -\cos t = -1 \text{ when } x = y = t = 0.$$

Consequently, $A_2 = \begin{bmatrix} -1 \\ -1 \end{bmatrix}$ and $-A_1^{-1}A_2 = \begin{bmatrix} 1 \\ 0 \end{bmatrix}$. This means $x'(0) = 1$ and $y'(0) = 0$.

The open set W of Theorem 4-3.2 is by no means the largest open set on which a function g of the required kind is defined, and one can seek to extend the domain of g to see if one can obtain a maximal domain. Such matters are discussed by Graves in [13].

Problem Set 4-3

4-3.P1. Determine whether the solvability near $(0,0,0,0)$ of the equations discussed in Example 4-3.6 for q and r follows from the implicit function theorem.

4-3.P2. Show that the system of equations:

$$3x + y - z - u^3 = 0$$
$$x - y + 2z + u = 0$$
$$2x + 2y - 3z + 2u = 0$$

can be solved for x, y, u in terms of z but not for x, y, z in terms of u.

4-3.P3. Let $f: \mathbb{R}^2 \times \mathbb{R}^2 \to \mathbb{R}^2$ be defined by

$$f(x,y) = (x_1 y_2 + x_2 y_1 - 1, \ x_1 x_2 - y_1 y_2),$$

where $x = (x_1, x_2)$ and $y = (y_1, y_2)$. Show that $f(1,0,0,1) = (0,0)$. Verify that the linear derivative $(D_y f)(1,0,0,1)$ is represented by the matrix $\begin{bmatrix} 0 & 1 \\ -1 & 0 \end{bmatrix}$. Use this and the implicit function theorem to show that y is a function of x near $(1,0)$. Compute the (matrix of the) linear derivative of this function at $(1,0)$.

4-3.P4. Suppose the equation $f(x,y,z) = 0$, where f is differentiable, can be solved for each of the three variables x,y,z as a differentiable function of the other two. Show that at any point where $f(x,y,z) = 0$ and at least one of the partial derivatives $\frac{\partial f}{\partial x}, \frac{\partial f}{\partial y}, \frac{\partial f}{\partial z}$ is nonzero, the other two are also nonzero, and we have

$$\frac{\partial x}{\partial y} \frac{\partial y}{\partial z} \frac{\partial z}{\partial x} = -1.$$

4-3.P5. The function $f(x,y,z,u) = x^2 + y^2 + z^2 + u^2 - 1$ satisfies $f(\frac12,\frac12,\frac12,\frac12) = 0$. Solve for each of the variables in terms of the other three in an open set containing $(\frac12,\frac12,\frac12)$ and check whether the four solutions satisfy $\frac{\partial x}{\partial y} \frac{\partial y}{\partial z} \frac{\partial z}{\partial u} \frac{\partial u}{\partial x} = -1$ at $(\frac12,\frac12,\frac12,\frac12)$.

4-3.P6. Let $J_F(x)$ denote the Jacobian of a map F at x. Suppose the maps $f:\mathbb{R}^n\to\mathbb{R}^n$ and $g_i:\mathbb{R}\to\mathbb{R}$ are all continuously differentiable. Define $h:\mathbb{R}^n\to\mathbb{R}^n$ by

$$h_i(x) = f_i(g_1(x_1),g_2(x_2),\dots,g_n(x_n)) \qquad \text{for } x = (x_1,\dots,x_n)\in\mathbb{R}^n, \ 1\le i\le n,$$

where $f_1,\dots,f_n$ are the components of f. Show that

$$J_h(x) = [J_f(g_1(x_1),g_2(x_2),\dots,g_n(x_n))]\cdot g_1{'}(x_1)\cdot g_2{'}(x_2)\cdot\ \cdots\ \cdot g_n{'}(x_n).$$

4-3.P7. Let $f(x,y_1,y_2) = x^2y_1 + e^x + y_2$ on $\mathbb{R}^3$. Show that there exists a differentiable function g on some open set containing $(1,-1)$ in $\mathbb{R}^2$ such that $g(1,-1) = 0$ and $f(g(y_1,y_2),y_1,y_2) = 0$ on that open set. Find the partial derivatives $(D_1g)(1,-1)$ and $(D_2g)(1,-1)$.

4-3.P8. Let f and g be functions on $\mathbb{R}$ with continuous derivatives, $f(0) = 0$ and $f'(0) \ne 0$. Consider the equation $f(x) = tg(x)$, $t \in \mathbb{R}$. Show that in a suitable interval $|t| < \delta$, there is a unique continuous function $x = x(t)$ that solves the equation and satisfies $x(0) = 0$. Find the derivative $x'(0)$. When $g(0) = 0$, what is the largest possible interval on which a solution is defined?

4-4 Implicit Function Theorem in Another Form

In applying the implicit function theorem in examples like (e), (f) and (g) in Example 4-3.4, the reference to an open set containing the given point is cumbersome and appears to complicate the matter rather than clarify it. However, it is unavoidable if the version given in Theorem 4-3.2 is to be used, because the uniqueness of the solution g is contingent upon $(g(y),y)$ belonging to that open set (denoted by U in the statement).

What one can do in order to avoid the reference to the open set is to establish another form of the theorem, in which the uniqueness of the solution g is contingent upon its being continuous and satisfying $g(b) = a$.

Establishing this version of the theorem requires Proposition 2-5.9.

4-4.1. Implicit Function Theorem. *Let $f:E\to\mathbb{R}^n$ be a continuously differentiable map from an open subset $E \subseteq \mathbb{R}^n\times\mathbb{R}^m$ into $\mathbb{R}^n$ such that $f(a,b) = 0$ for some $(a,b) \in E$. Let A_1 and A_2 be linear maps of $\mathbb{R}^n$ and $\mathbb{R}^m$, respectively, into $\mathbb{R}^n$ defined by $A_1h = f'(a,b)(h,0)$ and $A_2k = f'(a,b)(0,k)$, so that $f'(a,b)(h,k) = A_1h + A_2k \ \forall\ h\in\mathbb{R}^n, k\in\mathbb{R}^m$. Suppose A_1 is invertible. Then*

(a) *there exists an open ball $B \subseteq \mathbb{R}^m$ centred at b and a unique continuous map $G:B\to\mathbb{R}^n$ such that*

$$G(b) = a \quad \text{and} \quad f(G(y),y) = 0 \ \forall\ y\in B;$$

(b) *moreover, G is continuously differentiable and $G'(b) = -A_1^{-1}A_2$.*

Proof. By Theorem 4-3.2, there exist open sets $U \subseteq \mathbb{R}^n \times \mathbb{R}^m$ and $W \subseteq \mathbb{R}^m$ with $b \in W$ and $(a, b) \in U$ and such that there exists a unique map $g: W \to \mathbb{R}^n$ for which

$$(g(y), y) \in U, \quad f(g(y), y) = 0 \quad \forall \, y \in W.$$

Moreover, $g(b) = a$, g is continuously differentiable on W and $g'(b) = -A_1^{-1}A_2$. Since W is open and $b \in W$, therefore there exists an open ball $B \subseteq W$ centred at b. Let G be the restriction of g to B. Then, except possibly for the uniqueness part, all other statements in (a) and (b) are true.

To prove uniqueness, suppose G_1 is any continuous map from B to $\mathbb{R}^n$ for which $G_1(b) = a$ and $f(G_1(y), y) = 0 \, \forall \, y \in B$. Consider the subsets Y and N of B defined as

$$Y = \{y \in B : G_1(y) = G(y)\} \text{ and } N = \{y \in B : G_1(y) \neq G(y)\}.$$

It is sufficient to show that N is empty. Since $G_1(b) = a = G(b)$, we have $b \in Y$, so that Y is not empty. Also, $Y \cup N = B$ and $Y \cap N$ is empty. Since G_1 and G are both continuous, the set N is open. It will now be shown that Y is also open.

Consider any $y_0 \in Y$. Then $G_1(y_0) = G(y_0)$. But $(G(y_0), y_0) = (g(y_0), y_0) \in U$. Therefore, $(G_1(y_0), y_0) \in U$. By continuity of G_1, there exists an open ball $B_1 \subseteq B \subseteq W$ centred at y_0 such that $y \in B_1 \Rightarrow (G_1(y), y) \in U$. Now define $G_2: W \to \mathbb{R}^n$ to agree with G_1 on B_1 and to agree with g on the rest of W. (It does not matter that this function could be discontinuous.) Then $(G_2(y), y) \in U$ and $f(G_2(y), y) = 0 \, \forall \, y \in W$. By the uniqueness in (a) of the statement of Theorem 4-3.2, G_2 must agree with g on the whole of W. Hence, G_1 must agree with g on the set B_1 (G_2 was defined as agreeing with G_1 on B_1). But g agrees with G on B (G was defined as the restriction of g to B) and $B_1 \subseteq B$. Thus, g agrees with G as well as G_1 on the set B_1. Therefore, G agrees with G_1 on B_1, so that $B_1 \subseteq Y$. It has been shown that any $y_0 \in Y$ is the centre of an open ball contained in Y. In other words, Y is open.

This shows that the open ball B is the union of the disjoint open sets Y and N. By Proposition 2-5.9, it follows that one among the sets Y and N must be empty. However, Y is not empty (recall that $b \in Y$). Therefore, N must be empty. As observed earlier, this completes the proof. □

4-4.2. Remark. If the examples such as (f) or (g) of Section 4-3 are discussed in the light of Theorem 4-4.1, then the conclusion obtained is that *on a ball about the given point* there is a unique *continuous* solution g of the given equation *which takes the given value at the given point*; the solution is also continuously differentiable.

4-4.3. Remark. It is not possible to replace the ball B in Theorem 4-4.1 by a more general open set W that may not have the 'connectedness' property of a ball assured by Proposition 2-5.9. Consider $f(x, y) = x^2 + y^2 - 1$, $E = \mathbb{R}^2$, $(a, b) = (1, 0)$.

On the open set

$$W = (-\frac{1}{3}, \frac{1}{3}) \cup (\frac{2}{3}, 1),$$

(shown along the vertical axis in the figure below) which contains b, there are two continuous solutions $x = G(y)$ of the equation $f(x, y) = 0$ (shown along the circle), both satisfying the requirement that $G(b) = a$, namely,

$$G(y) = \begin{cases} \sqrt{1-y^2} & \text{if } y \in (-\frac{1}{3}, \frac{1}{3}) \\ \pm\sqrt{1-y^2} & \text{if } y \in (\frac{2}{3}, 1). \end{cases}$$

Note that $(-\frac{1}{3}, \frac{1}{3}) \cup (\frac{2}{3}, 1)$ is the union of disjoint open sets $(-\frac{1}{3}, \frac{1}{3})$ and $(\frac{2}{3}, 1)$ and neither of these open sets is empty.

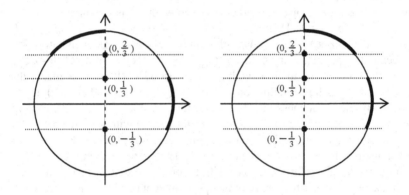

However, if we consider this example in the light of the previous version of the implicit function theorem (Theorem 4-3.2), the open sets $U = \{(x, y) : x > 0\}$ and $W = (-1, 1)$ together have the property that the only solution $x = g(y)$ of the equation $f(x, y) = 0$ which is defined for $y \in W$ and satisfies the requirement that $(g(y), y) \in U$ (i.e., that $g(y) > 0$) is given by $g(y) = \sqrt{(1 - y^2)}$. Moreover, it is continuously differentiable and satisfies $g(0) = 1$.

We now present the **implicit function theorem in two dimensions**.

4-4.4. Theorem. *Let $E \subseteq \mathbb{R}^2$ be open and $F: E \rightarrow \mathbb{R}$ be continuous with a partial derivative D_2F that is positive (or negative) everywhere. Suppose $(a, b) \in E$ and $F(a, b) = 0$. Then there exist intervals $(a - \delta, a + \delta), (b - \eta, b + \eta)$ with a unique function $f: (a - \delta, a + \delta) \rightarrow (b - \eta, b + \eta)$ such that*

$$F(x, f(x)) = 0 \text{ for all } x \in (a - \delta, a + \delta) \quad \text{and} \quad b = f(a). \tag{1}$$

This function f is continuous everywhere. Moreover, if c belongs to $(a - \delta, a + \delta)$ and F is differentiable at $(c, f(c))$, then f is differentiable at c and

$$(D_1F)(c, f(c)) + (D_2F)(c, f(c)) \cdot f'(c) = 0. \tag{2}$$

Proof. We shall work with the case when D_2F is positive because the contrary case is similar.

Since $(a, b) \in E$, an open subset of $\mathbb{R}^2$, there exists $\eta > 0$ such that

$$|x - a| < \eta, |y - b| \leq \eta \implies (x, y) \in E,$$

so that, for each $x \in (a - \eta, a + \eta)$, the function $y \to F(x, y)$ is defined on the interval $[b - \eta, b + \eta]$. Now, this function has derivative $(D_2F)(x, y)$ and therefore it has a positive derivative on its entire domain $[b - \eta, b + \eta]$. When $x = a$, it vanishes at b and it follows that it is negative at $b - \eta$ and positive at $b + \eta$:

$$F(a, b - \eta) < 0 < F(a, b + \eta).$$

By continuity of F on E, there must exist a positive $\delta < \eta$ such that

$$F(x, b - \eta) < 0 < F(x, b + \eta) \quad \text{for} \quad a - \delta < x < a + \delta.$$

In other words, for each $x \in (a - \delta, a + \delta)$, the function $y \to F(x, y)$ is negative at $b - \eta$ and positive at $b + \eta$. But this function is continuous and so, the intermediate value theorem yields at least one $y \in (b - \eta, b + \eta)$ where the function vanishes: $F(x, y) = 0$. However, there can be at most one $y \in \mathbb{R}$ satisfying $F(x, y) = 0$, because the derivative of the function $y \to F(x, y)$, namely $(D_2F)(x, y)$ is positive everywhere. Define $f(x)$ to be the unique such y and we have a function f on $(a - \delta, a + \delta)$ that fulfills (1). Since, the element $y \in \mathbb{R}$ where $F(x, y) = 0$ is unique, the function f is also unique.

Observe that, in view of the fact that the aforementioned y obtained by using the intermediate value theorem lies in $(b - \eta, b + \eta)$, we have

$$b - \eta < f(x) < b + \eta \quad \text{for} \quad a - \delta < x < a + \delta. \tag{3}$$

To prove the continuity of f, consider any $\alpha \in (a - \delta, a + \delta)$ and any $\varepsilon > 0$. We need to show that there exists a $\delta_1 > 0$ such that

$$f(\alpha) - \varepsilon < f(x) < f(\alpha) + \varepsilon \quad \text{for} \quad \alpha - \delta_1 < x < \alpha + \delta_1.$$

Since $\alpha \in (a - \delta, a + \delta)$, we have $b - \eta < f(\alpha) < b + \eta$ by (3). We may therefore assume that

$$\varepsilon < \min \{ f(\alpha) - (b - \eta), (b + \eta) - f(\alpha) \}$$

Then we have

$$b - \eta < f(\alpha) - \varepsilon < f(\alpha) + \varepsilon < b + \eta,$$

so that the function $y \to F(\alpha, y)$ is defined at $f(\alpha) - \varepsilon$ and at $f(\alpha) + \varepsilon$. Since $F(\alpha, f(\alpha)) = 0$, the function vanishes at $y = f(\alpha)$ and has a positive derivative everywhere on its domain. Therefore it is negative at $f(\alpha) - \varepsilon$ and positive at $f(\alpha) + \varepsilon$, i.e., $F(\alpha, f(\alpha) - \varepsilon) < 0 < F(\alpha, f(\alpha) + \varepsilon)$. As before, it follows by continuity of F that there exists a positive δ_1 such that $(\alpha - \delta_1, \alpha + \delta_1) \subseteq (a - \delta, a + \delta)$ and

$$F(x, f(\alpha) - \varepsilon) < 0 < F(x, f(\alpha) + \varepsilon) \quad \text{for} \quad \alpha - \delta_1 < x < \alpha + \delta_1.$$

In other words, for each $x \in (\alpha - \delta_1, \alpha + \delta_1)$, the function $y \rightarrow F(x, y)$ is negative at $f(\alpha) - \varepsilon$ and positive at $f(\alpha) + \varepsilon$. Therefore, the unique y where the function vanishes must lie between $f(\alpha) - \varepsilon$ and $f(\alpha) + \varepsilon$, i.e.,

$$f(\alpha) - \varepsilon < f(x) < f(\alpha) + \varepsilon \quad \text{for} \quad \alpha - \delta_1 < x < \alpha + \delta_1.$$

Continuity of f at α is thus established.

We proceed to prove (2). For any sufficiently small $h \neq 0$, the number $c + h$ belongs to $(a - \delta, a + \delta)$. Denoting $f(c + h) - f(c)$ by k, we have

$$0 = F(c + h, f(c + h)) - F(c, f(c)) = F(c + h, f(c) + k)) - F(c, f(c))$$
$$= h[(D_1 F)(c, f(c))] + k[(D_2 F)(c, f(c))] + u(h, k)(|h| + |k|),$$

where $u(h, k) \rightarrow 0$ as $(h, k) \rightarrow (0,0)$. Dividing by h and regrouping terms, we have

$$0 = (D_1 F)(c, f(c)) + \frac{k}{h}[(D_2 F)(c, f(c)) + u(h, k)\frac{|k|}{k}] + u(h, k)\frac{|h|}{h}, \qquad (4)$$

where $\frac{|k|}{k}$ can be taken to be any real number in the event that $k = 0$. By the continuity of f at c, we have $\lim_{h \rightarrow 0} k = 0$. It follows from this that the first and third terms in (4) each have a limit as $h \rightarrow 0$ and therefore so does the second term. But in the second term, the factor that is multiplied to $\frac{k}{h}$ has limit $(D_2 F)(c, f(c))$, which is nonzero. This implies that $\frac{k}{h}$ has a limit, so that $f'(c)$ exists. The existence of $f'(c)$, together with (4), leads to (2). $\qquad \square$

The following consequence is known by the same name as the foregoing theorem.

4-4.5. Corollary. *Let $E \subseteq \mathbb{R}^2$ be open and $F: E \rightarrow \mathbb{R}$ be differentiable with a positive (or negative) partial derivative $D_2 F$ everywhere. Suppose $F(a, b) = 0$, where $(a, b) \in E$. Then there exists an interval $(a - \delta, a + \delta)$ with a unique rea- valued function f defined on it such that*

$$F(x, f(x)) = 0 \text{ for all } x \in (a - \delta, a + \delta) \quad \text{and} \quad b = f(a). \qquad (1)$$

This function f is differentiable with derivative $f'(x)$ satisfying

$$(D_1 F)(x, f(x)) + (D_2 F)(x, f(x)) \cdot f'(x) = 0 \qquad (2)$$

everywhere on its domain.

4-4.6. Example. Let $E = \mathbb{R}^2$, $F(x, y) = ye^y - |x|^{\frac{1}{2}}$ and $(a, b) = (0, 0)$. Since F is not differentiable at (a, b), none among Theorem 4-3.2, Theorem 4-4.1 and Corollary 4-4.5 is applicable. However, Theorem 4-4.4 does show that a unique continuous solution $y = f(x)$ such that $b = f(a)$ must be valid on some interval containing 0.

Problem Set 4-4

4-4.P1. Given the equation $x^2 + y^2 = 1$ and the point $(1,0)$, find three open subsets W of $\mathbb{R}$, each containing 0, and satisfying the respective conditions:

(i) there exists a unique solution $x = g(y)$ of the given equation having domain W and satisfying the condition that $g(y) > 1/\sqrt{2}$;

(ii) there exist several solutions $x = g(y)$ of the given equation having domain W and satisfying the condition that $g(y) > -1/\sqrt{2}$; do they satisfy $g(0) = 1$?

(iii) there exists a unique solution $x = g(y)$ of the given equation having domain W and satisfying the condition that $g(y) > 0$, but four continuous solutions having domain W and satisfying the condition that $g(0) = 1$.

4-4.P2. State a theorem that includes Theorem 4-3.2 as well as Theorem 4-4.1, and then prove it, starting from the inverse function theorem.

4-4.P3. (a) The implicit function theorem 4-4.4 gives only a sufficient condition for solvability. Example 4-4-4 (h) shows that nonvanishing of the partial derivative is not necessary. Give an example of a function f on $\mathbb{R}^2$ such that $\frac{\partial f}{\partial y}(0,0) = 0$, f is not differentiable at $(0,0)$, but the equation $f(x,y) = 0$ has a unique solution $y = g(x)$ near 0 such that $g(0) = 0$.

(b) Give an example of a function f on $\mathbb{R}^2$ such that $\frac{\partial f}{\partial y}(0,0) = 0$ and the equation $f(x,y) = 0$ has two (not more) differentiable solutions $y = g(x)$ near 0 such that $g(0) = 0$.

(c) Give an example of a function f on $\mathbb{R}^2$ such that $\frac{\partial f}{\partial y}(0,0) = 0$ and the equation $f(x,y) = 0$ has at least four differentiable solutions $y = g(x)$ near 0 such that $g(0) = 0$.

4-4.P4. Prove the following variant of Theorem 4-4.4: Let $E \subseteq \mathbb{R}^2$ be open and $F : E \rightarrow \mathbb{R}$ be continuous with a positive partial derivative D_2F at $(a,b) \in E$, and let $F(a,b) = 0$. Then there exists an interval $(a - \delta, a + \delta)$ with at least one real valued function f defined on it such that

$$F(x, f(x)) = 0 \text{ for all } x \in (a - \delta, a + \delta) \quad and \quad b = f(a). \tag{1}$$

If f is continuous at $c \in (a - \delta, a + \delta)$ and F is differentiable at $(c, f(c))$ with $(D_2F)(c, f(c)) \neq 0$, then f is differentiable at c and

$$(D_1F)(c, f(c)) + (D_2F)(c, f(c)) \cdot f'(c) = 0. \tag{2}$$

4-4.P5. Prove the version of Theorem 4-4.1, in which (1) is altered as follows: $\exists$ an open set $W \subseteq \mathbb{R}^m$ containing b and such that, on any open subset B of W that contains b and is not a union of two nonempty disjoint open sets, there exists a unique continuous map $G: B \rightarrow \mathbb{R}^n$ for which $G(b) = a$ and $f(G(y), y) = 0$ $\forall$ $y \in B$.

5

Extrema

5-1 Necessary Conditions

In an optimisation problem, the objective is to locate a maximum or minimum (or extremum) of some function, often called the *objective function*. The techniques of solving problems where the objective function depends only on one variable are introduced in an elementary calculus course soon after the concept of derivative of a function of one variable is discussed. The optimisation of functions of several variables is discussed after the concept of partial derivatives of such functions has been introduced. Sometimes, the chosen variables may have one or more quantitative relations between them in the form of equations, usually referred to as *constraints*. The method of Lagrange multipliers, which is used to deal with such problems, is also discussed in multivariable calculus. The reader is presumed to be adept at implementing the method in specific instances.

The purpose of this chapter is to discuss from a theoretical perspective the methods of optimisation, constrained as well as unconstrained, of functions of several variables. We shall begin with a formal definition of local maximum and minimum.

5-1.1. Definition. *A real valued function f on a domain $S \subseteq \mathbb{R}^n$ is said to have a* **local maximum** *at a point $x \in S$ if there is some $\delta > 0$ such that*

$$\|y - x\| < \delta \quad \Rightarrow \quad y \in S, \ f(y) \le f(x).$$

If the stronger condition that

$$y \in S \quad \Rightarrow \quad f(y) \le f(x)$$

holds, then f is said to have an **absolute maximum** *at x. Similarly for* **local** *and* **absolute minimum**. As is customary, the word *extremum* will be understood to mean maximum or minimum.

In case the stricter condition that $0 < \|y - x\| < \delta \Rightarrow y \in S, \ f(y) < f(x)$ holds, we speak of a **local strict maximum** at x. Similarly for **local strict minimum**.

Suppose f has a local extremum at x and h is any nonzero element of $\mathbb{R}^n$, which we wish to take as a direction vector. The function $\phi:\left(-\delta/\|h\|, \delta/\|h\|\right) \to \mathbb{R}$ defined by $\phi(\xi) = f(x + \xi h)$ then has an extremum at $\xi = 0$. Now suppose further that f has a derivative at x in the direction h, which we shall denote by $(D_h f)(x)$. Since

S. Shirali, H.L. Vasudeva, *Multivariable Analysis*,
DOI 10.1007/978-0-85729-192-9_5, © Springer-Verlag London Limited 2011

$$\frac{\phi(t)-\phi(0)}{t} = \frac{f(x+th)-f(x)}{t} \qquad \text{for } 0 < |t| < \delta/\|h\|,$$

$\phi'(0)$ exists and equals $(D_h f)(x)$. By an elementary result about functions on intervals (see, e.g., Shirali and Vasudeva [23, Proposition 9-5.2] or Berberian [3, Theorem 8.4.4]), $\phi'(0)$ must be 0; therefore $(D_h f)(x)$ must also be 0.

In summary, if a function has a directional derivative at a point where it has a local extremum, then that derivative must be 0. In particular, a partial derivative that exists at a point of local extremum is always 0. The same is then true regarding the linear derivative.

Vanishing of partial derivatives is thus a necessary condition for a local extremum; however, it is far from being sufficient, as is illustrated by the function $f(x,y) = xy$, which has no local extremum at $(0,0)$ although its partial derivatives vanish there. Sufficient conditions are discussed in the next section.

We go on to consider what are called 'constrained extrema'.

Suppose it is required to find a point of extremum of a real valued function ϕ on some open set $S \subseteq \mathbb{R}^k$ subject to n 'constraints'

$$f_1(x_1,\dots,x_k) = 0, \ f_2(x_1,\dots,x_k) = 0, \dots, f_n(x_1,\dots,x_k) = 0.$$

What this means is that a point of extremum is to be found for the function obtained by restricting ϕ to the subset of S described by the n equations called constraints. The number of constraints n is taken to be less than the number of variables k, so that the subset described by them does not reduce to a single point or the empty set. Thus $n < k$.

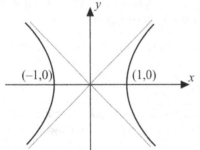

As an example with $k = 2$ and $n = 1$, we consider minimising $\phi(x,y) = x^2 + y^2$ subject to the single constraint $f_1(x,y) = x^2 - y^2 - 1 = 0$. In terms of analytic geometry, this asks for the point on the hyperbola $x^2 - y^2 - 1 = 0$ that is nearest to the origin. From the adjoining graph, the reader can see that the nearest points are $(\pm 1, 0)$. Note that the set described by the constraint, to which ϕ is restricted, has no interior points and therefore one cannot speak of a local maximum or minimum of ϕ in the sense discussed so far. A modified definition is as below.

5-1.2. Definition. *Let* $\phi: S \to \mathbb{R}$ *and* $f: S \to \mathbb{R}^n$ *be functions on a subset* $S \subseteq \mathbb{R}^k$ *and* $T = \{x \in S : f(x) = 0\}$. *Then* ϕ *is said to have a* **constrained local maximum** *at* $a \in T$ *if there is some* $\delta > 0$ *such that*

$$\|x - a\| < \delta, \ x \in T \quad \Rightarrow \quad \phi(a) \geq \phi(x).$$

Similarly for **constrained local minimum**. As is customary, the word *extremum* will be understood to mean maximum or minimum.

When we speak of a constrained local extremum, some constraint $f(x) = 0$ is taken as understood from the context even when none may be stated explicitly.

A glance at the graph of the hyperbola in the example above will convince the reader that a number δ as in Def. 5-1.2 indeed exists for each of the points $(1,0)$ and $(-1,0)$.

Presumably, the reader is aware of the method of Lagrange multipliers for solving such constrained extremum problems in calculus, e.g., as in Thomas and Finney [27], and can independently verify that a straightforward use of the method yields both the solutions of the above problem.

Our purpose here is to give a theoretical justification why the Lagrange multiplier equations constitute a necessary condition. Some observations would be in order before we proceed.

The constraint in the foregoing example requires that $x^2 = y^2 + 1$, which leads to $\phi(x,y) = 2y^2 + 1 = \Phi(y)$, say. Since the equality $x^2 = y^2 + 1$ puts no restriction on y, the function Φ is to be minimised on the domain $\mathbb{R}$. Therefore, if a minimum exists, it must be a local minimum and we can find it by setting $\Phi'(y)$ equal to 0, whereby we obtain $y = 0$. The constraint then shows that $x = \pm 1$.

At first sight, it may seem that we could just as well have rephrased the constraint as $y^2 = x^2 - 1$, which leads to $\phi(x,y) = 2x^2 - 1 = \Phi(x)$, say. The equality $y^2 = x^2 - 1$ restricts x^2 to be greater than or equal to 1 and Φ is therefore to be minimised on the domain $\{x \in \mathbb{R} : x^2 \geq 1\}$; the minimum occurs when $x = \pm 1$, which is not an interior point of the domain. Thus, we do not have a local minimum, and mindlessly setting the derivative equal to 0 results in the disaster that $x = 0$ is the critical value. The present rephrasing of the constraint has brought about a situation in which the first derivative test for a local extremum is not applicable.

Although the Lagrange multiplier avoids solving the constraint equations and obtaining the function Φ, its theoretical justification will nevertheless be in terms of this function. In order to ensure that we do not have the situation illustrated in the preceding paragraph, care has to be taken that the constrained local extremum of ϕ corresponds to a *local* extremum of Φ. In the next two paragraphs, we indicate how this is going to be done.

Suppose that from the n constraints

$$f_1(x_1,\ldots,x_k) = 0,\ f_2(x_1,\ldots,x_k) = 0,\ldots,\ f_n(x_1,\ldots,x_k) = 0,$$

one can express n variables, the first n say, in terms of the remaining $k - n$ variables as

$$x_1 = g_1(x_{n+1}, \ldots, x_k), \; x_2 = g_2(x_{n+1}, \ldots, x_k), \ldots, \; x_n = g_n(x_{n+1}, \ldots, x_k).$$

What this means is that the point

$$(g_1(x_{n+1}, \ldots, x_k), g_2(x_{n+1}, \ldots, x_k), \ldots, g_n(x_{n+1}, \ldots, x_k), x_{n+1}, \ldots, x_k)$$

always belongs to the set T defined by the constraints. As was seen in connection with the implicit function theorem, one can at best expect such a thing to be valid in some open set $W \subseteq \mathbb{R}^{k-n}$ containing $(a_{n+1}, \ldots, a_k)$, where $(a_1, \ldots, a_k)$ is a given point of T. If $g = (g_1, \ldots, g_n)$ is continuous, then so is the map

$$(x_{n+1}, \ldots, x_k) \to (g_1(x_{n+1}, \ldots, x_k), g_2(x_{n+1}, \ldots, x_k), \ldots, g_n(x_{n+1}, \ldots, x_k), x_{n+1}, \ldots, x_k).$$

This continuity has the consequence that for any open ball B centred at $(a_1, \ldots, a_k)$, a sufficiently small open subset of W that contains $(a_{n+1}, \ldots, a_k)$ is mapped into $B \cap T$. Hence, if B has the property that $\phi(x_1, \ldots, x_k) \geq \phi(a_1, \ldots, a_k)$ for every $(x_1, \ldots, x_k) \in B \cap T$, then the function Φ defined on W by

$$\Phi(x_{n+1}, \ldots, x_k) = \phi(g_1(x_{n+1}, \ldots, x_k), g_2(x_{n+1}, \ldots, x_k), \ldots, g_n(x_{n+1}, \ldots, x_k), x_{n+1}, \ldots, x_k)$$

has a local minimum at $(a_{n+1}, \ldots, a_k)$. This will play a role in the proof below.

Now suppose further that the ball B can be so chosen that every $(x_1, \ldots, x_k) \in B \cap T$ is of the form

$$(g_1(x_{n+1}, \ldots, x_k), g_2(x_{n+1}, \ldots, x_k), \ldots, g_n(x_{n+1}, \ldots, x_k), x_{n+1}, \ldots, x_k).$$

Part (b) of Theorem 4-3.2 assures us that B can indeed be chosen in this manner. Then the converse of the conclusion of the previous paragraph holds: if Φ has a local minimum at $(a_{n+1}, \ldots, a_k)$, it follows that $\phi(x_1, \ldots, x_k) \geq \phi(a_1, \ldots, a_k)$ for every $(x_1, \ldots, x_k) \in B \cap T$, which means ϕ has a constrained local minimum at $(a_1, \ldots, a_k)$. This will play a role in a subsequent proof [see Theorem 5-2.9].

It should be noted that corresponding features are true for a local strict minimum. Moreover, similar statements are true about Φ having a local (strict) maximum when the inequalities above are reversed.

The function g is going to be obtained by an application of the implicit function theorem and it will therefore be convenient to switch to the notation we have already used there. Accordingly, we denote $(x_1, \ldots, x_n) \in \mathbb{R}^n$ by x and $(x_{n+1}, \ldots, x_k) \in \mathbb{R}^{k-n}$ by y. Also, m will denote $k - n$ and we are free to use the symbol k to mean something else.

Before proceeding further, the reader would do well to review the statement of Implicit Function Theorem 4-3.2 and check out what A_1 and A_2 denote there, especially the relation between them. Remarks 4-3.1 and 4-3.3 are crucial to what follows.

5-1.3. Theorem. *Let $\phi : S \to \mathbb{R}$ and $f : S \to \mathbb{R}^n$ be continuously differentiable functions on an open subset $S \subseteq \mathbb{R}^n \times \mathbb{R}^m$ and $T = \{(x, y) \in S : f(x, y) = 0\}$. Suppose ϕ has a constrained local extremum at $(a, b) \in T$, the constraint being that $f(x, y) =$*

0. *Assume the linear derivative A_1 of f with respect to x at (a,b), i.e., the map $A_1: \mathbb{R}^n \to \mathbb{R}^n$ such that $A_1 h = f'(a,b)(h,0)$, to be invertible. Then there exist n real numbers $\lambda_1, \ldots, \lambda_n$ such that*

$$(D_j \phi)(a,b) + \sum_{p=1}^{n} \lambda_p (D_j f_p)(a,b) = 0 \quad \text{for } 1 \leq j \leq n + m. \tag{1}$$

Proof. We shall need not only the linear map $A_1: \mathbb{R}^n \to \mathbb{R}^n$ but also the associated map $A_2: \mathbb{R}^m \to \mathbb{R}^n$ defined as $A_2 k = f'(a,b)(0,k)$, and the corresponding maps $B_1: \mathbb{R}^n \to \mathbb{R}$ and $B_2: \mathbb{R}^m \to \mathbb{R}$ with ϕ in place of f. Besides, we shall need the following property of B_1 and B_2:

$$[\phi'(a,b)](h,k) = [\phi'(a,b)][(h,0) + (0,k)]$$

$$= B_1 h + B_2 k \qquad \text{for } (h,k) \in \mathbb{R}^n \times \mathbb{R}^m. \tag{2}$$

Since A_1 is invertible, the implicit function theorem provides a continuously differentiable function g on some open set W containing b such that

and
$$g(b) = a, \qquad y \in W \Rightarrow (g(y),y) \in T \tag{3}$$

$$g'(b) = -A_1^{-1} A_2. \tag{4}$$

By the chain rule, the map $\Phi: W \to \mathbb{R}$ given by

$$\Phi(y) = \phi(g(y),y)$$

has linear derivative at b given by $\Phi'(b) = \phi'(g(b),b)P$, where P is the linear derivative at b of the map $y \to (g(y),y)$. By Remark 4-3.1, P is given by

$$P(k) = (g'(b)k, k) \qquad \text{for } k \in \mathbb{R}^m,$$

and hence $\Phi'(b)$ is given by

$$\Phi'(b)(k) = \phi'(g(b),b)(g'(b)k, k)$$

$$= \phi'(a,b)(g'(b)k, k)$$

$$= B_1(g'(b)k) + B_2 k, \quad \text{in view of (2)}.$$

Thus, $\Phi'(b) = B_1 g'(b) + B_2$ and it follows from (4) that

$$\Phi'(b) = -(B_1 A_1^{-1})A_2 + B_2. \tag{5}$$

Now, it is a consequence of (3), the continuity of g and the hypothesis of a constrained local extremum at (a,b) that Φ has a local extremum at b, so that $\Phi'(b) = 0$. By (5), this means

$$-(B_1 A_1^{-1})A_2 + B_2 = 0. \tag{6}$$

If we set $\lambda = -B_1 A_1^{-1}$, we have

$$\lambda A_1 + B_1 = 0 \tag{7}$$

by the very definition of λ, and from (6), we also have

$$\lambda A_2 + B_2 = 0. \tag{8}$$

Since λ is a linear map from $\mathbb{R}^n$ to $\mathbb{R}$, it is represented by a $1 \times n$ matrix $[\lambda_1, \ldots, \lambda_n]$. Also, A_1 and A_2 are represented by the respective matrices [see Remark 4-3.3]

$$\begin{bmatrix} D_1 f_1 & \cdots & D_n f_1 \\ D_1 f_2 & \cdots & D_n f_2 \\ & \ddots & \\ D_1 f_n & \cdots & D_n f_n \end{bmatrix} \quad \text{and} \quad \begin{bmatrix} D_{n+1} f_1 & \cdots & D_{n+m} f_1 \\ D_{n+1} f_2 & \cdots & D_{n+m} f_2 \\ & \ddots & \\ D_{n+1} f_n & \cdots & D_{n+m} f_n \end{bmatrix},$$

while

$$\begin{bmatrix} D_1 \phi & \cdots & D_n \phi \end{bmatrix} \quad \text{and} \quad \begin{bmatrix} D_{n+1} \phi & \cdots & D_{n+m} \phi \end{bmatrix},$$

respectively represent B_1 and B_2, all partial derivatives being understood as taken at (a,b). Using these matrix representations, we find that (7) asserts the first n equations in (1) and (8) asserts the remaining ones.   □

5-1.4. Remarks. (a) Since the order of variables can be changed in an extremum problem, we can choose any n variables as constituting x in the above theorem, not necessarily the first n. If some other n variables are chosen, then the matrix of A_1 will consist not of the first n columns of the Jacobian matrix of $f'(a, b)$ as in the above proof, but the n columns corresponding to the variables chosen. Therefore the condition that A_1 is invertible effectively says that the $n \times n$ matrix formed by some n columns of the $n \times (n+m)$ Jacobian matrix of $f'(a, b)$ is invertible. It can happen that no matter which n columns we select, the $n \times n$ matrix formed by them fails to be invertible. One of the examples below will illustrate what can happen in that event.

   In the actual instances we discuss, the number of constraints will be either one or two, and correspondingly, n will be either 1 or 2. Therefore we shall be able to check the above condition on the Jacobian matrix without recourse to the methods that are available in linear algebra.

   In fact, when $n = 1$, the Jacobian matrix is $1 \times (1+m)$ and the condition is simply that some entry in the matrix, i.e., some partial derivative, is nonzero. When $n = 2$, the Jacobian matrix is $2 \times (2+m)$ and the condition is that some 2×2 'submatrix' is invertible; this can easily be seen to be equivalent to the two rows of the Jacobian matrix not being proportional to each other.

(b) Let us consider the example discussed just before Def. 5-1.2, namely, the problem of minimising $\phi(x,y) = x^2 + y^2$ subject to the constraint $f_1(x,y) = x^2 - y^2 - 1 = 0$. We can choose either x or y to play the role of x in the theorem, as long as the conditions are satisfied. At the points of minimum $(\pm 1, 0)$, we find that $\partial f_1 / \partial x$ is nonzero but $\partial f_1 / \partial y$ is zero. Therefore, the condition of invertibility is not fulfilled if the present y is chosen to play the role of x in the theorem. This is what was behind the failure of the attempt to find the extremum by expressing y in terms of x from the constraint, which had led to $\phi(x,y) = 2x^2 - 1 = \Phi(x)$.

When we discuss a concrete example or problem, it will be convenient to use the following standard terminology:

- The functions f and ϕ will be called the *constraint function* and *objective function*, respectively.

- The set T of the theorem will be called the *constraint set*.

- The equations $(D_j\,\phi)(x,y) + \sum_{p=1}^{n}\lambda_p(D_j f_p)(x,y) = 0$ $(1 \leq j \leq n + m)$ will be called the *Lagrange equations*.

- The equations $f_p(x,y) = 0$ $(1 \leq p \leq n)$ will be called the *constraint equations*.

- The λ_p will be called the *Lagrange multipliers*.

- The function $\phi + \sum_{p=1}^{n}\lambda_p f_p$ will be called the *Lagrangian*.

We shall use the phrase 'invertibility condition holds at' a point of the constraint set to mean that for *some choice* of x (i.e., for some n variables of the problem), the linear derivative of the constraint function with respect to x is invertible at the point. This means that the 'invertibility condition fails at' a point of the constraint set if, for *every choice* of n variables of the problem, the linear derivative of the constraint function with respect to x fails to be invertible at the point.

Using this terminology, Theorem 5-1.3 can be summarised as saying that if a local extremum occurs at a point where the invertibility condition holds, then the Lagrange equations are satisfied at that point. This leaves open the possibility that an extremum occurs at a point where the invertibility condition fails and the Lagrange equations are not satisfied. Then it would be futile to check for the condition *after* solving the Lagrange equations. We recommend that the points (if any) of the constraint set at which the invertibility condition fails, be checked first to see if any of them is an extremum. Usually textbook problems do not have such points, because authors are not so sadistic as to include problems that do. There are exceptions, though.

5-1.5. Examples. (a) Consider the problem of finding all constrained local extrema of $\phi(x,y) = 1 - (x+y) + xy$ (objective function) subject to the constraint $f(x,y) = x^2 + y^2 - 1 = 0$. The Jacobian matrix of the constraint function is $[2x \quad 2y]$. Since the entries cannot both become zero at any point where the constraint is satisfied, the invertibility condition holds on the entire constraint set. Therefore the Lagrange equations must be satisfied at any point of local extremum. Let us solve the Lagrange equations

$$-1 + y + \lambda(2x) = 0, \qquad -1 + x + \lambda(2y) = 0$$

together with the constraint

$$x^2 + y^2 - 1 = 0.$$

The Lagrange equations lead to $2\lambda(x-y)-(x-y)=0$, from which we deduce that $x = y$ or $\lambda = \frac{1}{2}$. In the former case, the constraint equation leads to $x = y = \pm 1/\sqrt{2}$. In the latter case, the Lagrange equations leads to $x + y = 1$, which along with the constraint equation, yields $xy = 0$, so that either $x = 1, y = 0$ or $x = 0, y = 1$. Thus we get the four points $P_1 = (1/\sqrt{2}, 1/\sqrt{2})$, $P_2 = (-1/\sqrt{2}, -1/\sqrt{2})$, $P_3 = (1,0)$ and $P_4 = (0,1)$. (It may be noted that P_1, P_2 correspond to $\lambda = \frac{1}{2}(\sqrt{2}-1)$ and P_3, P_4 correspond to $\lambda = \frac{1}{2}$.) Evaluation of ϕ at these points shows that ϕ attains its minimum at P_3 and P_4 and maximum at P_2. It remains to determine whether P_1 is a point of constrained local extremum. To this end, we note that P_1 satisfies $x \geq 0$, $y \geq 0$ and consider the related problem of finding extrema of $1 - (x + y) + xy$ subject to $y = \sqrt{(1 - x^2)}$, $0 \leq x \leq 1$. This is precisely equivalent to the problem of extrema of $\Phi(x) = 1 - [x + \sqrt{(1 - x^2)}] + x\sqrt{(1 - x^2)}$ on the interval $[0,1]$. Routine computations show that Φ attains its maximum value at $x = 1/\sqrt{2}$. While we know that, if ϕ has a constrained local maximum at $(1/\sqrt{2}, 1/\sqrt{2})$, then Φ has a local maximum at $1/\sqrt{2}$, we shall now argue the converse. [But see 5-1.P2.] The just proven fact that Φ attains its maximum value at $x = 1/\sqrt{2}$ means that $0 \leq x \leq 1 \Rightarrow \Phi(x) \leq \Phi(1/\sqrt{2})$. In view of the aforementioned equivalence,

$$0 \leq x \leq 1, \ y \geq 0, \ y = \sqrt{(1 - x^2)} \Rightarrow \phi(x,y) = 1 - (x + y) + xy \leq \Phi(1/\sqrt{2}).$$

When $y \geq 0$, the function $y = \sqrt{(1 - x^2)}$ provides the *unique* solution of the constraint equation. Therefore, the statement displayed above can be rephrased as

$$0 \leq x \leq 1, \ y \geq 0, \ x^2 + y^2 - 1 = 0 \Rightarrow \phi(x,y) = 1 - (x + y) + xy \leq \Phi(1/\sqrt{2}).$$

Now consider the open ball of radius $1 - 1/\sqrt{2}$ centred at $P_1 = (1/\sqrt{2}, 1/\sqrt{2})$. If (x,y) lies in this ball, it satisfies $0 \leq x \leq 1$, $y \geq 0$; if it also satisfies the constraint, then it follows that $\phi(x,y) \leq \Phi(1/\sqrt{2}) = \phi(1/\sqrt{2}, 1/\sqrt{2})$. Thus, ϕ has a constrained local maximum at P_1.

(b) The constraint $y^5z + z^5x + x^5y = 3(\frac{\pi}{4})^6$ holds at $(\frac{\pi}{4}, \frac{\pi}{4}, \frac{\pi}{4})$. If one attempts to maximise or minimise $\tan x + \tan y + \tan z$ (locally) subject to the constraint, the point $(\frac{\pi}{4}, \frac{\pi}{4}, \frac{\pi}{4})$ immediately presents itself as one solution of the constraint and Lagrange equations, with $\lambda = -1/3(\frac{\pi}{4})^5$. However, settling its status (whether maximum or minimum or neither) by the sort of procedure adopted in the preceding example is not a realistic option. More on this in the next section.

(c) The Lagrange multiplier method furnishes those points of extremum where the invertibility condition holds. However, an extremum can occur at a point where the condition does not hold, as the following example shows. Suppose we

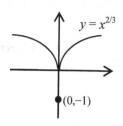

wish to minimise $x^2 + (y + 1)^2$ subject to $y^3 - x^2 = 0$. (Graphically, this amounts to minimising the distance from the point $(0,-1)$ on the y-axis to the curve $y = x^{2/3}$; see the accompanying figure to guess the solution right away.) The Jacobian matrix of the constraint function is $[-2x \quad 3y^2]$. There exists one point in the constraint set, namely $(0,0)$, where both entries of the matrix are zero. At any other point where a local extremum occurs, the Lagrange equations must hold: $2x - 2\lambda x = 0$ and $2(y + 1) + 3\lambda y^2 = 0$. It follows from these equations and the constraint that $x \neq 0$ and hence that $\lambda = 1$ and $3y^2 + 2y + 2 = 0$. However, there is no such real number y! Therefore, if at all there is a minimum, it must occur at $(0,0)$. Although the Lagrange equations did not yield this point, one can directly verify by elementary methods that an absolute minimum indeed occurs there. To wit, $y^3 - x^2 = 0 \Rightarrow y \geq 0 \Rightarrow x^2 + (y + 1)^2 = y^3 + (y + 1)^2 \geq 1 = 0^2 + (0 + 1)^2$. What the Lagrange multiplier method has done for us is to guarantee that there is no local extremum other than $(0,0)$.

Problem Set 5-1

5-1.P1. Vanishing of the first derivative is not a sufficient condition for an extremum, e.g., $y = x^3$ has no extremum at 0 although its derivative vanishes there. Use this to show that the Lagrange equations can hold when there is no local extremum.

5-1.P2. Let $S \subseteq \mathbb{R}^n \times \mathbb{R}^m$ be open and $T = \{(x,y) \in S : f(x,y) = 0\}$, where f is a map from S to $\mathbb{R}^n$. Suppose $(a,b) \in T$, $W \subseteq \mathbb{R}^m$ is open, $b \in W$ and $g:W \to \mathbb{R}^n$ is the unique map such that $(g(y),y) \in T$ whenever $y \in W$. Then prove the following:
(a) $y \in W, (x,y) \in T \Rightarrow x = g(y)$. In particular, $a = g(b)$.
(b) If $\phi:S \to \mathbb{R}$ has the property that the map $\Phi:W \to \mathbb{R}$ defined by $\Phi(y) = \phi(g(y),y)$ has a local minimum at b, then ϕ has a constrained local minimum at $(a,b) \in T$, the constraint being that $f(x,y) = 0$.

5-1.P3. Minimise $x^2 + y^2 + z^2$ subject to $x - y + z = 2$ and $2x + y + 4z = 16$, given that a minimum exists.

5-1.P4. Use the Lagrange multiplier method to find
(a) the point on the line $x + y = 4$ that is closest to the circle $x^2 + y^2 = 1$;
(b) the point on the circle $x^2 + y^2 = 1$ that is closest to the line $x + y = 4$.

5-1.P5. Find all solutions to the Lagrange equations for minimising the (square of the) distance between two points (x,y) and (u,v), subject to the two constraints that (x,y) lie on the circle $x^2+y^2=1$ and (u,v) lie on the line $u+v=4$. Check whether there are any points of extremum other than the solutions obtained.

5-1.P6. Find the absolute maximum and minimum values of $x^2+y^2+z^2$ subject to the constraints

$$\frac{x^2}{4}+\frac{y^2}{5}+\frac{z^2}{25}=1 \quad \text{and} \quad z=x+y.$$

5-1.P7. Solve the problem of finding all absolute minima of $x^2+y^2+(z-1)^2$ subject to the constraint $3x^2-2xy+2y^2-2x-6y+7=0$ by (a) the Lagrange multiplier method and (b) converting it to an unconstrained problem.

5-1.P8. Choose polar coordinates (r,θ) in the plane so as to have $-\frac{\pi}{2}<\theta\le\frac{\pi}{2}$ and r positive, zero or negative. Define $f:\mathbb{R}^2\to\mathbb{R}$ as

$$f(r,\theta)=\begin{cases} r^2 & \text{if } \theta=0 \\ r^2\cos\frac{r}{\theta} & \text{if } \theta\ne0 \end{cases}.$$

Show that
(a) f is continuous at the origin;
(b) f does not have a local minimum at the origin;
(c) the restriction of f to a line through the origin has a local strict minimum there.

5-1.P9. Define $f:\mathbb{R}^2\to\mathbb{R}$ as

$$f(x,y)=\begin{cases} x^2+y^2-2x^2y-\dfrac{4x^6y^2}{(x^4+y^2)^2} & \text{if } (x,y)\ne(0,0) \\ 0 & \text{if } (x,y)=(0,0). \end{cases}$$

(a) Show that f is continuous at $(0,0)$.
(b) For $-\frac{\pi}{2}<\theta\le\frac{\pi}{2}$ and $t\in\mathbb{R}$, define $g_\theta(t)=f(t\cos\theta,t\sin\theta)$. Show that $g_\theta(0)=0$, $g_\theta'(0)=0$ and $g_\theta''(0)=2$. Thus the restriction of f to a line through $(0,0)$ has a strict local minimum at $(0,0)$.
(c) Show that $(0,0)$ is not a local minimum for f by considering $f(x,x^2)$.

5-1.P10. Define $f:\mathbb{R}^2\to\mathbb{R}$ as

$$f(x,y) = \begin{cases} x^2 + y^2 & \text{if } y = 0 \\[2mm] (x^2 + y^2)\cos\dfrac{x^2 + y^2}{\arctan(y/x)} & \text{if } x \neq 0 \neq y \\[3mm] (x^2 + y^2)\cos\dfrac{x^2 + y^2}{\pi/2} & \text{if } x = 0 \neq y. \end{cases}$$

Show that
(a) f is continuous at $(0,0)$;
(b) the restriction of f to any line through $(0,0)$, i.e., $y = kx$ or $x = 0$, has a local strict minimum at $(0,0)$;
(c) f does not have a local minimum at $(0,0)$.

5-2 Sufficient Conditions

Recall the second derivative test for local extrema in a single variable. It asserts that a sufficient, but not necessary, condition for a function f to have a maximum (respectively, minimum) at a point x of its interval of definition is that f be differentiable on an open subinterval containing x with derivative 0 at x and that the second derivative at x be negative (respectively, positive) [see, e.g., Shirali and Vasudeva [23, Proposition 9-5.12]]. We begin by establishing the analogue for local extrema in $\mathbb{R}^n$.

5-2.1. Theorem. *Suppose that x is a point in the domain $S \subseteq \mathbb{R}^n$ of a real-valued function f such that the derivatives $D_j f$ $(1 \leq j \leq n)$ are differentiable at each point of some ball centred at x, while $D_j f(x) = 0$ $(1 \leq j \leq n)$ and the second partial derivatives $D_{ij} f$ are continuous at x. Let*

$$Q(h) = \sum_{j=1}^{n} h_j \left[\sum_{i=1}^{n} h_i D_{ij} f(x) \right] \quad \text{for } h \in \mathbb{R}^n.$$

(a) *If $Q(h) > 0$ for all nonzero $h \in \mathbb{R}^n$, then f has a local strict minimum at x.*
(b) *If $Q(h) < 0$ for all nonzero $h \in \mathbb{R}^n$, then f has a local strict maximum at x.*
(c) *If there exist nonzero h' and h'' in $\mathbb{R}^n$ such that $Q(h') > 0 > Q(h'')$, then f has neither a local minimum nor a local maximum at x.*

Proof. We begin by noting the property of Q that, for any $\lambda \in \mathbb{R}$ and any $h \in \mathbb{R}^n$,
$$Q(\lambda h) = \lambda^2 Q(h). \tag{1}$$

Since Q is continuous, it has a minimum value m and a maximum value M on the compact set $\{h \in \mathbb{R}^n : \|h\| = 1\}$. Then for any nonzero $h \in \mathbb{R}^n$, we have $m \leq Q(h/\|h\|) \leq M$, from which it follows by using (1) that

$$m\|h\|^2 \leq Q(h) \leq M\|h\|^2 \text{ for any } h \in \mathbb{R}^n. \tag{2}$$

Let r be the radius of the ball mentioned in the hypothesis. By Proposition 3-5.5, for any $h \in \mathbb{R}^n$ with $\|h\| < r$, there exists $\theta \in (0,1)$ such that

$$f(x+h) - f(x) = \sum_{j=1}^{n} h_j \cdot D_j f(x) + \frac{1}{2} \sum_{j=1}^{n} h_j [\sum_{i=1}^{n} h_i D_{ij} f(x+\theta h)]$$

$$= \frac{1}{2} \sum_{j=1}^{n} h_j [\sum_{i=1}^{n} h_i (D_{ij} f(x+\theta h) - D_{ij} f(x))] + \frac{1}{2} Q(h). \qquad (3)$$

(a) Under the hypothesis here, $m > 0$. By continuity of $D_{ij}f$, there exists a positive $\delta < r$ such that

$$\sum_{j=1}^{n} \sum_{i=1}^{n} |D_{ij} f(x+h) - D_{ij} f(x)| < \frac{1}{2} m \text{ whenever } \|h\| < \delta.$$

Since $\theta \in (0,1)$, we have $\|\theta h\| \leq \|h\|$ and hence

$$\sum_{j=1}^{n} \sum_{i=1}^{n} |D_{ij} f(x+\theta h) - D_{ij} f(x)| < \frac{1}{2} m \text{ whenever } \|h\| < \delta.$$

Now, $|h_i| \leq \|h\|$ for all i. Therefore

$$|\sum_{j=1}^{n} h_j [\sum_{i=1}^{n} h_i (D_{ij} f(x+\theta h) - D_{ij} f(x))]| \leq \frac{1}{2} m \|h\|^2$$

and hence from (3) and then (2), we obtain

$$f(x+h) - f(x) \geq \frac{1}{2} m \|h\|^2 - \frac{1}{4} m \|h\|^2 = \frac{1}{4} m \|h\|^2 \quad \text{for } \|h\| < \delta.$$

This shows that f has a local strict minimum at x.

(b) This argument is similar but $M < 0$ and $\frac{1}{2} m$ is to be replaced by $-\frac{1}{2} M$.

(c) In view of (1), we may assume that $\|h'\| = 1 = \|h''\|$. By continuity of $D_{ij}f$, there exists a positive $\delta < r$ such that $\sum_{j=1}^{n} \sum_{i=1}^{n} |D_{ij} f(x+h) - D_{ij} f(x)| < Q(h')$ whenever $\|h\| < \delta$. Then

$$|\sum_{j=1}^{n} h_j [\sum_{i=1}^{n} h_i D_{ij} f(x+h) - D_{ij} f(x)])| < Q(h') \|h\|^2 \quad \text{whenever } 0 < \|h\| < \delta.$$

Therefore, when $h = \lambda h'$, $0 < \lambda < \delta$, we have $\|h\|^2 = \lambda^2$ and it follows from (3) and (1) that

$$f(x+h) - f(x) > \frac{1}{2} Q(h) - \frac{1}{2} Q(h') \|h\|^2 = \frac{1}{2} \lambda^2 Q(h') - \frac{1}{2} Q(h') \lambda^2 = 0.$$

This shows that f does not have a local maximum at x. A similar argument shows that f also does not have a local minimum at x. □

5-2.2. Remark. In the preceding section we mentioned vanishing of first partial derivatives as a necessary condition for a local extremum; second partial derivatives were not involved and were not even assumed to exist. When the latter exist, the argument of part (c) of the above theorem can be used for obtaining an additional necessary condition for a local maximum (respectively, minimum), namely, that $Q(h) \leq 0$ (respectively, $Q(h) \geq 0$) for all nonzero $h \in \mathbb{R}^n$, provided of course that the conditions therein about differentiability and continuity are fulfilled. Indeed, if $Q(h') > 0$ for some nonzero h', then one can reason exactly as in (c) above that f does not have a local maximum at x.

In concrete applications, the conditions stipulated within (a), (b) or (c) have to be verified for the function at hand. Methods for doing so are available through linear algebra, but we shall not discuss them. The verification is easy to carry out in some instances, as we now illustrate:

Example. We seek the extrema of

$$f(x_1, x_2, x_3) = 3x_1^2 + 2x_2^2 + 4x_3^2 + 4x_3x_1 - 10x_1 - 4x_2 - 12x_3 \quad \text{on } \mathbb{R}^3.$$

The first partial derivatives are

$$D_1 f = 6x_1 + 4x_3 - 10, \qquad D_2 f = 4x_2 - 4, \qquad D_3 f = 8x_3 + 4x_1 - 12.$$

It is immediate that all three vanish at $(x_1, x_2, x_3) = (1,1,1)$ and nowhere else. So a local extremum, if any, must occur at this point. In order to apply the theorem, we compute second partial derivatives, which happen to be constants:

$$\begin{array}{lll}
D_{11}f = 6 & D_{21}f = 0 & D_{31}f = 4 \\
D_{12}f = 0 & D_{22}f = 4 & D_{32}f = 0 \\
D_{13}f = 4 & D_{23}f = 0 & D_{33}f = 8
\end{array}$$

Hence Q is given by

$$Q(h_1, h_2, h_3) = 6h_1^2 + 4h_2^2 + 8h_3^2 + 8h_3 h_1.$$

Upon recasting this as

$$Q(h_1, h_2, h_3) = 2(h_1 + 2h_3)^2 + 4h_1^2 + 4h_2^2,$$

we see that $Q > 0$ unless $(h_1, h_2, h_3) = (0,0,0)$. From the theorem we can now conclude that f has a strict local minimum at $(x_1, x_2, x_3) = (1,1,1)$.

We have obtained this conclusion by routine computation. A skillful but elementary computation shows that

$$f(x_1, x_2, x_3) = (x_1 - x_2)^2 + (x_1 + x_2 - 2)^2 + (x_1 + 2x_3 - 3)^2 - 13,$$

whereby the same conclusion can be obtained without any differentiation.

The verification is even easier if either $[D_{ij}f(x)]$ is a diagonal matrix (which means $D_{ij}f(x) = 0$ for $i \neq j$) or $n = 2$. In the former case, we have $Q(h) = \sum_{i=1}^{n} D_{ii}f(x)h_i^2$, from which it follows that the condition of (a) (respectively, (b)) is fulfilled if each diagonal entry $D_{ii}f(x)$ is positive (respectively, negative) and that the condition of (c) is fulfilled if some diagonal entry is positive and another is negative. We first present an instance when this situation occurs and then go on to discuss the case when $n = 2$.

Example. Let $a > b > c > 0$. For the function

$$f(x,y,z) = (ax^2 + by^2 + cz^2)\exp(-x^2 - y^2 - z^2) \quad \text{on } \mathbb{R}^3,$$

we show that there are seven points where all partial derivatives vanish. We also show that the function has a local maximum at two of the seven points, a minimum at one and neither a maximum nor a minimum at the remaining four.

Since $\frac{\partial f}{\partial x} = 2x\exp(-x^2 - y^2 - z^2)[a - (ax^2 + by^2 + cz^2)]$, and similarly for $\frac{\partial f}{\partial y}, \frac{\partial f}{\partial z}$, the points where all three partial derivatives vanish are precisely the solutions of (1)–(3) below:

$$x = 0 \text{ or } a = ax^2 + by^2 + cz^2 \tag{1}$$

$$y = 0 \text{ or } b = ax^2 + by^2 + cz^2 \tag{2}$$

$$z = 0 \text{ or } c = ax^2 + by^2 + cz^2. \tag{3}$$

Now $(0,0,0)$ is a solution, where f obviously has a minimum. We shall obtain six others. If a solution (x,y,z) has two nonzero coordinates, say y and z, then (2) and (3) lead to $b = c$, which contradicts the hypothesis. Therefore a solution of (1)–(3) can have at most one nonzero coordinate. So, a solution with $x \neq 0$ must satisfy $y = 0 = z$ and hence by (1) must also satisfy $a = ax^2$, so that $x = \pm 1$. We conclude that $(\pm 1,0,0)$, $(0,\pm 1,0)$, $(0,0,\pm 1)$ are the only solutions of (1)–(3) besides $(0,0,0)$, thereby making a total of seven solutions.

It will now be shown that $(\pm 1,0,0)$ are points of maximum while $(0,\pm 1,0)$, $(0,0,\pm 1)$ provide neither a maximum nor a minimum. The second partial derivatives are given by

$$\frac{\partial^2 f}{\partial x^2} = 2\exp(-x^2 - y^2 - z^2)[a - (5ax^2 + by^2 + cz^2) + 2x^2(ax^2 + by^2 + cz^2)],$$

$$\frac{\partial^2 f}{\partial y \partial x} = -4xy\exp(-x^2 - y^2 - z^2)[(a + b) - (ax^2 + by^2 + cz^2)],$$

and correspondingly for the others. Upon evaluating them at $(\pm 1,0,0)$, we find that the Hessian matrix at both points is a diagonal matrix with entries $-4a/e, 2(b - a)/e, 2(c - a)/e$. All these are negative in view of the hypothesis that $a > b > c > 0$. As noted above, the condition stipulated in (a) of Theorem 5-2.1 is fulfilled and thus $(\pm 1,0,0)$ are points of maximum. Upon evaluating at $(0,\pm 1,0)$ however, we find that the Hessian matrix at these points is a diagonal matrix with entries $2(a - b)/e, -4b/e, 2(c - b)/e$, which are, respectively, positive, negative and negative. Thus the condition stipulated in (c) of Theorem 5-2.1 is fulfilled and $(0,\pm 1,0)$ are points of neither maximum nor minimum. At the remaining points $(0,0,\pm 1)$, the Hessian matrix turns out to be a diagonal matrix with entries $2(a - c)/e, 2(b - c)/e, -4c/e$, which are respectively positive, positive and negative. Once again by (c) of Theorem 5-2.1, $(0,0,\pm 1))$ are points of neither maximum nor minimum.

We go on to discuss the case when $n = 2$.

First of all, $Q(h)$ simplifies to $a_{11}h_1^2 + (a_{12} + a_{21})\, h_1 h_2 + a_{22} h_2^2$, where $a_{ij} = D_{ij}f(x)$, $i,j = 1,2$. By Young's theorem (Theorem 3-5.4), $a_{12} = a_{21}$ and therefore $Q(h)$ further simplifies to $a_{11}h_1^2 + 2a_{12}h_1 h_2 + a_{22}h_2^2$. When $a_{11} \neq 0$, an elementary computation gives

$$a_{11}h_1^2 + 2a_{12}h_1 h_2 + a_{22}h_2^2 = a_{11}\Big[\big(h_1 + \tfrac{a_{12}}{a_{11}}h_2\big)^2 - \frac{a_{12}^2 - a_{11}a_{12}}{a_{11}^2}h_2^2\Big],$$

from which it follows that a sufficient condition for $Q(h)$ to be positive for all nonzero (h_1,h_2) is that $a_{12}^2 < a_{11}a_{22}$ and $a_{11} > 0$. Obviously, the condition $a_{11} > 0$ can be replaced by $a_{22} > 0$. Similarly, a sufficient condition for $Q(h)$ to be negative for all nonzero (h_1,h_2) is that $a_{12}^2 < a_{11}a_{22}$ and either $a_{11} < 0$ or $a_{22} < 0$.

Next, suppose $a_{12}^2 > a_{11}a_{22}$. Then if $a_{11} \neq 0$, the equality displayed above also shows that $Q(h)$ can take positive as well as negative values; this happens even if $a_{11} = 0$, because in this situation, $a_{12}^2 > 0$ and $Q(h) = h_2(2a_{12}h_1 + a_{22}h_2)$, where $a_{12} \neq 0$. Let us summarise all of this as a theorem, keeping in view that $a_{ij} = D_{ij}f(x)$.

5-2.3. Theorem. *Suppose that x is a point in the domain $S \subseteq \mathbb{R}^2$ of a real valued function f such that the derivatives $D_1 f$ and $D_2 f$ are differentiable at each point of some ball (disc) centred at x, while $D_1 f(x) = 0 = D_2 f(x)$ and the second partial derivatives $D_{11}f$, $D_{12}f$ and $D_{22}f$ are continuous at x.*
(a) *If $D_{12}f(x)^2 < [D_{11}f(x)] \cdot [D_{22}f(x)]$ and $D_{11}f(x) > 0$ (or $D_{22}f(x) > 0$), then f has a local strict minimum at x.*
(b) *If $D_{12}f(x)^2 < [D_{11}f(x)] \cdot [D_{22}f(x)]$ and $D_{11}f(x) < 0$ (or $D_{22}f(x) < 0$), then f has a local strict maximum at x.*
(c) *If $D_{12}f(x)^2 > [D_{11}f(x)] \cdot [D_{22}f(x)]$ then f has neither a local minimum nor a local maximum at x.*

In the sufficient condition of Theorem 5-2.1, a crucial role is played by $Q(h)$. Note that Q is a special kind of a map from $\mathbb{R}^n$ to $\mathbb{R}$, which is closely related to a linear map without itself being linear. Such maps have a name:

5-2.4. Definition. *A map $Q:\mathbb{R}^n \to \mathbb{R}$ is called a **quadratic form** if there exists an $n \times n$ matrix $[a_{ij}]$ such that*

$$Q(h) = \sum_{j=1}^{n} h_j \big[\sum_{i=1}^{n} h_i a_{ij} \big] \quad \text{for } h \in \mathbb{R}^n.$$

*Q is said to be **positive definite** if $Q(h) > 0$ whenever $h \neq 0$ and **positive semidefinite** if $Q(h) \geq 0$ for all h. Similarly for **negative definite** and **negative semidefinite**.*

*When f is a real-valued function having second partial derivatives $D_{ij}f(x)$, the quadratic form given by the matrix $[D_{ij}f(x)]$ is called the **Hessian form** of f at x. The matrix is called the **Hessian matrix** of f at x.*

Thus, what we have called $Q(h)$ in Theorem 5-2.1 is in fact the Hessian form of f at the point x. We recount in terms of the concepts just defined what the three parts of the theorem say under the stated hypotheses. Part (a) says that if the Hessian form is positive definite at the point in question, then there is a local strict minimum, while part (b) says that if the Hessian form is negative definite, then there is a local strict maximum. Part (c) says that if the Hessian form is neither positive semidefinite nor negative semidefinite, then there is neither a local minimum nor a local maximum. The necessary condition for a local maximum in Remark 5-2.2 was that the Hessian form be negative semidefinite.

5-2.5. Examples. (a) Let $[a_{i\,j}]$ be the 3×3 matrix whose rows are, respectively,

$$[2 \quad 1 \quad 4], \qquad [1 \quad 1 \quad 3], \qquad [4 \quad 3 \quad 11].$$

It is a simple computation that the associated quadratic form Q is given by

$$Q(h_1, h_2, h_3) = 2h_1^2 + h_2^2 + 11h_3^2 + 6h_2h_3 + 8h_3h_1 + 2h_1h_2.$$

One can verify that this can be put in the form

$$Q(h_1, h_2, h_3) = (h_1 + h_2 + 3h_3)^2 + (h_1 + h_3)^2 + h_3^2.$$

It follows that Q is positive definite. If the 11 is changed to 10, then the last term h_3^2 will have to be deleted and Q will be positive semidefinite.

(b) Let $[a_{i\,j}]$ be the 2×2 matrix whose rows are, respectively,

$$[2\lambda \quad 1] \quad \text{and} \quad [1 \quad 2\lambda].$$

The associated quadratic form Q maps (h_1, h_2) into $2\lambda(h_1^2 + h_2^2) + 2h_1h_2$. When the value of λ is $\frac{1}{2}(\sqrt{2} - 1)$, this becomes

$$Q(h_1, h_2) = (\sqrt{2} - 1)(h_1^2 + h_2^2) + 2h_1h_2.$$

We find that Q is neither positive semidefinite nor negative semidefinite. However, when restricted to those (h_1, h_2) for which $h_1 = h_2$, it will behave as though it is positive definite; when restricted to those (h_1, h_2) for which $h_1 + h_2 = 0$, it will behave as though it is negative definite.

Now suppose $\alpha:\mathbb{R} \rightarrow \mathbb{R}^2$ is a linear map such that every (h_1, h_2) in the range of α satisfies $h_1 + h_2 = 0$; for instance,

$$\alpha(k) = (k, -k).$$

Then a consequence of the fact observed above is that in case the composition $Q \circ \alpha$—which surely maps $\mathbb{R}$ into $\mathbb{R}$—turns out to be a quadratic form in $\mathbb{R}$, then it is negative definite. Similarly, for $\alpha(k) = (k, k)$, the composition $Q \circ \alpha$ is positive definite if it is a quadratic form.

(c) Consider the 3×3 matrix whose rows are, respectively,

$$[A \quad B \quad B], \qquad [B \quad A \quad B], \qquad [B \quad B \quad A],$$

where A and B are distinct real numbers. The associated quadratic form Q is

$$Q(h_1, h_2, h_3) = A(h_1^2 + h_2^2 + h_3^2) + 2B(h_2h_3 + h_3h_1 + h_1h_2)$$

$$= B(h_1 + h_2 + h_3)^2 + (A - B)(h_1{}^2 + h_2{}^2 + h_3{}^2).$$

This shows that Q need not be positive (or negative) definite. However, when restricted to those $(h_1, h_2, h_3) \in \mathbb{R}^3$ for which $h_1 + h_2 + h_3 = 0$, it will behave as though it is positive or negative definite according as $A > B$ or $A < B$.

Now suppose $\alpha : \mathbb{R}^2 \to \mathbb{R}^3$ is a linear map such that every (h_1, h_2, h_3) in the range of α satisfies $h_1 + h_2 + h_3 = 0$; for instance,

$$\alpha(k_1, k_2) = (k_1, k_2, -(k_1 + k_2)) \quad \text{or} \quad (2k_1, 3k_2, -(2k_1 + 3k_2)).$$

Then a consequence of the fact observed above is that in case the composition $Q \circ \alpha$, which surely maps $\mathbb{R}^2$ into $\mathbb{R}$, is a quadratic form in $\mathbb{R}^2$, then it is positive or negative definite according as $A > B$ or $A < B$. That the composition will always be a quadratic form follows from the next result.

5-2.6. Proposition. *Let Q be a quadratic form in $\mathbb{R}^n$ with matrix $[a_{ij}]$ and $\alpha : \mathbb{R}^m \to \mathbb{R}^n$ be a linear map with matrix $[b_{ip}]$, $1 \le i \le n$ and $1 \le p \le m$. Then the composition $Q \circ \alpha : \mathbb{R}^m \to \mathbb{R}$ is a quadratic form in $\mathbb{R}^m$ with matrix $[c_{pq}]$, $1 \le p, q \le m$, where*

$$c_{pq} = \sum_{j=1}^{n} b_{jq} \Big(\sum_{i=1}^{n} a_{ij} b_{ip} \Big). \tag{A}$$

Proof. Consider any $k \in \mathbb{R}^m$. Since α has matrix $[b_{iq}]$, the jth component of $\alpha(k)$ is

$$\alpha(k)_j = \sum_{q=1}^{m} b_{jq} k_q.$$

Therefore

$$(Q \circ \alpha)(k) = \sum_{j=1}^{n} \alpha(k)_j \Big[\sum_{i=1}^{n} \alpha(k)_i a_{ij} \Big] = \sum_{j=1}^{n} \Big(\sum_{q=1}^{m} b_{jq} k_q \Big) \Big[\sum_{i=1}^{n} \Big(\sum_{p=1}^{m} b_{ip} k_p \Big) a_{ij} \Big]$$

$$= \sum_{q=1}^{m} k_q \Big[\sum_{p=1}^{m} k_p \Big[\sum_{j=1}^{n} b_{jq} \Big(\sum_{i=1}^{n} a_{ij} b_{ip} \Big) \Big] \Big]. \qquad \square$$

The above proposition can be reformulated as saying that $Q \circ \alpha$ is a quadratic form given by the matrix product $[b_{ip}]^T [a_{ij}][b_{ip}]$, where the superscript T indicates transpose.

It may be recalled from the proof of the necessary conditions for a constrained extremum [Theorem 5-1.3] that we used the composed function called Φ in the argument, although the point of the Lagrange multiplier method is to circumvent an explicit computation of Φ. Since sufficient conditions for an unconstrained extremum involve the Hessian form, one may anticipate that any proof concerning sufficient conditions for a constrained extremum would involve the Hessian form of the composed function Φ. Therefore it is useful to express the second partial derivatives of a composed function $\phi \circ G$ in terms of second and first derivatives of ϕ and G. We do so in the next proposition, by employing the chain rule with respect to partial derivatives as explained in Section 3-4.

5-2.7. Proposition. *Let U and V be open subsets of $\mathbb{R}^m$ and $\mathbb{R}^n$, respectively, and $G:U \to V$, $\phi:V \to \mathbb{R}$ be functions such that ϕ as well as all the n component functions G_j $(1 \le j \le n)$ of G have differentiable first partial derivatives everywhere. Then $\phi \circ G$ has differentiable first partial derivatives and at any $x \in U$, and for any p,q with $1 \le p \le m$, $1 \le q \le m$, we have*

$$[D_{p\,q}(\phi \circ G)](x) = \sum_{j=1}^{n} \Big[\big[\sum_{i=1}^{n} (D_{i\,j}\,\phi)(G(x)) \cdot (D_p\,G_i)(x)) \big] \cdot (D_q\,G_j)(x))$$

$$+ (D_j\,\phi)(G(x)) \cdot (D_{p\,q}\,G_j(x)) \Big], \tag{1}$$

where all multiplications of numbers are indicated by a dot $\cdot$. In particular, if the second partial derivatives of ϕ as well as of all the n component functions G_j are continuous at $G(a)$ and a respectively, then the second partial derivatives of $\phi \circ G$ are continuous at a.

Proof. By the chain rule, we have

$$[D_q(\phi \circ G)](x) = \sum_{j=1}^{n} (D_j\,\phi)(G(x)) \cdot (D_q\,G_j)(x)).$$

Since all first partial derivatives on the right side have been assumed differentiable, those of $\phi \circ G$ are also differentiable; besides, we can apply the chain rule once again and a routine computation leads to (1), which then implies the last statement. $\qquad\qquad\Box$

5-2.8. Remark. Continuing with the notation of the above proposition, we note that:

(a) The Hessian form $H_{\phi \circ G}$ at x of the composition $\phi \circ G$ is given by the matrix with (p,q)th entry $[D_{p\,q}(\phi \circ G)](x)$, which is the left side of (1).

(b) The Hessian form Q at $G(x)$ of ϕ is given by the matrix with (i,j)th entry

$$a_{i\,j} = (D_{i\,j}\,\phi)(G(x)).$$

(c) Lastly, the linear derivative $\alpha = G'(x)$ at x of G is given by the matrix with (i,p)th entry

$$b_{i\,p} = (D_p\,G_i)(x).$$

Therefore it follows from Proposition 5-2.6 that on the right side of (1), the first term (after implementing the double summation) is the (p,q)th entry of a matrix that gives the quadratic form $Q \circ \alpha$. In other words, (1) asserts that the Hessian matrix $H_{\phi \circ G}$ at x of the composition $\phi \circ G$ differs from a matrix that gives the quadratic form $Q \circ \alpha$ by

$$\sum_{j=1}^{n} (D_j \phi)(G(x)) \cdot (D_{p\,q} G_j(x)).$$

In particular, if this sum happens to be 0 for all p and q, then the Hessian form of $\phi \circ G$ is the same as the composition $Q \circ \alpha$, where Q is the Hessian form of ϕ at $G(x)$ and α is the linear derivative of G at x.

It may be pertinent here to note that for a twice differentiable function L defined on an open subset $S \subseteq \mathbb{R}^n \times \mathbb{R}^m$, the Hessian form at a point $(a,b) \in S$ is the quadratic form given by the matrix $[(D_{i\,j} L)(a,b)]$ of dimension $(n+m) \times (n+m)$. Also, it may be recalled that the function $L = \phi + \sum_{p=1}^{n} \lambda_p f_p$, the Hessian form of which will play a crucial role in the next result, is called the Lagrangian.

5-2.9. Theorem. *Let $\phi:S \to \mathbb{R}$ and $f:S \to \mathbb{R}^n$ be differentiable functions on an open subset $S \subseteq \mathbb{R}^n \times \mathbb{R}^m$ such that all the partial derivatives of ϕ and of every component function of f are differentiable. Let $T = \{(x,y) \in S : f(x,y) = 0\}$. Assume that*

(a) *all second partial derivatives are continuous at $(a,b) \in T$;*

(b) *the linear derivative A_1 of f with respect to x at (a,b), i.e., the map $A_1:\mathbb{R}^n \to \mathbb{R}^n$ such that $A_1 h = f'(a,b)(h,0)$, is invertible;*

(c) *there exist n real numbers $\lambda_1, \dots, \lambda_n$ such that*

$$(D_j \phi)(a,b) + \sum_{r=1}^{n} \lambda_r (D_j f_r)(a,b) = 0 \qquad \text{for } 1 \le j \le n+m. \qquad (1)$$

If the Hessian form H at (a,b) of the function L defined on S by

$$L(x,y) = \phi(x,y) + \sum_{r=1}^{n} \lambda_r f_r(x,y)$$

satisfies

$$H(u) > 0 \qquad \text{whenever} \qquad 0 \ne u \in \mathbb{R}^n \times \mathbb{R}^m \text{ and } f'(a,b)(u) = 0,$$

then ϕ has a constrained local strict minimum at (a,b), the constraint being that $f(x,y) = 0$. If the inequality is reversed, then ϕ has a constrained local strict maximum at (a,b).

Proof. We shall need not only the linear map $A_1:\mathbb{R}^n \to \mathbb{R}^n$ but also the associated map $A_2:\mathbb{R}^m \to \mathbb{R}^n$ defined as $A_2 k = f'(a,b)(0,k)$, and the corresponding maps $B_1:\mathbb{R}^n \to \mathbb{R}$ and $B_2:\mathbb{R}^m \to \mathbb{R}$ with ϕ in place of f. We shall use the following property of B_1 and B_2:

$$[\phi'(a,b)](h,k) = [\phi'(a,b)][(h,0) + (0,k)]$$
$$= B_1 h + B_2 k \qquad \text{for } (h,k) \in \mathbb{R}^n \times \mathbb{R}^m. \qquad (2)$$

Since A_1 is invertible, the implicit function theorem provides a continuously differentiable function g on some open set W containing b such that

$$g(b) = a, \qquad y \in W \Rightarrow (g(y),y) \in T \tag{3}$$

and

$$g'(b) = -A_1^{-1}A_2. \tag{4}$$

According to part (b) of the theorem just quoted, g maps into an open subset U of $\mathbb{R}^n \times \mathbb{R}^m$ such that every $(x,y) \in T \cap U$ is of the form $(g(y),y)$ with $y \in W$. This has the consequence that if the map $\Phi : W \to \mathbb{R}$ given by

$$\Phi(y) = \phi(g(y),y)$$

has a local strict minimum at b, then ϕ has a constrained local strict minimum at (a,b), the constraint being that $f(x,y) = 0$.

Therefore we need only show that Φ has a local strict minimum at b. With this in mind, we show first that $\Phi'(b) = 0$.

By the chain rule, the linear derivative at b of the map Φ is given by $\Phi'(b) = \phi'(g(b),b)G'(b)$, where G is the map

$$G(y) = (g(y),y), \; \forall\, y \in W. \tag{5}$$

By Remark 4-3.1, $G'(b) : \mathbb{R}^m \to \mathbb{R}^n \times \mathbb{R}^m$ is given by

$$G'(b)(k) = (g'(b)k, k) \qquad \forall\, k \in \mathbb{R}^m.$$

Note that it follows from this equality that $G'(b)$ is injective, something that we shall need only towards the end of the proof. It also follows from this equality that $\Phi'(b)$ is given by

$$\begin{aligned}
\Phi'(b)(k) &= \phi'(g(b),b)(g'(b)k, k) \\
&= \phi'(a,b)(g'(b)k, k) \\
&= B_1(g'(b)k) + B_2 k, \quad \text{in view of (2).}
\end{aligned}$$

Thus $\Phi'(b) = B_1 g'(b) + B_2$ and it follows from (4) that

$$\Phi'(b) = -(B_1 A_1^{-1})A_2 + B_2. \tag{6}$$

Let λ be the linear map from $\mathbb{R}^n$ to $\mathbb{R}$ represented by an $1 \times n$ matrix $[\lambda_1, \dots, \lambda_n]$. We shall argue that

$$\lambda A_1 + B_1 = 0 \tag{7}$$

and

$$\lambda A_2 + B_2 = 0. \tag{8}$$

To see why, observe that A_1 and A_2 are represented [see Remark 4-3.3] by the respective matrices

$$\begin{bmatrix} D_1 f_1 & \cdots & D_n f_1 \\ D_1 f_2 & \cdots & D_n f_2 \\ & \ddots & \\ D_1 f_n & \cdots & D_n f_n \end{bmatrix} \quad \text{and} \quad \begin{bmatrix} D_{n+1} f_1 & \cdots & D_{n+m} f_1 \\ D_{n+1} f_2 & \cdots & D_{n+m} f_2 \\ & \ddots & \\ D_{n+1} f_n & \cdots & D_{n+m} f_n \end{bmatrix},$$

while

$$[D_1\phi \quad \cdots \quad D_n\phi] \quad \text{and} \quad [D_{n+1}\phi \quad \cdots \quad D_{n+m}\phi],$$

respectively, represent B_1 and B_2, all partial derivatives being understood as taken at (a,b). Using these matrix representations, we find that the first n equations in (1) assert (7) and the remaining ones assert (8).

Now (7) can be rephrased as $\lambda = -B_1 A_1^{-1}$. Together with (8), this yields

$$-(B_1 A_1^{-1})A_2 + B_2 = 0. \tag{9}$$

By virtue of (6), this means $\Phi'(b) = 0$, as claimed.

In the light of Theorem 5-2.1, it remains only to prove that Φ has differentiable first partial derivatives, its second derivatives are continuous at b and that its Hessian form H_Φ at b is positive definite.

Since Φ is the composition $\phi \circ G$, where $G(y) = (g(y),y)$, it follows by Proposition 5-2.7 that firstly, it has differentiable first partial derivatives with second derivatives continuous at b and secondly, that

$$[D_{p\,q}(\phi \circ G)](y) = \sum_{j=1}^{n+m} \left[\left[\sum_{i=1}^{n+m} (D_{i\,j}\,\phi)(G(y)) \cdot (D_p\,G_i)(y)) \right] \cdot (D_q\,G_j)(y)) + \right.$$

$$\left. (D_j\,\phi)(G(y)) \cdot (D_{p\,q}\,G_j(y)) \right].$$

The same argument applies with the n components f_r in place of ϕ. This leads to n more equations like the one above, which the reader may choose either to imagine or to write down (without defacing this book!). However, in these n equations, the left sides will be 0 because $f((g(y),y) = 0$ everywhere. Therefore, upon multiplying them by $\lambda_1,\ldots,\lambda_n$, respectively, and adding to the equation displayed above, we get

$$[D_{p\,q}(\phi \circ G)](y) = \sum_{j=1}^{n+m} \left[\left[\sum_{i=1}^{n+m} (D_{i\,j}\,L)(G(y)) \cdot (D_p\,G_i)(y)) \right] \cdot (D_q\,G_j)(y)) \right]$$

$$+ \sum_{j=1}^{n+m} (D_j\,L)(G(y)) \cdot (D_{p\,q}\,G_j(y)).$$

Now the given equality (1) of the hypothesis implies that the second summation on the right side here is 0 when $y = b$. Consequently,

$$[D_{p\,q}(\phi \circ G)](b) = \sum_{j=1}^{n+m} \left[\left[\sum_{i=1}^{n+m} (D_{i\,j}\,L)(G(b)) \cdot (D_p\,G_i)(b)) \right] \cdot (D_q\,G_j)(b)) \right]$$

$$= \sum_{j=1}^{n+m} [(D_q\,G_j)(b)) \cdot \left[\sum_{i=1}^{n+m} (D_{i\,j}\,L)(G(b)) \cdot (D_p\,G_i)(b)) \right].$$

By Proposition 5-2.6, this means that, at the point b, the Hessian form H_Φ of $\Phi = \phi \circ G$ is the composition $H \circ \alpha$, where the linear map $\alpha : \mathbb{R}^m \to \mathbb{R}^n \times \mathbb{R}^m$ is represented by the matrix having $(D_p G_i)(b)$ as its (i,p)th entry. But this matrix represents precisely the linear derivative $G'(b)$. Thus

$$H_\Phi = H \circ G'(b).$$

With a view to checking the positive definiteness of H_Φ, consider any $h \in \mathbb{R}^m$, $h \neq 0$. The element $u = G'(b)(h) \in \mathbb{R}^n \times \mathbb{R}^m$ has the property that

$$f'(a,b)(u) = f'(a,b)(G'(b)(h)) = 0,$$

because

$$f'(a,b)(G'(b)(h)) = [f'(G(b))G'(b)](h) = [(f \circ G)'(b)](h) \text{ and } f \circ G = 0 \text{ on } W.$$

As already noted above, $G'(b)$ is injective, and therefore $u \neq 0$. The hypothesis about H now implies that $H(u) > 0$, so that $H_\Phi(h) = [H \circ G'(b)](h) = H(u) > 0$, confirming the positive definiteness of H_Φ.   □

5-2.10. Examples. (a) In 5-1.5(a), we settled the status of the point $P_1 = (1/\sqrt{2}, 1/\sqrt{2})$ by using an explicit solution of the constraint equation, which seems such a shame because that is precisely what the Lagrange multiplier method is meant to avoid. By means of the above theorem we can now handle the matter without resorting to an explicit solution (which may be considered the shameless way to do it!). The Hessian matrix of the Lagrangian $L(x,y) = \phi(x,y) + \lambda f(x,y)$ has rows

$$[2\lambda \quad 1] \quad \text{and} \quad [1 \quad 2\lambda].$$

Therefore the Hessian form Q maps (h_1,h_2) into $2\lambda(h_1^2 + h_2^2) + 2h_1 h_2$. For P_1, the value of λ is $\frac{1}{2}(\sqrt{2} - 1)$. So, as seen in Example 5-2.5(b), $Q(h_1,h_2) = (\sqrt{2} - 1)(h_1^2 + h_2^2) + 2h_1 h_2$ and it is negative for all nonzero (h_1,h_2) satisfying $h_1 + h_2 = 0$. We wish to know whether it is positive (or negative) for all nonzero (h_1,h_2) satisfying $f'(1/\sqrt{2}, 1/\sqrt{2})(h_1,h_2) = 0$. The linear derivative $f'(1/\sqrt{2}, 1/\sqrt{2})$ has the 1×2 matrix

$$[(D_1 f)(1/\sqrt{2}, 1/\sqrt{2}) \quad (D_2 f)(1/\sqrt{2}, 1/\sqrt{2})] = [\sqrt{2} \quad \sqrt{2}].$$

Therefore, $f'(1/\sqrt{2}, 1/\sqrt{2})(h_1,h_2) = 0 \Rightarrow h_1 + h_2 = 0$. As already noted, the quadratic form Q is negative for all such nonzero (h_1,h_2). Since $(D_2 f)(1/\sqrt{2}, 1/\sqrt{2}) = \sqrt{2} \neq 0$, the invertibility condition of the above theorem holds, and it follows thereby that ϕ has a local (strict) maximum at P_1.

(b) Let f be a twice continuously differentiable function on an open subset U of $\mathbb{R}^3$ that includes the point (a,a,a), having the *cyclicity* property that

$$f(x,y,z) = f(y,z,x) = f(z,x,y) \quad \text{whenever} \quad (x,y,z), (y,z,x), (z,x,y) \in U.$$

Observe that if one among (x,y,z), (y,z,x), (z,x,y) belongs to any ball that is contained in U and centred at (a,a,a), then so do the other two. Suppose that $D_1 f(a,a,a) \neq 0$. Also, let F be a twice continuously differentiable on an open interval containing a. We shall show that, subject to the constraint

$$f(x,y,z) = f(a,a,a),$$

the function

$$\phi(x,y,z) = F(x) + F(y) + F(z)$$

has a local strict maximum or minimum at (a,a,a) according as

$$F''(a) < F'(a)\frac{f_{11}(a,a,a) - f_{12}(a,a,a)}{f_1(a,a,a)}$$

or

$$F''(a) > F'(a)\frac{f_{11}(a,a,a) - f_{12}(a,a,a)}{f_1(a,a,a)},$$

where the subscripts indicate partial differentiation (f_1 means $D_1 f$ and so on).

In view of the cyclicity property and the observation about it, $f_1(a,a,a) = f_2(a,a,a) = f_3(a,a,a)$ and $f_{11}(a,a,a) = f_{22}(a,a,a) = f_{33}(a,a,a)$; also all other second partial derivatives are equal to each other. Therefore, firstly, the linear derivative $f'(a,a,a)$ is represented by the 1×3 matrix with each entry equal to $f_1(a,a,a)$. As this is given to be nonzero, the invertibility condition of the above theorem is fulfilled. Since the entries of $f'(a,a,a)$ are equal and nonzero, the elements $(h_1,h_2,h_3) \in \mathbb{R}^3$ such that $f'(a,a,a)(h_1,h_2,h_3) = 0$ are those for which $h_1 + h_2 + h_3 = 0$. Secondly, the Hessian matrix of the Lagrangian $\phi + \lambda f$ at (a,a,a) is as in Example 5-2.5(c), with

$$A = F''(a) + \lambda f_{11}(a,a,a) \text{ and } B = \lambda f_{12}(a,a,a).$$

As discussed there, the associated quadratic form is negative or positive for the relevant nonzero elements $(h_1,h_2,h_3) \in \mathbb{R}^3$ according as $A < B$ or $A > B$. Now,

$$\phi_i(a,a,a) + \lambda f_i(a,a,a) = 0 \text{ for } i = 1,2,3 \text{ if } \lambda = -\frac{F'(a)}{f_1(a,a,a)}.$$

Hence $A = F''(a) - F'(a)f_{11}(a,a,a)/f_1(a,a,a)$ and $B = -F'(a)f_{12}(a,a,a)/f_1(a,a,a)$. This shows that $A < B$ or $A > B$ according as

$$F''(a) < F'(a)\frac{f_{11}(a,a,a) - f_{12}(a,a,a)}{f_1(a,a,a)}$$

or

$$F''(a) > F'(a)\frac{f_{11}(a,a,a) - f_{12}(a,a,a)}{f_1(a,a,a)}.$$

The theorem now yields the required conclusion.

We can now determine whether the point $(\pi/4, \pi/4, \pi/4)$ in Example 5-1.5(b) is a local strict maximum or minimum [see 5-2.P9].

For a treatment of the theorem that gives formal recognition to its differential geometry aspects, the reader may consult Edwards [10]. Another discussion is available in the Internet article by Cheng [6].

Problem Set 5-2

5-2.P1. Find all points of local maxima and minima of the function f defined on $\mathbb{R}^2$ by
$$f(x,y) = 2x^3 - 3x^2 + 2y^3 + 3y^2.$$

5-2.P2. (a) Find all local maxima and minima of
$$f(x_1,x_2,x_3) = x_1^4 + x_2^4 + x_3^4 - 4x_1x_2x_3 \quad \text{on } \mathbb{R}^3$$
by using Theorem 5-2.1.
(b) Obtain the final conclusion of part (a) without any differentiation by recasting $f(x_1,x_2,x_3)$ as a sum of squares plus a constant.
(c) Use your answer to (b) to suggest a fourth degree polynomial $g(x_1,x_2,x_3)$, for which the final conclusion is the same but g is not of the form $\alpha f + \beta$, with α and β constant.
(d) Show that a search for the local extrema of the function $F(x_1,x_2,x_3) = x_1^{10} + x_2^{10} + x_3^{10} - 10x_1x_2x_3$ on $\mathbb{R}^3$ leads to the same final conclusion as for the function f of part (a).

5-2.P3. Let $a > b > 0 > c$. For the function
$$f(x,y,z) = (ax^2 + by^2 + cz^2) \exp(-x^2 - y^2 - z^2) \quad \text{on } \mathbb{R}^3$$
show that there are seven points where all partial derivatives vanish. Also show that the function has a local maximum at two of the seven points, a minimum at one and neither a maximum nor a minimum at the remaining four.

5-2.P4. Find all extrema of $F(x,y,z) = xyz(x + y + z - 1)$ on $\mathbb{R}^3$.

5-2.P5. For the quadratic form Q in $\mathbb{R}^3$ associated with the matrix with rows
$$[1 \quad 2 \quad -1], [2 \quad 13 \quad -5] \quad \text{and} \quad [-1 \quad -5 \quad 4],$$
write $Q(h)$ as a sum of squares and determine whether Q is positive definite. You may denote h by (a,b,c) instead of (h_1,h_2,h_3).

5-2.P6. For the quadratic form Q in $\mathbb{R}^3$ associated with the matrix with rows
$$[1 \quad -1 \quad -1], [-1 \quad 5 \quad -5] \quad \text{and} \quad [-1 \quad -5 \quad 10],$$

write $Q(h)$ as a sum of squares (if possible) and determine whether Q is positive definite. You may denote h by (a,b,c) instead of (h_1,h_2,h_3).

5-2.P7. Let Q be the quadratic form in $\mathbb{R}^3$ associated with the matrix having rows

$$[A \quad B \quad -B], \ [B \quad A \quad -B] \quad \text{and} \quad [-B \quad -B \quad A], \quad \text{where } A \neq B.$$

Let $\alpha:\mathbb{R}^2{\rightarrow}\mathbb{R}^3$ be the map $\alpha(a,b) = (a,b,a+b)$. Determine whether $Q{\circ}\alpha$ is positive definite.

5-2.P8. Let Q be as in 5-2.P6 above and α be as in 5-2.P7. Determine whether $Q{\circ}\alpha$ is positive definite.

5-2.P9. The point $(\pi/4,\pi/4,\pi/4)$ satisfies the Lagrange equations for the function $\tan x + \tan y + \tan z$ subject to the constraint $y^5z + z^5x + x^5y = 3(\pi/4)^6$. Determine whether it is a point of constrained local strict maximum or minimum.

5-2.P10. The point $(\pi/6,\pi/6,\pi/6)$ satisfies the Lagrange equations for the function $\sin x + \sin y + \sin z$ subject to the constraint $yz + zx + xy = 3(\pi/6)^2$. Determine whether it is a point of constrained local strict maximum or minimum.

5-2.P11. The point $(\pi/4,\pi/4,\pi/4)$ satisfies the Lagrange equations for the function $\tan x + \tan y + \tan z$ subject to the constraint $yz + zx + xy = 3(\pi/4)^2$. Determine whether it is a point of constrained local strict maximum or minimum.

5-2.P12. If the hypothesis about H in Theorem 5-2.9 is changed to:

$$\exists \ u',u''\in \mathbb{R}^n{\times}\mathbb{R}^m \text{ such that } f'(a,b)(u') = f'(a,b)(u'') = 0 \text{ and } H(u') > 0 > H(u''),$$

will it be true that on every open set containing (a,b), ϕ takes values greater than as well as less than $\phi(a,b)$ while satisfying the constraint? Justify.

5-2.P13. Lagrange equations for the function $x^2 + y^2 + z^2$ subject to the constraint $z = xy + 2$ have $(x,y,z) = (0,0,2)$, $\lambda = -4$ as one solution. Determine whether this is a local extremum without converting it to an unconstrained problem.

5-2.P14. (This is the one-dimensional analogue of 5-2.P13.) Lagrange equations for the function $x^2 + y^2 + z^2 + u^2$ subject to the constraint $f(x,y,z,u) = u - xyz - 2 = 0$ have $(x,y,z,u) = (0,0,0,2)$, $\lambda = -4$ as one solution. Determine whether this is a local extremum without converting it to an unconstrained problem.

5-2.P15. Lagrange equations for the function $x^3 + y^3 + z^3 + u^3$ subject to the constraint $f(x,y,z,u) = u - (yz + zx + xy) - 1 = 0$ have $(x,y,z,u) = (0,0,0,1)$, $\lambda = -3$ as one solution. Determine whether this is a local extremum. Conversion to an unconstrained problem is acceptable.

5-2.P16. Let F and g be twice continuously differentiable functions on $\mathbb{R}$, $g(0) = 0 = F'(0) \neq F''(0)$. Lagrange's equations for a local extremum of $F(x) + F(y) + F(z)$ with constraint $f(x,y,z) = z - g(x)g(y) - C = 0$ (where C is some constant) have $(x,y,z) = (0,0,C)$, $\lambda = -F'(C)$ as one solution. Show that this solution is a local extremum if $|F''(0)| > |F'(C)| \cdot g'(0)^2$. Under what further condition is it a maximum?

5-2.P17. Lagrange equations for the function $x^2 + y^2 + z^2$ subject to the constraint $f(x,y,z) = ze^z - xy(x^2 + y^2) - e = 0$ have $(x,y,z) = (0,0,1)$, $\lambda = -1/e$ as one solution. Determine whether this is a local extremum.

5-2.P18. Lagrange equations for the function xyz subject to the constraint $f(x,y,z) = ze^z - xy(x^2 + y^2) - e = 0$ have $(x,y,z) = (0,0,1)$, $\lambda = 0$ as one solution. Determine whether this is a local extremum.

5-2.P19. In 5-1.P5, two solutions were found for the Lagrange equations. For each of them determine whether the point corresponds to a local maximum or minimum or neither.

5-2.P20. Find the local extrema of x subject to the constraint

$$x^2 + y^2 + z^2 - 2b\sqrt{(x^2 + y^2)} + (b^2 - a^2) = 0,$$

where a and b are given positive constants satisfying $b > a$. (The constraint represents a torus and the reader may find it instructive to visualise what the solution means.)

5-2.P21. Find the smallest possible area for a hexagon circumscribing a unit circle.

6

Riemann Integration in Euclidean Space

6-1 Cuboids and Pavings

A straightforward analogue of a closed interval in higher dimensions is a Cartesian product of closed intervals. Although many authors prefer to call them 'intervals', we shall refer to them as *cuboids*. They are best visualised as rectangles in $\mathbb{R}^2$ and as 'boxes' in $\mathbb{R}^3$.

In this chapter and the next, we shall use the norm $\|x\| = \|x\|_\infty = \max\{|x_i| : 1 \leq i \leq n\}$. Consequently, the closed ball of radius r centred at $a \in \mathbb{R}^n$ consists of all $x \in \mathbb{R}^n$ such that $\max\{|x_i - a_i| : 1 \leq i \leq n\} \leq r$. This is the same as $\{x \in \mathbb{R}^n : a_i - r \leq x_i \leq a_i + r$ for $1 \leq i \leq n\}$; in other words, the Cartesian product of the intervals $[a_i - r, a_i + r]$, each of which has length $2r$.

6-1.1. Definition. *A subset I of $\mathbb{R}^n$ is called a* **closed cuboid** *if there are closed intervals $[a_i, b_i]$, where $a_i < b_i$, $1 \leq i \leq n$, such that*

$$I = [a_1, b_1] \times \cdots \times [a_n, b_n].$$

An **open cuboid** *is defined analogously. The interval $[a_i, b_i]$ will be referred to as the ith* **edge** *and will sometimes be denoted by I_i. In case the lengths of all the edges are equal, we speak of a* **cube**. *The product of the lengths of the edges,*

$$(b_1 - a_1) \cdots (b_n - a_n)$$

is called the **volume** *of the cuboid (whether open or closed) and is denoted by* **vol(I)**.

When $n = 1$, a closed cuboid is just a closed bounded interval and its volume is the length.

There will be little occasion to work with open cuboids and we shall understand a cuboid to be closed unless specified as open. It is easy to check that what was called a cuboid in 3-3.P6 is the same as a closed cuboid in the sense just defined. This fact will play no role in our considerations.

6-1.2. Remarks. (a) A cuboid $[a_1, b_1] \times \cdots \times [a_n, b_n]$ with $b_i - a_i = r$ for each i is the same as a closed ball of radius $r/2$ centred at c, where $c_i = (a_i + b_i)/2$ for each i. However, we shall usually describe it as a cube, or even as a cuboid. Similarly, an open cuboid with each side equal to r is an open ball of radius $r/2$.

S. Shirali, H.L. Vasudeva, *Multivariable Analysis*,
DOI 10.1007/978-0-85729-192-9_6, © Springer-Verlag London Limited 2011

(b) For a cuboid $I = [a_1, b_1] \times \cdots \times [a_n, b_n]$, we have

$$I_i = [a_i, b_i] = \{x_i : x \in I\}$$

and

$$a_i = \min \{x_i : x \in I\} \text{ and } b_i = \max \{x_i : x \in I\}.$$

(c) The interior I° of the closed cuboid $I = [a_1, b_1] \times \cdots \times [a_n, b_n]$ is the open cuboid formed by the corresponding open intervals, namely, $(a_1, b_1) \times \cdots \times (a_n, b_n)$. To see why, consider any x in the latter (open) cuboid. By definition, $a_i < x_i < b_i$ for each i. If we take δ to be a positive number less than each of the $2n$ positive numbers $b_i - x_i, x_i - a_i$, then $\|y - x\| < \delta$ implies $y \in I$. This shows that $x \in I^\circ$. Conversely, suppose such a positive number δ exists for some point $x \in \mathbb{R}^n$, i.e., $x \in I^\circ$. Then the point y for which $y_i = x_i - \delta/2$ for each i satisfies $\|y - x\| < \delta$ and therefore $y \in I$, from which it follows that $a_i \leq x_i - \delta/2$, so that $a_i < x_i$ for each i. Similarly, we can show $x_i < b_i$ and thus $x \in (a_1, b_1) \times \cdots \times (a_n, b_n)$. The simple proof that the closure of the open cuboid $(a_1, b_1) \times \cdots \times (a_n, b_n)$ is the corresponding closed cuboid $[a_1, b_1] \times \cdots \times [a_n, b_n]$ is left as 6-1.P4.

(d) Given a closed cuboid $I = [a_1, b_1] \times \cdots \times [a_n, b_n]$ and $\varepsilon > 0$, there exists an open cuboid J such that $I \subset J$ and $\mathrm{vol}(J) < \mathrm{vol}(I) + \varepsilon$. Consider the cuboid $J = (a_1 - \delta, b_1 + \delta \times \cdots \times (a_n - \delta, b_n + \delta)$, where $\delta > 0$. Surely it contains I and its volume is $\mathrm{vol}(J) = (b_1 - a_1 + 2\delta) \cdots (b_n - a_n + 2\delta)$, which tends to $\mathrm{vol}(I)$ as $\delta \to 0$. Therefore, it is possible to choose δ small enough to ensure that $\mathrm{vol}(J) < \mathrm{vol}(I) + \varepsilon$.

For example, if $I = [1,2] \times [3,5] \subset \mathbb{R}^2$, which has volume $\mathrm{vol}(I) = (2 - 1)(5 - 3) = (1)(2) = 2$, then the open cuboid $J = (1 - \delta, 2 + \delta) \times (3 - \delta, 5 + \delta)$ has volume $\mathrm{vol}(J) = (1 + 2\delta)(2 + 2\delta) = 2 + 6\delta + 4\delta^2 < \mathrm{vol}(I) + 10\delta$ if $\delta < 1$; now, this is guaranteed to be less than $\mathrm{vol}(I) + \varepsilon$ provided that $\delta < \min\{1, \varepsilon/10\}$.

Similarly, given an open cuboid I and $\varepsilon > 0$, there exists a closed cuboid J such that

$$J \subset I \text{ and } \mathrm{vol}(I) < \mathrm{vol}(J) + \varepsilon.$$

Consider the cuboid

$$J = [a_1 + \delta, b_1 - \delta] \times \cdots \times [a_n + \delta, b_n - \delta],$$

where $0 < \delta < \frac{1}{2} \min\{b_i - a_i : 1 \leq i \leq n\}$. It is contained in the given cuboid I and its volume is $\mathrm{vol}(J) = (b_1 - a_1 - 2\delta) \cdots (b_n - a_n - 2\delta)$, which tends to $\mathrm{vol}(I)$ as $\delta \to 0$. Therefore, it is possible to choose δ small enough to ensure that $\mathrm{vol}(I) < \mathrm{vol}(J) + \varepsilon$.

For example, if $I = (1,2) \times (3,5) \subset \mathbb{R}^2$, which has volume $\mathrm{vol}(I) = (2 - 1)(5 - 3) = (1)(2) = 2$, then the closed cuboid

$$J = [1 + \delta, 2 - \delta] \times [3 + \delta, 5 - \delta], \quad \text{where } \delta < \tfrac{1}{2},$$

has volume $\text{vol}(J) = (1 - 2\delta)(2 - 2\delta) = 2 - 6\delta + 4\delta^2 > \text{vol}(I) - 6\delta$; now, this is guaranteed to be greater than $\text{vol}(I) - \varepsilon$ provided that $\delta < \min\{\frac{1}{2}, \frac{\varepsilon}{6}\}$.

(e) [Required in Proposition 6-1.11 below] For $I = [a_1, b_1] \times \cdots \times [a_n, b_n]$, we have

$$\sup\{\|y - x\| : y \in I, x \in I\} = \sup\{\|y - x\| : y \in I^\circ, x \in I^\circ\}$$
$$= \max\{b_i - a_i : 1 \le i \le n\} = M, \text{ say.}$$

It is clear that neither of these sups can exceed M. To prove the reverse, let $M = b_k - a_k$, where $1 \le k \le n$. Then for a given positive $\varepsilon < M/2$, any pair of points y and x of I° such that $x_k = a_k + \frac{\varepsilon}{2}$ and $y_k = b_k - \frac{\varepsilon}{2}$ satisfies $\|y - x\| \ge (b_k - a_k) - \varepsilon = M - \varepsilon$.

(f) Suppose $n > 1$. If we delete $[a_p, b_p]$ in the Cartesian product $I = [a_1, b_1] \times \cdots \times [a_n, b_n]$, we get a cuboid in $\mathbb{R}^{n-1}$ with volume $\text{vol}(I) / (b_p - a_p)$. We introduce no special name for this cuboid in $\mathbb{R}^{n-1}$ but shall denote it by $I \sim I_p$. Thus

$$\text{vol}(I \sim I_p) = \text{vol}(I) / (b_p - a_p).$$

When $n = 1$, there is only one edge to begin with, and deleting it results in an empty Cartesian product. So there is no such cuboid as $I \sim I_p$; however, $\text{vol}(I) / (b_p - a_p) = 1$ and it will be convenient to make the convention that the symbol $\text{vol}(I \sim I_p)$ means 1 when $n = 1$.

6-1.3. Definition. *For any closed cuboid $I = [a_1, b_1] \times \cdots \times [a_n, b_n]$, the $2n$ subsets*

$$\{x \in I : x_p = a_p\} \text{ and } \{x \in I : x_p = b_p\}, \quad 1 \le p \le n$$

*are called its **lower faces** and **upper faces**, respectively.*

When $n = 2$ or 3, a cuboid is a rectangle or a 'box' and its faces in the sense defined above are what are called *sides*. When $n = 1$, a cuboid is an interval and the left endpoint is the only lower face, while the right endpoint is the only upper face. When $n = 2$, the lower and left sides are the two lower faces and the upper and right sides are the two upper faces.

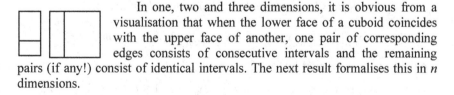 In one, two and three dimensions, it is obvious from a visualisation that when the lower face of a cuboid coincides with the upper face of another, one pair of corresponding edges consists of consecutive intervals and the remaining pairs (if any!) consist of identical intervals. The next result formalises this in n dimensions.

6-1.4. Proposition. *Suppose*

$$I = [a_1, b_1] \times \cdots \times [a_n, b_n] \quad and \quad J = [c_1, d_1] \times \cdots \times [c_n, d_n]$$

are cuboids and the lower face $\{x \in I : x_p = a_p\}$ of I is the same set as the upper face $\{x \in J : x_q = d_q\}$ of J. Then $p = q$, $a_p = d_p$ and $[a_i, b_i] = [c_i, d_i]$ for $i \neq p$. Also, an interior point of either cuboid cannot belong to the other. Lastly, $I \cup J$ is the closed cuboid $K_1 \times \cdots \times K_n$, where $K_p = [c_p, b_p]$ and $K_i = [a_i, b_i] = [c_i, d_i]$ for $i \neq p$.

Proof. If $p \neq q$, then the points x and x' with $x_i = x'_i = a_i$ for $i \neq q$ and $x_q = a_q$, $x'_q = b_q$ both belong to the first mentioned face; however they cannot both belong to the second, because otherwise $a_q = x_q = d_q = x'_q = b_q$, contradicting the stipulation [see Def. 6-1.1] that $a_q < b_q$. Thus $p = q$. The point x with each $x_i = a_i$ belongs to the first mentioned face and therefore to the second, which means $x_p = d_p$, and hence $a_p = d_p$. The proof that $[a_i, b_i] = [c_i, d_i]$ for $i \neq p$ is left as 6-1.P1.

An interior point x of I must satisfy $a_p < x_p$. Since $a_p = d_p$, it cannot satisfy $x_p \leq d_p$, which rules out the possibility that it belongs to J. A similar argument shows that an interior point of J cannot belong to I.

For the last part, suppose $x \in I \cup J$. Then $x_p \in [a_p, b_p] \cup [c_p, d_p]$ while $x_i \in [a_i, b_i]$ for $i \neq p$. Since $a_p = d_p$, we have $[a_p, b_p] \cup [c_p, d_p] = [c_p, b_p]$. Therefore, x has been shown to lie in the closed cuboid $K_1 \times \cdots \times K_n$, where $K_p = [c_p, b_p]$ and $K_i = [a_i, b_i] = [c_i, d_i]$ for $i \neq p$. The converse is just as easy. $\square$

The n-dimensional analogue of a partition will be defined in terms of partitions of intervals in one dimension, the properties of which we take for granted. In order to avoid the appearance of giving a circular definition, we prefer to call the n-dimensional analogue by another name, namely, *paving*.

6-1.5. Definition. *If $I = [a_1, b_1] \times \cdots \times [a_n, b_n]$ is a closed cuboid, any set P of n partitions*

$$P_i : a_i = x_{i\,0} < x_{i\,1} < \cdots < x_{i\,m_i} = b_i, \qquad 1 \leq i \leq n,$$

of the respective edges $[a_i, b_i]$ is called a **paving** *of the cuboid I. A cuboid J for which the ith edge J_i is one of the subintervals of $[a_i, b_i]$ formed by P_i, i.e.,*

$$J = J_1 \times \cdots \times J_n, \text{ where each } J_i \text{ is a subinterval formed by } P_i$$

or

$$J = [x_{1\,j_1-1}, x_{1\,j_1}] \times [x_{2\,j_2-1}, x_{2\,j_2}] \times \cdots \times [x_{n\,j_n-1}, x_{n\,j_n}], \quad 1 \leq j_i \leq m_i, \quad 1 \leq i \leq n,$$

is called a **cuboid formed by** *(or of) the paving P. An open cuboid*

$$(x_{1\,j_1-1}, x_{1\,j_1}) \times (x_{2\,j_2-1}, x_{2\,j_2}) \times \cdots \times (x_{n\,j_n-1}, x_{n\,j_n}), \quad 1 \leq j_i \leq m_i, \quad 1 \leq i \leq n,$$

is called an **open cuboid formed by** *(or of) the paving. For clarity, we may sometimes speak of the former as a* **closed cuboid formed by** *(or of) the paving.*

6-1.6. Definition. *A family of cuboids is said to be* **nonoverlapping** *if no interior point of any one of them belongs to any other cuboid of the family.*

A subfamily of a nonoverlapping family of cuboids is clearly nonoverlapping.

6-1.7. Remarks. (a) By Remark 6-1.2(c), the open cuboids formed by a paving are the interiors of the (closed) cuboids formed by it. Likewise, the (closed) cuboids formed by a paving are the closures of the open cuboids formed by it.

(b) Given a partition of an interval (in $\mathbb{R}$), it is trivial that every point of the interval belongs to some subinterval formed by the partition. In higher dimensions, this has the consequence that, given a paving of a cuboid, every point of the latter belongs to some cuboid formed by the paving. Thus a cuboid is contained in the union of the cuboids formed by a given paving of it. The reverse inclusion is obvious and we conclude that any cuboid is precisely equal to the union of the family of all the cuboids formed by a given paving.

(c) Let $n > 1$. Consider the cuboid $I{\sim}I_p$ in $\mathbb{R}^{n-1}$ obtained by deleting the pth edge I_p of a cuboid I in $\mathbb{R}^n$. In other words,

$$I{\sim}I_p = I_1 \times \cdots \times I_{p-1} \times I_{p+1} \times \cdots \times I_n.$$

If P is a paving of I consisting of the partitions $P_1, \ldots, P_n$ of the respective edges $I_1, \ldots, I_n$, then deleting P_p leads to a paving of $I{\sim}I_p$, which we shall denote by $P{\sim}P_p$. Fix any subinterval $[\alpha, \beta]$ of I_p formed by P_p and, as illustrated in the figure for two dimensions with $p = 1$, consider any cuboid K formed by P, for which the pth edge is $[\alpha, \beta]$.

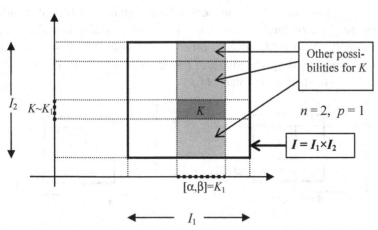

That is,

$K = K_1 \times \cdots \times K_n$, where each K_i is a subinterval formed by P_i and $K_p = [\alpha, \beta]$.

Obviously, $K{\sim}K_p$ is a cuboid formed by $P{\sim}P_p$ and this provides a bijection between such cuboids formed by P and *all* the cuboids formed by $P{\sim}P_p$. This is reflected in the figure by the fact that the vertical sides of all the four different shaded rectangles correspond to all the four different subintervals of the partition of the vertical side of the main big rectangle I.

(d) The total volume of the cuboids formed by a paving of a cuboid I is equal to $\mathrm{vol}(I)$. This follows by a straightforward induction on n using (c) above, the initial case $n = 1$ being simply the well known fact that when we have a partition of an interval in $\mathbb{R}$, the total length of the subintervals agrees with the length of the whole interval.

(e) The family of cuboids formed by a paving of a cuboid is nonoverlapping, and hence so is any subfamily. In order to prove this, it is sufficient to show that, if J and K are distinct cuboids formed by the same paving P, then an interior point of either of them cannot belong to the other. In symbols, $J^\circ \cap K = \varnothing = J \cap K^\circ$. By definition of cuboid, $J \neq K$ implies that $J_p \neq K_p$ for some p ($1 \leq p \leq n$). Since the intervals J_p and K_p are formed by the same partition of the pth edge, they can have only an endpoint in common. Thus if $x \in J^\circ$, its pth component, x_p belongs to the interior of J_p and therefore cannot belong to K_p, which implies that $x \notin K$.

It now follows from the definition of 'face' that any common point of J and K must belong to a face of J and also to a face of K.

The next proposition deals with a situation which is depicted in the two figures below for two dimensions. Suppose we start with a cuboid $I = I_1 \times I_2$, and partitions P_1, P_2 of the edges I_1, I_2 respectively [see the left figure]. Then P_1, P_2 constitute a paving of I. Consider a cuboid J (drawn shaded) such that the endpoints of its edges occur among points of the partitions P_1, P_2. Then J must have a paving Q [now see the figure on the right] such that the family $\mathcal{F}$ of all cuboids formed by Q is a subfamily of the family $\mathcal{G}$ of all the cuboids formed by the given paving P_1, P_2 of I. Moreover, J cannot be the union of any other family of cuboids formed by the given paving of I. And, of course, the total volume of the cuboids in $\mathcal{F}$ is equal to the volume of J.

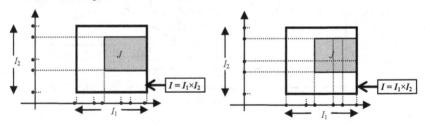

6-1.8. Proposition. *Suppose $I = [a_1, b_1] \times \cdots \times [a_n, b_n]$ is a closed cuboid, and the partitions*

$$P_i : a_i = x_{i\,0} < x_{i\,1} < \cdots < x_{i\,m_i} = b_i, \qquad 1 \leq i \leq n,$$

provide a paving of I. Let $J = [\alpha_1, \beta_1] \times \cdots \times [\alpha_n, \beta_n]$ be a cuboid such that each of the α_i and β_i are points of P_i. Then there exists a paving $Q_1, \ldots, Q_n$ of J such that the family $\mathcal{F}$ of all the cuboids formed by this paving is a subfamily of the *family $\mathcal{G}$ of all the cuboids formed by the given paving $P_1, \ldots, P_n$. Moreover, $\mathcal{F}$ is the unique subfamily of $\mathcal{G}$ having union J, the total volume of all the cuboids of $\mathcal{F}$ is equal to $\mathrm{vol}(J)$, and an interior point of J cannot belong to a cuboid of $\mathcal{G}$ that is not in $\mathcal{F}$.*

Proof. By hypothesis, for each i, there exist p_i and q_i such that

$$\alpha_i = x_{i \, p_i}, \quad \beta_i = x_{i \, q_i}, \quad p_i < q_i.$$

Now the points

$$Q_i: \alpha_i = x_{i \, p_i} < \cdots < x_{i \, q_i} = \beta_i$$

furnish a partition of $[\alpha_i, \beta_i]$, so that $Q_1, \ldots, Q_n$ is a paving of J. Besides, the subintervals of $[\alpha_i, \beta_i]$ formed by Q_i are among the subintervals of $[a_i, b_i]$ formed by P_i. It therefore follows from Def. 6-1.5 that the family $\mathcal{F}$ of all the cuboids formed by $Q_1, \ldots, Q_n$ is a subfamily of the family $\mathcal{G}$ of all the cuboids formed by the given paving $P_1, \ldots, P_n$.

By Remark 6-1.7(b), $\cup \mathcal{F} = J$. For the uniqueness, first observe that, given a partition of an interval (in $\mathbb{R}$), the midpoint of any subinterval formed by the partition belongs to none of the other subintervals. In higher dimensions, this has the consequence that the 'centre' of any one of the cuboids formed by a paving belongs to *none* of the other cuboids. Now consider any subfamily $\mathcal{F}_1$ of the family of all the cuboids formed by given paving such that $\cup \mathcal{F}_1 = J = \cup \mathcal{F}$. If some cuboid were to be in $\mathcal{F}$ but not in $\mathcal{F}_1$ (or vice versa), then its centre would belong to $\cup \mathcal{F}$ but not to $\cup \mathcal{F}_1$ (or vice versa), a contradiction. Consequently, $\mathcal{F}_1 = \mathcal{F}$.

Since $\mathcal{F}$ is the family of all the cuboids formed by a paving of J, it follows by Remark 6-1.7(d) that the total volume of all the cuboids of $\mathcal{F}$ is equal to $\mathrm{vol}(J)$.

It remains to prove that an interior point of J cannot belong to a cuboid of $\mathcal{G}$ that is not in $\mathcal{F}$. For this purpose, we first observe from the above definition of Q_i that, if a subinterval $[x_{i \, j}, x_{i \, j+1}]$ formed by P_i satisfies both the inequalities

$$x_{i j} < \beta_i \quad \text{and} \quad x_{i j+1} > \alpha_i, \tag{1}$$

then it is also a subinterval formed by Q_i. Now consider a cuboid $K = K_1 \times \cdots \times K_n$ in $\mathcal{G}$ that is not in $\mathcal{F}$, and an interior point $x = (x_1, \ldots, x_n)$ of J. We have to show that $x \notin K$. Since K is not in $\mathcal{F}$, then for some i, the interval $K_i = [x_{i \, j}, x_{i \, j+1}]$ is not a subinterval formed by the partition Q_i, so that (1) does not hold. Since x is an interior point of $J = [\alpha_1, \beta_1] \times \cdots \times [\alpha_n, \beta_n]$, we must have $\alpha_i < x_i < \beta_i$. If it were also the case that $x_{i j} \le x_i \le x_{i j+1}$, then (1) would hold. Therefore, $x_i \notin [x_{i j}, x_{i j+1}] = K_i$, from which it follows that $x \notin K$. $\qquad \square$

In $\mathbb{R}^2$, if one has a finite number of rectangles, possibly overlapping, inside a single rectangle I, the sides of the inner rectangles can be 'produced' to generate a paving of I. When this is done, each of the inner rectangles will be a union of some family of rectangles formed by the paving; this will continue to be so even if the paving is refined. There is of course a similar phenomenon in $\mathbb{R}^3$ but one has to 'produce' the faces. The next proposition describes the matter in $\mathbb{R}^n$ but without introducing any formal definition of 'producing'.

6-1.9. Proposition. *Let* $K_1, \ldots, K_m$ *be cuboids contained in a single cuboid* I *and let* $\delta > 0$ *be given. Then there exists a paving of* I *such that*
(a) *each* K_k *is the union of some subfamily* $\mathcal{F}_k$ *of the family of all the cuboids formed by the paving; moreover,* $\mathrm{vol}(K_k)$ *equals the total volume of all the cuboids of* $\mathcal{F}_k$;
(b) *the total volume of the cuboids in the subfamilies* $\mathcal{F}_k$ *does not exceed the total volume of the* K_k;
(c) *an interior point of* K_k *cannot be in any cuboid formed by the paving that is not in* $\mathcal{F}_k$;
(d) *for any cuboid* J *formed by the paving,* $\max\{\|x - x'\| : x, x' \in J\} < \delta$.

Proof. Since all the K_k are contained in I, their ith edges are subintervals of the ith edge of I. Therefore, the endpoints of these edges give rise to a partition of the ith edge of I. Take any refinement of it such that every subinterval has length less than δ, and denote it by P_i. Then $P_1, \ldots, P_n$ constitute a paving of I satisfying (d). By Proposition 6-1.8, for each k, there is some subfamily $\mathcal{F}_k$ of the family of all the cuboids formed by the paving such that (c) holds and K_k is the union of $\mathcal{F}_k$ while the total volume of the cuboids of $\mathcal{F}_k$ equals $\mathrm{vol}(K_k)$, so that (a) also holds. Taking the sum over all k and noting that some cuboids may occur in more than one $\mathcal{F}_k$, we get (b). □

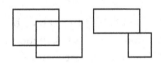

It is easy to conjecture on the basis of some figures that an intersection of two (closed) cuboids is either a cuboid or is a subset (perhaps empty) of a face of each. The next proposition establishes this in general.

6-1.10. Proposition. *Let* I *and* J *be* (*closed*) *cuboids such that* $I \cap J$ *is not a cuboid. Then* $I \cap J$ *is a subset* (*perhaps empty*) *of a face of* I *as well as of a face of* J.

Proof. Let $I = [a_1, b_1] \times \cdots \times [a_n, b_n]$ and $J = [\alpha_1, \beta_1] \times \cdots \times [\alpha_n, \beta_n]$, where $a_i < b_i$ and $\alpha_i < \beta_i$ for $1 \leq i \leq n$. Then $x \in I \cap J$ if and only if

$$\max\{a_i, \alpha_i\} \leq x \leq \min\{b_i, \beta_i\} \qquad \text{for } 1 \leq i \leq n.$$

If $\max\{a_i, \alpha_i\} < \min\{b_i, \beta_i\}$ for every i, then $I \cap J$ is a cuboid. So, suppose

$$\max\{a_i, \alpha_i\} \geq \min\{b_i, \beta_i\} \qquad \text{for some } i.$$

In case $\max\{a_i,\alpha_i\} > \min\{b_i,\beta_i\}$, the intersection $I \cap J$ is empty. We need consider only the case when

$$\max\{a_i,\alpha_i\} = \min\{b_i,\beta_i\}.$$

This equality implies that every $x \in I \cap J$ satisfies $x_i = \max\{a_i,\alpha_i\} = \min\{b_i,\beta_i\}$. Two possibilities can arise (not mutually exclusive): $\max\{a_i,\alpha_i\} = a_i$ and $\max\{a_i,\alpha_i\} = \alpha_i$.

Suppose $\max\{a_i,\alpha_i\} = a_i$. Then $I \cap J$ is a subset of the face of I given by $x_i = a_i$. Besides,

$$\min\{b_i,\beta_i\} = \max\{a_i,\alpha_i\} = a_i < b_i \qquad \text{so that} \qquad \min\{b_i,\beta_i\} = \beta_i.$$

Therefore, $I \cap J$ is a subset of the face of J given by $x_i = \beta_i$. The possibility that $\max\{a_i,\alpha_i\} = \alpha_i$ is argued along similar lines. $\qquad\square$

We conclude this section with a preview of some of the other tedious considerations that will be required in the sequel [e.g. Lemma 7-2.1, Proposition 7-2.2, Proposition 7-3.5]—but stay with us!

6-1.11. Proposition. *Let $H \subseteq W \subseteq \mathbb{R}^n$, where H is compact and W is open. If $\mathcal{F}$ is a finite family of closed cuboids that cover H (i.e., the union of the cuboids in the family contains H), then there exists a family of closed cuboids $\mathcal{G}$ that also covers H and:*
(a) each cuboid in $\mathcal{G}$ is contained in W;
(b) $\mathcal{G}$ is finite and the total volume of the cuboids in $\mathcal{G}$ is no greater than the total volume of the cuboids in the family $\mathcal{F}$;
(c) $\mathcal{G}$ is a nonoverlapping family.

Proof. [See figure below, where W is oval shaped, H is a segment and $\mathcal{F}$ consists of two cuboids.] First we argue that there is a positive lower bound to the distances between points of H and points of the complement W^c of W, which is to say, there exists $\delta > 0$ such that

$$h \in H, \; \|h - x\| < \delta \Rightarrow x \in W.$$

If this were not so, then there would exist sequences $\{h_n\}$ in H and $\{x_n\}$ in W^c such that $\|h_n - x_n\| \to 0$. Since H is compact, some subsequence $\{h_{n_k}\}$ would converge to a limit $h \in H$. Since $H \subseteq W$, we would have $h \in W$, an open set, and hence there would exist some $\eta > 0$ such that $\|h - x\| < \eta \Rightarrow x \in W$. By choosing $k \in \mathbb{N}$ large enough to make $\|h_{n_k} - x_{n_k}\| < \eta/2$ as well as $\|h_{n_k} - h\| < \eta/2$, we would have the contradiction that $\|h - x_{n_k}\| < \eta$ even though $x_{n_k} \notin W$. This establishes the existence of a positive δ of the kind indicated above.

Now let $\mathcal{F}$ be a finite family of closed cuboids that cover H and consider any one cuboid I in the family. There is a paving $P_1, \ldots, P_n$ of I,

$$P_i : a_i = x_{i\,0} < x_{i\,1} < \cdots < x_{i\,m_i} = b_i, \qquad 1 \le i \le n,$$

such that $\max\{\,|x_{i\,j_i} - x_{i\,j_i-1}|\ :\ 1\le j_i\le m_i\,,\ 1\le i\le n\} < \delta$. It follows by Remark 6-1.2(e) that for any cuboid

$$J = [x_{1\,j_1-1},x_{1\,j_1}]\times[x_{2\,j_2-1},x_{2\,j_2}]\times\cdots\times[x_{n\,j_n-1},x_{n\,j_n}]\,,\quad 1\le j_i\le m_i\,,\quad 1\le i\le n,$$

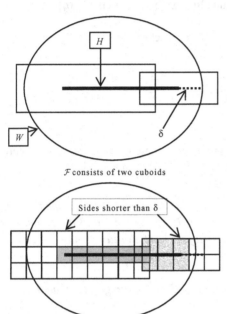

$\mathcal{F}$ consists of two cuboids

Sides shorter than δ

formed by the paving, we have

$$\sup\{\|y - x\|\ :\ y\in J, x\in J\} < \delta.$$

Now select only those cuboids formed by the paving which contain at least one point h of H. Clearly, they cover the intersection $I\cap H$. Also, every point x of such a cuboid satisfies $\|h - x\| < \delta$ and therefore, belongs to W. Thus all the selected cuboids are subsets of W. Moreover, by Remark 6-1.7(d), their total volume is no greater than that of I. If we form the family $\mathcal{H}$ of cuboids selected in this manner for *all* the various $I \in \mathcal{F}$, it will not only cover H but will also satisfy (a) and (b) with $\mathcal{H}$ in place of $\mathcal{G}$.

To obtain a family $\mathcal{G}$ that also satisfies (c), we apply Proposition 6-1.9 to the cuboids $K_1,\ldots,K_m$ of the finite family $\mathcal{H}$, and thereby obtain families $\mathcal{F}_1,\ldots,\mathcal{F}_m$ as in that proposition. By (a) and (b) therein, the family $\mathcal{G} = \mathcal{F}_1\cup\cdots\cup\mathcal{F}_m$ satisfies (a) and (b) of the present proposition. That it also satisfies (c) follows from Remark 6-1.7(e) and the fact that, according to Proposition 6-1.9, all the cuboids of $\mathcal{G}$ are formed by the same paving of a certain cuboid. $\square$

Problem Set 6-1

6-1.P1. Complete the proof that $[a_i,b_i] = [c_i,d_i]$ for $i\ne p$ in Proposition 6-1.4.

6-1.P2. Suppose $I = [a_1,b_1]\times\cdots\times[a_n,b_n]$ and $J = [c_1,d_1]\times\cdots\times[c_n,d_n]$ are cuboids and the lower face $\{x\in I: x_p = a_p\}$ of I is the same set as the lower face $\{x\in J: x_q = c_q\}$ of J. Then $p = q$, $a_p = c_p$ and $[a_i,b_i] = [c_i,d_i]$ for $i\ne p$. Also, either $I\subseteq J$ or $J\subseteq I$. Lastly, I and J have interior points in common.

6-1.P3. Prove the following converse of Proposition 6-1.4: If a union of two (closed) cuboids is a cuboid and they have no common interior points, then there is one dimension in which the lower face of one is the upper face of the other, and their edges in the remaining dimensions coincide.

6-1.P4. Prove that the closure of the open cuboid $I = (a_1, b_1) \times \cdots \times (a_n, b_n)$ is the corresponding closed cuboid $J = [a_1, b_1] \times \cdots \times [a_n, b_n]$.

6-1.P5. Suppose $I = [a_1, b_1] \times \cdots \times [a_n, b_n]$ is a closed cuboid, and the partitions

$$P_i : a_i = x_{i\,0} < x_{i\,1} < \cdots < x_{i\,m_i} = b_i, \qquad 1 \le i \le n,$$

provide a paving of I. Let $\mathcal{F}$ be a subfamily of the family of all the cuboids formed by the paving. If the union $\cup \mathcal{F}$ is a cuboid J, show that there exists a paving $Q_1, \ldots, Q_n$ of J such that the family of all the cuboids formed by this paving is none other than $\mathcal{F}$. Moreover, the total volume of all the cuboids belonging to $\mathcal{F}$ is equal to the volume of J.

6-2 Riemann Integral Over Cuboids

The definition of Riemann integral over a cuboid and the proofs of most basic properties are direct analogues of the one-dimensional case, with pavings playing the role of partitions. We shall therefore restrict ourselves to formal definitions and comment on some proofs.

6-2.1. Definition. *Let $I \subset \mathbb{R}^n$ be a cuboid and $f : I \to \mathbb{R}$ be a bounded function. Given a paving P of I, let $K_1, \ldots, K_m$ denote the cuboids formed by P and for $j = 1, \ldots, m$, let*

$$m_{K_j} = \inf\, \{f(x) : x \in K_j\} \quad and \quad M_{K_j} = \sup\, \{f(x) : x \in K_j\}.$$

*Then the **lower sum of** f over the paving P is*

$$L(f, P) = \sum_{j=1}^{m} m_{K_j} \operatorname{vol}(K_j)$$

*and the **upper sum** is*

$$U(f, P) = \sum_{j=1}^{m} M_{K_j} \operatorname{vol}(K_j).$$

It is trivial that $L(f, P) = -U(-f, P)$.

Sometimes one may need to take the sum only over *some* cuboids. It is then convenient to introduce a name, such as $\mathcal{F}$, for the family of cuboids involved in the sum and write $\sum_{j=1}^{m} m_j \operatorname{vol}(K_j)$ and $\sum_{j=1}^{m} M_j \operatorname{vol}(K_j)$ respectively as

$$\Sigma_{K \in \mathcal{F}}\, m_K \operatorname{vol}(K) \quad and \quad \Sigma_{K \in \mathcal{F}}\, M_K \operatorname{vol}(K).$$

It is easy to prove that $L(f + g, P) \geq L(f, P) + L(g, P)$ and $U(f + g, P) \leq U(f, P) + U(g, P)$. Also, $L(cf, P) = cL(f, P)$ and $U(cf, P) = cU(f, P)$ for any nonnegative constant c. If $c < 0$, then $L(cf, P) = cU(f, P)$ and $U(cf, P) = cL(f, P)$.

Recall that a partition Q of an interval is called a refinement of a partition P of the same interval when Q includes every point of P.

6-2.2. Definition. *A paving* Q *of* $I = [a_1, b_1] \times \cdots \times [a_n, b_n]$ *consisting of*

$$Q_i: a_i = \xi_{i\,0} < \xi_{i\,1} < \cdots < \xi_{i\,p_i} = b_i, \qquad 1 \leq i \leq n,$$

is a **refinement** *of a paving* P *consisting of*

$$P_i: a_i = x_{i\,0} < x_{i\,1} < \cdots < x_{i\,m_i} = b_i, \qquad 1 \leq i \leq n,$$

if each $x_{i\,j_i}$ *is also a point of* Q_i. *In other words, if each partition* Q_i *is a refinement of the partition* P_i *in the usual sense in one dimension.*

As with partitions, any two pavings have a common refinement. If the partition P_i is refined by adding a point t_i to it, which means $x_{i\,p-1} < t_i < x_{i\,p}$ for some p, $1 \leq p \leq m_i$, then we obtain a refinement of P; call it P'. Among the cuboids formed by P, those having $[x_{i\,p-1}, x_{i\,p}]$ as the ith edge are not among the ones formed by P'. However, each such cuboid C is the union of two cuboids formed by P', which have $[x_{i\,p-1}, t_i]$ and $[t_i, x_{i\,p}]$ as their respective ith edges and have all other edges the same as C. Those not having $[x_{i\,p-1}, x_{i\,p}]$ as the ith edge are among the cuboids formed by P'. Also, P' forms no cuboids besides the aforementioned ones.

Using what has been noted in the preceding paragraph, one can easily adapt the one-dimensional arguments about partitions to prove that refining a paving does not decrease the lower sum and does not increase the upper sum, and that $L(f, P_1) \leq U(f, P_2)$ for any pavings P_1 and P_2.

6-2.3. Definition. *For a bounded function* $f: I \rightarrow \mathbb{R}$, *where* I *is a cuboid in* $\mathbb{R}^n$, *the supremum of the set of all lower sums* $L(f, P)$ *is called the* **lower Riemann integral of** f *over* I *and is denoted by* $\underline{\int}_I f$. *The infimum of all upper sums* $U(f, P)$ *is called the* **upper Riemann integral of** f *over* I *and is denoted by* $\overline{\int}_I f$.

It is trivial to see that

$$\underline{\int}_I (f + g) \geq \underline{\int}_I f + \underline{\int}_I g,$$

$$\overline{\int}_I (f + g) \leq \overline{\int}_I f + \overline{\int}_I g$$

and that

$$\underline{\int}_I f = -\overline{\int}_I (-f).$$

From the fact that $L(f,P_1) \leq U(f,P_2)$ for any pavings P_1 and P_2, one can easily obtain the inequality $\underline{\int}_I f \leq \overline{\int}_I f$. Like the previous two, the next definition is also a direct analogue of the one in $\mathbb{R}$.

6-2.4. Definition. *A bounded function $f:I \to \mathbb{R}$, where I is a cuboid in $\mathbb{R}^n$, is called* **Riemann integrable** *if $\underline{\int}_I f = \overline{\int}_I f$. The common value of the upper and lower integrals is denoted by $\int_I f$ and is called the* **Riemann integral of** *the function f.*

We shall usually drop the word 'Riemann' in this connection, as we do not intend to discuss any other type of integral.

When $n = 1$, a paving is simply a partition and all the concepts defined in this Section so far reduce to the ones known by the same name in one dimension.

The following criterion of integrability is proved the same way as in $\mathbb{R}$ and is equally useful:

6-2.5. Proposition. *A bounded function $f:I \to \mathbb{R}$, where I is a cuboid in $\mathbb{R}^n$, is integrable if and only if, for every $\varepsilon > 0$, there exists a paving P of I such that*

$$U(f,P) - L(f,P) < \varepsilon. \tag{A}$$

Such a paving satisfies

$$\int_I f - \varepsilon < L(f,P) \leq U(f,P) < \int_I f + \varepsilon. \tag{B}$$

It can now be proved by imitating the one-dimensional case that, whenever functions f and g on a cuboid are both integrable, the following are also integrable: $f + g$, fg, $|f|$ and cf (where c is a constant); moreover the usual equalities and inequalities concerning them hold.

6-2.6. Example. Let $f:[0,1] \times [0,1] \to \mathbb{R}$ be defined as

$$f(x,y) = \begin{cases} 3 & \text{if } y \in \mathbb{Q} \\ x^2 & \text{if } y \notin \mathbb{Q}. \end{cases}$$

We shall show that f is not integrable. On any cuboid $K \subseteq [0,1] \times [0,1]$, we have $M_K = 3$ and also $m_K \leq 1$, the latter because $y \notin \mathbb{Q} \Rightarrow f(x,y) = x^2 \leq 1$. Consequently, $M_K - m_K \geq 2$ for every K. It follows that $U(f,P) - L(f,P) \geq 2 \cdot \text{vol}([0,1] \times [0,1]) = 2$ for every paving P. By Proposition 6-2.5, f cannot be integrable.

6-2.7. Proposition. *Suppose $J \subseteq I$, both cuboids, and $f:I \to \mathbb{R}$ is integrable. Then the restriction g of the function f to the cuboid J is also integrable.*

Proof. Denote the ith edges of I and J by I_i and J_i respectively. Since $J \subseteq I$, each J_i is a subinterval of I_i, i.e., has its endpoints lying in the latter. Therefore, given any paving of I, its partition of I_i has a refinement P_i which is obtained by adding *only* the endpoints of J_i (unless already present). Those points of P_i that lie in J_i form a partition Q_i of the latter and the subintervals formed by Q_i are precisely those among the ones formed by P_i that are subsets of J_i. From the definition of cuboids formed by a paving [Def. 6-1.5], it now follows that the family $\mathcal{G}$ of cuboids formed by the paving $Q = \{Q_1, \dots, Q_n\}$ of J is a subfamily of the family $\mathcal{F}$ of cuboids formed by the refinement $P = \{P_1, \dots, P_n\}$ of the given paving of I.

Now consider any $\varepsilon > 0$. Using Proposition 6-2.5, we find that there exists a paving P_1 of I such that $U(f, P_1) - L(f, P_1) < \varepsilon$. Note that any refinement must also satisfy the same inequality. In particular, if P is its refinement obtained by adding *only* the endpoints of J_i to the partition of the I_i, then we have $U(f, P) - L(f, P) < \varepsilon$. For Q, $\mathcal{F}$ and $\mathcal{G}$ as in the preceding paragraph,

$$0 \le U(g, Q) - L(g, Q) = \Sigma_{K \in \mathcal{G}} (M_K - m_K) \operatorname{vol}(K)$$

and

$$\Sigma_{K \in \mathcal{F}} (M_K - m_K) \operatorname{vol}(K) = U(f, P) - L(f, P) < \varepsilon.$$

Since $\mathcal{G} \subseteq \mathcal{F}$ and $M_K - m_K \ge 0$, we also have

$$\Sigma_{K \in \mathcal{G}} (M_K - m_K) \operatorname{vol}(K) \le \Sigma_{K \in \mathcal{F}} (M_K - m_K) \operatorname{vol}(K).$$

Together with the above two inequalities, this implies that $0 \le U(g, Q) - L(g, Q) < \varepsilon$. The existence of such a paving Q of J proves the integrability of g by Proposition 6-2.5.   □

6-2.8. Example. Let Let $f : [0,3] \times [0,1] \to \mathbb{R}$ be defined by the same scheme as in Example 6-2.6, which is that

$$f(x,y) = \begin{cases} 3 & \text{if } y \in \mathbb{Q} \\ x^2 & \text{if } y \notin \mathbb{Q}. \end{cases}$$

The restriction of this function to $[0,1] \times [0,1]$ is the function of Example 6-2.6, which was shown not to be integrable. It now follows by Proposition 6-2.7 that the function here is also not integrable.

6-2.9. Remark. Returning to the bijection mentioned in Remark 6-1.7(c) when $n > 1$, the total volume of cuboids K with pth edge $[\alpha, \beta]$ must be the product of $\beta - \alpha$ with the total volume of the cuboids $K \sim K_p$. Since the latter are precisely the cuboids formed by the paving $P \sim P_p$ of $I \sim I_p$, it follows from Remark 6-1.7(d) that their total volume is $\operatorname{vol}(I \sim I_p)$. Therefore, the total volume of all the cuboids having $[\alpha, \beta]$ as the pth edge is $(\beta - \alpha) \operatorname{vol}(I \sim I_p)$. This is trivially true when $n = 1$ because of our convention in this case that $\operatorname{vol}(I \sim I_p) = 1$ despite the fact that there is no such cuboid as $I \sim I_p$. In view of Remark 6-1.2(f), this also equals

$(\beta - \alpha)\,\mathrm{vol}(I)\big/(b_p - a_p)$, where $I_p = [a_p, b_p]$. What this means in terms of the figure shown earlier in this connection is that the total area of the shaded rectangles is the product of their common width $\beta - \alpha$ with the ratio of the area of I to the length of its horizontal edge. We emphasise that the above conclusion is valid even when $n = 1$, so that we may use it below [in Proposition 6-2.10 and Proposition 6-2.11] without assuming $n > 1$.

6-2.10. Proposition. *Let I be a cuboid and both $f:I \to \mathbb{R}$ $g:I \to \mathbb{R}$ be bounded. If $f(x) = g(x)$ for $x \in I^\circ$, then*

$$\underline{\int}_I f = \underline{\int}_I g \quad and \quad \overline{\int}_I f = \overline{\int}_I g.$$

In particular, if one of the functions is integrable, so is the other and has the same integral.

Proof. Consider any $\varepsilon > 0$. By definition of lower integral, there exists a paving Q of I such that

$$\underline{\int}_I g - \tfrac{\varepsilon}{3} < L(g, Q). \tag{1}$$

By definition of paving [see Def. 6-1.5], Q consists of some partitions $Q_1, \ldots, Q_n$ of the edges of I. Let δ be *any* positive number less than the lengths of *all* the subintervals formed by *all* the Q_i. (We are free to require at a later stage that δ be even smaller.) For $1 \leq i \leq n$, each Q_i has a refinement P_i that has just two additional points, one in the first subinterval at a distance less than δ from the left endpoint and one in the last subinterval at a distance less than δ from the right endpoint. Now let P be the paving of I consisting of the refinements $P_1, \ldots, P_n$. Then (1) holds with P in place of Q.

The family of cuboids formed by P can be subdivided into two disjoint subfamilies, one consisting of those having an edge that shares an endpoint with an edge of I and those having no such edge. Name these subfamilies as $\mathcal{B}$ and $\mathcal{I}$ respectively. Recall that, by definition, a cuboid K formed by P has edges that are subintervals formed by the partitions P_i, so that an edge K_p that may share an endpoint with an edge of I must have length less than δ. Since the edge of I has two endpoints, it follows by Remark 6-2.9 that the total volume of such cuboids is at most $2\delta \cdot \mathrm{vol}(I \sim I_p)$ and hence the total volume of the cuboids in $\mathcal{B}$ is at most $2\delta \cdot \sum_{p=1}^{n} \mathrm{vol}(I \sim I_p)$, i.e.,

$$\Sigma_{K \in \mathcal{B}}\, \mathrm{vol}(K) \leq 2\delta \cdot \sum_{p=1}^{n} \mathrm{vol}(I \sim I_p). \tag{2}$$

We now require that $\delta < \varepsilon \big/ \big(6M \sum_{p=1}^{n} \mathrm{vol}(I \sim I_p)\big)$, where M is a common upper bound of $|f|$ and $|g|$. Then, with obvious meanings for $m_K{}^f$ and $m_K{}^g$, it follows from (2) regarding f that

$$|\Sigma_{K\in B}\, m_K{}^f\,\mathrm{vol}(K)| \le M\Sigma_{K\in B}\,\mathrm{vol}(K) < 2M\delta\cdot\sum_{p=1}^{n}\mathrm{vol}(I{\sim}I_p) < \tfrac{\varepsilon}{3}$$

and similarly, regarding g, that

$$|\Sigma_{K\in B}\, m_K{}^g\,\mathrm{vol}(K)| < \tfrac{\varepsilon}{3}.$$

These two inequalities imply that

$$L(f,P) = \Sigma_{K\in B}\, m_K{}^f\,\mathrm{vol}(K) + \Sigma_{K\in \mathcal{I}}\, m_K{}^f\,\mathrm{vol}(K) > -\tfrac{\varepsilon}{3} + \Sigma_{K\in \mathcal{I}}\, m_K{}^f\,\mathrm{vol}(K) \quad (3)$$

and

$$L(g,P) = \Sigma_{K\in B}\, m_K{}^g\,\mathrm{vol}(K) + \Sigma_{K\in \mathcal{I}}\, m_K{}^g\,\mathrm{vol}(K) < \tfrac{\varepsilon}{3} + \Sigma_{K\in \mathcal{I}}\, m_K{}^g\,\mathrm{vol}(K). \quad (4)$$

Now, it is immediate from the definition of $\mathcal{I}$ that the cuboids in it are subsets of I°. Therefore, $m_K{}^f = m_K{}^g$ for $K\in \mathcal{I}$ and the summations on the right sides of (3) and (4) are equal. It follows that

$$L(f,P) - L(g,P) > 2(-\tfrac{\varepsilon}{3}).$$

As already noted, (1) holds with P in place of Q. Hence, we obtain from the above inequality that $L(f,P) > \int_I g - \varepsilon$ and hence that $\int_I f \ge L(f,P) > \int_I g - \varepsilon$. As this has been established for every positive ε, we have $\underline{\int}_I f \ge \underline{\int}_I g$. The reverse inequality follows by a similar argument with the roles of f and g interchanged.

The equality for upper integrals follows either by an analogous argument or from the observation that $\underline{\int}_I \phi = -\overline{\int}_I(-\phi)$ for any bounded ϕ. □

6-2.11. Proposition. *Let $J \subseteq I$, both cuboids, and let $g:J\to\mathbb{R}$ and $f:I\to\mathbb{R}$ both be bounded. Suppose $f(x) = 0$ for $x \notin J$ while $f(x) = g(x)$ for $x \in J^\circ$. Then*

$$\underline{\int}_I f = \underline{\int}_J g \quad \text{and} \quad \overline{\int}_I f = \overline{\int}_J g.$$

In particular, if one of the functions is integrable, so is the other and has the same integral.

Proof. By Proposition 6-2.10, the function $g_1:J\to\mathbb{R}$ such that $g_1 = g$ on J° and $g_1 = f$ elsewhere on J has the same lower and upper integrals as g. It is sufficient therefore, to prove that f has the same upper and lower integrals as g_1. Since there will be no further occasion to refer to g_1, we may as well denote g_1 by g. Then the function f agrees with g on J and is 0 for $x \notin J$.

In view of the equality $\underline{\int}_K \phi = -\overline{\int}_K(-\phi)$ for any bounded ϕ and any cuboid K, one need consider only lower integrals.

We shall first prove $\underline{\int}_I f \ge \underline{\int}_J g$.

   Consider any $\varepsilon > 0$. By definition of lower integral, there exists a paving Q of J such that

$$\underline{\int_J} g - \frac{\varepsilon}{2} < L(g,Q). \qquad (1)$$

By definition of paving [see Def. 6-1.5], Q consists of some partitions $Q_1,\ldots,Q_n$ of the edges of J. In the next paragraph, we describe a paving P of I to be obtained from Q.

   Denote the edges of J and I by $J_i = [a_i,b_i]$ and $I_i = [\alpha_i,\beta_i]$, respectively. Since $J \subseteq I$, we have $J_i \subseteq I_i$, and hence $\alpha_i \le a_i < b_i \le \beta_i$, for each i. If $J = I$, there is nothing to prove. So we assume $J \subset I$. This has the consequence that for some i, we must have either $\alpha_i < a_i$ or $b_i < \beta_i$. Let δ be *any* positive number less than all those differences $a_i - \alpha_i$ and $\beta_i - b_i$ that are positive. (We are free to require at a later stage that δ be even smaller.) Then δ has the property that, in case $\alpha_i < a_i$, the point $a_i - \delta$ belongs to the interval (α_i, a_i), and in case $b_i < \beta_i$, the point $b_i + \delta$ belongs to (b_i, β_i). Thus the partition Q_i of J_i gives rise to a partition P_i of I_i that consists all the points of Q_i and also α_i, $a_i - \delta$ in case $\alpha_i < a_i$ and $b_i + \delta$, β_i in case $b_i < \beta_i$. Note that P_i includes at most four additional points besides those of Q_i. Furthermore,

(a) subintervals formed by P_i that have both endpoints in J_i are among those already formed by Q_i
(b) and conversely;
(c) subintervals formed by P_i that have exactly one endpoint in J_i are at most two in number and have length δ, while
(d) those that have neither endpoint in J_i are disjoint from J_i.

We shall now work with the paving P of I consisting of the partitions $P_1,\ldots,P_n$.

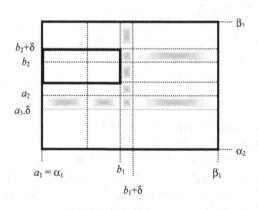

$b_2+\delta$
b_2

a_2
$a_3.\delta$

$a_1 = \alpha_1$ b_1 β_1
$b_1+\delta$

Cuboids of B are partially shaded

   The family of cuboids formed by P can be subdivided into three disjoint subfamilies $\mathcal{I}$, $\mathcal{B}$ and $\mathcal{O}$ as follows. $\mathcal{I}$ consists of those cuboids K for which every edge K_i has both endpoints in J_i (so that $K \subseteq J$ and is one of the cuboids formed by Q); $\mathcal{B}$ consists of those cuboids K for which some edge K_p has exactly one endpoint in J_p (so that K_p has length δ by (c) above);

$\mathcal{O}$ consists of the remaining cuboids K, each of which must have an edge with both endpoints outside J_i (so that K is disjoint from J by (d) above). Therefore, denoting the infimum of f over any cuboid K by m_K, we have

$$L(f,P) = \Sigma_{K \in \mathcal{I}}\, m_K \operatorname{vol}(K) + \Sigma_{K \in \mathcal{B}}\, m_K \operatorname{vol}(K) + \Sigma_{K \in \mathcal{O}}\, m_K \operatorname{vol}(K).$$

By (a) and (b) above, $\mathcal{I}$ is precisely the family of cuboids formed by Q; moreover f agrees with g on these cuboids. So, the foregoing equality leads to

$$L(f,P) = L(g,Q) + \Sigma_{K \in \mathcal{B}}\, m_K \operatorname{vol}(K) + \Sigma_{K \in \mathcal{O}}\, m_K \operatorname{vol}(K). \tag{2}$$

By (c), each $K \in \mathcal{B}$ has at least one edge of length δ and it therefore follows from Remark 6-2.9 that the total volume of all the cuboids in $\mathcal{B}$ cannot exceed $2\delta \cdot \sum_{p=1}^{n} \operatorname{vol}(I \sim I_p)$, i.e.

$$\Sigma_{K \in \mathcal{B}}\, \operatorname{vol}(K) \le 2\delta \cdot \sum_{p=1}^{n} \operatorname{vol}(I \sim I_p).$$

We now require that $\delta < \varepsilon / 4M \sum_{p=1}^{n} \operatorname{vol}(I \sim I_p)$, where M is an upper bound of $|f|$. It then follows from the inequality just noted that

$$|\Sigma_{K \in \mathcal{B}}\, m_K \operatorname{vol}(K)| \le M\Sigma_{K \in \mathcal{B}}\, \operatorname{vol}(K) < 2M\delta \cdot \sum_{p=1}^{n} \operatorname{vol}(I \sim I_p) < \frac{\varepsilon}{2}. \tag{3}$$

By (d), each $K \in \mathcal{O}$ has an edge K_p disjoint from the corresponding edge J_p of J, and therefore, is itself disjoint from J, so that f is 0 on K. Hence, (2) becomes

$$L(f,P) = L(g,Q) + \Sigma_{K \in \mathcal{B}}\, m_K \operatorname{vol}(K),$$

which, in conjunction with (3), implies

$$|L(f,P) - L(g,Q)| < \frac{\varepsilon}{2}. \tag{4}$$

Consequently,

$$L(f,P) - L(g,Q) > -\frac{\varepsilon}{2}.$$

Therefore, in view of (1), we have $L(f,P) > \int_J g - \varepsilon$ and hence $\underline{\int}_I f \ge L(f,P) > \int_J g - \varepsilon$. Since this is true for every positive ε, it follows that

$$\underline{\int}_I f \ge \underline{\int}_J g,$$

as required.

The proof of the reverse inequality is similar in parts but we shall present full details of everything that is slightly different.

Again, consider an arbitrary $\varepsilon > 0$ and let $P' = \{P'_1, \ldots, P'_n\}$ be an arbitrary paving of I. As before, denote the edges of J and I by $J_i = [a_i, b_i]$ and $I_i = [\alpha_i, \beta_i]$ respectively. Since $J \subseteq I$, we have $J_i \subseteq I_i$, and hence $\alpha_i \le a_i < b_i \le \beta_i$, for each i. If $J = I$, there is nothing to prove. So we assume $J \subset I$. This has the consequence that for some i, we must have either $\alpha_i < a_i$ or $b_i < \beta_i$. Let δ be *any* positive number less than all those differences $a_i - \alpha_i$ and $\beta_i - b_i$ that are positive. (We are free to require at a later stage that δ be even smaller.) Then δ has the property that, in case $\alpha_i < a_i$, the point $a_i - \delta$ belongs to the interval (α_i, a_i), and in case $b_i < \beta_i$, the point $b_i + \delta$ belongs to (b_i, β_i). Let P_i be the refinement of P'_i obtained by including the points $a_i, a_i - \delta$ (unless $\alpha_i = a_i$) and also the points $b_i, b_i + \delta$ (unless $b_i = \beta_i$). Note that P_i includes at most four additional points besides those of P'_i. (It is worth bearing in mind that the subinterval of P_i that begins at b_i may end before $b_i + \delta$, in which case its length is less than δ; similarly for a subinterval that ends at a_i.) Let the partition Q_i of J_i consist of those points of the partition P_i of I_i that belong to $[a_i, b_i]$. Then

(a) subintervals formed by P_i that have both endpoints in J_i are among those already formed by Q_i

(b) and conversely;

(c) subintervals formed by P_i that have exactly one endpoint in J_i are at most two in number and have length δ or less, while

(d) those that have neither endpoint in J_i are disjoint from J_i.

We now argue exactly as above with the paving P of I consisting of the partitions $P_1, \ldots, P_n$ and the paving Q of J consisting of the partitions $Q_1, \ldots, Q_n$, thereby obtaining (4). Keeping in mind that P is a refinement of P', it then follows that

$$L(f, P') \le L(f, P) \le L(g, Q) + \tfrac{\varepsilon}{2} \le \underline{\int}_J g + \tfrac{\varepsilon}{2}.$$

Since this has been proved for an arbitrary paving P' of I, it follows that $\underline{\int}_I f \le \underline{\int}_J g + \tfrac{\varepsilon}{2}$. Here $\varepsilon > 0$ is arbitrary and we therefore conclude that $\underline{\int}_I f \le \underline{\int}_J g$. $\square$

Problem Set 6-2

6-2.P1. Let $I = [0,1] \times [0,2] \subset \mathbb{R}^2$ and the paving P consist of the partitions P_1 : $0, 1$ of $[0,1]$ and P_2 : $0, 1, 2$ of $[0,2]$. If $f(x_1, x_2) = x_1 + x_2$, find $m(K)$ and $M(K)$ for each cuboid K formed by the paving.

6-2.P2. Let $I = [0,1] \times [0,2] \subset \mathbb{R}^2$, n be a positive integer, and the paving P consist of the partitions

$$P_1 : \frac{j}{n}, 0 \le j \le n \qquad \text{and} \qquad P_2 : \frac{2k}{n}, 0 \le k \le n.$$

Compute $L(f, P)$ for $f(x_1, x_2) = x_1 + x_2$.

6-2.P3. Let $I = [0,1] \times [0,2] \times [2,3] \times [1,5] \subset \mathbb{R}^4$, n be a positive integer, and the paving P consist of the partitions

$$P_1 : \frac{j_1}{n}, \, 0 \le j_1 \le n, \qquad P_2 : \frac{j_2}{n}, \, 0 \le j_2 \le 2n,$$

$$P_3 : 2 + \frac{j_3}{n}, \, 0 \le j_3 \le n, \qquad P_4 = 1 + \frac{j_4}{n}, \, 0 \le j_4 \le 4n.$$

For $f(x_1, x_2, x_3, x_4) = x_1 + x_2 + x_3 + x_4$, compute $U(f,P) - L(f,P)$.

6-2.P4. Suppose in the proof of Proposition 6-2.11, one subdivides the family of cuboids formed by P as follows: $\mathcal{I}$ is as before; $\mathcal{O}$ consists of the cuboids K for which some edge K_p has both endpoints outside J_p; and $\mathcal{B}$ consists of the remaining cuboids. Can one still carry out the proof (or are these subfamilies perhaps the same as before)?

6-2.P5. Suppose $J \subseteq I$, both cuboids, and $g : J \to \mathbb{R}$ is the restriction to J of $f : I \to \mathbb{R}$. Define $\chi : I \to \mathbb{R}$ as being 1 on J and 0 elsewhere on I (usually called the 'characteristic function' of J). Show that g is integrable if, and only if the product $f\chi$ is, in which case, $\int_J g = \int_I (f\chi)$.

6-2.P6. Let $f : [0,1] \times [a,b] \to \mathbb{R}$ be defined as

$$f(x,y) = \begin{cases} 0 & \text{if } x \notin \mathbb{Q} \text{ or } y \notin \mathbb{Q} \\ 1/q & \text{if } y \in \mathbb{Q} \text{ and } x = \frac{p}{q} \text{ with minimal } q \in \mathbb{N}. \end{cases}$$

Prove that f is integrable and has integral 0.

6-2.P7. (a) If $f : I \to \mathbb{R}$ is integrable, where I is a cuboid in $\mathbb{R}^n$, and if $f(x) \ge 0$ on I, show that $\int_I f \ge 0$.

(b) If $f : I \to \mathbb{R}$ is integrable, where I is a cuboid in $\mathbb{R}^n$, show that $|f|$ is integrable and that $|\int_I f| \le \int_I |f|$.

6-2.P8. Suppose $f : [a_1, b_1] \times \cdots \times [a_n, b_n] \to \mathbb{R}$ is increasing in each variable ($n \ge 2$). Show that f is integrable.

6-2.P9. Let $f : I \to \mathbb{R}$ and $g : I \to \mathbb{R}$ be integrable, where $I \subseteq \mathbb{R}^m$ is a cuboid. Show that the product function $fg : I \to \mathbb{R}$ is integrable.

6-2.P10. Let $f : I \to \mathbb{R}$ be integrable, where $I \subseteq \mathbb{R}^m$ is a cuboid and let $\phi : I \times \mathbb{R}^n \to \mathbb{R}$ be defined by

$$\phi(x,y) = f(x) \text{ for } x \in I, \, y \in \mathbb{R}^n.$$

Show that ϕ is integrable over $I \times J$, where J is any cuboid in $\mathbb{R}^n$.

6-2.P11. If f_1 and f_2 are Riemann integrable over intervals $I_1 = [a_1, b_1]$ and $I_2 = [a_2, b_2]$ respectively and $(f_1 f_2)(x,y) = f_1(x) f_2(y)$ for $(x,y) \in I = I_1 \times I_2$, prove that

$$\int_I f_1 f_2 = (\int_{I_1} f_1)(\int_{I_2} f_2), \quad \text{i.e.,} \quad \int_I f_1(x) f_2(y) \, dx \, dy = (\int_{I_1} f_1(x) \, dx)(\int_{I_2} f_2(y) \, dy).$$

6-3 Iterated Integral Over Cuboids

Computation of integrals over cuboids in $\mathbb{R}^n$ is usually carried out by n iterations, a procedure from multivariable calculus, with which the reader is doubtless familiar. Although the procedure works for all continuous functions, it can break down for more general functions. We shall illustrate this presently, but first we take note of two examples in $\mathbb{R}$ that we shall need. One is the *Dirichlet* function, which takes only two different values, one at rational numbers and the other at irrational numbers. It is well known from analysis in $\mathbb{R}$ that this function is not integrable over any interval. The second function on $\mathbb{R}$ that we wish to note is the *Thomae* function, which is 0 at irrational numbers and equals $1/q$ at any rational number p/q, where q is the minimal possible positive integer in any such representation of the latter. Such a function has already been mentioned in 6-2.P6, where $f(x,y)$ becomes the Thomae function for each fixed rational y. The solution of that problem essentially consists in showing that the Thomae function is integrable over $[0,1]$. Since all lower sums are 0, it follows that the integral is 0.

Let I and J be cuboids in $\mathbb{R}^n$ and $\mathbb{R}^m$ respectively. Then $I\times J$ is a cuboid in $\mathbb{R}^{n+m}$. If $f:I\times J\to\mathbb{R}$ is bounded, then for each $y \in J$, the function $\phi:I\to\mathbb{R}$ defined by $\phi(x) = f(x,y)$ is also bounded. Therefore, it has upper and lower integrals, which we shall denote by

$$\overline{\int}_I f(x,y)\,dx \quad \text{and} \quad \underline{\int}_I f(x,y)\,dx \quad \text{respectively.}$$

Each is a bounded function of y and we shall denote their upper and lower integrals by

$$\overline{\int}_J dy\,\overline{\int}_I f(x,y)\,dx, \quad \underline{\int}_J dy\,\overline{\int}_I f(x,y)\,dx \quad \text{and} \quad \overline{\int}_J dy\,\underline{\int}_I f(x,y)\,dx,$$

$$\underline{\int}_J dy\,\underline{\int}_I f(x,y)\,dx$$

respectively. Similarly, we have the following functions of x:

$$\overline{\int}_J f(x,y)\,dy \quad \text{and} \quad \underline{\int}_J f(x,y)\,dy \quad \text{respectively}$$

and their upper and lower integrals

$$\overline{\int}_I dx\,\overline{\int}_J f(x,y)\,dy, \quad \underline{\int}_I dx\,\overline{\int}_J f(x,y)\,dy \quad \text{and} \quad \overline{\int}_I dx\,\underline{\int}_J f(x,y)\,dy,$$

$$\underline{\int}_I dx\,\underline{\int}_J f(x,y)\,dy$$

respectively. This notation dispenses with explicit mention of ϕ or its analogue $y\to f(x,y)$, and we need not introduce any letters to denote them.

6-3.1. Examples. (a) Let $f: [0,1] \times [a,b] \to \mathbb{R}$ be defined [as in 6-2.P6] thus:

$$f(x,y) = \begin{cases} 0 & \text{if } x \notin \mathbb{Q} \text{ or } y \notin \mathbb{Q} \\ 1/q & \text{if } y \in \mathbb{Q} \text{ and } x = \frac{p}{q} \text{ with minimal } q \in \mathbb{N}. \end{cases}$$

Here $I = [0,1]$ and $J = [a,b]$. In order to consider $\int_J f(x,y)\,dy$, fix any $x \in I = [0,1]$. If $x \notin \mathbb{Q}$ then $f(x,y) = 0 \ \forall \ y \in J$ and therefore, the integral exists and is 0. But if $x \in \mathbb{Q}$, say $x = p/q$ with q minimal positive, then $f(x,y) = 0$ for irrational y and $1/q$ for rational y; in other words, Dirichlet's function with the two values 0 and $1/q$. This means $\int_J f(x,y)\,dy$ does not even exist. And this despite the fact that $\int_{I \times J} f$ does exist and is 0, as seen in 6-2.P6.

Note however that $\underline{\int}_J f(x,y)\,dy = 0$ for all x, so that $\int_I dx \underline{\int}_J f(x,y)\,dy = 0$. It is also true [see Problem 6-3.P1] that $\int_I dx \overline{\int}_J f(x,y)\,dy = 0$.

In order to consider $\int_I f(x,y)\,dx$, fix any $y \in J = [a,b]$. If $y \notin \mathbb{Q}$ then $f(x,y) = 0 \ \forall \ x \in I$ and therefore, the integral exists and is 0. But if $y \in \mathbb{Q}$, then $f(x,y)$ is Thomae's function and therefore, has integral 0. Thus $\int_I f(x,y)\,dx = 0 \ \forall \ y \in J$. So, $\int_J dy \int_I f(x,y)\,dx = 0 = \int_{I \times J} f$.

(b) Let $f: [0,3] \times [0,1] \to \mathbb{R}$ be as in Example 6-2.8:

$$f(x,y) = \begin{cases} 3 & \text{if } y \in \mathbb{Q} \\ x^2 & \text{if } y \notin \mathbb{Q}. \end{cases}$$

As seen before, $\int_{[0,3] \times [0,1]} f$ does not exist. In order to consider $\int_{[0,3]} f(x,y)\,dx$, fix any $y \in [0,1]$. If $y \in \mathbb{Q}$ then $f(x,y) = 3 \ \forall \ x \in [0,3]$ and so $\int_{[0,3]} f(x,y)\,dx = 9$. If $y \notin \mathbb{Q}$, then $f(x,y) = x^2 \ \forall \ x \in [0,3]$ and so $\int_{[0,3]} f(x,y)\,dx = \frac{1}{3}(27-0) = 9$. So, $\int_{[0,3]} f(x,y)\,dx = 9 \ \forall \ y \in [0,1]$ and we therefore have $\int_{[0,1]} dy \int_{[0,3]} f(x,y)\,dx = 9$. And this despite the fact that $\int_{[0,3] \times [0,1]} f$ does not even exist.

We shall now show that what was noted regarding upper and lower integrals in the first of the above two examples actually typifies what happens in general.

6-3.2. Theorem. *Let I and J be cuboids in $\mathbb{R}^n$ and $\mathbb{R}^m$ respectively and suppose that $f: I \times J \to \mathbb{R}$ is integrable. Then the functions Φ and Ψ defined on J by*

$$\Phi(y) = \underline{\int}_I f(x,y)\,dx \quad \text{and} \quad \Psi(y) = \overline{\int}_I f(x,y)\,dx$$

are integrable and both have integral equal to $\int_{I \times J} f$, which is to say,

$$\int_J dy \underline{\int}_I f(x,y)\,dx = \int_{I \times J} f = \int_J dy \overline{\int}_I f(x,y)\,dx.$$

The analogous functions of x given by

$$\underline{\int}_J f(x,y)\,dy \quad and \quad \overline{\int}_J f(x,y)\,dy$$

are also integrable and

$$\int_I dx \underline{\int}_J f(x,y)\,dy = \int_{I\times J} f = \int_I dx \overline{\int}_J f(x,y)\,dy.$$

Proof. Let P be a paving of $I\times J$. Then P is comprised of a paving of I and a paving of J. Denote the families of cuboids formed by the latter two by $\mathcal{F}$ and $\mathcal{G}$ respectively. Then the cuboids formed by P are precisely those that are of the form $K\times L$, where $K\in\mathcal{F}$ and $L\in\mathcal{G}$.

Set

$$M_{K\times L} = \sup\{f(x,y):(x,y)\in K\times L\}, \qquad N_L = \sup\{\Psi(y):y\in L\}.$$

Also, for any $y\in J$, set

$$M_K(y) = \sup\{f(x,y):x\in K\}.$$

Then $y\in L \Rightarrow M_{K\times L} \ge M_K(y)$. It follows for each $y\in L$ that

$$\Sigma_{K\in\mathcal{F}}\, M_{K\times L}\,\mathrm{vol}(K) \ge \Sigma_{K\in\mathcal{F}}\, M_K(y)\,\mathrm{vol}(K) \ge \overline{\int}_I f(x,y)\,dx = \Psi(y).$$

Since this is true for all $y\in L$, it further follows that

$$\Sigma_{K\in\mathcal{F}}\, M_{K\times L}\,\mathrm{vol}(K) \ge N_L.$$

If we multiply both sides by $\mathrm{vol}(L)$ and take the summation over all $L\in\mathcal{G}$, while taking into account that $\mathrm{vol}(K)\cdot\mathrm{vol}(L) = \mathrm{vol}(K\times L)$, we get

$$\Sigma_{L\in\mathcal{G}}\,\Sigma_{K\in\mathcal{F}}\, M_{K\times L}\,\mathrm{vol}(K\times L) \ge \Sigma_{L\in\mathcal{G}}\, N_L\,\mathrm{vol}(L) \ge \overline{\int}_J \Psi.$$

As already noted, the family of cuboids formed by P is precisely $\{K\times L : K\in\mathcal{F}, L\in\mathcal{G}\}$. Therefore, the left side in the above inequality is nothing but $U(f,P)$. Thus

$$U(f,P) \ge \overline{\int}_J \Psi$$

for any paving P. It follows that

$$\int_{I\times J} f \ge \overline{\int}_J \Psi = \overline{\int}_J dy \overline{\int}_I f(x,y)\,dx.$$

An analogous argument shows that

$$\int_{I\times J} f \le \underline{\int}_J \Phi = \underline{\int}_J dy \underline{\int}_I f(x,y)\,dx.$$

Consequently,

$$\int_{I\times J} f \le \underline{\int}_J dy \underline{\int}_I f(x,y)\,dx \le \underline{\int}_J dy \,\overline{\int}_I f(x,y)\,dx \le \overline{\int}_J dy \,\overline{\int}_I f(x,y)\,dx \le \int_{I\times J} f$$

and also

$$\int_{I\times J} f \le \underline{\int}_J dy \underline{\int}_I f(x,y)\,dx \le \overline{\int}_J dy \underline{\int}_I f(x,y)\,dx \le \overline{\int}_J dy \,\overline{\int}_I f(x,y)\,dx \le \int_{I\times J} f.$$

The first of these quadruple inequalities implies that $\overline{\int}_I f(x,y)\,dx$, which is $\Psi(y)$, is integrable with integral $\int_{I\times J} f$, and the second one implies the corresponding thing about $\underline{\int}_I f(x,y)\,dx$.

The rest follows by an analogous argument.   □

6-3.3. Remark. When f is continuous, it is continuous as a function of x alone (for each fixed y) and therefore, in the above theorem we have

$$\underline{\int}_I f(x,y)\,dx = \overline{\int}_I f(x,y)\,dx = \int_I f(x,y)\,dx$$

and furthermore,

$$\int_J dy \int_I f(x,y)\,dx = \int_{I\times J} f.$$

Similarly, f is a continuous function of y alone (for each fixed x) and therefore

$$\underline{\int}_J f(x,y)\,dy = \overline{\int}_J f(x,y)\,dy = \int_J f(x,y)\,dy$$

and furthermore,

$$\int_I dx \int_J f(x,y)\,dy = \int_{I\times J} f.$$

The integrals $\int_J dy \int_I f(x,y)\,dx$ and $\int_I dx \int_J f(x,y)\,dy$ are called **iterated** (or **repeated**) **integrals**. The double equality

$$\int_J dy \int_I f(x,y)\,dx = \int_{I\times J} f = \int_I dx \int_J f(x,y)\,dy,$$

which has been shown to hold when f is continuous, is sometimes known as **Fubini's Theorem.**

Problem Set 6-3

6-3.P1. For the function f of Example 6-3.1(a), show that $\int_I dx \,\overline{\int}_J f(x,y)\,dy = 0$.

6-3.P2. For the function f of Example 6-3.1(b), determine whether $\int_{[0,3]} dx \int_{[0,1]} f(x,y)\,dy = 9$.

6-3.P3. For the function f of Example 6-3.1(b), where $I = [0,3]$ and $J = [0,1]$, compute whichever of the following exist

$$\int_I dx \underline{\int}_J f(x,y)\,dy, \quad \int_I dx \,\overline{\int}_J f(x,y)\,dy, \quad \int_J dy \underline{\int}_I f(x,y)\,dx,$$

$$\int_J dy \,\overline{\int}_I f(x,y)\,dx$$

and determine which (if any) are equal.

6-3.P4. For the function $f: [0,\pi] \times [0,1] \to \mathbb{R}$ defined by

$$f(x,y) = \begin{cases} \cos x & \text{if } y \text{ is rational} \\ 0 & \text{if } y \text{ is irrational} \end{cases},$$

show that $\int_{[0,1]} dy \int_{[0,\pi]} f(x,y)\,dx$ exists but that $\int_{[0,\pi]} dx \int_{[0,1]} f(x,y)\,dy$ and $\int_{[0,\pi] \times [0,1]} f$ do not.

6-3.P5. For the function $f: [0,1] \times [0,1] \to \mathbb{R}$ defined by

$$f(x,y) = \begin{cases} 1/y^2 & 0 < x < y < 1 \\ -1/x^2 & 0 < y < x < 1 \;, \\ 0 & \text{elsewhere} \end{cases}$$

show that $\int_{[0,1]} dx \int_{[0,1]} f(x,y)\,dy$ and $\int_{[0,1]} dy \int_{[0,1]} f(x,y)\,dx$ both exist but are unequal.

6-3.P6. Using Fubini's theorem and the positivity of the integral of a positive continuous function on a rectangle with nonempty interior, give a simple proof of the equality $\frac{\partial^2 f}{\partial x \partial y} = \frac{\partial^2 f}{\partial y \partial x}$ for mixed partial derivatives, assuming that both are continuous.

6-3.P7. For each $i \in \mathbb{N}$, let ϕ_i be continuous on $\mathbb{R}$, vanish outside $I_i = (2^{-i}, 2^{1-i}]$ and satisfy $\int_{I_i} \phi_i = 1$. Put

$$f(x,y) = \sum_{i=1}^{\infty} [\phi_i(x) - \phi_{i+1}(x)] \phi_i(y).$$

Show that the function is well defined on all of $\mathbb{R}^2$ (i.e., the series always converges) and

$$\int_0^1 dy \int_0^1 f(x,y)\,dx = 0 \quad \text{but} \quad \int_0^1 dx \int_0^1 f(x,y)\,dy = 1.$$

6-4 Riemann Integral Over Other Sets

Suppose we want the Riemann integral of a bounded function over a set that is not a cuboid. Right at the outset, we require the set be bounded, which implies that it is contained in some cuboid. We select any one such cuboid I, say, extend the function to be zero outside the set and then take the integral (if it exists) over I. Before proceeding any further, we must show that it does not matter which cuboid containing the set is selected. The formal statement will be easier if we introduce the following notation for the extension just referred to: Suppose E is a bounded nonempty subset of $\mathbb{R}^n$; for any cuboid $I \supseteq E$ and any function $f:E\rightarrow\mathbb{R}$, we denote by f_I the real valued function on I defined as

$$f_I(x) = \begin{cases} f(x) & \text{if } x \in E \\ 0 & \text{if } x \notin E. \end{cases}$$

6-4.1. Proposition. *Suppose E is a bounded subset of $\mathbb{R}^n$ and $f:E\rightarrow\mathbb{R}$ is bounded. For any cuboids I_1 and I_2 containing E,*

$$\underline{\int}_{I_1} f_{I_1} = \underline{\int}_{I_2} f_{I_2} \quad \text{and} \quad \overline{\int}_{I_1} f_{I_1} = \overline{\int}_{I_2} f_{I_2}.$$

In particular, the function f_{I_1} is integrable if, and only if f_{I_2} is, in which case,

$$\int_{I_1} f_{I_1} = \int_{I_2} f_{I_2}.$$

Proof. Let $J = I_1 \cap I_2$ and suppose J is a cuboid. Then both the following are true: (1) the restrictions of f_{I_1} and f_{I_2} to J are both equal to f_J and (2) $f_{I_1}(x) = f_{I_2}(x) = f_J(x)$ for $x \in J^\circ$ while $f_{I_1}(x) = f_{I_2}(x) = 0$ for $x \notin J$. Therefore, the equality of lower and upper integrals follows immediately from Proposition 6-2.11.

If J is not a cuboid, then by Proposition 6-1.10, it is a subset of a face of I_1 as well as of a face of I_2. Therefore, f_{I_1} vanishes on the interior of I_1 while f_{I_2} vanishes on the interior of I_2. By Proposition 6-2.10, it follows that

$$\underline{\int}_{I_1} f_{I_1} = \underline{\int}_{I_2} f_{I_2} = 0 = \overline{\int}_{I_1} f_{I_1} = \overline{\int}_{I_2} f_{I_2}. \qquad \square$$

With the above proposition in hand, we are almost ready to define $\int_E f$ as $\int_I f_I$, except that we need to ensure that the constant function 1 will turn out to be integrable! We restrict ourselves to sets E for which this happens. The function on a set $X \supseteq E$, which equals 1 on E and 0 on the complement (which may be empty) is called the **characteristic function** of E. It will be denoted by χ_E without explicit mention of X, which will usually be understood from the context. When f is the constant function 1 on E, any extension f_I to a cuboid $I \supseteq E$ is obviously the characteristic function of E on I. The constant function 1 on E will turn out to be integrable if we restrict E to have the property that, for some cuboid $I \supseteq E$, the characteristic function χ_E is integrable. The same will then be true for any cuboid $I \supseteq E$ by Proposition 6-4.1.

6-4.2. Definition. *A bounded set $E \subseteq \mathbb{R}^n$ is said to* **have content** *if for some cuboid $I \supseteq E$, the characteristic function of E on I is integrable. Its integral is called the* **content of E** *and will be denoted by $c(E)$.*

In view of Proposition 6-4.1, E has content if and only if for every cuboid $I \supseteq E$, the characteristic function χ_E of E on I is integrable. Moreover the integral $\int_I \chi_E$ is the same regardless of the particular cuboid; so the content is independent of I.

Let E be the set of all those points in the cuboid $I = [0,1] \times [0,1]$ that have rational components. It is quickly seen that E does not have content. Indeed, every cuboid of every paving of I contains points of E as well as points of its complement, so that the upper and lower sums of χ_E are 1 and 0, respectively. This means the upper and lower integrals are also 1 and 0, respectively.

6-4.3. Definition. *Suppose $E \subseteq \mathbb{R}^n$ has content and $f : E \to \mathbb{R}$ is bounded. Then the* **lower** *and* **upper (Riemann) integrals** $\underline{\int}_E f$ *and* $\overline{\int}_E f$ *of f are the lower and upper Riemann integrals $\underline{\int}_I f_I$ and $\overline{\int}_I f_I$ for any cuboid $I \supseteq E$. The function f is said to be* **integrable** *if $\underline{\int}_I f = \overline{\int}_I f$ and the common value of the lower and upper integrals, denoted by $\int_E f$, is called the* **(Riemann) integral** *of f.*

We emphasise once again that the values of the lower and upper integrals, and hence also the integrability and value of the integral, are independent of the choice of the cuboid containing E [Proposition 6-4.1].

6-4.4. Remarks. (a) If $c(E) = 0$, then $\int_I \chi_E = 0$ for any cuboid $I \supseteq E$. So, for every $\varepsilon > 0$, there exists a paving P of I such that $U(\chi_E, P) < \varepsilon$. But $U(\chi_E, P) = \Sigma_{K \in \mathcal{F}} \operatorname{vol}(K)$, where $\mathcal{F}$ is the family of those cuboids formed by P that have a nonempty intersection with E. For any bounded function $f : E \to \mathbb{R}$, the absolute values of the lower and upper sums cannot exceed $(\sup |f|) \Sigma_{K \in \mathcal{F}} \operatorname{vol}(K) < (\sup |f|) \varepsilon$. It follows that

$$\underline{\int}_E f = \overline{\int}_E f = 0 \quad \text{when} \quad c(E) = 0.$$

(b) It can be proved by using the corresponding results for integrals over cuboids that, whenever functions f and g on E are both integrable, the following are also integrable: $f + g$, fg, $|f|$ and αf (where α is a constant); moreover the usual equalities and inequalities concerning them hold.

(c) Let $f : E \to \mathbb{R}$ be integrable and $A \subseteq E \subseteq I$, where I is a cuboid. Denote the restriction of f to A by $f|_A$. By definition of the extension to I, it follows that

$(f|_A)_I = f_I \cdot \chi_A$. Suppose both E and its subset A have content. Then f_I and χ_A are both integrable and hence their product $(f|_A)_I$ is integrable by part (b) above. This implies by Def. 6-4.3 that $f|_A$ is integrable. In future, we shall avoid introducing the symbol $f|_A$ by using notation as in Def. 6-4.5 below.

(d) Let $f: E \rightarrow \mathbb{R}$ be bounded and $A \subseteq E \subseteq I$, where I is a cuboid. Suppose f is zero outside A. Then $(f|_A)_I = f_I$. If both A and E have content, then it follows that

$$\underline{\int}_I f_I = \underline{\int}_I (f|_A)_I = \underline{\int}_I f_I \cdot \chi_A \text{ and } \overline{\int}_I f_I = \overline{\int}_I (f|_A)_I = \overline{\int}_I f_I \cdot \chi_A$$

for any cuboid $I \supseteq E$.

(e) As in the case of integrals over a cuboid, for any set E having content, we have

$$\underline{\int}_E (f+g) \geq \underline{\int}_E f + \underline{\int}_E g,$$

$$\overline{\int}_E (f+g) \leq \overline{\int}_E f + \overline{\int}_E g$$

and

$$\underline{\int}_E f = -\overline{\int}_E (-f).$$

6-4.5 Definition. *Let $A \subseteq E \subseteq \mathbb{R}^n$, where A has content, and suppose $f: E \rightarrow \mathbb{R}$ is bounded on A. Then the* **lower** *and* **upper (Riemann) integrals** $\underline{\int}_A f$ *and* $\overline{\int}_A f$ *of f on A are the lower and upper Riemann integrals of the restriction $f|_A$, i.e.,* $\underline{\int}_I (f|_A)_I$ *and* $\overline{\int}_I (f|_A)_I$ *for any cuboid $I \supseteq A$. The function f is said to be* **integrable on A** *if $\underline{\int}_A f = \overline{\int}_A f$ and the common value of the lower and upper integrals, denoted by $\int_A f$, is called the* **(Riemann) integral** *of f on A.*

Using this terminology, we can rephrase Remark 6-4.4(c) as saying that if both E and its subset A have content and a function f is integrable on E, then it is also integrable on A. Similarly, we can rephrase Remark 6-4.4(d) as saying that if $f: E \rightarrow \mathbb{R}$ is bounded and $A \subseteq E \subseteq I$, where I is a cuboid and both A and E have content, and if f is zero outside A, then $\overline{\int}_A f = \overline{\int}_E f$ and $\underline{\int}_A f = \underline{\int}_E f$.

It may be noted that in the above definition, E need not be bounded, but A has to be, because it is assumed to have content.

6-4.6. Remarks. (a) It is trivial to show for characteristic functions that

$$E \subseteq F \Rightarrow \chi_E \leq \chi_F;$$

$$\chi_{E \cap F} = \chi_E \cdot \chi_F \quad \text{and} \quad \chi_{E \cup F} = \chi_E + \chi_F - \chi_E \cdot \chi_F.$$

Also, denoting the complement of a set F by F^c, and the set difference $E \cap F^c$ by $E \backslash F$, we have

$$\chi_{E \setminus F} = \chi_{E \cap F^c} = \chi_E - \chi_{E \cap F} \, .$$

Taken in conjunction with Remark 6-4.4, these properties have the following straightforward consequences, which will be used in this chapter without reference:

(b) If E and F both have content, then: $E \subseteq F \Rightarrow \chi_E \le \chi_F \Rightarrow c(E) \le c(F)$.

(c) If $c(F) = 0$ and $E \subseteq F$, then E has content and $c(E) = 0$.

(d) If E and F both have content, then so do $E \cap F$ and $E \cup F$; moreover

$$c(E \cup F) \le c(E) + c(F) \text{ in general,}$$

while

$$c(E \cup F) = c(E) + c(F) \text{ in case } c(E \cap F) = 0.$$

(e) If $c(E) = c(F) = 0$, then $c(E \cap F) = c(E \cup F) = 0$.

(f) If $f : E \to \mathbb{R}$ is bounded on the subset $A \subseteq E$ having content zero, then f is integrable over A and $\int_A f = 0$.

(g) If $c(A \cap B) = 0$ and $f : A \cup B \to \mathbb{R}$ is bounded, then by Remark 6-4.4(e),

$$\underline{\int}_{A \cup B} f \ge \underline{\int}_A f + \underline{\int}_B f,$$

$$\overline{\int}_{A \cup B} f \le \overline{\int}_A f + \overline{\int}_B f.$$

If furthermore A and B have content and f is integrable over A as well as B, then $A \cup B$ has content, f is integrable over $A \cup B$ and

$$\int_{A \cup B} f = \int_A f + \int_B f.$$

(h) If E and F have content, then so does the difference $E \setminus F$, and $c(E \setminus F) = c(E) - c(E \cap F)$.

Any closed cuboid has content equal to its volume, because the integral of the constant function 1 is the volume. For an open cuboid, the same can be seen to be true by applying Proposition 6-2.11. We shall therefore use the terms 'content' and 'volume' interchangeably in connection with cuboids.

6-4.7. Proposition. *A finite union of cuboids has content and its content does not exceed the total volume of the cuboids.*

Proof. This is immediate from Remark 6-4.6(d) and the fact that the content of a cuboid is the same as its volume.   □

6-4.8. Proposition. *A face of a cuboid has content zero.*

Proof. A face $\{x \in I : x_p = a_p\}$ of a cuboid $I = [a_1, b_1] \times \cdots \times [a_n, b_n]$ is contained in the cuboid $J = J_1 \times \cdots \times J_n$, where $J_p = [a_p - \varepsilon, a_p + \varepsilon]$, $\varepsilon > 0$, and $J_i = [a_i, b_i]$ for $i \ne p$. The volume (or content) of J is $2\varepsilon \cdot \mathrm{vol}(I)/(b_p - a_p)$. By Remark 6-4.6(b), the content of the face cannot exceed the content (or volume) of J, which is $2\varepsilon \cdot \text{vol}(I)/(b_p - a_p)$. Since this holds for an arbitrary $\varepsilon > 0$, we conclude that the content of a face is always 0. $\qquad\square$

6-4.9. Proposition. *Let $\mathcal{F}$ be a nonoverlapping family of cuboids. Then*
(a) *the intersection of a cuboid of $\mathcal{F}$ with any union of other cuboids of $\mathcal{F}$ has content zero;*
(b) *the content of the union $\cup \mathcal{F}$ equals the sum of the contents (volumes) of the cuboids of $\mathcal{F}$.*

Proof. (a) By definition of nonoverlapping [see Def. 6-1.6], an interior point of a cuboid K of the family cannot belong to any union K' of other cuboids of the family. Therefore, $K \cap K'$, if nonempty, must consist of points belonging to the faces of K. Since there are only finitely many faces and each of them has content zero by Proposition 6-4.8, it follows by Remark 6-4.6(d) that the union of all the faces has content zero. Therefore, the subset $K \cap K'$ must have content zero.

(b) This follows from what has just been proved and Remark 6-4.6(d). $\qquad\square$

6-4.10. Proposition. *Let $E \subseteq \mathbb{R}^n$ be bounded. If E has content, then for any $\varepsilon > 0$, there exists a finite family $\mathcal{F}$ of closed cuboids such that*
(i) *$\mathcal{F}$ covers E;*
(ii) *the total volume of all the cuboids of $\mathcal{F}$ is less than $c(E) + \varepsilon$;*
(iii) *$\mathcal{F}$ is nonoverlapping.*
Moreover, the following are equivalent:
(α) *E has content zero;*
(β) *for any $\varepsilon > 0$, there exist finitely many closed cuboids which cover E, have total volume less than ε and form a nonoverlapping family;*
(γ) *for any $\varepsilon > 0$, there exist finitely many closed cuboids which cover E and have total volume less than ε;*
(δ) *for any $\varepsilon > 0$, there exist finitely many open cuboids which cover E and have total volume less than ε.*

Proof. Throughout this argument, whenever I is a cuboid containing E, the symbol M_K will denote the sup of χ_E on a cuboid $K \subseteq I$. Thus, M_K is 1 or 0 depending on whether K contains a point of E or not.

Suppose E has content and $\varepsilon > 0$. Let $E \subseteq I$, a cuboid. Since $c(E) = \int_I \chi_E$ by definition, there exists a paving P of I such that $0 \le U(\chi_E, P) < c(E) + \varepsilon$. Let $\mathcal{F}$ be the family of those cuboids formed by P that contain a point of E, and let $\mathcal{G}$ consist of the rest. Then $\mathcal{F}$ covers E. Furthermore, since M_K is 0 or 1 according as $K \in \mathcal{G}$ or $K \in \mathcal{F}$, we have

$$\Sigma_{K\in\mathcal{F}} \mathrm{vol}(K) = \Sigma_{K\in\mathcal{F}} M_K \mathrm{vol}(K) + \Sigma_{K\in\mathcal{G}} M_K \mathrm{vol}(K) = U(\chi_E,P) < c(E) + \varepsilon.$$

Thus, $\mathcal{F}$ fulfills (i) and (ii). It follows from Remark 6-1.7(e) that it also fulfills (iii).

$(\alpha)\Rightarrow(\beta)$. This is immediate from what has just been proved.

$(\beta)\Rightarrow(\gamma)$. Trivial.

$(\gamma)\Rightarrow(\delta)$. Let $\varepsilon > 0$ and consider any m closed cuboids that cover E and have total volume less than $\frac{\varepsilon}{2}$. By Remark 6-1.2(d), each closed cuboid K is contained in an open cuboid with volume less than $\mathrm{vol}(K) + \frac{\varepsilon}{2m}$. Therefore, we have m open cuboids which cover E and have total volume less than $\frac{\varepsilon}{2} + \frac{\varepsilon}{2m}m = \varepsilon$.

$(\delta)\Rightarrow(\alpha)$. Let $\varepsilon > 0$ and consider any m open cuboids which cover E and have total volume less than ε. Then the corresponding closed cuboids (their closures) $K_1,\ldots,K_m$ do the same. Besides, any point of E belongs to the interior of one or more of them. Let I be a cuboid that contains their union and hence also E. By Proposition 6-1.9, there exists a paving P of I such that (a)–(c) of that proposition hold. In terms of the notation there, (c) ensures that $M_K = 0$ whenever $K \notin \mathcal{F}_k$ for every k. Therefore, it follows from (a) and (b) that

$$U(\chi_E,P) \le \Sigma_k(\Sigma_{K\in\mathcal{F}_k} \mathrm{vol}(K)) \le \Sigma_k \mathrm{vol}(K_k) < \varepsilon.$$

Since this is true for every $\varepsilon > 0$, χ_E is integrable with $\int_I \chi_E = 0$. $\square$

Recall from Def. 2-4.12 that the boundary ∂A of a subset A is the set of those points that belong to its closure $\overline{A}$ but not to its interior A°. For a point x of a closed cuboid $[a_1,b_1]\times\cdots\times[a_n,b_n]$, the meaning of not belonging to its interior $(a_1,b_1)\times\cdots\times(a_n,b_n)$ is simply that for some p, one of the inequalities $a_p \le x_p \le b_p$ should be an equality, which is the same as saying that x belongs to a face. Thus, the boundary of a cuboid is the union of all its faces.

As noted immediately after Def. 6-4.2, the set $\{(x,y)\in[0,1]\times[0,1] : x,y\in\mathbb{Q}\}$ has no content. Its interior is empty and therefore, the boundary is the same as its closure $[0,1]\times[0,1]$.

6-4.11. Corollary. *Suppose a bounded set $E \subseteq \mathbb{R}^n$ does not have content zero. Then there exists $\eta > 0$ such that, given any finitely many cuboids covering E, those among them that have a point of E in their interior have total volume greater than η.*

Proof. Since E does not have content zero, it follows by Proposition 6-4.10(γ) that there exists $\eta > 0$ such that any finite family of cuboids covering E has total volume greater than 2η.

Suppose, if possible, that $\mathcal{F}$ is a finite family of cuboids covering E such that the subfamily $\mathcal{G}$ of those cuboids of $\mathcal{F}$ that have points of E in their interior have total volume η', where $\eta' < 2\eta$. The cuboids of $\mathcal{F}$ that are not in the subfamily can have points of E only on their boundary. But the boundaries have content zero by Proposition 6-4.8 and there are only finitely many of them. Therefore, their union has content zero and by Proposition 6-4.10(γ), there exists a finite family $\mathcal{H}$ of cuboids covering the union and with total volume less than $2\eta - \eta'$. Now $\mathcal{H} \cup \mathcal{G}$ is a finite family of cuboids covering E and having total volume less than $(2\eta - \eta') + \eta' = 2\eta$. This is a contradiction, which shows that $\mathcal{F}$ cannot have the supposed property.   □

The next proof uses the simple fact that if two functions agree on an open ball, then either one of them is continuous at the centre if and only if the other one is.

6-4.12. Lemma. *Let $E \subseteq J$, where J is a cuboid, and $f : E \to \mathbb{R}$ be any function. If the extension f_J (see description above) is discontinuous at $x \in J$, then $x \in \overline{E}$; moreover, if $x \notin \partial E$, then $x \in E^\circ$ and f is discontinuous at x.*

Proof. Suppose, if possible, that $x \notin \overline{E}$. Since $\overline{E}$ is closed, some open ball centred at x is disjoint from E and therefore, f_J is zero everywhere on the intersection of that ball with J. So, f_J cannot be discontinuous at x. Next, suppose $x \notin \partial E$. Then, by definition of boundary, $x \in E^\circ$. Therefore, some open ball centred at x is a subset of E. So f_J agrees with f everywhere on that ball. If f were to be continuous at x, so would f_J.   □

6-4.13. Theorem. *Suppose $I \subseteq \mathbb{R}^n$ is a cuboid and the set E of points of discontinuity of a bounded function $f : I \to \mathbb{R}$ has content zero. Then f is integrable.*

Proof. Let J be a closed cuboid such that $I \subseteq J^\circ$. Denote by f_J the extension of f to J obtained by setting $f_J(x) = 0$ for $x \notin I$. By Lemma 6-4.12, the discontinuities of f_J are contained in $E \cup \partial I$. If we can prove f_J integrable, it will follow by Proposition 6-2.7 that f is also integrable. Since ∂I has content zero, so does $E \cup \partial I$ and therefore, it is sufficient to prove the result when $E \subseteq I^\circ$.

As usual, we denote by M_K and m_K, respectively, the supremum and infimum of f on a cuboid K. Also, M will denote $\sup \{ |f(x)| : x \in I \}$.

Let $\varepsilon > 0$. Since E has content zero, by Proposition 6-4.10 there exists a finite family $\mathcal{U}$ of open cuboids which cover E and have total volume less than ε. Since $E \subseteq I^\circ$, we may replace them with their intersections with I° and assume that all of them are subsets of I°. This ensures that their closures are subsets of I, so that Proposition 6-1.9 is applicable.

Now the union of all cuboids of $\mathcal{U}$ is an open subset of I and therefore, its complement in I, which we shall denote by H, is a compact set. Since $\mathcal{U}$ covers E, which is the set of discontinuities of f, the latter is continuous at each point of the compact set H. Therefore, it is uniformly continuous on H and there exists $\delta > 0$ such that

$$x, x' \in H, \ \|x - x'\| < \delta \Rightarrow |f(x) - f(x')| < \varepsilon. \tag{1}$$

Applying Proposition 6-1.9 to the closures $K_1, \ldots, K_m$ of the cuboids of $\mathcal{U}$, we get a paving P of I satisfying (a)–(d) of that proposition. In terms of the notation there, denote $\bigcup_{k=1}^m \mathcal{F}_k$ by $\mathcal{F}$, and let $\mathcal{G}$ consist of the remaining cuboids formed by P. By (c), a cuboid K of $\mathcal{G}$ cannot contain an interior point of any K_k and is therefore disjoint from the union of the cuboids of $\mathcal{U}$, which means it is a subset of H. By (d) and (1), it follows that $x, x' \in K \Rightarrow |f(x) - f(x')| < \varepsilon$. Therefore

$$0 < M_K - m_K \le \varepsilon \qquad \text{whenever} \quad K \in \mathcal{G}. \tag{2}$$

By (b), the total volume of all the cuboids of $\mathcal{F}$ is no greater than the total volume of all the cuboids of $\mathcal{U}$ and hence no greater than ε. That is,

$$\Sigma_{K \in \mathcal{F}} \operatorname{vol}(K) \le \varepsilon. \tag{3}$$

Using (2) and (3), we get

$$0 \le U(f, P) - L(f, P) = \Sigma_{K \in \mathcal{F}} (M_K - m_K) \operatorname{vol}(K) + \Sigma_{K \in \mathcal{G}} (M_K - m_K) \operatorname{vol}(K)$$

$$< 2M \cdot \Sigma_{K \in \mathcal{F}} \operatorname{vol}(K) + \varepsilon \cdot \Sigma_{K \in \mathcal{G}} \operatorname{vol}(K) \le 2M\varepsilon + \operatorname{vol}(I) \cdot \varepsilon = \varepsilon \cdot [2M + \operatorname{vol}(I)].$$

Since this has been shown to hold for any $\varepsilon > 0$, it follows that f is integrable. $\square$

Consider again the subset of $[0,1] \times [0,1]$ consisting of points with rational coordinates. We have noted already that it does not have content and that its boundary is $[0,1] \times [0,1]$. Since the boundary has content 1, in particular, it does not have content zero. The equivalence between a set not having content and its boundary not having content zero (which includes the possibility that the boundary has no content) is what the next proposition is about.

6-4.14. Proposition. *A bounded set $E \subseteq \mathbb{R}^n$ has content if and only if its boundary ∂E has content zero.*

Proof. Suppose ∂E has content zero and $I \supseteq E$ is a cuboid. The function χ_E is continuous on E° and therefore, its points of discontinuity must lie on ∂E by Lemma 6-4.12. Therefore, by Proposition 6-4.13, χ_E is integrable, which means E has content.

For the converse, suppose ∂E does not have content zero. As usual, we denote by M_K and m_K, respectively, the supremum and infimum of f on a cuboid K. Now, there exists $\eta > 0$ as in Proposition 6-4.11. It is sufficient to show for any cuboid $I \supseteq E$ and any paving P of I that $U(\chi_E, P) - L(\chi_E, P) > \eta$, because it will immediately follow that χ_E is not integrable, which means E does not have content. So, consider any paving P of any cuboid $I \supseteq E$. Let $\mathcal{F}$ be the family of those cuboids formed by P which have points of ∂E in their interior. By Proposition 6-4.11,

$$\Sigma_{K \in \mathcal{F}} \operatorname{vol}(K) > \eta.$$

Since each $K \in \mathcal{F}$ has a point of ∂E in its interior, it follows that K contains points of E as well as of its complement. Consequently, χ_E takes the value 1 as well as 0 on K and hence $M_K - m_K = 1$. Together with the inequality displayed above, this implies

$$U(\chi_E, P) - L(\chi_E, P) \geq \Sigma_{K \in \mathcal{F}} (M_K - m_K) \operatorname{vol}(K) = \Sigma_{K \in \mathcal{F}} \operatorname{vol}(K) > \eta.$$

As already noted, the result follows from here. □

6-4.15. Theorem. *Suppose $E \subseteq \mathbb{R}^n$ is a bounded set having content and the set of points of discontinuity of a bounded function $f: E \to \mathbb{R}$ has content zero. Then f is integrable.*

Proof. Let F be the set of points of discontinuity of f, and let $I \supseteq E$ be a cuboid. By Lemma 6-4.12, the set of discontinuities of the extension f_I is a subset of $F \cup \partial E$. Since E has content, ∂E has content zero by Proposition 6-4.14, while F has content zero by hypothesis. Therefore, $F \cup \partial E$ has content zero and hence, so does the set of discontinuities of f_I. It follows by Theorem 6-4.13 that f_I is integrable. Hence f is integrable by Def. 6-4.3. □

6-4.16. Proposition. *Suppose $F \subseteq \mathbb{R}^n$ is a bounded set having content and $f: F \to \mathbb{R}$ is a bounded function such that the set E of points where $f(x) \neq 0$ has content zero. Then f is integrable and $\int_F f = 0$.*

Proof. Let $M = \sup\{|f(x)| : x \in F\}$. Now proceed as in the proof of '$(\delta) \Rightarrow (\alpha)$' in Proposition 6-4.10 but with ε replaced by $\frac{\varepsilon}{M}$. The inequality displayed there will now turn out as

$$|U(f_I, P)| \leq \Sigma_k (\Sigma_{K \in \mathcal{F}_k} M \cdot \operatorname{vol}(K)) \leq M \cdot \Sigma_k \operatorname{vol}(K_k) < \varepsilon.$$

Since $-f$ satisfies the same hypotheses, it follows that the lower sum $L(f_I, P)$ satisfies a similar inequality. The conclusion now follows. □

It is now straightforward to show that, if two functions on a set with content differ only on a subset of content zero, then one of them is integrable if and only if the other one is, in which case their integrals agree. Sometimes it happens that a function f is defined only on a subset of a convenient set F and the subset E on which it is undefined has content zero. For example, $F = \{(x,y) \in \mathbb{R}^2 : 0 \le x \le 1, x \le y \le 1\}$ and $f(x,y) = \exp(-x/y)$. This function is undefined on the subset $E = \{(x,y) \in F : y = 0\}$, which consists of the single point $(0,0)$ and therefore, has content zero. Since f is bounded, it can be extended to all of F so as to continue being bounded; one can then work with the extended function on the set F. The first thing to check is whether F has content. We shall leave the matter to the next section [see Example 6-5.2(b)], where a general sufficient condition for the purpose will be discussed.

Problem Set 6-4

6-4.P1. Suppose a bounded set $E \subseteq \mathbb{R}^n$ has positive content. Show that it has an interior point.

6-4.P2. If a set E has content, show that its closure $\overline{E}$ must also have content and that $c(\overline{E}) = c(E)$.

6-4.P3. Let $1 \le p \le n$ and s be a fixed real number. Define a map $\alpha : \mathbb{R}^n \to \mathbb{R}^n$ as $\alpha(x)_i = x_i$ if $i \ne p$ and $\alpha(x)_p = x_p + s$. If $E \subseteq \mathbb{R}^n$ has content, show that $\alpha(E)$ also has content and $c(\alpha(E)) = c(E)$.

6-4.P4. If $E \subseteq \mathbb{R}^n$ has content and $f : E \to \mathbb{R}$ is integrable, show that $|f|$ is integrable and that $|\int_E f| \le \int_E |f|$.

6-4.P5. Justify Remark 6-4.6(f).

6-4.P6. Let $f : E \to \mathbb{R}$ be bounded, where $E \subseteq \mathbb{R}^n$ is also bounded. Suppose there exists a sequence $\{X_k : k \in \mathbb{N}\}$ of subsets of E such that f is integrable over each $E \backslash X_k$ (i.e., $E \cap X_k^c$) and $c(X_k) \to 0$ as $k \to \infty$. Show that f is integrable over E and that

$$\int_{E \backslash X_k} f \to \int_E f \text{ as } k \to \infty.$$

6-4.P7. (A mean value theorem) Suppose $E \subseteq \mathbb{R}^n$ has content and $f : E \to \mathbb{R}$ is integrable. Prove that there exists $\mu \in [m, M]$, where $m = \inf\{f(x) : x \in E\}$ and $M = \sup\{f(x) : x \in E\}$, such that $\int_E f = \mu \cdot c(E)$. If E is also closed and connected and if f is continuous, show that $\mu = f(\xi)$ for some $\xi \in E$.

6-4.P8. Let $E \subset \mathbb{R}^n$ with $c(E) \ne 0$, and $E = \overline{E^\circ}$. If f is a continuous nonnegative function on E and M denotes $\sup\{f(x) : x \in E\}$, show that

$$\lim_{m \to \infty} (\int_E f^m)^{1/m} = M.$$

Show that the condition that $E = \overline{E^\circ}$ cannot be omitted.

6-4.P9. Let $E \subseteq \mathbb{R}^n$ have content and suppose $f:E \to \mathbb{R}$ and $g:E \to \mathbb{R}$ are integrable over E. Show that fg is integrable over E.

6-4.P10. Let K be a cuboid and $f:K \to \mathbb{R}$ be integrable. For each $x \in K$, define the cuboid J_x to be $[a_1,x_1] \times \cdots \times [a_n,x_n]$ and let $F(x) = \int_{J_x} f$. Show that the function $F:K \to \mathbb{R}$ is continuous.

6-4.P11. Let $E = \{(x,y) \in \mathbb{R}^2 : 0 \le x \le 1,\ 0 \le y \le 1,\ (x,y) \ne (0,0)\}$, and define $f:E \to \mathbb{R}$ by $f(x,y) = 2xy/(x^2 + y^2)$. Show that $\int_E f$ exists.

6-5 Iterated Integral Over Other Sets

One frequently needs integrals over sets bounded by parts of spherical or plane surfaces described by an equation such as $z = (1 - x^2 - y^2)^{1/2}$ on some compact set $E \subseteq \mathbb{R}^2$. In order to ensure that the sets have content, it is necessary to know that a part of the boundary described as

$$z = \phi(x,y),\ (x,y) \in E$$

has content zero. We begin by establishing this.

6-5.1. Proposition. *Suppose $E \subseteq \mathbb{R}^{n-1}$ has content and $f:E \to \mathbb{R}$ is uniformly continuous. Then the subset $F = \{(x,y) \in \mathbb{R}^n : x \in E,\ y = f(x)\}$ of $\mathbb{R}^n$ has (n-dimensional) content zero.*

Proof. Let $\varepsilon > 0$. By uniform continuity, there exists $\delta > 0$ such that

$$|f(x) - f(x')| < \tfrac{1}{2}\varepsilon/(c(E) + 1) \quad \text{whenever } x,x' \in E \text{ and } \|x - x'\| < \delta.$$

By Proposition 6-4.10, there exist finitely many closed cuboids that cover E and have total volume less than $c(E) + 1$. Each of them has a paving such that every cuboid K formed by the paving satisfies $x,x' \in K \Rightarrow \|x - x'\| < \delta$. Let $\mathcal{G}$ be the (finite) family of all cuboids formed by all pavings. Then each $K \in \mathcal{G}$ satisfies:

$$x,x' \in K \cap E \ \Rightarrow\ |f(x) - f(x')| < \tfrac{1}{2}\varepsilon/(c(E) + 1). \qquad (1)$$

Moreover,
$$\mathcal{G} \text{ covers } E \qquad (2)$$
and
$$\Sigma_{K \in \mathcal{G}}\, \mathrm{vol}(K) < c(E) + 1. \qquad (3)$$

By (1), for each $K \in \mathcal{G}$, there exists an interval $I \subset \mathbb{R}$ with length no greater than $\varepsilon/(c(E) + 1)$ such that $x \in K \cap E \Rightarrow f(x) \in I$; in other words, $\{(x,y) \in \mathbb{R}^n : x \in K \cap E,\ y = f(x)\} \subseteq K \times I$. Now $K \times I$ is a cuboid in $\mathbb{R}^n$ with $\mathrm{vol}(K \times I) \le \mathrm{vol}(K) \cdot \varepsilon/(c(E) + 1)$.

From (2), it follows that $F \subseteq \bigcup_{K \in \mathcal{G}} (K \times I)$, and from (3) that $\Sigma_{K \in \mathcal{G}} \text{vol}(K \times I) < \varepsilon$. Proposition 6-4.10 now implies that F has content zero.   □

6-5.2. Examples. (a) Suppose we wish to show that the disc $D = \{(x,y) : x^2 + y^2 \le a^2\} \subseteq \mathbb{R}^2$ has content ($a > 0$). Its boundary can be described in two parts as $y = \sqrt{(a^2 - x^2)}$ and $y = -\sqrt{(a^2 - x^2)}$ both on the compact set $[-a, a]$. By Proposition 6-5.1, each part has content zero. Therefore, the disc D has content. For any $f : D \to \mathbb{R}$ which is known to be integrable, one can now invoke Theorem 6-3.2 to evaluate $\int_D f$ as the iterated (i.e., repeated) integral

$$\int_{-a}^{a} dx \int_{-\sqrt{a^2-x^2}}^{\sqrt{a^2-x^2}} f(x,y)\,dy .$$

In particular, the content $c(D)$ is given by this iterated integral with $f(x,y) = 1$. The usual elementary evaluation confirms that this works out to be πa^2.

(b) We return to the example briefly mentioned at the end of the preceding section:

$$F = \{(x,y) \in \mathbb{R}^2 : 0 \le x \le 1,\ x \le y \le 1\} \text{ and } f(x,y) = \exp(-x/y).$$

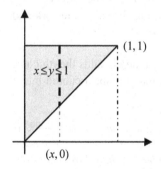

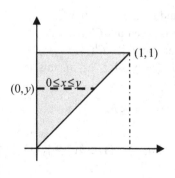

(F is the shaded triangle in the upper figure shown alongside.) The boundary of F is the union of the three subsets (not disjoint)

$$\{(x,y) \in \mathbb{R}^2 : x = 0,\ 0 \le y \le 1\},$$
$$\{(x,y) \in \mathbb{R}^2 : 0 \le x \le 1,\ y = 1\},$$
$$\{(x,y) \in \mathbb{R}^2 : 0 \le x \le 1,\ y = x\}.$$

The first is described by the function $x = 0$ on the compact domain $0 \le y \le 1$; the second and the third by the functions $y = 1$ and $y = x$ on the compact domain $0 \le x \le 1$. It follows by Proposition 6-5.1 that each of the three subsets has content zero and hence F has content. The function f is undefined at $(0,0)$ and we may set it equal to any value at that point. The only discontinuity will be at $(0,0)$ and therefore, $\int_F f$ exists by Theorem 6-4.13. The pair of inequalities that define F can easily be seen to be equivalent to the pair $0 \le y \le 1$, $0 \le x \le y$ (see the figure). It follows that the set F can alternatively be described as

$$F = \{(x,y) \in \mathbb{R}^2 : 0 \leq y \leq 1,\ 0 \leq x \leq y\}.$$

By Theorem 6-3.2, $\int_F f$ can therefore be evaluated as

$$\int_0^1 dy \int_0^y \exp(-x/y)\, dx.$$

It is left to the reader to check that this works out to be $(e-1)/2e$.

Let $\alpha:\mathbb{R}^n \to \mathbb{R}^n$ be the (linear) map given by

$$\alpha(x_1, \ldots, x_n) = (x_1 + x_2, x_2, \ldots, x_n)$$

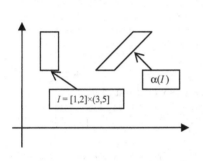

and I be a cuboid $[a_1, b_1] \times \cdots \times [a_n, b_n]$. Then $\alpha(I)$ need not be a cuboid. In the language of elementary geometry, it is a 'parallelogram' when $n = 2$ and a 'parallelepiped' when $n = 3$. We shall use these terms only informally, without using the concepts in any proof. A visualisation, as in the figure shown on the left, suggests that the parallelogram has the same base and height as the cuboid. Therefore, on the basis of elementary geometry, one would expect it to have the same content. The assertion that $c(\alpha(I)) = c(I)$ makes sense within our formal framework and we now prove that it is indeed true in all dimensions $n \geq 2$.

The next proposition says that a linear map which merely adds the jth component to the ith component ($i \neq j$) has the property that it maps a cuboid into a set having the same content as the cuboid.

6-5.3. Proposition. *Let $n \geq 2$ and $1 \leq i \leq n$, $1 \leq j \leq n$, $i \neq j$. Suppose $\alpha:\mathbb{R}^n \to \mathbb{R}^n$ is the (linear) map given by $\alpha(x_1, \ldots, x_n) = (y_1, \ldots, y_n)$, where $y_k = x_k$ for $k \neq i$ and $y_k = x_k + x_j$ for $k = i$ (i.e., $y_i = x_i + x_j$). Then for any cuboid I, the set $\alpha(I)$ has content and $c(\alpha(I)) = c(I)$.*

Proof. For ease of notation, we shall assume $i = 1$ and $j = 2$, so that $\alpha(x_1, \ldots, x_n) = (x_1 + x_2, \ldots, x_n)$. The map α has an inverse, which is given by $\alpha^{-1}(x_1, \ldots, x_n) = (x_1 - x_2, x_2, \ldots, x_n)$. Now, $(y_1, \ldots, y_n) \in \alpha(I)$ if and only if $\alpha^{-1}(y_1, \ldots, y_n) = (y_1 - y_2, y_2, \ldots, y_n) \in I$. Denoting the edges of I by $[a_i, b_i]$, $1 \leq i \leq n$, this is equivalent to

$$(y_2 \ldots, y_n) \in [a_2, b_2] \times \cdots \times [a_n, b_n] \quad \text{and} \quad a_1 + y_2 \leq y_1 \leq b_1 + y_2.$$

So,

$$\alpha(I) = \{(y_1, \ldots, y_n) \in \mathbb{R}^n : (y_2, \ldots, y_n) \in [a_2, b_2] \times \cdots \times [a_n, b_n], a_1 + y_2 \leq y_1 \leq b_1 + y_2\}. \quad (1)$$

In particular, $\alpha(I)$ is a subset of the cuboid

$$J = [a_1 + a_2, b_1 + b_2] \times [a_2, b_2] \times \cdots \times [a_n, b_n].$$

Among the boundary points of $\alpha(I)$, those that satisfy $y_i = a_i$ or b_i, $2 \le i \le n$, lie in a face of J and therefore form a set of content zero [Proposition 6-4.8]. The remaining boundary points of $\alpha(I)$ satisfy

$$\text{either } (y_2, \ldots, y_n) \in [a_2, b_2] \times \cdots \times [a_n, b_n] \quad \text{and} \quad y_1 = a_1 + y_2$$

$$\text{or } (y_2, \ldots, y_n) \in [a_2, b_2] \times \cdots \times [a_n, b_n] \quad \text{and} \quad y_1 = b_1 + y_2.$$

Since the equalities $y_1 = a_1 + y_2$ and $y_1 = b_1 + y_2$ both define uniformly continuous functions on $[a_2, b_2] \times \cdots \times [a_n, b_n]$, Proposition 6-5.1 shows that these boundary points also form a set of content zero. Therefore, the entire boundary of $\alpha(I)$ has content zero, which implies that $\alpha(I)$ has content by Proposition 6-4.14. It remains to show that $c(\alpha(I)) = c(I)$.

Let f denote the characteristic function of $\alpha(I)$ on the cuboid J, which contains it. By Def. 6-4.3, $c(\alpha(I)) = \int_J f$. By Theorem 6-3.2, $\int_J f$ is the integral over the cuboid $[a_2, b_2] \times \cdots \times [a_n, b_n] \subseteq \mathbb{R}^{n-1}$ of

$$g(y_2, \ldots, y_n) = \int_{[a_1 + a_2, b_1 + b_2]} f(y_1, \ldots, y_n) \, dy_1 .$$

For any $(y_2, \ldots, y_n) \in [a_2, b_2] \times \cdots \times [a_n, b_n]$, it follows from (1) that

$$(y_1, \ldots, y_n) \in \alpha(I) \quad \Leftrightarrow \quad a_1 + y_2 \le y_1 \le b_1 + y_2.$$

Since f is the characteristic function of $\alpha(I)$, the above equivalence implies that the value of $f(y_1, \ldots, y_n)$ is 1 or 0 depending on whether $a_1 + y_2 \le y_1 \le b_1 + y_2$ or not. Therefore,

$$g(y_2, \ldots, y_n) = \int_{[a_1 + y_2, b_1 + y_2]} f(y_1, \ldots, y_n) \, dy_1 = \int_{[a_1 + y_2, b_1 + y_2]} 1 \, dy_1$$

$$= \int_{[a_1 + y_2, b_1 + y_2]} 1 \, dy_1 = b_1 - a_1 ,$$

which is a constant function. Hence

$$c(\alpha(I)) = \int_J f = \int_{[a_2, b_2] \times \cdots \times [a_n, b_n]} (b_1 - a_1) = (b_1 - a_1) \cdots (b_n - a_n) = \mathrm{vol}(I) = c(I).$$

$\square$

The above proposition can be proved without resorting to Theorem 6-3.2 on repeated integrals. See 6-5.P1.

Problem Set 6-5

6-5.P1. Suppose $n \ge 2$, $I = [a_1, b_1] \times \cdots \times [a_n, b_n]$, where $b_1 - a_1 > b_2 - a_2$ and

$S = \{(x_1,\dots,x_n) \in \mathbb{R}^n : (x_2,\dots,x_n) \in [a_2,b_2]\times\cdots\times[a_n,b_n],\ a_1 + x_2 \le x_1 \le b_1 + x_2\}.$

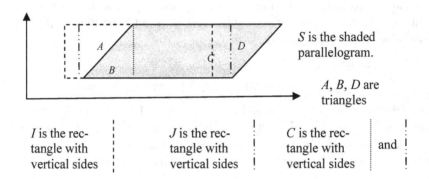

S is the shaded parallelogram.

A, B, D are triangles

I is the rectangle with vertical sides	J is the rectangle with vertical sides	C is the rectangle with vertical sides	and

Prove the following without using Proposition 6-5.3:

(a) $S = B \cup C \cup D$, where

$B = \{(x_1,\dots,x_n) \in \mathbb{R}^n : (x_2,\dots,x_n) \in [a_2,b_2]\times\cdots\times[a_n,b_n],\ a_1 + x_2 \le x_1 \le a_1 + b_2\}$
$C = \{(x_1,\dots,x_n) \in \mathbb{R}^n : (x_2,\dots,x_n) \in [a_2,b_2]\times\cdots\times[a_n,b_n],\ a_1 + b_2 \le x_1 \le b_1 + a_2\}$
$D = \{(x_1,\dots,x_n) \in \mathbb{R}^n : (x_2,\dots,x_n) \in [a_2,b_2]\times\cdots\times[a_n,b_n],\ b_1 + a_2 \le x_1 \le b_1 + x_2\}.$

(b) The cuboid

$$J = [a_1 + a_2, b_1 + a_2]\times[a_2,b_2]\times\cdots\times[a_n,b_n]$$

is the union $A \cup B \cup C$, where

$A = \{(x_1,\dots,x_n) \in \mathbb{R}^n : (x_2,\dots,x_n) \in [a_2,b_2]\times\cdots\times[a_n,b_n],\ a_1 + a_2 \le x_1 \le a_1 + x_2\}.$

(c) A, B, C, D all have content and that $A \cap B$, $B \cap C$, $C \cap D$ all have content zero.

(d) S has content and $c(S) = (b_1 - a_1)\cdots(b_n - a_n)$.

(e) If the hypothesis that $b_1 - a_1 > b_2 - a_2$ is dropped, S need not be the union of B, C, D but the equality of part (d) is nevertheless valid.

6-5.P2. Let T denote the triangular region $\{(x,y) \in \mathbb{R}^2 : x \ge 0,\ y \ge 0,\ 0 \le \frac{x}{a} + \frac{y}{b} \le 1\}$, where $a > 0$ and $b > 0$. Assume that f is continuous on T and has continuous partial derivative $D_{2\,1} f$ on the interior of T. Prove that there is a point (x_0, y_0) on the segment joining $(a, 0)$ to $(0, b)$ such that

$\int_T D_{2\,1} f = f(0,0) - f(a,0) + a D_1 f(x_0, y_0).$

6-5.P3. If $A > 0$ and $0 < \varepsilon < 1$, show that the subset

$\{(x,y) \in \mathbb{R}^2 : x \ge 0,\ y \ge 0,\ 0 \le \frac{x^2}{x^2+y^2} < \varepsilon,\ 0 < x^2+y^2 < A\}$

of the first quadrant of $\mathbb{R}^2$ (see the figure) has content that does not exceed $A\varepsilon^{1/2}$.

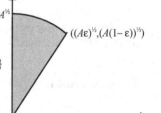

$A^{1/2}$

$((A\varepsilon)^{1/2}, (A(1-\varepsilon))^{1/2})$

7

Transformation of Integrals

7-1 Special Cuboids

So far we have presumed a minimal knowledge of linear algebra on the part of the reader. However in this chapter, we shall use basic properties of determinants and the fact that any invertible matrix is a product of 'elementary' matrices.

The transformation formula that justifies the so called 'substitution' or 'change of variables' rule for evaluating a Riemann integral is fairly easy to establish in $\mathbb{R}$. In higher dimensions however, the corresponding formula is far more difficult to prove. This is the task we take up in this chapter.

One hurdle that can be foreseen right at the outset is that even so simple a transformation of variables as

$$u = x + y \qquad v = x - y$$

need not map a cuboid into a cuboid. More generally, one would need to investigate whether a transformation of variables maps a set having content into a set of the same kind.

The *diameter* of a nonempty bounded subset $E \subseteq \mathbb{R}^n$ is understood to be $\sup\{\|x - y\| : x, y \in E\}$. Using this terminology, Remark 6-1.2(e) says that the cuboid $I = [a_1, b_1] \times \cdots \times [a_n, b_n]$ and its interior I° both have diameter $\max\{b_i - a_i : 1 \le i \le n\}$, where it is understood that the norm $\|\ \|$ is $\|\ \|_\infty$. We now introduce the notation **diam E** for the diameter of a set $E \subseteq \mathbb{R}^n$.

For the diameter of a nonempty bounded set to be zero, it is necessary and sufficient that it contain one and only one point. In fact if the diameter is zero, then $\|x - y\| = 0$ for x, y belonging to the set; i.e., $x = y$ for every pair x, y of points of the set.

For any cuboid I, it is easy to see that $\mathrm{vol}(I) \le (\mathrm{diam}\ I)^n$. Equality holds when all edges are of equal length, not otherwise. Now suppose $n > 1$. Then the ratio $(\mathrm{diam}\ I)^n / \mathrm{vol}(I)$ can be arbitrarily large: for instance, take $I = [0, a] \times [0, 3]$, where $0 < a < 3$, so that the diameter is 3, the volume is $3a$ and the ratio is $3/a$. As suggested by this instance as well as by other instances that the reader can surely come up with, the ratio may be intuitively taken as a measure of how far

S. Shirali, H.L. Vasudeva, *Multivariable Analysis*,
DOI 10.1007/978-0-85729-192-9_7, © Springer-Verlag London Limited 2011

the cuboid is from being a cube (all edges equal in length). If we require that no edge be longer than double the shortest, then the cuboid has to be reasonably close to being a cube and, as will be seen at the beginning of the forthcoming proof, the ratio cannot exceed 2^n.

Example. The cuboid $[0,1] \times [0,\pi]$ in $\mathbb{R}^2$ has its 'vertical' edge longer than double the other. However, it can be 'broken up' into subcuboids with vertical edge longer but not longer than twice the other. One way is to partition the vertical edge with points 0, $\pi/2$ and π. Since $1 < \pi/2 < 2$, this will serve the purpose. Instead of partitioning the vertical edge into two equal subintervals, we could have partitioned it into three equal subintervals. We cannot go beyond 3—unless we also partition the horizontal edge—the reason being that $\pi/4 < 1$.

Similarly, if the cuboid $[0,1] \times [0,\sqrt{29}]$ is to be broken up as above, one may partition the vertical edge into anywhere from 3 to 5 equal subintervals, because these are the only integers n for which $1 < \frac{1}{n}\sqrt{29} < 2$. The fact that this can always be done in at least one way is the crux of the next proof. Before reading it, the reader would do well to solve 7-1.P1.

7-1.1. Proposition. *A cuboid in $\mathbb{R}^n$ always has a paving P such that every cuboid K formed by it satisfies* $(\operatorname{diam} K)^n < 2^n \operatorname{vol}(K)$.

Proof. When $n = 1$, this is trivial, because a cuboid is now an interval and its diameter equals the length, which is the same as its volume. So we need consider only $n > 1$.

To begin with, note that for a cuboid K to have the property in question, it is sufficient that there exist $l > 0$ such that the length L of every side satisfies $l \leq L < 2l$. For if this obtains then $\operatorname{diam} K < 2l$ and $\operatorname{vol}(K) \geq l^n$, which implies $(\operatorname{diam} K)^n < 2^n \operatorname{vol}(K)$.

Consider a cuboid $[a_1, b_1] \times \cdots \times [a_n, b_n]$; suppose

$$\min\{b_i - a_i : 1 \leq i \leq n\} = b_j - a_j, \quad \text{where } 1 \leq j \leq n.$$

For each i ($1 \leq i \leq n$), let m_i be the largest positive integer (easily seen to exist) such that $b_j - a_j \leq (b_i - a_i)/m_i$. Then $(b_i - a_i)/2m_i < b_j - a_j$ and hence

$$b_j - a_j \leq \frac{b_i - a_i}{m_i} < 2(b_j - a_j). \tag{1}$$

For each i, let P_i be the partition of $[a_i, b_i]$ that subdivides the interval into subintervals of length $(b_i - a_i)/m_i$ each. Then the paving consisting of the partitions $P_1, \ldots, P_n$ has the property that each cuboid K formed by it has edges with lengths $(b_i - a_i)/m_i$. In view of (1) and the remark in the preceding paragraph, K has the required property.   □

Given a nonempty bounded subset of $\mathbb{R}^2$, it is intuitively clear what is meant by 'circumscribing' it in a rectangle with horizontal and vertical sides. It is a consequence of the fact that we are using the max norm $\| \ \|_\infty$ that the circumscribing rectangle, generally a cuboid, has the same diameter as the given set. If all points of the set lie on the same horizontal or vertical line, then the rectangle reduces to a segment, but it is nevertheless contained in a proper rectangle having the same diameter. There is cause for some discomfort when the set has diameter zero, because no cuboid can have diameter zero. But even in this extreme case, one can enclose the set in a cuboid having arbitrarily small diameter. We now go on to formulate and prove the analogue of all this in n dimensions. The concept of circumscribing will not be required.

7-1.2. Proposition. [Needed in Propositions 7-2.2, 7-4.2] *Let $E \subseteq \mathbb{R}^n$ be nonempty and bounded. If* diam $E = 0$ *and* $\varepsilon > 0$, *then there exists a closed cuboid $K \supseteq E$ such that* diam $K = \varepsilon$. *If* diam $E > 0$, *then there exists a closed cuboid $K \supseteq E$ such that* diam $K =$ diam E.

Proof. Denote diam E by δ. If s is some point in E, then any $x \in E$ satisfies $|x_i - s_i| \le \|x - s\| \le \delta$ for every i $(1 \le i \le n)$, so that $s_i - \delta \le x_i \le s_i + \delta$. Therefore,

$$a_i = \inf\{x_i : x \in E\} \quad \text{and} \quad b_i = \sup\{x_i : x \in E\}. \tag{1}$$

exist.

Suppose $\delta = 0$. Then there is only one point in E and $a_i = b_i$ for every i. The cuboid K with edges $[a_i - \frac{1}{2}\varepsilon, a_i + \frac{1}{2}\varepsilon]$, $1 \le i \le n$, therefore has the required property.

Now suppose $\delta > 0$. Then there are at least two points in E and therefore, it cannot happen that $a_i = b_i$ for every i; otherwise $(a_1, \ldots, a_n)$ would be the only point in E and the diameter would be zero. Define

$$a_i' = a_i \quad \text{and} \quad b_i' = b_i \qquad \text{if } a_i < b_i \tag{2}$$

and

$$a_i' = a_i - \frac{\delta}{2} \quad \text{and} \quad b_i' = b_i + \frac{\delta}{2} = a_i + \frac{\delta}{2} \qquad \text{if } a_i = b_i. \tag{3}$$

Then $K = [a_1', b_1'] \times \cdots \times [a_n', b_n']$ is a closed cuboid. Besides, it contains E and therefore diam $K \ge$ diam E. It remains to prove the reverse inequality: diam $K \le$ diam $E = \delta$.

For any i such that $a_i = b_i$, we have $b_i' - a_i' = \delta$ by (3). It is sufficient to prove $b_i' - a_i' \le \delta$ for those i for which $a_i < b_i$. In view of (2), the inequality to be proved is equivalent to $b_i - a_i \le \delta$. In order to arrive at this, consider any $\varepsilon > 0$. It follows from (1) that there exist $x, y \in E$ such that

$$x_i < a_i + \frac{\varepsilon}{2} \quad \text{and} \quad b_i - \frac{\varepsilon}{2} < y_i.$$

It follows that $y_i - x_i > b_i - a_i - \varepsilon$. Since $y_i - x_i \leq \|y - x\| \leq \delta$, we conclude that $\delta > b_i - a_i - \varepsilon$. But this is true for any $\varepsilon > 0$ and therefore $b_i - a_i \leq \delta$. As already noted, this completes the proof.   □

We close this section with a sharpening of Proposition 7-1.1, which we shall need later. The sharpening consists in replacing the factor 2 by an arbitrary $\mu > 1$. As in Proposition 7-1.1, a cuboid K will satisfy $(\text{diam } K)^n < \mu^n \text{vol}(K)$ if there exists $l > 0$ such that the length L of every edge satisfies $l \leq L < \mu l$. Suppose, for instance, that we want such a paving of $[0,1] \times [0,\sqrt{29}]$ with $\mu = \frac{3}{2}$. As before, we seek to take the length of the shorter edge as l and subdivide only the longer edge into n equal subintervals of length L each. In the present case, $l = 1$ and $L = \sqrt{29}/n$. The inequality $l \leq L < \mu l$ now becomes $1 \leq \sqrt{29}/n < \frac{3}{2}$. This holds when $n = 4$ or 5. If we choose $n = 4$ (or 5), the required paving is made up of the (trivial) partition of $[0,1]$ consisting only of its endpoints and the partition of $[0,\sqrt{29}]$ that subdivides it into 4 (or 5) equal subintervals. In this example as well as in the two described earlier, the search for a paving in which the partition of the shorter edge consists only of its endpoints turned out to be successful. But it can happen that we have to subdivide the shorter edge also, as we now illustrate.

Suppose we want a paving of $[0,5] \times [0,6]$ with $\mu = \frac{11}{10}$. Since $6 > 5(\frac{11}{10})$, we must subdivide the longer edge into 2 or more subintervals. If we subdivide only the longer edge into n subintervals of length $\frac{6}{n}$ each $(n \geq 2)$, then the longer edge of each resulting subcuboid is of length $5 > \frac{11}{10}\frac{6}{n}$. This makes it necessary to subdivide both edges. In fact, if we subdivide $[0,5]$ into 11 equal parts and $[0,6]$ into 13 equal parts, we shall have obtained the paving we want, because $\frac{5}{11} < \frac{6}{13} < \frac{5}{11}\frac{11}{10}$. In order to see how to arrive at 11 and 13, read the next proof!

7-1.3. Proposition. [Used in Proposition 7-3.5] *Given $\mu > 1$ and a paving of a cuboid in $\mathbb{R}^n$, there exists a refinement of the paving such that every cuboid K formed by the refinement satisfies $(\text{diam } K)^n < \mu^n \text{vol}(K)$.*

Proof. When $n = 1$, this is trivial. So we need consider only $n > 1$. For a cuboid K to have the property in question, it is sufficient that there exist $l > 0$ such that the length L of every edge satisfies $l \leq L < \mu l$. For, if this obtains, then diam $K < \mu l$ and $\text{vol}(K) \geq l^n$, which implies $(\text{diam } K)^n < \mu^n \text{vol}(K)$.

Choose a positive integer N such that $1 + \frac{1}{N} < \mu$. Let k_0 be the total number of subintervals formed by all the n partitions that make up the given paving and denote their lengths by λ_i, $1 \le i \le k_0$. Let j be an index ($1 \le j \le k_0$) such that λ_j is least among all the finitely many positive numbers λ_i. Using the integer part function [ ], set

$$m_i = \left[N \frac{\lambda_i}{\lambda_j} \right] \qquad \text{for each } i, \ 1 \le i \le k_0.$$

Then $m_j = N$, and by minimality of λ_j, we have $m_i \ge N = m_j$. Also, $m_i \le N \frac{\lambda_i}{\lambda_j} < m_i + 1$, so that

$$\frac{\lambda_j}{N} \le \frac{\lambda_i}{m_i} < \frac{1}{m_i} \frac{m_i + 1}{N} \lambda_j = \left(1 + \frac{1}{m_i} \right) \frac{\lambda_j}{N} \le \left(1 + \frac{1}{N} \right) \frac{\lambda_j}{N} < \mu \frac{\lambda_j}{N}.$$

Thus, if $l = \lambda_j/N$, we have $l \le \lambda_i/m_i < \mu l$. Now, refine the given partitions of the edges by subdividing the ith subinterval into m_i equal subintervals ($1 \le i \le k_0$), which must then have length λ_i/m_i each. Then the length L of any subinterval satisfies $l \le L < \mu l$. It follows that the length of the edge of any cuboid K formed by the refinement satisfies the same inequality. As observed at the beginning, this is sufficient to ensure that $(\operatorname{diam} K)^n < \mu^n \operatorname{vol}(K)$. $\square$

Problem Set 7-1

7-1.P1. For each of the following cuboids, we want a paving in which the first edge (normally called *horizontal*) has the partition consisting only of its endpoints. What is the minimum and maximum number of equal subintervals for a partition of the vertical edge if each cuboid formed by the paving is to have a vertical edge longer than the horizontal but not longer than twice as much?

(a) $[0,1] \times [0,2e]$; (b) $[0,2] \times [0,\sqrt{99}]$

7-1.P2. For the cuboid $[0,1] \times [0,2e]$, we want a paving in which the first edge (normally called *horizontal*) has the partition consisting only of its endpoints. What is the minimum and maximum number of equal subintervals for a partition of the other edge if each cuboid formed by the paving is to have a vertical edge longer than the horizontal but not longer than $\frac{3}{2}$ times the horizontal?

7-1.P3. Name a triplet n_1, n_2, n_3 of positive integers such that, if the edges of the cuboid $[0,6] \times [0,9] \times [0,11]$ are respectively subdivided into n_1, n_2, n_3 equal subintervals, each cuboid K of the resulting paving will satisfy $(\operatorname{diam} K)^3 < (\frac{4}{3})^3 \operatorname{vol}(K)$.

7-1.P4. (This Problem explains why we switched to $\frac{11}{10}$ from $\frac{3}{2}$ in the example just before Proposition 7-1.3.) For $0 < a < b$, let P be a paving of $[0,a] \times [0,b]$ consisting of the trivial partition of $[0,a]$, i.e., only one subinterval, and a partition of $[0,b]$ consisting of n equal subintervals; each cuboid formed by P thus has edges of length a and b/n. Let l be the smaller of these and L the bigger.

(a) If $\mu > \sqrt{2}$, show that it is possible to choose n in such a way that $l \leq L < \mu l$.

(b) If $1 < \mu \leq \sqrt{2}$, show that for certain values of a and b it is not possible to choose n as above.

7-2 Transformation of Content

As a step towards the transformation formula, we need to establish what relation there is, if any, between the contents of E and $\alpha(E)$ when α is a map that serves for changing variables in an integral. In other words, when α is an injective continuously differentiable function defined on some set containing E. The present section addresses this matter of 'transformation of content'.

Examples. Let $E = [1,3] \times [1,2] \subseteq \mathbb{R}^2$ and $\alpha : E \to \mathbb{R}^2$ be defined as $\alpha(u,v) = (x,y) =$

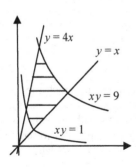

$(\frac{u}{v}, uv)$. The usual way to describe α is to write $x = \frac{u}{v}$, $y = uv$. Then all four of u,v,x and y are positive and we find that $u = \sqrt{(xy)}$ and $v = \sqrt{\frac{y}{x}}$. In other words, $\alpha^{-1}(x,y) = (\sqrt{(xy)}, \sqrt{\frac{y}{x}})$. We conclude from this that $(u,v) \in E$ if and only if (x,y) satisfies the pair of inequalities $1 \leq \sqrt{(xy)} \leq 3$ and $1 \leq \sqrt{\frac{y}{x}} \leq 2$; or equivalently, (x,y) lies in the first quadrant between the hyperbolas $xy = 1$, $xy = 9$ and the straight lines $y = x, y = 4x$. Thus, $\alpha(E)$ can be expressed as (see adjoining figure)

$$\alpha(E) = \{(x,y) \in \mathbb{R}^2 : x > 0,\ y > 0,\ 1 \leq xy \leq 9,\ x \leq y \leq 4x\}.$$

As another illustration, consider $E = \{(u,v) \in \mathbb{R}^2 : 0 < u < A$ and $0 < v < 2\pi\}$ $= (0,A) \times (0,2\pi)$, an open rectangular subset of $\mathbb{R}^2$. For $0 < \delta < \min\{A, \frac{\pi}{2}\}$, the set $E_\delta = [\delta, A] \times [\delta, 2\pi - \delta]$ is a closed rectangle. The transformation α defined on E by $\alpha(u,v) = (u\cos v, u\sin v)$ is injective. Setting $(x,y) = \alpha(u,v)$ renders u,v into polar coordinates in the (x,y)-plane. So, one can guess that α maps E *bijectively* onto the subset

$$G = \{(x,y) \in \mathbb{R}^2 : 0 < x^2 + y^2 < A^2, \text{ either } x < 0 \text{ or } y \neq 0\}.$$

A precise justification of why α maps E bijectively onto G is left as 7-2.P3. One can also sketch $\alpha(E_\delta)$ in the (x,y)-plane as a subset of the disc centred at the origin and having radius A, but with an appropriate subset removed (the latter is shaped like a keyhole; see the figure below).

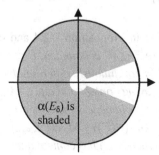

$\alpha(E_\delta)$ is shaded

Our first result here is concerned purely with open and compact subsets of $\mathbb{R}^n$ and its statement (i) will be used in the proposition immediately after it as well as elsewhere. Statement (ii) will be used in Proposition 7-2.4 and Proposition 7-2.5 only. We are about to use Theorem 2-5.7, according to which, a subset of $\mathbb{R}^n$ is compact if and only if it is closed and bounded.

7-2.1. Lemma. *Let $H \subseteq W$, where W is an open subset of $\mathbb{R}^n$ and H is compact. Then there exists an open set V such that*

(i) $H \subseteq V \subseteq \overline{V} \subseteq W$ *and $\overline{V}$ is compact*

and

(ii) *there exists $\eta > 0$ such that $h \in H$, $\|h - x\| < \eta \Rightarrow x \in V$.*

Proof. As in the proof of Proposition 6-1.11, let $\delta > 0$ have the property that

$$h \in H, \|h - x\| < \delta \Rightarrow x \in W.$$

Define V to be the union of some finitely many open $\frac{\delta}{2}$-balls covering the compact set H. Since there are only finitely many balls in the union, the closure $\overline{V}$ is contained in the union of the closures, i.e., the corresponding closed balls. Therefore,

$$x \in \overline{V} \Rightarrow \|h - x\| \leq \frac{\delta}{2} \text{ for some } h \in H$$

$$\Rightarrow x \in W.$$

Thus $\bar{V} \subseteq W$. Furthermore, $x \in \bar{V} \Rightarrow \|x\| \le \|h - x\| + \|h\| \le \frac{\delta}{2} + \|h\|$, which shows that $\bar{V}$ is bounded. Since it is closed, it must be compact. This proves (i).

Since H is a compact subset of the open set V, we can once again proceed as in the proof of Proposition 6-1.11 and obtain $\eta > 0$ such that $h \in H$, $\|h - x\| < \eta$ $\Rightarrow x \in V$. This proves (ii). $\square$

7-2.2. Proposition. [Used in Proposition 7-2.4 and the transformation formula (Theorem 7-4.4)] *Suppose $E \subseteq \mathbb{R}^n$ is bounded, $\bar{E} \subseteq W$, where W is an open subset of $\mathbb{R}^n$, and $\alpha : W \rightarrow \mathbb{R}^n$ is continuously differentiable. Then there exists $M_0 > 0$ such that for any $F \subseteq E$ having content and any $\varepsilon > 0$, $\alpha(F)$ is covered by a finite family of cuboids with total volume less than $M_0 \cdot c(F) + \varepsilon$. In particular,*
(a) if $\alpha(F)$ has content, then $c(\alpha(F)) \le M_0 \cdot c(F)$, and
(b) if F has content zero, then $\alpha(F)$ has content and $c(\alpha(F)) = 0$.

Proof. Since $\bar{E}$ is compact, it follows by Lemma 7-2.1 that there exists an open set V such that $\bar{E} \subseteq V \subseteq \bar{V} \subseteq W$ and $\bar{V}$ is compact. Since α is continuously differentiable on W, the real-valued function which maps each $x \in W$ into $\|\alpha'(x)\|$ is continuous, and is therefore bounded on the compact subset $\bar{V}$ of W. So, there exists $M > 0$ such that

$$\|\alpha'(x)\| \le M \qquad \text{whenever} \quad x \in \bar{V}.$$

We shall show that $M_0 = (2M)^n$ has the required property.

Consider any $F \subseteq E$ having content and any $\varepsilon > 0$. By Proposition 6-4.10, there exists a finite family of closed cuboids that cover F and have total volume less than $c(F) + \varepsilon/(2M)^n$. Since their union is closed, they also cover $\bar{F}$. Now $\bar{F}$ is a compact subset of the open set V. So, by Proposition 6-1.11, there exists a finite family $\mathcal{F}$ of closed cuboids which cover $\bar{F}$, have total volume less than $c(F) + \varepsilon/(2M)^n$ and are contained in V and hence also in $\bar{V}$. Each cuboid may now be replaced by the cuboids formed by any paving of it; call the resulting family $\mathcal{H}$. By selecting each paving as in Proposition 7-1.1, we may assume that each cuboid K of $\mathcal{H}$ satisfies

$$(\text{diam } K)^n < 2^n \text{vol}(K).$$

Considering that a cuboid is convex and every cuboid of $\mathcal{H}$ is a subset of $\bar{V}$, it follows by Corollary 3-3.4 that $\|\alpha(x) - \alpha(y)\| \le M\|x - y\|$ whenever x and y belong to the same cuboid of the family. Thus, $\text{diam } \alpha(K) \le M \cdot \text{diam } K$ for each $K \in \mathcal{H}$. By Proposition 7-1.2, $\alpha(K)$ is contained in a closed cuboid $\alpha(K)'$ having diameter at most $M \cdot \text{diam } K$. Now

$$\text{vol}(\alpha(K)') \leq (M \cdot \text{diam} K)^n < M^n 2^n \text{vol}(K).$$

This shows that the finite family $\{\alpha(K)' : K \in \mathcal{H}\}$ of closed cuboids, which certainly covers $\alpha(F)$, has total volume less than $(2M)^n[c(F) + \varepsilon/(2M)^n] = M_0 \cdot c(F) + \varepsilon$, where $M_0 = (2M)^n$. This completes the proof of the existence of M_0 of the required kind.

Statement (a) is now an easy consequence. For the proof of statement (b), note that when F has content zero, the finite family $\{\alpha(K)' : K \in \mathcal{H}\}$ has total volume less than ε. Since it covers $\alpha(F)$, it follows upon using the equivalences in Proposition 6-4.10 that $\alpha(F)$ has content and that $c(\alpha(F)) = 0$. $\qquad\square$

Example. Let W be the open first quadrant in the plane and let α be the continuously differentiable function on W defined by $\alpha(x,y) = (1/x, 1/y)$. Then α maps $E = \{(x,y) \in W : 0 < x < 1, 0 < y < 1\}$ into a subset of $\mathbb{R}^2$ that is not even bounded and therefore has no content. But for any $\delta > 0$ and $A > 0$, it maps $\{(x,y) \in W : \delta < x < \delta + A, \delta < y < \delta + A \}$ into a bounded rectangle, which does have content. The closure of the latter set is contained in W, while that of E is not. However, the transformation given by $\alpha(x,y) = (\sqrt{x}, \sqrt{y})$ does map E into a set with content, in fact, into itself. Note that, unlike the first transformation, this one has a continuous extension to the closure of W.

Note that the above proposition makes no claim that if F has positive content then $\alpha(F)$ has content. We shall need the fact that under certain circumstances, such a thing does happen. Before we prove the result in this direction, we establish the following corollary to the inverse function theorem.

7-2.3. Proposition. [Used in Proposition 7-4.2] *Let W be an open subset of $\mathbb{R}^n$ and $\alpha: W \rightarrow \mathbb{R}^n$ be a continuously differentiable injective map with an invertible derivative α' everywhere on W. Then*
(a) α maps open subsets of W onto open subsets of $\mathbb{R}^n$;
(b) α^{-1} is continuously differentiable and

   (i) $(\alpha^{-1})'(y) = \alpha'(\alpha^{-1}(y))^{-1}$ for every $y \in \alpha(W)$;
   (ii) $(\alpha^{-1})'(\alpha(x)) = \alpha'(x)^{-1}$ for every $x \in W$.

Proof. Let V be an open subset of W. By Inverse Function Theorem 4-2.1, every $x \in V$ belongs to an open set $W_x \subseteq V$ such that firstly, $\alpha(W_x) \subseteq \alpha(V)$ is open in $\mathbb{R}^n$ and secondly, the restriction of α to W_x has a continuously differentiable inverse β satisfying $(\beta)'(y) = \alpha'(\alpha^{-1}(y))^{-1}$ for every $y \in \alpha(W_x)$. Now

$$y \in \alpha(V) \Rightarrow y = \alpha(x) \text{ for some } x \in V \Rightarrow y \in \alpha(W_x)$$

$$\text{for some } x \in V \Rightarrow y \in \bigcup_{x \in V} \alpha(W_x)$$

and

$$y \in \bigcup_{x \in V} \alpha(W_x) \Rightarrow y \in \alpha(V), \quad \text{because } \alpha(W_x) \subseteq \alpha(V).$$

Thus,

$$\alpha(V) = \bigcup_{x \in V} \alpha(W_x).$$

(a) Since $\alpha(V)$ is the union of all the open sets $\alpha(W_x)$, $x \in V$, it must be open in $\mathbb{R}^n$.

(b) Since β must agree with α^{-1} on its own domain, i.e., on $\alpha(W_x)$, it follows that α^{-1} has the two properties: being continuously differentiable and satisfying $(\alpha^{-1})'(y) = \alpha'(\alpha^{-1}(y))^{-1}$ on $\alpha(W_x)$. Since $\alpha(V)$ is the union of sets $\alpha(W_x)$, it further follows that α^{-1} has the two properties on $\alpha(V)$. But this is true for an arbitrary open subset V of W. Consequently, α^{-1} has the two properties on the whole of $\alpha(W)$. Since the injectivity of α ensures that every $x \in W$ is of the form $\alpha^{-1}(y)$, $y \in \alpha(W)$, it now follows that $(\alpha^{-1})'(\alpha(x)) = \alpha'(x)^{-1}$ for every $x \in W$.   □

The result about $\alpha(F)$ having content when F does is as follows:

7-2.4. Proposition. [Used in Proposition 7-4.2] *Let W be an open subset of $\mathbb{R}^n$ and $\alpha:W \rightarrow \mathbb{R}^n$ be a continuously differentiable injective map with an invertible derivative α' everywhere on W. Then*

(a) *if F has content and $\overline{F} \subseteq W$, then $\alpha(F)$ has content;*

(b) *if $H \subseteq W$ is compact, then there exists $\eta > 0$ such that, for any subset $F \subseteq W$ with diameter less than η, having content and containing a point of H, the set $\alpha(F)$ has content*

Proof. By 2-6.P8(c), continuity and injectivity of α together imply that it maps the boundary of a set E satisfying $\overline{E} \subseteq W$ into the boundary of $\alpha(E)$, which is to say $\alpha(\partial E) \subseteq \partial(\alpha(E))$. By 4-2.P5 [or by Proposition 7-2.3(a)], $\alpha(W)$ is open; in case $\overline{E}$ is compact, $\alpha(\overline{E})$ is also compact and hence $\overline{\alpha(E)} \subseteq \alpha(\overline{E}) \subseteq \alpha(W)$. Since the inverse of α is also continuous (by Inverse Function Theorem 4-2.1) and injective, it follows when $\overline{E}$ is compact that $\partial(\alpha(E)) \subseteq \alpha(\partial E)$, so that $\alpha(\partial E) = \partial(\alpha(E))$.

Let $H \subseteq W$ be compact. According to Lemma 7-2.1, there exists an open set V such that

(i) $H \subseteq V \subseteq \overline{V} \subseteq W$ and $\overline{V}$ is compact;

(ii) there exists $\eta > 0$ such that: $h \in H$, $\|h - x\| < \eta \Rightarrow x \in V$.

Consider a subset E of $\overline{V}$ that has content. Its boundary ∂E must have content zero by Proposition 6-4.14. Therefore, by Proposition 7-2.2(b), $\alpha(\partial E)$ also has content zero. But $\overline{E} \subseteq \overline{V} \subseteq W$ and therefore $\alpha(\partial E) = \partial(\alpha(E))$, as observed at the beginning. Hence, by Proposition 6-4.14 again, it follows that $\alpha(E)$ has content. Thus, α maps a subset of $\overline{V}$ that has content into one that also has content.

(a) Now let F have content and $\overline{F} \subseteq W$. Since a set having content is bounded, so is its closure, which must then be compact. Thus $\overline{F}$ is compact and hence we may choose H to be $\overline{F}$ in what has just been proved, whereupon we can conclude that $\alpha(F)$ has content.

(b) Let the subset $F \subseteq W$ contain a point of H and have diameter less than η. By (ii) above, F is a subset of V, so that $\overline{F} \subseteq \overline{V} \subseteq W$. It now follows by (a) that $\alpha(F)$ has content. $\square$

The next proposition is needed in Proposition 7-4.2.

7-2.5. Proposition. *Let W be an open subset of $\mathbb{R}^n$ and $\alpha:W\to\mathbb{R}^n$ be a continuously differentiable map. Suppose H is a compact subset of W. Then for any $\varepsilon > 0$, there exists $\delta > 0$ such that, for any cuboid K containing a point $x \in H$ and having diameter less than δ,*

$$K \subseteq W \qquad \text{and} \qquad \|\alpha(b) - \alpha(a) - \alpha'(x)(b - a)\| < \varepsilon\|b - a\| \quad \forall\, a,b \in K.$$

Proof. Let V and η be as in Lemma 7-2.1. By compactness of $\overline{V}$, the map $\alpha':W\to\mathscr{L}(\mathbb{R}^n,\mathbb{R}^n)$ is uniformly continuous on $\overline{V}$. So there exists $\delta > 0$ such that

$$x,y \in \overline{V},\ \|x - y\| < \delta \quad \Rightarrow \quad \|\alpha'(x) - \alpha'(y)\| < \frac{\varepsilon}{2n}. \tag{1}$$

(As elsewhere in our discussion of integration, the norm in $\mathbb{R}^n$ is $\|\ \|_\infty$, with the corresponding norm understood in $\mathscr{L}(\mathbb{R}^n,\mathbb{R}^n)$.) If we replace δ by $\min\{\delta,\eta\}$, then we have

$$x \in H,\ \|x - y\| < \delta \quad \Rightarrow \quad y \in V. \tag{2}$$

Now let K be any cuboid containing $x \in H$ and having diameter less than δ, and consider any $a,b \in K$. By (2), $K \subseteq V \subseteq \overline{V}$. Therefore, $K \subseteq W$, and moreover by (1),

$$y \in K \quad \Rightarrow \quad \|\alpha'(y) - \alpha'(a)\| < \frac{\varepsilon}{2n}. \tag{3}$$

From this inequality and convexity of a cuboid, we obtain by Corollary 3-3.7 that

$$\|\alpha(b) - \alpha(a) - \alpha'(a)(b - a)\| \le \frac{\varepsilon}{2}\|b - a\|. \tag{4}$$

Since $x \in K$, it follows from (3) that $\|\alpha'(x) - \alpha'(a)\| < \frac{\varepsilon}{2n}$ and hence that

$$\|\alpha'(x)(b - a) - \alpha'(a)(b - a)\| < \frac{\varepsilon}{2n}\|b - a\| \le \frac{\varepsilon}{2}\|b - a\|.$$

Combining this with (4), we get

$$\|\alpha(b) - \alpha(a) - \alpha'(x)(b - a)\| < \varepsilon \|b - a\|. \qquad \square$$

Problem Set 7-2

7-2.P1. Give an example of a subset $F \subseteq W \subseteq \mathbb{R}$, where W is open, $\alpha:W\rightarrow\mathbb{R}$ is continuously differentiable with an invertible derivative (which means nonzero in dimension 1) everywhere, F has content but $\alpha(F)$ does not.

7-2.P2. Show that the result of Proposition 7-2.2 holds with
$$M_0 = (\sup\{\|\alpha'(x)\| : x \in \overline{E}\})^n.$$

7-2.P3. Show that the transformation α defined on E by $\alpha(u,v) = (u\cos v, u\sin v)$ is bijective and maps $E = \{(u,v) \in \mathbb{R}^2 : 0 < u < A \text{ and } 0 < v < 2\pi\}$ onto $G = \{(x,y) \in \mathbb{R}^2 : 0 < x^2 + y^2 < A^2, \text{ either } x < 0 \text{ or } y \neq 0\}$.

7-3 Set Functions

The path that we take in proving the transformation formula involves studying what are called *set functions*, which are functions with a class (family) of subsets of a set as their domain. As is often the case in mathematics, their usefulness stems from the properties of the domain. Of particular importance to us is the class of subsets having content, and we begin by proving some of its properties that we shall need.

Throughout this chapter, the symbol A^c will denote the complement of a subset A of $\mathbb{R}^n$.

7-3.1. Proposition. *Suppose $E \subseteq \mathbb{R}^n$. Then the class $\mathfrak{C}_E$ of subsets of E having content has the properties that*
(a) $\varnothing \in \mathfrak{C}_E$;
(b) $A, B \in \mathfrak{C}_E \implies A \cup B, A \cap B, A^c \cap B \in \mathfrak{C}_E$.

Proof. (a) is trivial, because $\varnothing$ is a subset of a cuboid of arbitrarily small volume.

(b) Let $A, B \in \mathfrak{C}_E$. By the sufficiency condition of Proposition 6-4.14, ∂A and ∂B both have content zero and hence by Remark 6-4.6(e), $\partial A \cup \partial B$ also has content zero. But the boundaries of $A \cup B, A \cap B, A^c \cap B$ are all subsets of $\partial A \cup \partial B$ and therefore have content zero by Remark 6-4.6(c). It follows by the necessity condition of Proposition 6-4.14 that $A \cup B, A \cap B, A^c \cap B \in \mathfrak{C}_E$. $\square$

7-3.2. Definition. *A **set function in a set** $E \subseteq \mathbb{R}^n$ is a real-valued function on the class $\mathfrak{C}_E$ of all subsets of E that have content. A set function σ in E is said to be* **additive** *if $\sigma(A \cup B) = \sigma(A) + \sigma(B)$ whenever $c(A \cap B) = 0$.*

7-3.3. Examples. (a) A trivial example of an additive set function is $\sigma(A) = 0$ for all $A \in \mathfrak{C}_E$. The most obvious nontrivial example of a set function is simply c. That it is additive is the substance of Remark 6-4.6(d).

(b) Let $f: E \rightarrow \mathbb{R}$ be integrable on every subset $A \in \mathfrak{C}_E$. Then $\sigma_f(A) = \int_A f$ defines an additive set function in E by Remark 6-4.6(g).

(c) Suppose $E \subseteq \mathbb{R}^n$ and $\alpha: E \rightarrow \mathbb{R}^n$ maps every subset of E that has content into a subset of $\alpha(E)$ that also has content. Let $f: \alpha(E) \rightarrow \mathbb{R}$ be integrable on every subset of $\alpha(E)$ that has content. Then $\sigma(A) = \int_{\alpha(A)} f$ defines a set function in E. If α is bijective, then it maps $A \cap B$ onto $\alpha(A) \cap \alpha(B)$; if α also maps a set with content zero onto a set with content zero, then it follows that σ is additive. For this example to be of the same kind as the preceding one, there needs to be an integrable function $g: E \rightarrow \mathbb{R}$ such that $\int_A g = \sigma(A) = \int_{\alpha(A)} f$ for all $A \subseteq E$ having content. Whether such a function g exists or not naturally depends upon α and f. Some of our initial results pertain to this matter, as it will turn out to be crucial for the transformation formula.

7-3.4. Proposition. *Let σ be an additive set function in E. If $\mathcal{F}$ is a finite non-overlapping family of cuboids contained in E then $\sigma(\cup \mathcal{F}) = \Sigma_{K \in \mathcal{F}} \, \sigma(K)$. In particular, if $f: E \rightarrow \mathbb{R}$ is integrable, then $\int_{\cup \mathcal{F}_k} f = \Sigma_{K \in \mathcal{F}} \int_K f$.*

Proof. Since $\mathcal{F}$ is finite, an induction argument using Proposition 6-4.9 leads to the desired equality. The last part follows upon using Remark 6-4.6(g). $\square$

If Φ is a function on an interval $[a, b] \subseteq \mathbb{R}$, then one can regard $\Phi(v) - \Phi(u)$ as defining something akin to a set function σ on subintervals $[u, v]$ of $[a, b]$. Then

$$[\Phi(v) - \Phi(u)]/(v - u) = \frac{\sigma(I)}{c(I)},$$

where $I = [u, v]$, $v - u \neq 0$, $\sigma(I) = \Phi(v) - \Phi(u)$. Its limit as $v - u \rightarrow 0$, $u \leq x \leq v$, is the derivative $\Phi'(x)$, and by the fundamental theorem of calculus, $\sigma(I) = \int_I \Phi'$, provided of course the derivative exists everywhere. If $\Phi' \leq f$, then $\sigma(I) \leq \int_I f$. The next proposition establishes such an inequality in higher dimensions under a hypothesis that falls short of assuming that a derivative exists and does not exceed f; what it does say has the rough consequence that 'if a derivative were to exist, it would not exceed f'. The quaint looking hypothesis is satisfied in a situation we shall encounter during the proof of the transformation formula (Theorem 7-4.4).

7-3.5. Proposition. *Let $W \subseteq \mathbb{R}^n$ be open, σ an additive set function in W that has nonnegative values, and let $f: W \to \mathbb{R}$ be integrable over every subset of W that has content. Suppose H is a compact subset of W and for every $\varepsilon > 0$, there exists a $\delta > 0$ such that, for any $x \in H$ and any cuboid I containing x and having* diam $I < \delta$,

$$I \subseteq W \quad \text{and} \quad \sigma(I) \leq [\sup \{f(z) : z \in I\} + \varepsilon](1+\varepsilon)^{n+1}(\text{diam } I)^n.$$

Then

$$\sigma(A) \leq \int_A f$$

whenever A has content and $\bar{A} \subseteq H^\circ$.

Proof. We shall first prove the inequality when $A = \bar{A} \subseteq H^\circ$ is a cuboid. Consider any $\varepsilon > 0$ and any $\mu > 1$. Since f is integrable over A, there exists a paving P_1 of A such that $U(f,P_1) < \int_A f + \varepsilon$. Let P be a refinement of P_1 such that every cuboid K formed by P has diameter less than δ. Then

$$U(f,P) \leq U(f,P_1) < \int_A f + \varepsilon,$$

and by Proposition 7-1.3, we can ensure that each cuboid K also satisfies (diam $K)^n < \mu^n \text{vol}(K)$. Denote by $\mathcal{F}$ the family of these cuboids. For each cuboid $K \in \mathcal{F}$, let $M_K = \sup \{f(z): z \in K\}$.

Since $K \subseteq A \subseteq H$, it contains points of H; moreover, diam $K < \delta$. Therefore by hypothesis,

$$\sigma(K) \leq (1+\varepsilon)^{n+1}(M_K + \varepsilon)(\text{diam } K)^n$$

and hence

$$\sigma(K) \leq (1+\varepsilon)^{n+1} \mu^n (M_K + \varepsilon)\text{vol}(K).$$

Therefore by Proposition 7-3.4,

$$\sigma(A) = \Sigma_{K \in \mathcal{F}} \sigma(K) \leq (1+\varepsilon)^{n+1} \mu^n \Sigma_{K \in \mathcal{F}} (M_K + \varepsilon)\text{vol}(K)$$

$$= (1+\varepsilon)^{n+1} \mu^n U(f,P) + (1+\varepsilon)^{n+1} \mu^n \varepsilon[\text{vol}(A)]$$

$$< (1+\varepsilon)^{n+1} \mu^n (\int_A f + \varepsilon) + (1+\varepsilon)^{n+1} \mu^n \varepsilon[\text{vol}(A)].$$

Since this has been proved for every $\varepsilon > 0$ and every $\mu > 1$, the required inequality holds for a cuboid A.

In view of Proposition 7-3.4, it follows that the inequality also holds when A is a union of a finite nonoverlapping family of cuboids.

We go on to the general case of an arbitrary set A having content and such that $\overline{A} \subseteq H^\circ$.

Consider any $\varepsilon > 0$. Since $|f|$ is integrable on the set H, it has a finite supremum, which we denote by M. Since A has content, it is bounded by definition; therefore $\overline{A}$ is compact. By Proposition 6-4.10, there exists a union F of a finite nonoverlapping family of cuboids such that

$$\overline{A} \subseteq F \text{ and } c(F) < c(\overline{A}) + \frac{\varepsilon}{M}.$$

By Proposition 6-1.11, we can ensure that $F \subseteq H^\circ$, so that, by what has been proved above, the required inequality holds for each cuboid in F. Now, $c(A) = c(\overline{A})$ [see 6-4.P3]. Therefore,

$$A \subseteq F \subseteq H^\circ \quad \text{and} \quad c(F) < c(A) + \frac{\varepsilon}{M}.$$

Using Proposition 7-3.1, and Remark 6-4.6(d) and 6-4.6(g), we have

$$c(F \cap A^c) = c(F) - c(A) < \frac{\varepsilon}{M} \quad \text{and} \quad \int_F f = \int_A f + \int_{F \cap A^c} f.$$

Therefore, recalling that $M = \sup\{|f(x)| : x \in H\}$, we have

$$\int_F f < \int_A f + M\frac{\varepsilon}{M} = \int_A f + \varepsilon.$$

Since σ is additive as well as nonnegative, it follows from the inclusion $A \subseteq F$ that

$$\sigma(A) \leq \sigma(F).$$

As F is a union of a finite nonoverlapping family of cuboids and the required inequality holds for each cuboid in the union, Proposition 7-3.4 and the above inequality lead to

$$\sigma(A) \leq \int_F f.$$

Together with the preceding inequalities about integrals, this implies that

$$\sigma(A) < \int_A f + \varepsilon.$$

This has been shown to hold for an arbitrary $\varepsilon > 0$ and therefore $\sigma(A) \leq \int_A f$. $\quad\square$

7-4 The Transformation Formula

We shall soon come to the main result of this chapter. The first proposition of this section can be regarded as the special case of the tansformation formula when the transformation of variables is linear and the integrand is identically 1, although no integral is explicitly mentioned.

When $\alpha:\mathbb{R}^n \to \mathbb{R}^n$ is a linear map, $\det \alpha$ will denote the determinant of the matrix representation of α as described in Chapter 2. Since α is its own linear derivative everywhere, $\det \alpha$ is also the Jacobian of α at each point of $\mathbb{R}^n$.

7-4.1. Proposition. *If $A \subseteq \mathbb{R}^n$ is a cuboid and $\alpha:\mathbb{R}^n \to \mathbb{R}^n$ is an invertible linear map, then*

$$c(\alpha(A)) = |\det \alpha| \, c(A).$$

(This is true even when α is not invertible, but the case will not arise in our considerations.)

Proof. If α is of the type that merely multiplies the kth component by some non-zero a, i.e.,

$$\alpha(x_1, \dots, x_n) = (y_1, \dots, y_n),$$

where

$$y_j = \begin{cases} x_j & \text{for } j \neq k \\ ax_j & \text{for } j = k , \end{cases} \tag{1}$$

then it maps the cuboid A onto a cuboid with kth edge of length $|a|$ times that of A and all other edges the same as those of A; the volume of the latter is $|a|\mathrm{vol}(A)$. Since the content of a cuboid is the volume and $\det \alpha = a$, the desired equality holds in this case.

If α is of the type that merely interchanges two components, i.e.

$$\alpha(x_1, \dots, x_n) = (y_1, \dots, y_n),$$

where

$$y_j = \begin{cases} x_\ell & \text{if } j = k \\ x_k & \text{if } j = l \\ x_j & \text{if } k \neq j \neq l , \end{cases} \tag{2}$$

then it only interchanges the kth and lth edges of any cuboid and the volume therefore remains unaltered; also, $\det \alpha = -1$. So, the desired equality holds in this case.

If α is of the type that merely adds one row to another row, i.e., there exist distinct indices k, l such that

$$\alpha(x_1, \dots, x_n) = (y_1, \dots, y_n),$$

where

$$y_j = \begin{cases} x_j & \text{for } j \neq k \\ x_j + x_l & \text{for } j = k , \end{cases} \tag{3}$$

(so that $y_k = x_k + x_l$), then $c(\alpha(A)) = c(A)$ by Proposition 6-5.3 while $\det \alpha = 1$. Therefore the desired equality holds in this case too.

By Remark 2-3.4, every invertible linear map is a composition of linear maps of the type (1), (2) and (3), i.e., having an elementary matrix. Also, if the equality in question holds for two linear maps, then it surely holds for their composition. Therefore it holds for all invertible linear maps.   □

7-4.2. Proposition. *Let W be an open subset of $\mathbb{R}^n$ and $\alpha:W \to \mathbb{R}^n$ be a continuously differentiable injective map with an invertible derivative α' everywhere on W. Suppose $H \subseteq W$ is a compact subset and $\varepsilon > 0$. Then there exists $\delta > 0$ such that, for any cuboid I that contains a point $x \in H$ and has diameter less than δ, we have $I \subseteq W$ and*

$$c(\alpha(I)) \leq |\det \alpha'(x)|(1+\varepsilon)^n (\operatorname{diam} I)^n.$$

Proof. Let the open set V be as in Lemma 7-2.1. By Proposition 7-2.3, $\alpha(W)$ is open, α^{-1} is continuously differentiable and $(\alpha^{-1})'(\alpha(x)) = \alpha'(x)^{-1}$ for every $x \in W$. Denoting the linear map $\alpha'(x)$ by β_x, this can be written as

$$\beta_x^{-1} = (\alpha^{-1})'(\alpha(x)) \qquad \text{for every } x \in W.$$

Therefore, as a map from W to $\mathfrak{L}(\mathbb{R}^n, \mathbb{R}^n)$, β_x^{-1} is continuous. Now $\overline{V}$ is compact and therefore there exists a real number $M > 0$ such that $\|\beta_x^{-1}\| \leq M$ for every $x \in \overline{V}$. This means that

$$\|\beta_x^{-1}(h)\| \leq M\|h\| \qquad \text{for every } x \in \overline{V} \text{ and every } h \in \mathbb{R}^n. \tag{1}$$

Now consider any $\varepsilon > 0$. By Proposition 7-2.5, there exists $\delta > 0$ such that, for any cuboid I containing a point $x \in H$ and having diameter less than δ,

$$I \subseteq W \qquad \text{and} \qquad \|\alpha(b) - \alpha(a) - \beta_x(b-a)\| < \frac{\varepsilon}{M}\|b - a\| \quad \forall\, a, b \in I.$$

Since I is closed, it follows by Proposition 7-2.4(a) that $\alpha(I)$ has content. Now consider any such cuboid I and any $a, b \in I$. Then the above inequality holds and therefore it follows from (1) that

$$\|\beta_x^{-1}(\alpha(b)) - \beta_x^{-1}(\alpha(a)) - (b-a)\| < M\frac{\varepsilon}{M}\|b - a\| = \varepsilon\|b - a\|.$$

This implies that

$$\|\beta_x^{-1}(\alpha(b)) - \beta_x^{-1}(\alpha(a))\| < (1+\varepsilon)\|b - a\|.$$

Therefore, $\beta_x^{-1}(\alpha(I))$ has diameter no greater than $(1+\varepsilon)\operatorname{diam} I$. By Proposition 7-1.2, $\beta_x^{-1}(\alpha(I))$ is contained in some cuboid J of the same diameter. So,

$$\alpha(I) \subseteq \beta_x(J) \tag{2}$$

and

$$\operatorname{diam} J \leq (1+\varepsilon)\operatorname{diam} I. \tag{3}$$

Now by Proposition 7-4.1, $c(\beta_x(J)) = |\det \beta_x|\operatorname{vol}(J)$ and, as already noted, $\alpha(I)$ has content. The inclusion (2) implies $c(\alpha(I)) \leq c(\beta_x(J))$. Consequently by (3),

$$c(\alpha(I)) \le |\det\beta_x|\,\text{vol}(J) \le |\det\beta_x|(\text{diam } J)^n \le |\det\alpha'(x)|(1+\varepsilon)^n(\text{diam } I)^n. \quad \square$$

7-4.3. Proposition. *Let $E \subseteq F \subseteq \mathbb{R}^n$, where E has content and $f:E \rightarrow \mathbb{R}$ is integrable. Let $g:F \rightarrow \mathbb{R}$ be the extension of f to F obtained by setting it equal to zero outside E. Then g is integrable on any subset [see Def. 6-4.5] of F that has content.*

Proof. Consider a subset A of F that has content and let I be a cuboid that contains F and hence also E and A. Denote by χ_A the characteristic function of A on I.

Since $f:E \rightarrow \mathbb{R}$ is integrable, its extension f_I to the cuboid I is integrable, i.e., $\int_I f_I$ exists, where

$$f_I(x) = \begin{cases} f(x) & \text{if } x \in E \\ 0 & \text{if } x \notin E \end{cases} \qquad \text{for any } x \in I.$$

Since A has content, the characteristic function χ_A (domain I) is integrable. Hence, the product $f_I\,\chi_A$ is integrable.

For the integrability of g on A, what we need is [by Def. 6-4.5] that the restriction $g|_A$ of g to the subset A should be integrable when extended as $(g|_A)_I$ to I by setting it equal to zero outside A. We shall prove the required integrability by arguing that $(g|_A)_I$ is, in fact, the same as the product $f_I\,\chi_A$, for which integrability has already been established in the preceding paragraph.

By definition of g, we have

$$g(x) = \begin{cases} f(x) & \text{if } x \in E \\ 0 & \text{if } x \notin E \end{cases} \qquad \text{for any } x \in F,$$

$$(g|_A)_I(x) = \begin{cases} g(x) & \text{if } x \in A \\ 0 & \text{if } x \notin A \end{cases} \qquad \text{for any } x \in I$$

and

$$\chi_A(x) = \begin{cases} 1 & \text{if } x \in A \\ 0 & \text{if } x \notin A \end{cases} \qquad \text{for any } x \in I.$$

From the above descriptions of the four functions, we deduce for any $x \in I$ that

$$x \in E \cap A \quad \Rightarrow \quad f_I(x)\chi_A(x) = f_I(x) = f(x) \quad \text{and} \quad (g|_A)_I(x) = g(x) = f(x);$$

$$x \notin A \quad \Rightarrow \quad f_I(x)\chi_A(x) = 0 \quad \text{and} \quad (g|_A)_I(x) = 0;$$

$$x \in A, x \notin E \quad \Rightarrow \quad f_I(x)\chi_A(x) = f_I(x) = 0 \quad \text{and} \quad (g|_A)_I(x) = g(x) = 0.$$

This covers all possibilities for $x \in I$ and therefore $f_I(x)\chi_A(x) = (g|_A)_I(x)$ everywhere on I. $\quad \square$

We now come to the main result. The proof begins with a special case which is adequate for most purposes, especially when the integrand is continuous. The reader has the option of focusing attention only on this case.

In the statement of the theorem, $\det \alpha'$ is the Jacobian of α and one may replace it by some such symbol as J_α if desired.

7-4.4. Theorem. Transformation Formula: *Let $W \subseteq \mathbb{R}^n$ be open and $\overline{E} \subseteq W$, where E has content. Suppose $\alpha : W \to \mathbb{R}^n$ is a continuously differentiable injective map with an invertible derivative α' everywhere on W. Then, for any integrable function $f : \alpha(\overline{E}) \to \mathbb{R}$, the function $(f \circ \alpha) |\det \alpha'| : \overline{E} \to \mathbb{R}$ is also integrable and*

$$\int_{\alpha(E)} f = \int_E (f \circ \alpha) |\det \alpha'|.$$

Proof. We shall first prove this under the additional hypothesis that $(f \circ \alpha) |\det \alpha'| : \overline{E} \to \mathbb{R}$ is integrable.

Since any integrable function f is a difference of integrable nonnegative-valued functions, namely, $\frac{1}{2}(|f| + f)$ and $\frac{1}{2}(|f| - f)$, it suffices to prove the formula for nonnegative f. Also, we need prove only the inequality

$$\int_{\alpha(E)} f \leq \int_E (f \circ \alpha) |\det \alpha'|,$$

because the reverse inequality will follow from it in view of the following:

(i) $\alpha^{-1} : \alpha(W) \to W$ enjoys the same properties as α
(ii) $(f \circ \alpha) |\det \alpha'|$ is integrable and nonnegative on E and
(iii) $|\det(\alpha' \circ \alpha^{-1})| \, |\det (\alpha^{-1})'| = 1$ on $\alpha(W)$.

The last mentioned equality results from the chain rule together with the fact that $\alpha \circ \alpha^{-1}$ is the identity map on $\alpha(W)$.

Since E has content, it is bounded and hence $\overline{E}$ is compact. By hypothesis, $\overline{E} \subseteq W$ and therefore by Lemma 7-2.1 (applied twice), there exists open sets V_1 and V_2 such that

$$\overline{E} \subseteq V_1 \subseteq \overline{V_1} \subseteq V_2 \subseteq \overline{V_2} \subseteq W$$

and $\overline{V_2}$ is compact (hence also $\overline{V_1}$). If f is extended to $\alpha(\overline{V_2})$ by setting it equal to zero outside $\alpha(\overline{E})$, then by Proposition 7-4.3, the extension is integrable on every subset of $\alpha(\overline{V_2})$ that has content.

It is sufficient to prove the inequality in question for the extended function and from here onwards we shall denote the extended function by f.

Let A be any subset of V_2 having content. Then $\overline{A} \subseteq \overline{V_2} \subseteq W$ and by Proposition 7-2.4(a), $\alpha(A)$ has content. Also, $\alpha(A) \subseteq \alpha(\overline{V_2})$, so that f is integrable over $\alpha(A)$. Define a nonnegative set function σ in V_2 by

$$\sigma(A) = \int_{\alpha(A)} f.$$

We shall argue that σ is additive. To see why, let $c(A_1 \cap A_2) = 0$. Then $c(\alpha(A_1 \cap A_2)) = 0$ by Proposition 7-2.2(b). However, the injectivity of α implies that $\alpha(A_1 \cap A_2) = \alpha(A_1) \cap \alpha(A_2)$. So, $c(\alpha(A_1) \cap \alpha(A_2)) = 0$. It follows from Remark 6-4.6(g) that σ is additive.

Now let H be the compact subset $\overline{V}_1$ of V_2. Consider any $\varepsilon > 0$.

By Lemma 7-2.1, there exists $\delta_1 > 0$ such that if a cuboid I contains some $x \in H$ and has diameter less than δ_1, then

$$I \subseteq V_2. \tag{1}$$

Since $\overline{V}_2$ is compact and $|\det \alpha'|$ does not vanish anywhere, it has a positive lower bound on $\overline{V}_2$, which we shall call m. Also, α' and hence $\det \alpha'$ are uniformly continuous on $\overline{V}_2$ and, accordingly, there exists $\delta_2 > 0$ such that whenever the diameter of a cuboid I is less than δ_2 and $I \subseteq \overline{V}_2$, we have

$$|\det \alpha'(y_1) - \det \alpha'(y_2)| < m\varepsilon \qquad \text{whenever} \quad y_1, y_2 \in I.$$

This implies for all $x \in H \cap I$ that

$$|\det \alpha'(x)| \le |\det \alpha'(y_1)| + m\varepsilon \le |\det \alpha'(y_1)|(1 + \varepsilon) \quad \text{whenever } y_1 \in I. \tag{2}$$

By Proposition 7-4.2, there exists $\delta_3 > 0$ such that, for any cuboid I that contains some $x \in H$ and has diameter less than δ_3, we have $I \subseteq W$ and

$$c(\alpha(I)) \le |\det \alpha'(x)|(1 + \varepsilon)^n (\text{diam } I)^n. \tag{3}$$

Now consider any cuboid I that contains x and has diameter less than $\delta = \min \{\delta_1, \delta_2, \delta_3\}$. Then I satisfies $(1), (2)$ as well as (3). From (1), it follows that $\sigma(I)$ is defined. Besides,

$$\sigma(I) = \int_{\alpha(I)} f \le \sup \{f(z) : z \in \alpha(I)\} \cdot c(\alpha(I))$$

$$\le \sup \{(f \circ \alpha)(u) : u \in I\} \cdot c(\alpha(I)).$$

Now $|\det \alpha'|$ is continuous on the compact set $\overline{V}_2$ and therefore bounded on it. Let M be an upper bound. Choose $y \in I$ such that $\sup \{(f \circ \alpha)(u) : u \in I\} \le (f \circ \alpha)(y) + \frac{\varepsilon}{M}$. Then the above inequality becomes

$$\sigma(I) \le [(f \circ \alpha)(y) + \frac{\varepsilon}{M}] \cdot c(\alpha(I)). \tag{4}$$

From (2) and (3), we get

$$c(\alpha(I)) \le |\det \alpha'(y)|(1 + \varepsilon)^{n+1} (\text{diam } I)^n.$$

Combining this with (4), we get

$$\sigma(I) \le [(f \circ \alpha)(y) + \frac{\varepsilon}{M}] |\det \alpha'(y)|(1 + \varepsilon)^{n+1} (\text{diam } I)^n$$

$$\leq \sup\left\{\left[(f\circ\alpha)(z) + \frac{\varepsilon}{M}\right]|\det\alpha'(z)| : z\in I\right\}\cdot(1+\varepsilon)^{n+1}(\operatorname{diam} I)^n$$

$$\leq \sup\left\{(f\circ\alpha)(z)|\det\alpha'(z)| + \varepsilon : z\in I\right\}\cdot(1+\varepsilon)^{n+1}(\operatorname{diam} I)^n.$$

By Proposition 7-3.5, we now have $\sigma(A) \leq \int_A (f\circ\alpha)|\det\alpha'|$ whenever $\bar{A}\subseteq H^\circ$. Since $\bar{E}\subseteq V_1\subseteq \bar{V}_1 = H$ and V_1 is open, then $\bar{E}\subseteq H^\circ$. So,

$$\sigma(E) \leq \int_E (f\circ\alpha)|\det\alpha'|,$$

which is the same as

$$\int_{\alpha(E)} f \leq \int_E (f\circ\alpha)|\det\alpha'|.$$

As noted at the beginning, this inequality is all we needed to prove in order to establish the result under the additional hypothesis that $(f\circ\alpha)|\det\alpha'| : \bar{E}\to\mathbb{R}$ is integrable.

By Theorem 6-4.15, the additional hypothesis is fulfilled when f is continuous. Therefore, the result has been established for continuous f, and in particular, for any constant function.

Now let $f:\alpha(\bar{E})\to\mathbb{R}$ be an integrable function. As before, we may assume f to be nonnegative. Since $\alpha(\bar{E})\subseteq\alpha(W)$, where $\alpha(\bar{E})$ is compact and $\alpha(W)$ is open, Lemma 7-2.1 yields $\eta > 0$ such that $h\in\alpha(\bar{E})$, $\|h-x\| < \eta \Rightarrow x\in\alpha(W)$.

Consider any $\varepsilon > 0$ and any cuboid I containing $\alpha(\bar{E})$. Then the integral of f over $\alpha(E)$ is the same as its integral over I, with the understanding that f is extended to be zero outside $\alpha(E)$. This means $f\circ\alpha$ is extended to be zero outside E. Now, there exists a paving P of I such that

$$\int_{\alpha(E)} f-\varepsilon < L(f,P) \leq U(f,P) < \int_{\alpha(E)} f+\varepsilon. \qquad (5)$$

By refining P if necessary, we may assume that the diameter of each cuboid formed by it is no greater than η. This guarantees that any cuboid K formed by P that intersects $\alpha(E)$ lies within $\alpha(W)$. This has two consequences for the family $\mathcal{F}$ of such cuboids. One is that

$$c(\alpha^{-1}(K)\cap\alpha^{-1}(K')) = c(\alpha^{-1}(K\cap K')) = 0 \text{ whenever } K\in\mathcal{F}, K'\in\mathcal{F}, K\neq K'. \qquad (6)$$

This follows from Proposition 7-2.2(b) upon noting that $c(K\cap K') = 0$ by Proposition 6-4.9(a). Another consequence is that $\alpha^{-1}(K)\subseteq W$, and hence for any constant function k on a cuboid K,

$$\int_K k = \int_{\alpha^{-1}(K)} (k\circ\alpha)|\det\alpha'|, \qquad (7)$$

keeping in mind that the result has been shown to hold for a constant function.

Denoting by F the union of all $\alpha^{-1}(K)$ with $K\in\mathcal{F}$, we have $E\subseteq F$. As $f\circ\alpha$ has been extended to be zero outside E, according to the observation made immediately after Def. 6-4.5,

$$\underline{\int}_F (f \circ \alpha)|\det \alpha'| = \underline{\int}_E (f \circ \alpha)|\det \alpha'|. \tag{8}$$

Also,

$$L(f,P) = \Sigma_{K \in \mathcal{F}} (\inf\{f(x) : x \in K\})\mathrm{vol}(K).$$

Denote by m_K the function on I that equals the constant $\inf\{f(x) : x \in K\}$ on K and is extended to be 0 outside K. Then $m_K \circ \alpha \le f \circ \alpha$ on $\alpha^{-1}(K)$. By (7), we have

$$(\inf\{f(x) : x \in K\})\mathrm{vol}(K) = \int_K m_K = \int_{\alpha^{-1}(K)} (m_K \circ \alpha)|\det \alpha'|.$$

It follows that

$$L(f,P) = \Sigma_{K \in \mathcal{F}} \int_{\alpha^{-1}(K)} (m_K \circ \alpha)|\det \alpha'| \le \Sigma_{K \in \mathcal{F}} \underline{\int}_{\alpha^{-1}(K)} (f \circ \alpha)|\det \alpha'|,$$

using the fact that $m_K \circ \alpha \le f \circ \alpha$ on $\alpha^{-1}(K)$. Therefore by (6) and Remark 6-4.6(g),

$$L(f,P) \le \underline{\int}_F (f \circ \alpha)|\det \alpha'|$$

$$= \underline{\int}_E (f \circ \alpha)|\det \alpha'| \quad \text{by (8)}.$$

A similar argument shows that $U(f,P) \ge \overline{\int}_E (f \circ \alpha)|\det \alpha'|$. Therefore by (5),

$$\int_{\alpha(E)} f - \varepsilon < \underline{\int}_E (f \circ \alpha)|\det \alpha'| \le \overline{\int}_E (f \circ \alpha)|\det \alpha'| < \int_{\alpha(E)} f + \varepsilon.$$

The required integrability and equality are now immediate.   □

The theorem above differs from the versions in Burkill and Burkill [5] and in Protter and Morrey [20] in that f is not assumed continuous. It also differs from the version in Spivak [26] in that it assumes that f is defined on a closed set rather than on an open set.

7-4.5. Example. Evaluate the following integral:

$$\int_F \left(\sqrt{\frac{y}{x}} + \sqrt{xy}\right) dx\, dy,$$

where F is the region in the first quadrant of the xy-plane between the hyperbolas $xy = 1$, $xy = 9$ and the lines $y = x$ and $y = 4x$ (as in the figure at the beginning of Section 7-2).

The way the evaluation would be presented in calculus would be with an introduction of new variables $u = \sqrt{(xy)}$ and $v = \sqrt{\frac{y}{x}}$, or equivalently, $x = \frac{u}{v}$ and $y = uv$. The 'uv-region corresponding to F' would then be given by the inequalities

$1 \leq u \leq 3$, $1 \leq v \leq 2$. This procedure fits into the framework of the preceding theorem, as we now describe.

First of all, $F = \{(x,y) \in \mathbb{R}^2 : x > 0,\ y > 0,\ 1 \leq xy \leq 9,\ 1 \leq \frac{y}{x} \leq 4\}$. As noted at the beginning of Section 7-2, by taking $u = \sqrt{(xy)}$ and $v = \sqrt{\frac{y}{x}}$, we can express F as $\alpha(E)$, where $E = \{(u,v) \in \mathbb{R}^2 : 1 \leq u \leq 3,\ 1 \leq v \leq 2\}$ and α is given by $\alpha(u,v) = (\frac{u}{v},\ uv)$. This map α is injective on $W = \{(u,v) \in \mathbb{R}^2 : u > 0,\ v > 0\}$. Also, α' has matrix

$$\begin{bmatrix} \dfrac{1}{v} & -\dfrac{u}{v^2} \\[2ex] v & u \end{bmatrix},$$

which is continuous and invertible on W and the Jacobian is $\det \alpha' = 2\frac{u}{v} > 0$. With $f(x,y) = \sqrt{\frac{y}{x}} + \sqrt{(xy)}$, we have $(f \circ \alpha)(u,v) = u + v$. The given integral and the integral

$$\int_E (f \circ \alpha)|\det \alpha'| = \int_E (u+v)2\tfrac{u}{v}\,du\,dv$$

both exist. Therefore, they are equal by the transformation formula (Theorem 7-4.4). Now the second of the integrals is easily found to be $8 + \frac{52}{3}\ln 2$, which is therefore the required value.

The most frequently used transformation in $\mathbb{R}^2$ is the introduction of polar coordinates to convert an integral over a disc of radius A centred at $(0,0)$ to an integral over an (r,θ)-rectangle $[0,A] \times [0,2\pi]$ or its interior E. However, the polar transformation $\alpha(r,\theta) = (r\cos\theta, r\sin\theta)$ is not injective on any open set containing $\bar{E}$. In order to accommodate the transformation to polar coordinates, we make use of some further ideas.

7-4.6. Definition. *For any nonempty subset E of $\mathbb{R}^n$, a **balloon** is a sequence $\{E_m\}$ of subsets such that*
   (i)   *each E_m has content,*
   (ii) *$E_m \subseteq E_{m+1} \subseteq E$ for each m,*
   (iii) *$\bigcup_{m \in \mathbb{N}} E_m = E$.*

7-4.7. Examples. (a) With $n = 1$, the subset $E = \mathbb{R}$ has no content; but the sequence of subsets $[-m,m]$, $m = 1,2,3,\ldots$ is a balloon. The subsets $[-m,\infty)$ do not provide a balloon because they do not have content.

(b) Let $n = 2$. The sequence of subsets $E_m = \{(x,y) \in \mathbb{R}^n : x^2 + y^2 < 1 - \frac{1}{m}\}$ constitutes a balloon for $E = \{(x,y) \in \mathbb{R}^2 : x^2 + y^2 < 1\}$ and so do their closures. Note that we cannot select a finite number of E_m that can cover E. In this connection, let

$$V_\varepsilon = \{(x,y) \in \mathbb{R}^n : (1-\varepsilon)^2 < x^2 + y^2 < (1+\varepsilon)^2\} \text{ , where } 0 < \varepsilon < 1.$$

Fom Example 6-5.2(a), we have $c(V_\varepsilon) = \pi[(1+\varepsilon)^2 - (1-\varepsilon)^2] = 4\pi\varepsilon$, which approaches zero with ε. No matter how small ε may be, we can select a finite number of E_m such that, when taken together with V_ε, they can cover E. What is at play here is the compactness of the closure of E, which is a consequence of the boundedness of E.

(c) Let $E = \{(u,v) \in \mathbb{R}^2 : 0 < u < A \text{ and } 0 < v < 2\pi\} = (0,A) \times (0,2\pi)$, an open rectangular subset of $\mathbb{R}^2$. If $E_m = [\frac{1}{m}, A] \times [\frac{1}{m}, 2\pi - \frac{1}{m}]$ for $m > \max\{1/A, \frac{1}{\pi}\}$ then the sequence of closed rectangles $\{E_m\}$ is a balloon for E, consisting of compact sets. The transformation α defined on E by $\alpha(u,v) = (u\cos v, u\sin v)$ is injective and continuously differentiable, its linear derivative α' being given by the matrix

$$\begin{bmatrix} \cos v & -u\sin v \\ \sin v & u\cos v \end{bmatrix}.$$

Thus $\alpha'(u,v)$ is an invertible linear map for all $(u,v) \in E$. By Proposition 7-2.4(a), the sets $\alpha(E_m)$ have content; also, they are compact. As noted at the beginning of Section 7-2, α maps E *bijectively* onto the subset $G = \{(x,y) \in \mathbb{R}^2 : 0 < x^2 + y^2 < A^2$, either $x < 0$ or $y \neq 0\}$ and therefore the sequence $\{\alpha(E_m)\}$ is a balloon for $\alpha(E)$, consisting of compact sets.

Recall from 2-4.P9 that, for any $F \subseteq \mathbb{R}^n$, *we have* $\overline{F} = F \cup \partial F = F^\circ \cup \partial F$. We shall use this fact presently.

7-4.8. Proposition. *Let $F \subset \mathbb{R}^n$ have content. Then for any $\varepsilon > 0$, there exists an open set $V \supseteq \partial F$ such that $F \cup V$ is open and $c(V) < \varepsilon$. In particular, $c(F \cup V) < c(F) + \varepsilon$.*

Proof. Since F has content, the boundary ∂F has content 0. It follows by the equivalence of (α) and (δ) in Proposition 6-4.10 that there exists an open set $V \supseteq \partial F$ such that $c(V) < \varepsilon$. We need only prove that $F \cup V$ is open. This follows from the computation

$$F \cup V = F \cup (\partial F \cup V) = (F \cup \partial F) \cup V = (F^\circ \cup \partial F) \cup V$$
$$= F^\circ \cup (\partial F \cup V) = F^\circ \cup V. \qquad \square$$

7-4.9. Proposition. *Let $\{E_m\}$ be a balloon for E. If E has content, then $\lim_{m \to \infty} c(E_m) = c(E)$.*

Proof. Since $c(E_m) \leq c(E_{m+1}) \leq c(E)$, the limit must exist and not exceed $c(E)$.

To prove the reverse inequality, consider any $\varepsilon > 0$. Since E has content, the preceding proposition yields an open set V containing ∂E such that $E \cup V$ is open and $c(V) < \varepsilon$. Since each E_m has content, the proposition also yields open sets V_m containing ∂E_m such that $E_m \cup V_m$ is open and $c(V_m) < \varepsilon / 2^m$. Now,

$$E \subseteq \bigcup_{m \in \mathbb{N}} (E_m \cup V_m) \text{ and } V \supseteq \partial E,$$

from which we get

$$\bar{E} = E \cup \partial E \subseteq V \cup \bigcup_{m \in \mathbb{N}} (E_m \cup V_m).$$

Thus, the sets V and $E_m \cup V_m$ ($m \in \mathbb{N}$) constitute an open cover of $\bar{E}$. Since E has content, it is bounded and therefore so is its closure $\bar{E}$. But this means $\bar{E}$ is compact and the foregoing open cover contains a finite subcover. Thus there exists $k \in \mathbb{N}$ such that V and $\{E_m \cup V_m : 1 \le m \le k\}$ constitute a cover of $\bar{E}$. But $E_m \subseteq E_k$ for $1 \le m \le k$ and therefore $\bar{E}$ is covered by V, $\{V_m : 1 \le m \le k\}$ and E_k. Consequently,

$$c(E) = c(\bar{E}) \le c(V) + \sum_{1 \le m \le k} c(V_m) + c(E_k) < 2\varepsilon + \lim_{m \to \infty} c(E_m).$$

This implies the required reverse inequality.   □

7-4.10. Examples. (a) Returning to the transformation $\alpha(u, v) = (u \cos v, u \sin v)$ discussed above [see Example 7-4.7(c)], we note that it is easy to see why $\lim_{m \to \infty} c(E_m) = c(E)$, because E_m and E are rectangles and both $c(E_m)$ and $c(E)$ can be computed directly. However, regarding the associated balloon $\alpha(E_m)$, the disc with a keyhole removed, the computation of the content calls for some effort. Nevertheless, Proposition 7-4.9 allows us to conclude painlessly that $\lim_{m \to \infty} c(\alpha(E_m)) = c(\alpha(E))$, which is of course πA^2.

(b) Consider $E = (0,1) \times (0, A)$ and $\alpha(u, v) = (\sqrt{(uv)}, \sqrt{[(1 - u)v]})$ on E. The sequence of sets $E_m = [\frac{1}{m}, 1 - \frac{1}{m}] \times [\frac{1}{m}, A - \frac{1}{m}]$, starting from a sufficiently large m, is a balloon for E. A simple computation shows that $\alpha'(u, v)$ is represented by the matrix

$$\frac{1}{2} \begin{bmatrix} \sqrt{\dfrac{v}{u}} & \sqrt{\dfrac{u}{v}} \\[2ex] -\sqrt{\dfrac{v}{1-u}} & \sqrt{\dfrac{1-u}{v}} \end{bmatrix},$$

which is invertible for every $(u, v) \in E$ (the determinant is $\frac{1}{4} [\sqrt{\frac{1-u}{u}} + \sqrt{\frac{u}{1-u}}]$, i.e., $\frac{1}{4} \sqrt{\frac{1}{u(1-u)}}$). The equation $(x, y) = (\sqrt{(uv)}, \sqrt{(v - uv)})$ is equivalent to the pair

$$x = \sqrt{(uv)}, \qquad y = \sqrt{(v - uv)},$$

which can be handled by elementary computation, leading to

$$u = \frac{x^2}{x^2 + y^2}, \qquad v = x^2 + y^2.$$

This shows that α maps E onto the subset

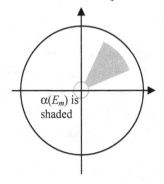

$\alpha(E_m)$ is shaded

$$\alpha(E) = \{(x,y) \in \mathbb{R}^2 : x > 0,\ y > 0,\ x^2 + y^2 < A\}$$

of the open first quadrant of $\mathbb{R}^2$ and has an inverse given by

$$\alpha^{-1}(x,y) = (x^2/(x^2 + y^2),\ x^2 + y^2).$$

For interested readers, we present a figure that depicts a representative $\alpha(E_m)$. Without computing $c(\alpha(E_m))$, one can deduce painlessly from Proposition 7-4.9 that $\lim_{m \to \infty} c(\alpha(E_m)) = c(\alpha(E))$. In terms of the figure, it means that the area of the shaded pizza slice (once bitten) approaches $\frac{1}{4}\pi A^2$.

7-4.11. Proposition. *Let $\{E_m\}$ be a balloon for E. If E has content and $f : E \to \mathbb{R}$ is integrable (on E), then f is integrable on each E_m and*

$$\lim_{m \to \infty} \int_{E_m} f = \int_E f.$$

Proof. That f is integrable on each E_m is a consequence of the observation recorded immediately after Def. 6-4.5. The same observation guarantees that f is integrable on $E \backslash E_m$. Since E is assumed to have content, we have $c(E \backslash E_m) = c(E) - c(E_m)$, which approaches zero by Proposition 7-4.9. It now follows by 6-4.P6 (wherein, we take $X_m = E \backslash E_m$) that the equality in question holds. $\square$

What makes this result useful is that transformations such as the one from polar to rectangular coordinates, or the one in Example 7-4.10(b), do not satisfy the hypotheses of the transformation formula on some desired domain of integration but do satisfy them on each set of an appropriately selected balloon for the domain. This has already been illustrated in the discussions above. We proceed to ensure our freedom to select a balloon according to our convenience provided the function is nonnegative-valued.

7-4.12. Proposition. *Let $f : E \to \mathbb{R}$ be nonnegative-valued and $\{E_m\}$ be a balloon for E (not assumed to have content). If $L = \lim_{m \to \infty} \int_{E_m} f$ exists, then for any other* *balloon consisting of sets on which f is integrable, the corresponding limit exists and equals L.*

Proof. Let $\{F_p\}$ be any other balloon for E such that f is integrable on each set F_p. Then f is integrable on each intersection $E_m \cap F_p$ [see observation immediately after Def. 6-4.5] and the inequality $\int_{E_m \cap F_p} f \leq \int_{E_m} f$ holds for each p and m, because $f \geq 0$. Moreover, the sequence $\int_{F_p} f$ is increasing (recall that $F_p \subseteq F_{p+1}$ by definition of a balloon).

Now, the sequence of sets $\{E_m \cap F_p\}$, $m = 1, 2, \ldots$, is a balloon for F_p and by Proposition 7-4.11 and the inequality noted at the beginning, we have

$$\int_{F_p} f = \lim_{m \to \infty} \int_{E_m \cap F_p} f \leq \lim_{m \to \infty} \int_{E_m} f = L.$$

Thus, L is an upper bound for the increasing sequence $\int_{F_p} f$. It follows that $\lim_{m \to \infty} \int_{F_p} f$ exists and does not exceed L. A similar argument establishes the reverse inequality. $\square$

The next proof uses the result of 2-6.P17.

7-4.13. Proposition. *Any open subset of $\mathbb{R}^n$ has a balloon consisting of compact sets.*

Proof. To begin with, consider a bounded open subset W. The first step is to construct a sequence $\{H_m\}$ of compact sets contained in W such that $H_m \subseteq H_{m+1}$ and $W = \bigcup_{m \in \mathbb{N}} H_m$. The second step is to construct a sequence $\{E_m\}$ of sets possessing the same properties and also having content.

Since W is assumed bounded, its complement W^c is nonempty and therefore, by 2-6.P17, the distance $d(x, W^c)$ of a point $x \in \mathbb{R}^n$ from W^c is a continuous function of x. Hence, the subsets $H_m = \{x \in \mathbb{R}^n : d(x, W^c) \geq \frac{1}{m}\}$, $m \in \mathbb{N}$, are closed in $\mathbb{R}^n$ and are contained in W. Since W is bounded, the sets H_m are also bounded and hence compact. Also, $H_m \subseteq H_{m+1}$ because $\frac{1}{m} > \frac{1}{m+1}$. Because W is open, each $x \in W$ satisfies $d(x, W^c) > 0$ and therefore belongs to H_m for some m. Consequently, $W = \bigcup_{m \in \mathbb{N}} H_m$. This completes the first step.

In view of compactness, each H_m is contained in a finite union Q_m of closed cuboids contained in W. Such a union of cuboids is compact and has content. Since a finite union of compact sets having content is again compact and has content, the sets $E_m = \bigcup_{k=1}^m Q_k \subseteq W$ are likewise. Moreover,

$$W = \bigcup_{m \in \mathbb{N}} H_m \subseteq \bigcup_{m \in \mathbb{N}} Q_m \subseteq \bigcup_{m \in \mathbb{N}} E_m \subseteq W.$$

This means $\{E_m\}$ is a balloon for W consisting of compact sets. The second step is now also complete and it has been shown that a *bounded* open set has a balloon consisting of compact sets.

Finally, suppose W is not bounded and for each $p \in \mathbb{N}$, let $W_p = \{x \in W : \|x\| < p\}$. Then each W_p is bounded as well as open, while $W = \bigcup_{p \in \mathbb{N}} W_p$. According to what has already been proved, for each p there is a balloon $\{E_{p,m}\}_{m \in \mathbb{N}}$ for W_p. Define

$$E_m = \bigcup_{p \le m} E_{p,m}. \tag{1}$$

Being a finite union of compact sets having content, each E_m is compact and has content. Since $E_{p,m} \subseteq E_{p,m+1}$, we have $E_m \subseteq E_{m+1} \subseteq W$. It remains only to prove that every $x \in W$ belongs to some E_m. With this in mind, consider any $x \in W$. Then $x \in E_{p,m}$ for some p and some m. But $E_{p,m} \subseteq E_{p,m+1}$ and hence $x \in E_{p,m'}$ for some $m' \ge p$. It now follows from our definition of E_m in (1) that $x \in E_{m'}$. □

7-4.14. Theorem. *Let $W \subseteq \mathbb{R}^n$ be open and $\alpha: W \to \mathbb{R}^n$ be a continuously differentiable injective map with an invertible derivative α' everywhere on W. Suppose that $f: \alpha(W) \to \mathbb{R}$ is a function that is nonnegative everywhere and integrable over every subset of its domain that has content and that the same is true of $(f \circ \alpha)|\det \alpha'| : W \to \mathbb{R}$. Then for any balloons $\{E_m\}$ and $\{F_p\}$ for $\alpha(W)$ and W, respectively, if either of the limits*

$$\lim_{m \to \infty} \int_{E_m} f \quad and \quad \lim_{p \to \infty} \int_{F_p} (f \circ \alpha)|\det \alpha'| \tag{1}$$

exists, then so does the other and the limits are equal. In case either of the integrals $\int_{\alpha(W)} f$ and $\int_W (f \circ \alpha)|\det \alpha'|$ exists (possibly both), it is equal to both limits. In particular, if both the integrals exist, then they are equal to each other.

Proof. By Props. 7-4.12 and 7-4.13, we may assume that $\{F_p\}$ consists of compact sets. Then $\{\alpha(F_p)\}$ consists of compact sets that also have content by Proposition 7-2.4(a). It is trivial to deduce that $\alpha(F_p) \subseteq \alpha(F_{p+1})$ and $\alpha(W) = \bigcup_{p \in \mathbb{N}} \alpha(F_p)$. Therefore $\{\alpha(F_p)\}$ is a balloon for $\alpha(W)$ consisting of compact sets. By Transformation Formula 7-4.4, we have

$$\int_{\alpha(F_p)} f = \int_{F_p} (f \circ \alpha)|\det \alpha'| \qquad \text{for each } p. \tag{2}$$

If the second limit mentioned in (1) exists, then (2) implies that $\lim_{p \to \infty} \int_{\alpha(F_p)} f$ exists and is equal to it. Since $\{\alpha(F_p)\}$ is a balloon for $\alpha(W)$, it follows by Proposition 7-4.12 that the first limit in (1) also exists and is equal to it. On the other hand, if the first limit exists, it follows by Proposition 7-4.12 that $\lim\limits_{p\to\infty}\int_{\alpha(F_p)}f$ exists and is equal to it. Therefore by (2), the second limit exists and is equal to it.

The last part follows by virtue of Proposition 7-4.11.   □

7-4.15. Corollary. *Suppose D is the disc* $\{(x,y)\in\mathbb{R}^2 : 0\le x^2+y^2\le A^2\}$, *where A > 0, and f:D→$\mathbb{R}$ is continuous. Let E denote the rectangle* $\{(r,\theta)\in\mathbb{R}^2 : 0\le r\le A,\ 0\le\theta\le 2\pi\}$, *i.e.,* $[0,A]\times[0,2\pi]$. *Then*

$$\int_D f = \int\int_E f(r\cos\theta, r\sin\theta)r\,dr\,d\theta = \int_0^{2\pi} d\theta \int_0^A f(r\cos\theta, r\sin\theta)r\,dr.$$

$$= \int_0^A dr \int_0^{2\pi} f(r\cos\theta, r\sin\theta)r\,d\theta.$$

Proof. Since f is continuous, the integrals all exist, while the second, third and fourth integrals are equal by Fubini's theorem [Remark 6-3.3]. So, we need only prove that the first and second are equal. Since $|f|$ is also continuous, the corresponding integrals with $|f|$ in place of f also exist and hence, as in the proof of Transformation Formula 7-4.4, we may restrict our considerations to nonnegative f.

Let $W = \{(u,v)\in\mathbb{R}^2 : 0<u<A$ and $0<v<2\pi\} = (0,A)\times(0,2\pi)$, an open subset of E. Since $E\backslash W$ has content zero, the integral over E is equal to that over W. Then, as seen in Example 7-4.7(c), the transformation $\alpha:W\to\mathbb{R}^2$ defined by $\alpha(r,\theta) = (r\cos\theta, r\sin\theta)$ satisfies the hypotheses of Theorem 7-4.14 and $\alpha(W) = \{(x,y)\in\mathbb{R}^2 : 0<x^2+y^2<A^2,$ either $x<0$ or $y\ne 0\}\subseteq D$. Again, $D\backslash\alpha(W)$ has content zero and so, the integral over D is equal to that over $\alpha(W)$. So it is sufficient to show that $\int_{\alpha(W)} f = \int\int_W f(r\cos\theta, r\sin\theta)r\,dr\,d\theta$. But since we have restricted considerations to nonnegative f, the foregoing equality is guaranteed by the last part of Theorem 7-4.14.   □

A similar argument justifies the usual procedure of evaluating the integral over the part of the disc in the first quadrant, i.e., $\{(x,y)\in\mathbb{R}^2 : 0\le x^2+y^2\le A^2, x\ge 0, y\ge 0\}$, via polar coordinates in the above manner, the integration with respect to θ being taken over $[0,\frac{\pi}{2}]$.

7-4.16. Examples. (a) We shall show that $\int_0^\infty \exp(-x^2)dx = \frac{1}{2}\sqrt{\pi}$. For any $A > 0$, let S_A denote the square $\{(x,y)\in\mathbb{R}^2 : 0\le x\le A,\ 0\le y\le A\}$. Then

$$[\int_0^A \exp(-x^2)dx\,]^2 = [\int_0^A \exp(-x^2)dx\,][\int_0^A \exp(-y^2)dy\,]$$

$$= \int_{S_A} \exp(-x^2-y^2)dx\,dy,$$

where we have used Fubini's theorem [6-4.P11 is an alternative] in the last step. Therefore, we need only prove that

$$\int_{S_A} \exp(-x^2 - y^2)\,dx\,dy \to \tfrac{\pi}{4} \text{ as } A \to \infty.$$

Let

$$Q_A = \{(x,y) \in \mathbb{R}^2 : 0 \le x^2 + y^2 \le A^2,\, x \ge 0,\, y \ge 0\},$$

the part in the first quadrant of the disc of radius A centred at $(0,0)$. Observe that $Q_A \subseteq S_A \subseteq Q_{A\sqrt2}$ and $\exp(-x^2 - y^2) > 0$ everywhere. It follows that

$$\int_{Q_A} \exp(-x^2 - y^2)\,dx\,dy \le \int_{S_A} \exp(-x^2 - y^2)\,dx\,dy \le \int_{Q_{A\sqrt2}} \exp(-x^2 - y^2)\,dx\,dy.$$

Introducing polar coordinates and applying the observation recorded after the proof of Cor. 7-4.15, we get

$$\int_{Q_A} \exp(-x^2 - y^2)\,dx\,dy = \int_0^{\pi/2} d\theta \int_0^A \exp(-r^2)r\,dr = \tfrac{\pi}{2}[\tfrac{1}{2}(1 - \exp(-A^2))]$$

$$= \tfrac{\pi}{4}[1 - \exp(-A^2)]$$

and similarly,

$$\int_{Q_{A\sqrt2}} \exp(-x^2 - y^2)\,dx\,dy = \tfrac{\pi}{4}[1 - \exp(-2A^2)].$$

Using these values in the double inequality displayed above, we get

$$\tfrac{\pi}{4}[1 - \exp(-A^2)] \le \int_{S_A} \exp(-x^2 - y^2)\,dx\,dy \le \tfrac{\pi}{4}[1 - \exp(-2A^2)].$$

It is immediate from this double inequality that $\int_{S_A} \exp(-x^2 - y^2)\,dy \to \tfrac{\pi}{4}$ as $A \to \infty$.

(b) The evaluation obtained above can also be expressed as

$$[\textstyle\int_0^\infty \exp(-x^2)\,dx\,]^2 = \tfrac{1}{4}\lim \int_{1/m}^{1-1/m} \frac{1}{\sqrt{s(1-s)}}\,ds \text{ as } m \to \infty, \qquad (A)$$

because the limit on the right here is in fact π. This form of the equality makes it possible to pretend (if one is so inclined) that trigonometric functions are still waiting for humans to discover them. While the virtues of avoiding trigonometric functions can be mathematically challenged [see the Internet article by Gilsdorf [11]], a proof of (A) can be differently enabled via the rational polar coordinates advocated in Wildberger [28]. For a point (x,y) in the first quadrant excluding $(0,0)$, the rational polar coordinates (s, Q) are given by

$$Q = x^2 + y^2, \qquad s = \frac{x^2}{x^2 + y^2} \qquad [= \cos^2\theta \text{ — of course!}].$$

Note that $s \in [0,1]$ and $Q \in (0,\infty)$. Every $(s,Q) \in [0,1]\times(0,\infty)$ uniquely determines a point $(x,y) \neq (0,0)$ in the first quadrant in accordance with the following transformation α:

$$x = \sqrt{(sQ)}, \qquad y = \sqrt{[(1-s)Q]}.$$

Though injective on $[0,1]\times(0,\infty)$, it is continuously differentiable only on the proper subset $(0,1)\times(0,\infty)$, which is easily seen to be mapped onto the *interior* of the first quadrant [see Example 7-4.10(b)]. Now Transformation Formula 7-4.4 requires the coordinate transformation α to be continuously differentiable on an open set W that contains the *closure* of the domain of integration. Therefore, it does not justify the use of rational polar coordinates for transforming an integral over any subset $\alpha(E)$ of the first quadrant whose closure includes a point of an axis. However, we can use Theorem 7-4.14, as we now show.

Since the functions involved are continuous, their integrals over bounded domains with content always exist. The domain of integration that was denoted by Q_A in (a) above can be replaced by its subset $\{(x,y) \in \mathbb{R}^2 : x > 0,\ y > 0,\ x^2 + y^2 \leq A^2\}$. This domain is, in turn, the image $\alpha(W)$ of the set $W = (0,1)\times(0,A^2]$. If we set $E_m = [\frac{1}{m}, 1-\frac{1}{m}]\times[\frac{1}{m}, A^2]$, then $\{E_m\}$ is a balloon for W. Since the functions involved are nonnegative, it follows by Theorem 7-4.14 that

$$\int_{Q_A} \exp(-x^2 - y^2)\,dx\,dy = \lim_{m\to\infty} \int_{E_m} \exp(-Q) \cdot \left| \det \alpha'(s,Q) \right| ds\,dQ$$

$$= \lim_{m\to\infty} \int_{E_m} \exp(-Q) \left[\frac{1}{4}\sqrt{\frac{1}{s(1-s)}} \right] ds\,dQ$$

$$= \lim_{m\to\infty} \int_{1/m}^{1-1/m} \left[\int_{1/m}^{A^2} \exp(-Q)\,dQ \right] \left[\frac{1}{4}\sqrt{\frac{1}{s(1-s)}} \right] ds$$

$$= \lim_{m\to\infty} \left[\exp(-1/m) - \exp(-A^2) \right] \frac{1}{4} \int_{1/m}^{1-1/m} \frac{1}{\sqrt{s(1-s)}}\,ds$$

$$= \left[1 - \exp(-A^2) \right] \cdot \lim_{m\to\infty} \frac{1}{4} \int_{1/m}^{1-1/m} \frac{1}{\sqrt{s(1-s)}}\,ds \,.$$

The required equality follows upon letting $A\to\infty$.

The equality claimed in Wildberger [28, p. 268, Example 27.3] is now established.

Problem Set 7-4

7-4.P1. Prove the following: Let $E \subseteq \overline{W}_1 \subseteq W \subseteq \mathbb{R}^n$, where E and W_1 have content, and W and W_1 are both open. Suppose $\alpha{:}W{\to}\mathbb{R}^n$ is a continuously differentiable map that is invertible on W_1 and such that $\{x \in W_1 : \det \alpha'(x) = 0\}$ has content zero. Then, for any integrable function $f{:}\alpha(\overline{E}){\to}\mathbb{R}$ such that $(f \circ \alpha)|\det \alpha'|{:}\overline{E}{\to}\mathbb{R}$ is also integrable, we have

$$\int_{\alpha(E)} f = \int_E (f \circ \alpha)|\det \alpha'|.$$

7-4.P2. Prove Cor. 7-4.15 by using 7-4.P1 instead of Theorem 7-4.14.

7-4.P3. Evaluate $\int_E \tan^{-1}(x + y)\, dx\, dy$, where $E = \{(x,y) \in \mathbb{R}^2 : x \geq 0, \ y \geq 0, \ x + y \leq 1\}$.

7-4.P4. Let $f(x,y) = 1/(x^2 + y^2)^{\alpha}$ and $E = \{(x,y) \in \mathbb{R}^2 : 0 < x^2 + y^2 \leq 1\}$. Consider a balloon $\{E_m\}_{m \in \mathbb{N}}$ for E such that f is bounded and hence integrable on each E_m. Show that $\lim_{m \to \infty} \int_{E_m} f$ exists for $\alpha < 1$ and it does not exist for $\alpha \geq 1$.

7-4.P5. Show that the hypothesis that f is nonnegative in Proposition 7-4.12 cannot be omitted.

7-4.P6. Determine real numbers a_1, b_1, c_1 and a_2, b_2, c_2 such that the transformation $T\colon \mathbb{R}^2 {\to} \mathbb{R}^2$ given by $T(x,y) = (u,v) = (a_1 + b_1 x + c_1 y, a_2 + b_2 x + c_2 y)$ maps $(0,0)$ into $(2,-1)$, $(1,0)$ into $(5,0)$ and $(0,1)$ into $(3,-2)$. Compute the Jacobian of T and use it to compute $\int_D \exp(2u - v)\, du\, dv$, where D is the image of $[0,1]\times[0,1]$ under T.

8

The General Stokes Theorem

Written by Harkrishan L. Vasudeva with help from Satish Shirali

8-1 Heuristic Background

The most important formula of analysis is the fundamental theorem of calculus. The formulas of Green, Gauss and Stokes are an extension of this theorem. They also constitute the extensively used part of the machinery of integral calculus. A far reaching generalisation of the above said theorems is the Stokes Theorem. In order to prove the theorem in its general form, we need to develop a good deal of material, known as *differential forms*. Much care has been taken to give clear definitions, examples and transparent proofs to tehnical challenging results. Differential forms also provide better insight into vector calculus, as is illustrated by the material covered in Section 8-8. A less formal and more intuitive introduction to the material covered in this chapter is available in Crowin and Szczarba [8], Lang [18] and Protter and Morrey [20].

One version of the substitution rule for Riemann integrals in one dimension is as follows (Pugh [21; p. 177]):

8-1.1. Theorem. *Let f:$[a,b]\to\mathbb{R}$ be Riemann integrable and Φ:$[\alpha,\beta]\to[a,b]$ be a bijection with a continuous positive first derivative. Then*

$$\int_{[a,b]} f(t)\,dt = \int_{[\alpha,\beta]} (f\circ\Phi)(s)\Phi'(s)\,ds.$$

Since Riemann integrability of f on $[a,b]$ implies that of $f(-t)$ on $[-b,-a]$ and also implies the equality

$$\int_{[a,b]} f(t)\,dt = \int_{[-b,-a]} f(-t)\,dt$$

(both integrals have the same sets of lower and upper sums), we have the following consequence of Theorem 8-1.1:

8-1.2. Theorem. *Let f:$[a,b]\to\mathbb{R}$ be Riemann integrable and Φ:$[\alpha,\beta]\to[a,b]$ be a bijection with a continuous negative first derivative. Then*

$$\int_{[a,b]} f(t)\,dt = -\int_{[\alpha,\beta]} (f\circ\Phi)(s)\Phi'(s)\,ds.$$

Combining the above two theorems, we have

8-1.3. Theorem. *Let f:$[a,b]\to\mathbb{R}$ be Riemann integrable and Φ:$[\alpha,\beta]\to[a,b]$ be a bijection with a continuous first derivative that vanishes nowhere. Then*

$$\int_{[a,b]} f(t)\,dt = \int_{[\alpha,\beta]} (f\circ\Phi)(s)|\Phi'(s)|\,ds.$$

S. Shirali, H.L. Vasudeva, *Multivariable Analysis*,
DOI 10.1007/978-0-85729-192-9_8, © Springer-Verlag London Limited 2011

It is this form of the substitution rule that the transformation formula of Chapter 7 generalises.

Theorem 8-1.3 does not use the fundamental theorem of calculus (FTC). However, there is another version of the substitution rule which is scarcely different from the FTC [Shirali and Vasudeva [23; p. 378]]:

8-1.4. Theorem. *Let* $F{:}[a,b]{\to}\mathbb{R}$ *and* $\Phi{:}[\alpha,\beta]{\to}[a,b]$ *both be differentiable. If* F' *and* $(F'\circ\Phi)\Phi'$ *are both Riemann integrable, then*

$$\int_{\Phi(\alpha)}^{\Phi(\beta)} F'(t)\,dt = \int_\alpha^\beta (F'\circ\Phi)(s)\Phi'(s)\,ds.$$

The proof of this theorem consists in merely observing that both sides are equal to $F(\Phi(\beta)) - F(\Phi(\alpha))$ by the FTC. The equality of the theorem can therefore also be expressed as

$$F(\Phi(\beta)) - F(\Phi(\alpha)) = \int_\alpha^\beta (F'\circ\Phi)(s)\Phi'(s)\,ds.$$

Written in this form, it looks more like the FTC than the substitution rule and we shall call it the 'substitution form of the FTC'.

Some features that distinguish Theorem 8-1.4 from Theorem 8-1.3 are worth noting. Perhaps the most obvious one is that $\Phi'(s)$ appears without absolute value. Precisely for this reason, it is difficult to state the equality in terms of integrals over intervals without distinguishing the lower and upper limits of integration. Another feature worth noting is that the interval of integration on the left side of Theorem 8-1.4 *need not be* the range of the substitution function Φ.

Having extended the substitution rule in the form of Theorem 8-1.3 to higher dimensions, it is natural to ask whether the substitution form of the FTC can also be extended. It can, but only for the case when F' has a continuous derivative. This is the task we take up in this chapter. The n-dimensional version is called the *general Stokes theorem* and resembles the well known calculus theorems of Green, Stokes and Gauss when $n = 2$ or 3.

A comparison of the equalities in Theorem 8-1.4, Theorem 8-1.3 and its extension to higher dimensions in Theorem 7-4.4 suggests that the role of Φ' should be played by a Jacobian but without absolute value. This will be the motivation for Jacobians without absolute value occurring as factors within integrands in the formulation of Def. 8-2.4 below.

Let us look at the equality in the substitution form of the FTC from a heuristic viewpoint, leaving it to later sections to make the ideas mathematically precise. First of all, the one-dimensional integral on the right side means that 'action' takes place over an interval in one direction, but action on the left side takes place at points, namely $\Phi(\beta)$ and $\Phi(\alpha)$. We can declare evaluation at a point to be zero-dimensional integration and write the left side as a difference of zero-dimensional integrals

$$\int_{\{\Phi(\beta)\}} F - \int_{\{\Phi(\alpha)\}} F.$$

By a further sleight of hand, we can then rewrite it as

$$\int_{\{\Phi(\beta)\} - \{\Phi(\alpha)\}} F.$$

It is to be kept in mind that whatever meaning we ultimately assign to the hither-to meaningless symbol $\{\Phi(\beta)\} - \{\Phi(\alpha)\}$, it is certainly not the algebraic difference of $\Phi(\beta)$ and $\Phi(\alpha)$ nor even a set theoretic difference. All this may sound like a joke, but precise meanings can be given to these rough ideas and a generalisation of the substitution form of the FTC can be formulated in terms of them and proved.

8-2 Differential Forms

In this section we give a formal mathematical meaning to the kind of expressions that appear as integrands of line and surface integrals in calculus and discuss their addition. The formal meaning will be essentially that they produce numbers out of 'parametrised' paths and surfaces. It should be noted that paths and surfaces differ only in the dimension of their parameter domains and will be collectively named as 'surfaces'; moreover, we shall define them as coming with a parametrisation and there will be no question of choosing a parametrisation or carrying out a reparametrisation. These terms will be used only for heuristic descriptions of a link-up with what the reader already knows from calculus.

8-2.1. Definition. *A* **k-surface** *in an open set* $U \subseteq \mathbb{R}^n$ *is the restriction to the cuboid* $[0,1]^k = \{(u_1, \ldots, u_k) : 0 \le u_i \le 1 \text{ for } 1 \le i \le k\}$ *of a* C^1 *map from an open set* $\subseteq \mathbb{R}^k$ *containing* $[0,1]^k$ *into* U.

If the map is C^2, we shall speak of a C^2 *k-surface* or of a *k-surface of class* C^2.

Remarks. (a) It may be emphasised that a k-surface is a map into an open subset U of a Euclidean space and not a subset of U.

(b) The domain of a k-surface can be taken to be any closed cuboid in $\mathbb{R}^k$ with nonempty interior. We have chosen the so called 'unit cuboid' in order to avoid introducing symbols for the endpoints of the edges. In specific examples, we may use other cuboids and leave it to the reader to rephrase matters in terms of the unit cuboid.

(c) It is not required that the map in the definition be injective on the interior of $[0,1]^k$; thus it is not necessarily a 'parametrisation' of its range.

(d) It is convenient to regard a map from the set $\{0\}$ consisting of a single point into an open set $U \subseteq \mathbb{R}^n$ as a **0-surface**.

8-2.2. Examples. (a) A 1-surface in $\mathbb{R}^2$ or $\mathbb{R}^3$ is the same as a continuously differentiable curve or path, as understood in calculus, provided that the continuously differentiable function involved can be extended to an open set containing the domain. The next two examples are special cases. It is left to the reader to verify that the extension is possible.

(b) The function that maps every $t \in [0,1]$ into $(\cos \pi t, \sin \pi t) \in \mathbb{R}^2$ is a 1-surface in $\mathbb{R}^2$, usually visualised as the upper semicircle with radius 1 and centre at the origin, traversed anticlockwise. It may seem more familiar to think of this as mapping every $t \in [0,\pi]$ into $(\cos t, \sin t)$. The reader will probably regard the function that maps $t \in [-1,1]$ into $(-t, \sqrt{(1 - t^2)}) \in \mathbb{R}^2$ as providing an 'equivalent parametrisation' of the same curve; however, this map is not a 1-surface because $\sqrt{(1 - t^2)}$ is not differentiable at the points $\pm1 \in [-1,1]$.

(c) For fixed $a > 0$ and $b > 0$, define

$$\gamma(t) = (a\cos t, b\sin t) \in \mathbb{R}^2, \qquad\qquad 0 \le t \le 2p\pi.$$

Then γ gives rise to a 1-surface, namely,

$$\lambda(t) = (a\cos(2p\pi t), b\sin(2p\pi t)) \in \mathbb{R}^2, \qquad\qquad 0 \le t \le 1.$$

Its range is an ellipse whenever $p \ge 1$ or $p \le -1$. When p is an integer, it is said to be a 'closed' curve, meaning thereby that $\lambda(0) = \lambda(1)$.

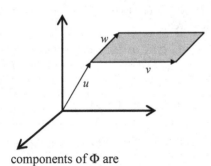

(d) Let u, v, w be vectors in $\mathbb{R}^3$ with v and w linearly independent. As usual, we shall denote points of $\mathbb{R}^2$ by (x,y) instead of (x_1, x_2). Put

$$\Phi(x,y) = u + xv + yw,$$
$$0 \le x \le 1, 0 \le y \le 1.$$

Then Φ is a 2-surface in $\mathbb{R}^3$ and its range is a parallelogram with one vertex at u and sides represented by v and w. The components of Φ are

$$\Phi_i(x,y) = u_i + xv_i + yw_i, \qquad\qquad i = 1,2,3.$$

(e) The function Φ that maps

$$(r,\theta) \in [0,1] \times [0,1] \quad \text{into} \quad (r\cos(2\pi\theta), r\sin(2\pi\theta)) \in \mathbb{R}^2$$

is a 2-surface in any open subset of $\mathbb{R}^2$ that contains its range, the unit disc (i.e., with radius 1, centre at the origin).

(f) Define a function Φ from $[0,1] \times [0,1]$ into $\mathbb{R}^3$ as

$$\Phi(r,\theta) = \left(\frac{2r\cos(2\pi\theta)}{1+r^2}, \frac{2r\sin(2\pi\theta)}{1+r^2}, \frac{1-r^2}{1+r^2} \right),$$

This is a 2-surface in $\mathbb{R}^3$ and its range is the upper unit hemisphere described by the equation $z = \sqrt{(1 - x^2 - y^2)}$. Indeed, the point $\Phi(r,\theta)$ is the point of intersection of the line joining $(0,0,-1)$ and $(r\cos 2\pi\theta, r\sin 2\pi\theta, 0)$ with the upper hemisphere. Similarly,

$$\Psi(r,\theta) = \left(\frac{2r\cos(2\pi\theta)}{1+r^2}, \frac{2r\sin(2\pi\theta)}{1+r^2}, -\frac{1-r^2}{1+r^2} \right)$$

maps onto the lower hemisphere. Both are 2-surfaces in $\mathbb{R}^3$, because the component functions have C^1 extensions to an open set containing $[0,1]\times[0,1]$. We note that

$$\Xi(r,\theta) = (r\cos(2\pi\theta), r\sin(2\pi\theta), \sqrt{(1 - r^2)})$$

also maps $[0,1]\times[0,1]$ onto the upper unit hemisphere, but is not a 2-surface for lack of a partial derivative when $r = \pm 1$. On the other hand,

$$\Lambda(\psi,\theta) = (\cos(2\pi\theta)\sin(\tfrac{1}{2}\pi\psi), \sin(2\pi\theta)\sin(\tfrac{1}{2}\pi\psi), \cos(\tfrac{1}{2}\pi\psi)), \quad (\psi,\theta)\in [0,1]^2$$

does define a 2-surface with the upper unit hemisphere as its range. Here $2\pi\theta$ and $\tfrac{1}{2}\pi\psi$ are the three-dimensional polar coordinates of the point $\Lambda(\psi,\theta)$.

(g) The function that maps $t \in [0,1]$ into $(\cos(2\pi t), \cos(4\pi t)) \in \mathbb{R}^2$ is a 1-surface or 'path' in $\mathbb{R}^2$. It is closed because it maps 0 and 1 into the same point of $\mathbb{R}^2$. Its range is the part of the parabola $y = 2x^2 - 1$ between $(-1,1)$ and $(1,1)$. The function that maps $t \in [0,1]$ into $(\cos(\pi t), \cos(2\pi t)) \in \mathbb{R}^2$ is also a 1-surface with the same range but is not a closed path.

(h) Let $\Phi:[0,1]^2 \to \mathbb{R}^2$ be the map given by

$$\Phi(r,\theta) = (\tfrac{1}{2}(3r - 1)\cos(2\pi\theta), \tfrac{1}{2}(3r - 1)\sin(2\pi\theta)).$$

The reader may verify that the range is the same as in Example (e) above. Nonetheless, they are two different 2-surfaces. This one is not injective on the interior of $[0,1]^2$ and therefore does not correspond to what is understood as a 'parametrisation' in calculus.

The definition of a differential form involves formidable looking integrals which are not all that difficult to set up. We begin with evaluation of similar integrals before coming to the formal definition.

8-2.3. Examples. (a) Consider the 2-surface of Example 8-2.2(d). In the expression $\omega = x_1 dx_1 dx_2$, to which we have yet to assign any mathematical meaning, we substitute $x_1 = \Phi_1(x,y)$ and $dx_1 dx_2 = \partial(\Phi_1,\Phi_2)/\partial(x,y)\, dx\, dy$ to get

$$\Phi_1(x,y)\frac{\partial(\Phi_1,\Phi_2)}{\partial(x,y)}\, dx\, dy = (u_1 + xv_2 + yw_3)(v_1 w_2 - v_2 w_1)\, dx\, dy.$$

The product on the right side describes a function of (x,y) on $[0,1]\times[0,1]$, which is continuous and therefore has an integral. The symbol $\int_{\Phi}\omega$ will mean the value of the integral. Note that, although the range of Φ is a parallelogram, we do not mention it in the symbol for the integral. This is partly because the parallelogram is implicit when we mention Φ and partly because we may want to make the same substitutions using a different Φ with some other range. Note that according to the scheme by which we have substituted for $dx_1 dx_2$, the order of x_1 and x_2 matters. Therefore, it is better to denote $dx_1 dx_2$ by dx_{12} and agree to distinguish it from dx_{21}.

The final integration will always be over the domain $[0,1] \times [0,1]$ of Φ. Thus

$$\int_\Phi x_1 \, dx_{12} \qquad \text{means} \qquad \int (u_1 + xv_2 + yw_3)(v_1w_2 - v_2w_1) \, dx \, dy,$$

where the latter integration is over $[0,1] \times [0,1]$. The value of the integral is easily seen to be $(u_1 + \tfrac{1}{2}v_2 + \tfrac{1}{2}w_3)(v_1w_2 - v_2w_1)$.

(b) In the case of a 1-surface, the Jacobian reduces to the derivative. For the 1-surface λ of Example 8-2.2(c) with $p = 1$, and $\omega = x \, dy$, the integral $\int_\lambda \omega$ will mean $\int_{[0,1]} a\cos 2\pi t(\frac{d}{dt} b \sin 2\pi t) \, dt$. The reader will find that this integral evaluates to πab. Similarly, $\int_\lambda y \, dx$ is $\int_{[0,1]} b \sin 2\pi t(\frac{d}{dt} a\cos 2\pi t) \, dt$, which evaluates to $-\pi ab$.

In the definition we are about to enunciate, ordered k-tuples $\langle i_1, i_2, \ldots, i_k \rangle$ are not expected to have distinct entries.

8-2.4. Definition. *A* **simple differential form of order** $k \geq 1$ *(or* **simple k-form** *for short) in an open set* $U \subseteq \mathbb{R}^n$ *is a real-valued function on the set of all k-surfaces in U for which there exists an ordered k-tuple* $\langle i_1, i_2, \ldots, i_k \rangle$ *with entries from among* $1, 2, \ldots, n$ *and a continuous function f on U such that the k-surface Φ is mapped into*

$$\int_{[0,1]^k} f(\Phi(u)) \frac{\partial(\Phi_{i_1}, \Phi_{i_2}, \ldots, \Phi_{i_k})}{\partial(u_1, u_2, \ldots, u_k)} \, du \, .$$

The simple k-form is then denoted by $f \, dx_{i_1 i_2 \cdots i_k}$ *and the above integral by*

$$\int_\Phi f \, dx_{i_1 i_2 \cdots i_k} \, .$$

If f equals 1 everywhere, then we write the simple k-form as simply $dx_{i_1 i_2 \cdots i_k}$.

If the integers $i_1, i_2, \ldots, i_k$ *are distinct, the k-form* $dx_{i_1 i_2 \cdots i_k}$ *is called a* **basic k-form.**

When $k = 1$, the Jacobians are understood to be the derivatives $\frac{d\Phi_r}{du}$ [see Example 8-2.3(b) above].

8-2.5. Remarks. (a) The definition requires the function f to be only continuous. However, what we want to do with differential forms will not work unless the function is at least of class C^1 and often C^2. It makes little difference if one requires them to have continuous partial derivatives of all orders.

(b) It is *not* assumed that the function f is unique. If the order of indices in $\langle i_1, i_2, \ldots, i_k \rangle$ is changed by a permutation, then the Jacobian is affected to the extent that it gets multiplied by the sign of the permutation. The multiplication is undone by multiplying by the sign of the permutation again. Thus

$$f\,dx_{i_1 i_2 \cdots i_k} \quad \text{and} \quad (\text{sign } \sigma) f\, dx_{j_1 j_2 \cdots j_k}$$

are the same differential form provided $\langle i_1, i_2, \ldots, i_k \rangle$ is obtained by applying σ to $\langle j_1, j_2, \ldots, j_k \rangle$ or vice versa. In particular, interchanging two among the dx_i reverses the sign. Thus when $k \geq 3$, we have $dx_{12} = (-1)dx_{21}$ and so on, while $dx_{123} = dx_{231} = dx_{312}$. Furthermore, when the indices $i_1, i_2, \ldots, i_k$ are not k distinct integers, the Jacobian vanishes and therefore the integral becomes 0. This means the function f can be replaced by 0.

(c) When $n \leq 3$, we denote x_1, x_2, x_3 by x, y, z respectively and write dx_{32} as $d(zy)$, dx_1 as $d(x)$ and so on. Accordingly, the integral $\int_\lambda y\, dx$ in Example 8-2.3(b) should have been denoted by $\int_\lambda y\, d(x)$.

(d) If the function f is zero everywhere on U, then $f\, dx_{i_1 i_2 \cdots i_k}$ maps every k-surface into the real number 0. When $k > 1$, the same is true of any simple k-form for which the indices $i_1, i_2, \ldots, i_k$ are not k distinct integers. In all these cases, we have the **zero simple k-form**. In particular, when $k > n$, the indices $i_1, i_2, \ldots, i_k$ cannot be distinct and hence we have the zero simple k-form.

(e) A **0-form** in an open set $U \subseteq \mathbb{R}^n$ is defined to be a continuous function on U. If f is a 0-form (that is, a continuous function on U) and Φ a 0-surface, then the zero-dimensional integral $\int_\Phi f$ is understood to mean $f(\Phi(0))$. The observations made under (b) above obviously do not apply to 0-forms.

8-2.6. Examples. (a) Consider the simple 1-form $x\,d(y)$ and the 1-surface λ in $\mathbb{R}^2$ of Example 8-2.2(c) with $p = 1$. The latter is given by

$$\lambda(t) = (a\cos(2\pi t), b\sin(2\pi t)) \in \mathbb{R}^2, \qquad 0 \leq t \leq 1.$$

Here $a > 0$ and $b > 0$. By Def. 8-2.4,

$$\int_\lambda x\,d(y) = \int_{[0,1]} (a\cos(2\pi t)) \frac{d}{dt}(b\sin(2\pi t))\, dt = \pi a b.$$

Note that the value of the integral has turned out to be the Jordan content of the subset of $\mathbb{R}^2$ described by the inequality $\frac{x^2}{a^2} + \frac{y^2}{b^2} \leq 1$. This may be familiar to the reader from calculus.

(b) Consider the simple 3-form $d(xyz)$ in $\mathbb{R}^3$ and the 3-surface

$$\Phi(r,\phi,\theta) = (ar\cos(2\pi\theta)\sin(\pi\phi), br\sin(2\pi\theta)\sin(\pi\phi), cr\cos(\pi\phi)),$$
$$(r,\theta,\phi) \in [0,1]^3.$$

Observe that Φ maps the cuboid $[0,1]^3$ onto the subset of $\mathbb{R}^3$ described by $\frac{x^2}{a^2} + \frac{y^2}{b^2} + \frac{z^2}{c^2} \leq 1$, called ellipsoid. The Jacobian of Φ is

$$\frac{\partial(\Phi_1, \Phi_2, \Phi_3)}{\partial(r,\phi,\theta)} = 2abc\pi^2 r^2 \sin \pi\phi.$$

Therefore by Def. 8-2.4,

$$\int_\Phi d(xyz) = \int_{[0,1]^3} 2abc\pi^2 r^2 \sin \pi\phi \, dr \, d\phi \, d\theta,$$

which works out to be $\frac{4}{3}\pi abc$. Note that the value of the integral has turned out to be the Jordan content of the ellipsoid mentioned above. Moreover, $\int_\Phi d(xzy) = -\frac{4}{3}\pi abc$.

We proceed to define a general differential form.

Like any two real-valued functions on a common domain, simple k-forms can be added and multiplied by constants. Thus, if ω_1 and ω_2 are simple k-forms in an open set U, their **sum** $\omega_1 + \omega_2$ is the function that maps every k-surface Φ in U into the real number $\int_\Phi \omega_1 + \int_\Phi \omega_2$. This is not to say that $\omega_1 + \omega_2$ is a simple k-form! A general differential form, called just *differential form*, is understood to be a (finite) sum of simple diferential forms. Naturally, the same differential form can be written as a sum of simple forms in various ways, just as a vector can be written as a sum of vectors in various ways. However, a vector can be written as a sum of specially chosen vectors (scalar multiples of vectors of a standard basis, for instance) in a unique way. We shall prove that a general differential form can analogously be written as a sum of specially chosen simple forms in a unique way.

A sum of 0-forms is a 0-form, and there is no distinction between simple and general 0-forms.

Throughout the rest of this section and the next two, we shall be working in an open subset U of $\mathbb{R}^n$ for some n. However, they will not always be mentioned explicitly.

We begin with the formal definition of a differential form in general, having order 1 or higher. A differential form of order 0 has already been defined in Remark 8-2.5(e).

8-2.7. Definition. *A* **differential form of order** $k \geq 1$ (or k**-form** for short) *in an open set* $U \subseteq \mathbb{R}^n$ *is a sum of simple k-forms in* U.

Thus a k-form ω ($k \geq 1$) can be represented as $\omega = \Sigma_I f_I dx_I$, where the summation ranges over some k-indices I and each f_I is a continuous function. If f

is a continuous function, then $f\omega$ denotes the k-form $\Sigma_I(ff_I)dx_I$. Thus $f(\Sigma_I f_I dx_I) = \Sigma_I(ff_I)dx_I$ by definition of the left side.

8-2.8. Remark. When $k = n-1 > 0$, only one among the indices $1, 2, \ldots, n$ fails to occur in any given k-index I. It is then easier to index the functions f_I by the single missing index. Thus, for example, when $n = 3$, it is easier to write a 2-form as

$$f_3 dx_{12} + f_2 dx_{13} + f_1 dx_{23} \quad \text{instead of} \quad f_{12} dx_{12} + f_{13} dx_{13} + f_{23} dx_{23}.$$

When $n = 2$, this hardly offers any advantage, but the fact that it can be done for $n \geq 2$ helps with notation in the proof of the main theorem of this chapter.

There will be occasions when we want every f_I to be of class C^1 or higher. We shall then describe the differential form as being of that class.

It is obvious that sums and multiples of k-forms by continuous functions f are again k-forms. Also, such familiar looking rules as

$$\omega_1 + \omega_2 = \omega_2 + \omega_1, \ \omega_1 + (\omega_2 + \omega_3) = (\omega_1 + \omega_2) + \omega_3,$$
$$f(\omega_1 + \omega_2) = f\omega_1 + f\omega_2, \ (f_1 + f_2)\omega = f_1\omega + f_2\omega, \ f_1(f_2\omega) = (f_1 f_2)\omega$$

are easily seen to hold.

A simple k-form ω is itself a k-form, because $\omega = \omega_0 + \omega$, where ω_0 denotes the zero simple k-form. Moreover, ω_0 satisfies $\omega = \omega_0 + \omega$ for *any* k-form ω. Therefore, we shall henceforth call it the **zero k-form** and denote it by 0. It will be clear from the context whether the symbol stands for a real number, a vector or the zero k-form. For any k-form ω, its constant multiple $(-1)\omega$ satisfies $\omega + (-1)\omega = 0$. Therefore we denote it by $-\omega$. In terms of this notation, the last part of Remark 8-2.5(b) can be expressed as

(a) $dx_{12} = -dx_{21}$ and so on when $k \geq 3$;

(b) $f dx_{i_1 i_2 \cdots i_k} = 0$ when the indices $i_1, i_2, \ldots, i_k$ are not k distinct integers (i.e., when one of the indices is repeated).

As with simple k-forms, the number that a k-form ω maps a k-surface Φ into is denoted by $\int_\Phi \omega$. Thus,

$$\int_\Phi (\omega_1 + \omega_2) = \int_\Phi \omega_1 + \int_\Phi \omega_2$$

and

$$\int_\Phi (c\omega) = c\int_\Phi \omega,$$

where c is any real number.

Since the ordered k-tuple $\langle i_1, i_2, \ldots, i_k \rangle$ occurs as an index, we call it a **k-index**; if we do not wish to specify k, then we speak of simply a **multi-index**. The integers i_j must satisfy $1 \leq i_j \leq n$, where n is the dimension of the space we are working with. When we are working with Euclidean spaces of different dimensions simultaneously, it may become necessary to specify the range of the integers in a k-index, in which case we shall speak of a k-index **in** or **from** $\langle 1, 2, \ldots, n \rangle$.

8-2.9. Notation. If the k-index $\langle i_1, i_2, \ldots, i_k \rangle$ is denoted by I, then dx_I denotes $dx_{i_1 i_2 \ldots i_k}$.

To every simple k-form ω there corresponds a k-index I and a continuous function f such that $\omega = f \, dx_I$. It follows from the definition of $\int_\Phi \omega$ that

$$f \, dx_I + g \, dx_I = (f + g) \, dx_I.$$

Hence every k-form ω is a sum

$$\omega = f_{I_1} \, dx_{I_1} + f_{I_2} \, dx_{I_2} + \cdots + f_{I_p} \, dx_{I_p}$$

with distinct k-indices $I_1, I_2, \ldots, I_p$, none of which is a permutation of any other; for instance,

$$y^2 \, d(xz) + \sin z \, d(yz) + 7 \, d(xx);$$

or what is the same thing

$$\sin z \, d(yz) - y^2 \, d(zx)$$

or

$$0 \, d(xy) + \sin z \, d(yz) - y^2 \, d(zx).$$

Also, the zero 2-form can be written as

$$0 = 2 d(xx) + 0 d(xz) = 7 d(xx) + 0 d(xz) + 0 d(xy), \text{ and so on.}$$

If ω is not 0, then the terms with k-indices containing a repeated entry can all be omitted.

The 3-form $5 \, d(xyz) + 8 \, d(xyx) - 4 \, d(yzx)$ is a basic 3-form, because it is the same as $d(yzx)$.

Suppose $k \geq 1$. We noted in Remark 8-2.5(b) above that even a basic k-form can have several representations. As a first step towards having a standard representation we note that there always exists a unique permutation that rearranges the ordered k-tuple $\langle i_1, i_2, \ldots, i_k \rangle$ of distinct entries $i_1, i_2, \ldots, i_k$ in ascending order.

A k-index $\langle i_1, i_2, \ldots, i_k \rangle$ is said to be **ascending** if $i_1 < i_2 < \ldots < i_k$. The integers $i_1, i_2, \ldots, i_k$ in an ascending k-index are necessarily distinct. Now, a set of k distinct integers can be arranged in increasing order in one and only one way. Therefore, for a simple k-form $f \, dx_{i_1 i_2 \ldots i_k}$ which is nonzero, so that $i_1, i_2, \ldots, i_k$ are distinct, there is a unique rearrangement of $\langle i_1, i_2, \ldots, i_k \rangle$ as an ascending k-index $J = \langle j_1, j_2, \ldots, j_k \rangle$ and

$$f \, dx_{i_1 i_2 \ldots i_k} = (\text{sign } \sigma) f \, dx_J,$$

where σ is the permutation that rearranges $\langle i_1, i_2, \ldots, i_k \rangle$ in ascending order as J. We shall soon show that $g = (\text{sign } \sigma)f$ is the only function such that the above equality holds.

Since every nonzero simple k-form can be written as $f\,dx_I$ with an ascending k-index I, it follows that any k-form ω, the *zero k-form included*, can be written as

$$\omega = f_{I_1}\,dx_{I_1} + f_{I_2}\,dx_{I_2} + \cdots + f_{I_p}\,dx_{I_p}$$

with distinct ascending k-indices $I_1, I_2, \ldots, I_p$. The possibility that some or all of the functions f_{I_r} are zero everywhere is not ruled out. Since there can be only finitely many distinct ascending k-indices, it further follows that an arbitrary k-form ω can be written as above, with all possible (distinct) ascending k-indices $I_1, I_2, \ldots$, included in the sum. In symbols,

$$\omega = \sum_I f_I\,dx_I,$$

where it is understood that I ranges over all ascending k-tuples. Thus the zero 2-form in $\mathbb{R}^3$ is represented as

$$0 = 0\,d(xy) + 0\,d(xz) + 0\,d(yz)$$

and in $\mathbb{R}^4$ as

$$0 = 0\,dx_{12} + 0\,dx_{13} + 0\,dx_{14} + 0\,dx_{23} + 0\,dx_{24} + 0\,dx_{34}.$$

If $\langle i_1, i_2, \ldots, i_k \rangle$ is different from $\langle i'_1, i'_2, \ldots, i'_k \rangle$, both ascending, then the two sets of k distinct integers $\{i_1, i_2, \ldots, i_k\}$ and $\{i'_1, i'_2, \ldots, i'_k\}$ must be different from each other and hence some i_j must be different from *all* the i'_j and vice versa. This too is a consequence of the fact that a set of k distinct integers can be arranged in increasing order in one and only one way. We shall use it in the next proof. The reader is reminded that the assumption $k \geq 1$ is still in force.

8-2.10. Proposition. *If the continuous function f does not vanish everywhere and I is an ascending k-index, then there exists a k-surface Φ such that*

(a) $\int_\Phi f\,dx_I \neq 0$ *and*

(b) $\int_\Phi g\,dx_{I'} = 0$ *for any continuous function g and any ascending k-index $I' \neq I$.*

Proof. Let $I = \langle i_1, i_2, \ldots, i_k \rangle$. Suppose f does not vanish at some $\xi \in U$. Without loss of generality, we may further suppose that $f(\xi) > 0$. Since the function is continuous, there exists a real number $h > 0$ such that every $x \in \mathbb{R}^n$ satisfying $|x_i - \xi_i| < 2h$ for $i = 1, \ldots, n$ belongs to U and satisfies $f(x) > 0$. Let $e_1, \ldots, e_n$ denote the standard basis of $\mathbb{R}^n$ and define a function Ψ on the open set $(-1,2)^k$ as

$$\Psi(u) = \xi + \sum_{j=1}^k (hu_j)e_{i_j}.$$

Its restriction Φ to $[0,1]^k$ is a k-surface in U and $f(\Phi(u)) > 0$ for every $u \in [0,1]^k$. Observe that the Jacobian

$$\frac{\partial(\Phi_{i_1}, \Phi_{i_2}, \ldots, \Phi_{i_k})}{\partial(u_1, u_2, \ldots, u_k)}$$

equals h^k everywhere and, consequently, by definition we have

$$\int_\Phi f \, dx_I = h^k \int_{[0,1]^k} f(\Phi(u)) \, du > 0.$$

Let $I' = \langle i'_1, i'_2, \ldots, i'_k \rangle \ne I$ be another ascending k-index. Some entry i'_j must be different from all the i_j, call it i'_{j_0}. Then we have $\Phi_{i'_{j_0}} = \xi_{i'_{j_0}}$, a constant, and hence the Jacobian

$$\frac{\partial(\Phi_{i'_1}, \Phi_{i'_2}, \ldots, \Phi_{i'_k})}{\partial(u_1, u_2, \ldots, u_k)}$$

vanishes everywhere, so that by definition,

$$\int_\Phi g \, dx_{I'} = 0. \qquad \qquad \square$$

8-2.11. Corollary. *Any k-form ω, where $k \ge 1$, has a unique representation as a sum $\omega = \sum_I f_I \, dx_I$, where the summation ranges over all ascending k-tuples I.*

Proof. The existence of such a representation has already been argued above. It remains only to prove that, if $\omega = \sum_I f_I \, dx_I = \sum_I g_I \, dx_I$, where both summations range over all possible ascending k-tuples I, then $f_I = g_I$ for each I.

Suppose there is some ascending k-tuple J such that $f = f_J - g_J$ does not vanish everywhere. By Proposition 8-2.10, there exists some k-surface Φ such that $\int_\Phi f \, dx_J \ne 0$, i.e., $\int_\Phi f_J \, dx_J \ne \int_\Phi g_J \, dx_J$, and also $\int_\Phi f_{I'} \, dx_{I'} = \int_\Phi g_{I'} \, dx_{I'} = 0$ for every $I' \ne J$. It follows that $\int_\Phi \sum_I f_I \, dx_I = \int_\Phi f_J \, dx_J \ne \int_\Phi g_J \, dx_J = \int_\Phi \sum_I g_I \, dx_I$. This implies $\sum_I f_I \, dx_I \ne \sum_I g_I \, dx_I$. $\qquad \square$

The unique representation of a k-form as $\sum_I f_I \, dx_I$, as guaranteed by the above Corollary, is called its **standard representation**.

The only possible ascending n-index is $\langle 1, 2, \ldots, n \rangle$. Therefore, any n-form in an open subset of $\mathbb{R}^n$ has standard representation consisting of a single term; thus every n-form is simple.

8-2.12. Remark. If $\omega = \sum_I f_I \, dx_I$ in standard representation, then for any continuous function f, the standard representation of $f\omega$ is $\sum_I (ff_I) dx_I$. If $\omega = \sum_I f_I \, dx_I$ and $\psi = \sum_I g_I \, dx_I$ both in standard representation, then $\omega + \psi = \sum_I (f_I + g_I) dx_I$ in standard representation. This simple observation will be crucial for some of our computations below.

Problem Set 8-2

8-2.P1. Let $\omega = \sum_{i=1}^{n} f_i dx_i$ and $\Phi:[0,1]\rightarrow\mathbb{R}^n$ be the restriction of a C^1 map. Express $\int_\Phi \omega$ as a Riemann integral over an interval. Calculate $\int_\Phi \omega$, when $\Phi(u) = (u, u^2, u^3)$ and $\omega = dx + dz$.

8-2.P2. Let $\omega = f_1 dx_{23} + f_2 dx_{31} + f_3 dx_{12}$, where f_1, f_2, f_3 are continuous on $\mathbb{R}^3$ and $\Phi:[0,1]^2\rightarrow\mathbb{R}^3$, $\Phi(u) = (\Phi_1(u), \Phi_2(u), \Phi_3(u))$ be a 2-surface in $\mathbb{R}^3$. Express $\int_\Phi \omega$ as a Riemann integral over a subset of $\mathbb{R}^2$.

8-2.P3. For the 2-surface Φ of Example 8-2.2(d) and the 2-form

$$\omega = (v_1 w_2 - v_2 w_1)dx_{12} + (v_1 w_3 - v_3 w_1)dx_{13} + (v_2 w_3 - v_3 w_2)dx_{23},$$

evaluate $\int_\Phi \omega$.

8-2.P4. Consider the 2-surface Φ in $\mathbb{R}^3$ of Example 8-2.2(f). Evaluate $\int_\Phi \omega$ when

$$\omega = x_1 dx_{23} + x_2 dx_{31} + x_3 dx_{12}.$$

8-2.P5. Give an example of a 1-surface Φ in $\mathbb{R}^n$ such that for any 1-form ω in $\mathbb{R}^n$, the value of $\int_\Phi \omega$ is 0.

8-2.P6. Give an example of a 3-surface of class C^2 in $\mathbb{R}^2$.

8-2.P7. Let φ be a C^1 function on $[0,1]$. Show that it can be extended to a C^1 function on an open interval containing $[0,1]$. Show that the mapping $\Phi:[0,1]^2\rightarrow\mathbb{R}^2$ defined by $\Phi(u,v) = (u, v\cdot\varphi(u))$ is a 2-surface and describe its range. Show also that the mapping $\Psi:[0,1]\rightarrow\mathbb{R}^2$ defined by $\Psi(t) = \Phi(t,1)$ is a 1-surface and describe its range.

8-3 Wedge Products

The collection of differential forms has a built-in multiplication process, called the *wedge product* or *exterior product*, and it is denoted by $\wedge$. We multiply a p-form in $\mathbb{R}^n$ by a q-form in $\mathbb{R}^n$ and obtain a $(p+q)$-form, which is 0 by definition if $p+q > n$. It suffices to define the wedge product of forms in standard representation and then show that the same formula works for forms that are in other representations.

Until further notice, only forms of order 1 or higher will be under consideration.

8-3.1. Notation. If $I = \langle i_1, i_2, \ldots, i_p\rangle$ is an ascending p-index and $J = \langle j_1, j_2, \ldots, j_q\rangle$ an ascending q-index, we denote by $\langle I,J\rangle$ the $(p+q)$-index

$$\langle I,J\rangle = \langle i_1, i_2, \ldots, i_p, j_1, j_2, \ldots, j_q\rangle.$$

8-3.2. Remark. Observe that $\langle I, J \rangle$ need not be an ascending index even when I and J are. Examples: $I = \langle 147 \rangle$ and $J = \langle 135 \rangle$ gives $\langle I, J \rangle = \langle 147135 \rangle$, and $I = \langle 147 \rangle$ and $J = \langle 235 \rangle$ gives $\langle I, J \rangle = \langle 147235 \rangle$, which is an odd permutation of the ascending index $\langle 123457 \rangle$. Also, $\langle 123457 \rangle$ is an even permutation of $\langle J, I \rangle = \langle 235147 \rangle$.

8-3.3. Definition. *Let* $\alpha = \Sigma_I f_I \, dx_I$ *and* $\beta = \Sigma_J g_J \, dx_J$ *be a p-form and a q-form, respectively, both in standard representation. Their* (**wedge**) **product** *is the* $(p + q)$*-form given by*

$$\alpha \wedge \beta = \Sigma_{I,J} f_I g_J \, dx_{\langle I, J \rangle}.$$

In $\mathbb{R}^3$ for example, we have

$$dx_1 \wedge dx_3 = (1dx_1 + 0dx_2 + 0dx_3) \wedge (0dx_1 + 0dx_2 + 1dx_3) = dx_{13}.$$

If we drop the terms in the standard representations of α and β for which f_I and g_J are zero everywhere, the computation of $\alpha \wedge \beta$ will not be affected. However, this option is available only when neither α nor β is the zero form. We do not wish to consider separate cases when one among them is the zero form. But in a specific situation, we may drop the zero terms, as we now illustrate.

$$dx_{15} \wedge dx_{2348} = dx_{152348} = -dx_{123458},$$

$$dx_{15} \wedge dx_{16} = dx_{1516} = 0 \text{ (the zero 4-form)}.$$

Also,

$$(x_1 dx_{13} + x_5 dx_{12}) \wedge (2dx_{245} - x_2 dx_{345}) = 2x_1 dx_{13245} - x_1 x_2 dx_{13345} + 2x_5 dx_{12245}$$

$$-x_2 x_5 dx_{12345}$$

$$= -2x_1 dx_{12345} - x_2 x_5 dx_{12345}$$

$$= -(2x_1 + x_2 x_5) dx_{12345}.$$

We draw attention to the fact that the right side of the defining equality for the wedge product need not be a standard representation for two reasons. One is that $\langle I, J \rangle$ may contain a repeated index. A second is that, if each $\langle I, J \rangle$ containing no repeated index is subjected to a permutation that rearranges its entries in ascending order, the ascending multi-indices obtained after rearrangement may not be distinct, as noted in Remark 8-3.2.

This makes it difficult to use the sum $\Sigma_{I,J} f_I g_J \, dx_{\langle I, J \rangle}$ for a further computation of a wedge product such as

$$(\Sigma_{I,J} f_I g_J \, dx_{\langle I, J \rangle}) \wedge (\Sigma_K h_K \, dx_K).$$

We shall overcome this difficulty by showing that the defining equality for the wedge product is actually valid even when the differential forms on the left side are not in standard representation. However, in order to do this, we first work with the definition as above in terms of standard representations.

8-3.4. Proposition. *Suppose I is a p-index and J is a q-index (neither one assumed to be ascending) and f_I, g_J are continuous functions. Then*

$$(f_I\,dx_I)\wedge(g_J\,dx_J) = (f_I\,g_J)dx_{\langle I,J\rangle}.$$

Proof. It is clear from Def. 8-3.3 that $0\wedge\beta = \alpha\wedge 0 = 0$ for any α and β.

If I contains a repeated index, then so does $\langle I,J\rangle$ and what is required to be proved reduces to $0\wedge(g_J\,dx_J) = 0$, which is true. Same when J contains a repeated index. So, assume that neither I nor J contains a repeated index.

Let ρ and σ be the permutations that rearrange I and J, respectively, as ascending indices ρI and σJ. Then by Remark 8-2.5(b),

$$f_I\,dx_I = (\text{sign}\,\rho)f_I\,dx_{\rho I} \quad \text{and} \quad g_J\,dx_J = (\text{sign}\,\sigma)g_J\,dx_{\sigma J}.$$

Moreover, the standard representations of $f_I\,dx_I$ and $g_J\,dx_J$ are

$$f_I\,dx_I = \textstyle\sum_{K\neq\rho I} 0\,dx_K + (\text{sign}\,\rho)f_I\,dx_{\rho I} \text{ and } g_J\,dx_J = \textstyle\sum_{L\neq\sigma J} 0\,dx_L + (\text{sign}\,\sigma)g_J\,dx_{\sigma J}.$$

Therefore, it follows from Def. 8-3.3 that

$$(f_I\,dx_I)\wedge(g_J\,dx_J) = (\text{sign}\,\rho)(\text{sign}\,\sigma)(f_I\,g_J)dx_{\langle \rho I,\,\sigma J\rangle}. \tag{1}$$

Now we consider two cases.

Case 1. $\langle I,J\rangle$ contains a repeated index.

In this case, $\langle \rho I,\ \sigma J\rangle$ also contains a repeated index. Consequently, both $(\text{sign}\,\rho)(\text{sign}\,\sigma)(f_I\,g_J)dx_{\langle \rho I,\,\sigma J\rangle}$ and $(f_I\,g_J)dx_{\langle I,J\rangle}$ are the zero form and it follows from (1) that $(f_I\,dx_I)\wedge(g_J\,dx_J) = 0 = (f_I\,g_J)dx_{\langle I,J\rangle}$.

Case 2. $\langle I,J\rangle$ contains no repeated index.

This means no index occurring in I occurs in J and we can unambiguously define τ to be the permutation of $\langle I,J\rangle$ that agrees with ρ on I and agrees with σ on J. Then τ is the product of the permutations ρ and σ (in either order) and sign $\tau = (\text{sign}\,\rho)(\text{sign}\,\sigma)$. Also, $\tau\langle I,J\rangle = \langle \rho I,\ \sigma J\rangle$. Therefore

$$(f_I\,g_J)dx_{\langle \rho I,\,\sigma J\rangle} = (f_I\,g_J)dx_{\tau\langle I,J\rangle} = (\text{sign}\,\tau)(f_I\,g_J)dx_{\langle I,J\rangle}.$$

Substituting this in (1), we get

$$(f_I\,dx_I)\wedge(g_J\,dx_J) = (\text{sign}\,\rho)(\text{sign}\,\sigma)(\text{sign}\,\tau)(f_I\,g_J)dx_{\langle I,J\rangle}$$
$$= (\text{sign}\,\tau)^2(f_I\,g_J)dx_{\langle I,J\rangle} = (f_I\,g_J)dx_{\langle I,J\rangle}. \qquad \square$$

8-3.5. Proposition. *Both distributive laws hold: If α and β are p-forms and γ is a q-form, then*

$$(\alpha + \beta)\wedge\gamma = \alpha\wedge\gamma + \beta\wedge\gamma \quad \text{and} \quad \gamma\wedge(\alpha + \beta) = \gamma\wedge\alpha + \gamma\wedge\beta.$$

Proof. Suppose

$$\alpha = \textstyle\sum_I f_I\,dx_I,\ \beta = \textstyle\sum_I g_I\,dx_I \text{ and } \gamma = \textstyle\sum_J h_J\,dx_J,$$

all being standard representations. Then by Def. 8-3.3,

$$\alpha\wedge\gamma = \textstyle\sum_{I,J} f_I h_J\,dx_{\langle I,J\rangle}, \qquad\qquad \beta\wedge\gamma = \textstyle\sum_{I,J} g_I h_J\,dx_{\langle I,J\rangle}.$$

Besides, the standard representation of $\alpha + \beta$ is $\Sigma_I (f_I + g_I) dx_I$. Therefore we can apply Def. 8-3.3 to obtain

$$(\alpha + \beta) \wedge \gamma = \Sigma_{I,J} (f_I + g_I) h_J dx_{\langle I,J \rangle}$$
$$= \Sigma_{I,J} f_I h_J dx_{\langle I,J \rangle} + \Sigma_{I,J} g_I h_J dx_{\langle I,J \rangle}.$$

The preceding three equalities lead immediately to $(\alpha + \beta) \wedge \gamma = \alpha \wedge \gamma + \beta \wedge \gamma$. The proof that $\gamma \wedge (\alpha + \beta) = \gamma \wedge \alpha + \gamma \wedge \beta$ is similar.   □

We now use the two propositions above to establish that the defining equality in Def. 8-3.3 for the wedge product is valid even when $\alpha = \Sigma_I f_I dx_I$ and $\beta = \Sigma_J g_J dx_J$ are not standard representations.

8-3.6. Proposition. *Let* $\alpha = \Sigma_I f_I dx_I$ *and* $\beta = \Sigma_J g_J dx_J$, *not necessarily standard representations. Then*

$$\alpha \wedge \beta = \Sigma_{I,J} f_I g_J dx_{\langle I,J \rangle}.$$

In other words,

$$(\Sigma_I f_I dx_I) \wedge (\Sigma_J g_J dx_J) = \Sigma_{I,J} f_I g_J dx_{\langle I,J \rangle}.$$

Proof. By repeated application of distributivity (Proposition 8-3.5), we have

$$\alpha \wedge \beta = \Sigma_I ((f_I dx_I) \wedge (\Sigma_J g_J dx_J)) = \Sigma_I \Sigma_J ((f_I dx_I) \wedge (g_J dx_J))$$
$$= \Sigma_{I,J} f_I g_J dx_{\langle I,J \rangle} \qquad \text{by Proposition 8-3.4.} \qquad □$$

8-3.7. Proposition. *The wedge product is associative: If* α, β, γ *are forms of any orders, then*

$$(\alpha \wedge \beta) \wedge \gamma = \alpha \wedge (\beta \wedge \gamma).$$

Proof. Straightforward computation using the above two propositions.   □

8-3.8. Proposition. *If* α *is a p-form and* β *is a q-form, then*

$$\alpha \wedge \beta = (-1)^{pq} (\beta \wedge \alpha).$$

Proof. In view of the distributivity proved in Proposition 8-3.5, it is sufficient to prove the equality only for simple forms $\alpha = f_I dx_I$ and $\beta = g_J dx_J$. Proposition 8-3.4 and the property noted just after Def. 8-2.7 that $f_1(f_2 \omega) = (f_1 f_2) \omega$ further reduces the matter to the case when $\alpha = dx_I$ and $\beta = dx_J$. Thus we need only prove that

$$dx_I \wedge dx_J = (-1)^{pq} (dx_J \wedge dx_I),$$

where I is a p-index and J is a q-index. We also know from Proposition 8-3.4 that

$$dx_I \wedge dx_J = dx_{\langle I,J \rangle} \text{ and } dx_J \wedge dx_I = dx_{\langle J,I \rangle}.$$

Therefore, we need only prove that

$$dx_{\langle I,J \rangle} = (-1)^{pq} dx_{\langle J,I \rangle}.$$

This will follow if we show that the sign of the permutation σ that rearranges $\langle I,J \rangle$ as $\langle J,I \rangle$ is $(-1)^{pq}$, because we know from Remark 8-2.5(b) that $dx_{\langle I,J \rangle} =$ (sign σ)$dx_{\langle J,I \rangle}$. To see why sign $\sigma = (-1)^{pq}$, let

$$I = \langle i_1, i_2, \dots, i_p \rangle \quad \text{and} \quad J = \langle j_1, j_2, \dots, j_q \rangle.$$

Then

$$\langle I,J \rangle = \langle i_1, i_2, \dots, i_p, j_1, j_2, \dots, j_q \rangle \quad \text{and} \quad \langle J,I \rangle = \langle j_1, j_2, \dots, j_q, i_1, i_2, \dots, i_p \rangle.$$

It is easily seen from here that permuting $\langle I,J \rangle$ into $\langle J,I \rangle$ can be achieved by successively interchanging each of $i_p, i_{p-1}, \dots, i_1$, in that order, with its immediate neighbours to the right $j_1, j_2, \dots, j_q$ (again, in that order), one after the other. This calls for q interchanges to be carried out for each of the p indices $i_p, i_{p-1}, \dots, i_1$. Therefore, sign $\sigma = (-1)^{pq}$. □

The above considerations did not take into account **forms of order 0**, which are defined to be continuous functions. We complete the picture by setting $f \wedge \alpha = \alpha \wedge f = f\alpha$, where f is a continuous function and thus a 0-form. It is left to the reader to verify that the properties of the wedge product that have been shown to hold continue to be valid when one or more of the forms involved are of order 0.

8-3.9. Proposition. *If h is a continuous function and α, β are forms of any orders, then $(h\alpha) \wedge \beta = h(\alpha \wedge \beta) = \alpha \wedge (h\beta)$.*

Proof. Immediate from Proposition 8-3.6. □

Note that, since $h\alpha = h \wedge \alpha$ and $h(\alpha \wedge \beta) = h \wedge (\alpha \wedge \beta)$, the first equality in Proposition 8-3.9 can also be obtained as a consequence of Proposition 8-3.7.

8-3.10. Remark. It is a consequence of Proposition 8-3.4 and Proposition 8-3.7 that, for any k-index $\langle i_1, i_2, \dots, i_k \rangle$, the equality

$$dx_{i_1 i_2 \cdots i_k} = dx_{i_1} \wedge dx_{i_2} \wedge \cdots \wedge dx_{i_k}$$

holds. We shall often write $dx_{i_1} \wedge dx_{i_2} \wedge \cdots \wedge dx_{i_k}$ for $dx_{i_1 i_2 \cdots i_k}$ or, what is the same thing, for dx_I, where $I = \langle i_1, i_2, \dots, i_k \rangle$.

Problem Set 8-3

8-3.P1. For 1-forms α, β, show by direct computation (without using Proposition 8-3.8) that $\alpha \wedge \beta = -\beta \wedge \alpha$.

8-3.P2. Show that there is a 2-form α in $\mathbb{R}^4$ such $\alpha \wedge \alpha \neq 0$.

8-4 The Exterior Derivative

In this section we define the *exterior derivative* of any differential form. More specifically, given any k-form ω of class C^1 in an open set $U \subseteq \mathbb{R}^n$, we produce a $(k+1)$-form in U, called the exterior derivative. A characterisation is provided at the end.

We remind the reader that all our discussion pertains to an open subset of $\mathbb{R}^n$.

8-4.1. Definition. *The* **exterior derivative** *of a 0-form f of class C^1 is the 1-form*

$$df = \sum_{j=1}^{n} (D_j f) dx_j$$

and that of a k-form $(k \geq 1)$ of class C^1 in standard representation $\omega = \sum_I f_I\, dx_I$ is the $(k+1)$-form

$$d\omega = \sum_I (df_I) \wedge dx_I.$$

It may appear at first sight that the symbol 'dx_j' now has two meanings, one in the sense of the definition of a differential form and another as the exterior derivative of the 0-form f given by $f(x) = x_j$. However, the exterior derivative of this 0-form is nothing but what we have called dx_j in the former sense. So the two meanings turn out to be the same.

8-4.2. Examples. (a) If $f: \mathbb{R}^3 \to \mathbb{R}$ is given by $f(x_1, x_2, x_3) = x_1^2 + x_2 x_3$, then $df = 2x_1 dx_1 + x_3 dx_2 + x_2 dx_3$.

(b) If $f_i: \mathbb{R}^n \to \mathbb{R}$ is the function that maps any point in $\mathbb{R}^n$ onto its ith coordinate, i.e., $f_i(x_1, x_2, \dots, x_n) = x_i$, then $D_j f_i = 1$ or 0 according as $j = i$ or $j \neq i$. Therefore, $df_i = dx_i$.

(c) Suppose $n = 2$ and ω is a 1-form given in xy-notation as

$$\omega = f dx + g dy.$$

Then

$$d\omega = (df) \wedge (dx) + (dg) \wedge (dy)$$

$$= \left(\frac{\partial f}{\partial x} dx + \frac{\partial f}{\partial y} dy \right) \wedge dx + \left(\frac{\partial g}{\partial x} dx + \frac{\partial g}{\partial y} dy \right) \wedge dy$$

$$= \frac{\partial f}{\partial y} dy \wedge dx + \frac{\partial g}{\partial x} dx \wedge dy$$

$$= \left(\frac{\partial g}{\partial x} - \frac{\partial f}{\partial y} \right) dx \wedge dy,$$

because the terms involving $dx \wedge dx$ and $dy \wedge dy$ vanish.

(d) Suppose $n = 4$ and ω is a 2-form given in terms of x_1, x_2, x_3, x_4 by

$$\omega = x_4 dx_1 \wedge dx_2 + x_2 x_3 dx_1 \wedge dx_3 .$$

Then

$$d\omega = dx_4 \wedge dx_1 \wedge dx_2 + (x_3 dx_2 + x_2 dx_3) \wedge dx_1 \wedge dx_3$$
$$= dx_4 \wedge dx_1 \wedge dx_2 - x_3 dx_1 \wedge dx_2 \wedge dx_3$$

since $dx_3 \wedge dx_2 \wedge dx_3 = 0$.

(e) The 1-form in $\mathbb{R}^2$ given by $\omega = x \, dy$ is not the (exterior) derivative of any 0-form. Suppose, if possible, that $\omega = df$, where f is a C^1 function on $\mathbb{R}^2$. Since by definition, $df = (D_x f)dx + (D_y f)dy$, for any closed curve γ with component functions γ_1 and γ_2, i.e., a 1-surface with $\gamma(0) = \gamma(1)$, we have

$$\int_\gamma \omega = \int_{[0,1]} \left[(D_x f)(\gamma_1(t), \gamma_2(t)) \left(\frac{d\gamma_1}{dt} \right) + (D_y f)(\gamma_1(t), \gamma_2(t)) \left(\frac{d\gamma_2}{dt} \right) \right] dt$$

$$= \int_{[0,1]} \frac{d}{dt} f(\gamma_1(t), \gamma_2(t)) \, dt = f(\gamma_1(1), \gamma_2(1)) - f(\gamma_1(0), \gamma_2(0))$$

$$= 0,$$

because $\gamma(0) = \gamma(1)$. On the other hand, for the closed curve given by $\gamma(t) = (\cos(2\pi t), \sin(2\pi t))$, we have $\int_\gamma \omega = \pi$ as seen in Example 8-2.6(a). This contradiction shows that $\omega = x \, dy$ is not the (exterior) derivative of any 0-form.

8-4.3. Theorem. *If α is a p-form and β a q-form, both of class C^1, then $\alpha \wedge \beta$ is also of class C^1 and*

$$d(\alpha \wedge \beta) = (d\alpha) \wedge \beta + (-1)^p \alpha \wedge (d\beta). \tag{1}$$

Moreover, d is linear. In particular, if $\omega = \sum_I f_I \, dx_I$, not necessarily in standard form, then $d\omega = \sum_I (df_I) \wedge dx_I$.

Proof. It is straightforward to see from the definition of wedge product that $\alpha \wedge \beta$ is also of class C^1.

The linearity of d on 0-forms is a trivial consequence of the linearity of partial differentiation. For higher order forms, when $\sum_I f_I \, dx_I$ and $\sum_I g_I \, dx_I$ are standard representations, their sum has standard representation $\sum_I (f_I + g_I) dx_I$ and the multiple $c \sum_I f_I \, dx_I$, where $c \in \mathbb{R}$, has standard representation $\sum_I (cf_I) dx_I$. This allows us to check linearity by a routine computation. It is then straightforward to verify that $d\omega = \sum_I (df_I) \wedge dx_I$ even if $\omega = \sum_I f_I \, dx_I$, not necessarily in standard form.

We begin by proving (1) when one among α and β is a 0-form. If both are, then (1) is essentially the product formula for derivatives. It can be summarised as

$$d(fg) = (df)g + f(dg),$$

268   The General Stokes Theorem

when f and g are C^1 functions. We shall use this special case shortly.

Suppose α is a 0-form and β is a q-form, $q > 0$. In view of the distributivity of the wedge product and linearity of d, we need prove (1) in this case only when $\beta = g\,dx_J$, a simple q-form. In order to avoid losing sight of the hypothesis that α is a 0-form of class C^1, which means simply a C^1 function, we shall denote it by f. Then

$$d\beta = dg \wedge dx_J.$$

Using the result of the preceding paragraph and the distributivity of the wedge product, we obtain

$$d(\alpha \wedge \beta) = d(f \wedge \beta) = d((fg)\,dx_J) = d(fg) \wedge dx_J = ((df)g + f(dg)) \wedge dx_J$$

$$= ((df)g)) \wedge dx_J + (f(dg)) \wedge dx_J.$$

Now, by Proposition 8-3.9, we have

$$((df)g)) \wedge dx_J = (df) \wedge (g\,dx_J) = (d\alpha) \wedge \beta$$

and

$$(f(dg)) \wedge dx_J = f(dg \wedge dx_J) = \alpha \wedge (d\beta).$$

Therefore $d(\alpha \wedge \beta) = (d\alpha) \wedge \beta + \alpha \wedge (d\beta)$. Since $p = 0$ in the present case, the foregoing equality is the same as (1).

If α is a p-form, $p > 0$, and β a 0-form, we can prove analogously that (1) holds, or alternatively, appeal to the fact that $\alpha \wedge \beta = \beta \wedge \alpha$ when one of these is 0-form, and apply the case just established in the preceding paragraph.

Finally, consider the case when both α and β are higher order forms. Again, in view of the distributivity of the wedge product and linearity of d, we need prove (1) only when both are simple forms. Therefore we take

$$\alpha = f\,dx_I \quad \text{and} \quad \beta = g\,dx_J,$$

where f and g are C^1 functions and dx_I, dx_J are basic forms with ascending indices I, J. Then

$$\alpha \wedge \beta = fg\,dx_{\langle I,J \rangle} = fg\,(dx_I \wedge dx_J).$$

Therefore,

$$d(\alpha \wedge \beta) = d(fg) \wedge dx_{\langle I,J \rangle} = ((df)g + f(dg)) \wedge dx_{\langle I,J \rangle}$$

$$= ((df)g) \wedge dx_{\langle I,J \rangle} + (f\,dg) \wedge dx_{\langle I,J \rangle}$$

Now, by Proposition 8-3.9,

$$((df)g) \wedge dx_{\langle I,J \rangle} = ((df)g) \wedge (dx_I \wedge dx_J) = df \wedge (dx_I \wedge g\,dx_J)$$

$$= (df \wedge dx_I) \wedge (g\,dx_J) = (d\alpha) \wedge \beta.$$

Also, by Proposition 8-3.8, $dg \wedge dx_I = (-1)^p dx_I \wedge dg$ and hence

$$(f\,dg) \wedge dx_{\langle I,J \rangle} = (f\,dg) \wedge (dx_I \wedge dx_J) = f(dg \wedge (dx_I \wedge dx_J))$$

$$= f((dg \wedge dx_I) \wedge dx_J) = f((-1)^p dx_I \wedge dg) \wedge dx_J)$$

$$= (-1)^p f \wedge dx_I \wedge (dg \wedge dx_J) = (-1)^p \alpha \wedge (d\beta).$$

Thus,

$$d(\alpha \wedge \beta) = (d\alpha) \wedge \beta + (-1)^p \alpha \wedge (d\beta). \qquad \square$$

Since d is defined only on forms of class C^1, then $d(d\omega)$ makes sense only for forms ω of class C^2.

8-4.4. Corollary. *Let ω be a C^2 form of order p in $U \subseteq \mathbb{R}^n$. Then $d(d\omega) = 0$.*

Proof. Observe that $d(dx_I) = d(1 dx_I) = 0$ because every partial derivative of the constant function 1 is 0.

Let $\omega = f \in C^2(U)$. Then

$$d(d\omega) = d\left(\sum_{j=1}^{n} (D_j f) dx_j\right)$$

$$= \sum_{j=1}^{n} d(D_j f) \wedge dx_j$$

$$= \sum_{i,j=1}^{n} (D_{i,j} f) dx_i \wedge dx_j.$$

$$= \sum_{1 \le i < j \le n} (D_i D_j f - D_j D_i f) dx_i \wedge dx_j$$

$$= 0,$$

because $D_{i,j} f = D_{j,i} f$ by Schwarz' Theorem 3-5.3 (f is of class C^2). Thus, $d(d\omega) = 0$ when ω is a C^2 form f of order 0. This may be written as $d^2 f = 0$.

If $\omega = f dx_I$, then by Def. 8-4.1, $d\omega = df \wedge dx_I$. Consequently, by (1) of Theorem 8-4.3, $d(d\omega) = d^2 f \wedge dx_I + (-1)^1 df \wedge d(dx_I) = 0$ since $d^2 f = 0$ and $d(dx_I) = 0$. This proves the result for simple forms. Additivity of d now implies it for general forms. $\qquad \square$

8-4.5. Remark. The 1-form in $\mathbb{R}^2$ given by $\omega = x\,dy$ is not the (exterior) derivative of any C^2 0-form. Indeed, if $x\,dy = df$, where f is a C^2 function, then $d^2 f = 0$, whereas $d(x\,dy) = dx \wedge dy \neq 0$. [See also Example 8-4.2(e).]

We have shown in the above paragraphs that the exterior derivative possesses the following properties:

(i) d is additive: $d(\alpha + \beta) = d\alpha + d\beta$;

(ii) $d(\alpha \wedge \beta) = (d\alpha) \wedge \beta + (-1)^p \alpha \wedge (d\beta)$, where p is the order of α;

(iii) $d(d\alpha) = 0$, where α is a C^2 form of any order.

These properties together with the fact that $df = \sum_{j=1}^{n} (D_j f) dx_j$ when f is a 0-form of class C^1 characterise d on the family of all C^1 forms. Applying the last mentioned fact to $f = x_i$, we obtain $df = \sum_{j=1}^{n} (D_j f) dx_j = 1 dx_i$, because all other terms in the summation have $D_j f = 0$. Thus $df = dx_i$. Since f is of class C^2, (iii) now leads to $0 = d(df) = d(dx_i)$. Hence, repeated application of (ii) yields

$$d(\, dx_{i_1} \wedge dx_{i_2} \wedge \cdots \wedge dx_{i_k}\,) = 0.$$

But by Remark 8-3.10 $dx_{i_1} \wedge dx_{i_2} \wedge \cdots \wedge dx_{i_k} = dx_I$, where $I = \langle i_1, i_2, \ldots, i_k \rangle$. Thus $d(dx_I) = 0$ for any k-index I. Now, if $\omega = \Sigma_I f_I dx_I$ is any k-form of class C^1, by (i) and (ii), we have $d\omega = \Sigma_I (df_I \wedge dx_I + f_I\, d(dx_I)) = \Sigma_I df_I \wedge dx_I$ since $d(dx_I) = 0$.

Problem Set 8-4

8-4.P1. Write the standard representation of the 1-form df, where $f(x_1, \ldots, x_n) = \sum_{i=1}^{n} x_i^2$ and show directly from the definition that its exterior derivative is 0.

8-4.P2. Let ω be a k-form in $U \subseteq \mathbb{R}^n$. If there is a $(k-1)$-form λ such that $d\lambda = \omega$, then ω is said to be **exact** in U. If $d\omega = 0$, then ω is said to be **closed** in U. Let $U = \mathbb{R}^2 \setminus \{(0,0)\}$ be the plane with origin removed. Show that the 1-form

$$\eta = \frac{x\,dy - y\,dx}{x^2 + y^2}$$

is closed but not exact.

8-5 Induced Mappings on Forms

Let us consider in detail what happens to functions (0-forms) under a mapping of their domain.

Suppose that U is an open subset of $\mathbb{R}^n$, V an open subset of $\mathbb{R}^m$. If a real valued function f is defined on V, then a map $T:U \to V$ naturally generates a related function

$$T^*f = f \circ T$$

on U. Thus, if T maps the open set U into the open set V, then the set of real valued functions on V is mapped (in the opposite direction) to the set of functions on U under the correspondence $f \to T^*f$ in the manner described above. If f and T are both continuous, then the same is true of T^*f.

In other words, we have shown that a mapping T^* of 0-forms on V into 0-forms on U arises naturally from a continuous map $T:U \to V$.

The definition below extends this idea to forms of higher order when T is a C^1 map from U to V.

We shall use x for points of U and y for points of V. Let $t_1, \ldots, t_m$ be the component functions of T. Note that for each i ($1 \le i \le m$), $dt_i = \sum_{j=1}^{n} (D_j t_i) dx_j$ is a 1-form in U. They will be mentioned in the forthcoming definition.

8-5.1. Definition. *With notation as above, the mapping T^*, which maps each k-form*

$$\omega = \Sigma_I b_I dy_I$$

in standard representation into

$$T^*\omega = \sum_I (b_I \circ T)\, dt_{i_1} \wedge dt_{i_2} \wedge \cdots \wedge dt_{i_k}\,,$$

where $\langle i_1, i_2, \ldots, i_k \rangle = I$, *is called the mapping (of forms)* **induced** *by the map* $T:U \to V$.

Example. Let $\omega = y_1\, dy_1 \wedge dy_2 + y_3{}^2\, dy_1 \wedge dy_3 + y_1 y_2\, dy_2 \wedge dy_3$ be a 2-form in $\mathbb{R}^3$, and let $T:\mathbb{R}^2 \to \mathbb{R}^3$ be defined by

$$T(x_1, x_2) = (x_1 + x_2, x_2{}^2, x_1 x_2).$$

The component functions of T are $t_1(x_1, x_2) = x_1 + x_2$, $t_2(x_1, x_2) = x_2{}^2$, $t_3(x_1, x_2) = x_1 x_2$. So, $dt_1 = dx_1 + dx_2$, $dt_2 = 2x_2\, dx_2$, $dt_3 = x_2\, dx_1 + x_1\, dx_2$. Therefore,

$$\begin{aligned}
T^*\omega &= (x_1 + x_2)(dx_1 + dx_2) \wedge (2x_2\, dx_2) + (x_1 x_2)^2 (dx_1 + dx_2) \wedge (x_2\, dx_1 + x_1\, dx_2) \\
&\quad + (x_1 + x_2)x_2{}^2 (2x_2\, dx_2) \wedge (x_2\, dx_1 + x_1\, dx_2) \\
&= \left(2x_2(x_1 + x_2) + x_1{}^3 x_2{}^2 - x_1{}^2 x_2{}^3 - 2x_2{}^4 (x_1 + x_2) \right) dx_1 \wedge dx_2.
\end{aligned}$$

8-5.2. Remarks. (a) If any of the terms in the standard representation $\omega = \sum_I b_I\, dy_I$ are the zero form of order k, then so is the corresponding term in $T^*\omega$. Therefore, in computing $T^*\omega$ from ω, we may omit the zero terms.

(b) Combined with this observation, the definition yields $T^*(dy_{i_\mu}) = dt_{i_\mu}$ for $\mu = 1, \ldots, m$.

(c) In the case of a simple k-form,

$$\omega = f_I\, dy_I = f_I\, dy_{i_1} \wedge dy_{i_2} \wedge \cdots \wedge dy_{i_k}\,,$$

the observation and Definition together yield

$$\begin{aligned}
T^*\omega &= (f_I \circ T)\, dt_{i_1} \wedge dt_{i_2} \wedge \cdots \wedge dt_{i_k} \\
&= (T^* f_I)\, T^*(dy_{i_1}) \wedge T^*(dy_{i_2}) \wedge \cdots \wedge T^*(dy_{i_k}).
\end{aligned}$$

The last equality holds irrespective of whether $\langle i_1, i_2, \ldots, i_k \rangle$ is an ascending index or not. Indeed, the same permutation is needed on each side to produce an ascending index.

(d) By the second part of Remark 8-2.12, $T^*(\omega_1 + \omega_2) = T^*(\omega_1) + T^*(\omega_2)$.

(e) It follows from (c) and (d) that, for an arbitrary k-form $\omega = \sum_I b_I\, dy_I$ in V, *whether in standard representation or not*, we have

$$T^*\omega = \sum_I (T^* b_I)\, T^*(dy_{i_1}) \wedge T^*(dy_{i_2}) \wedge \cdots \wedge T^*(dy_{i_k}).$$

(f) If ω_1 and ω_2 are forms of any orders, then

$$T^*(\omega_1 \wedge \omega_2) = T^*(\omega_1) \wedge T^*(\omega_2).$$

This is trivial if one is a 0-form, and follows from (d) and (e) for other cases.

(g) A k-surface Φ is, by definition, the restriction of a C^1 map defined on an open set containing the cuboid $[0,1]^k$. Therefore it makes sense to speak of Φ^*, thereby meaning Φ_1^*, where Φ_1 is any C^1 function of which Φ is the restriction. In case there are more than one such Φ_1, then whatever we say about Φ^* will be valid with any choice of Φ_1.

The following is a natural property of induced mappings on forms. The symbol ST denotes the composition $S \circ T$, and T^*S^* denotes $T^* \circ S^*$.

8-5.3. Proposition. *Let U, V, W be open sets in $\mathbb{R}^n, \mathbb{R}^m$ and $\mathbb{R}^r$, respectively. Suppose $T: U \to V$ and $S: V \to W$ are C^1 maps. If ω is a k-form in W, then $S^*\omega$ is a k-form in V, $T^*S^*\omega$ and $(ST)^*\omega$ are k-forms in U and*

$$T^*S^*\omega = (ST)^*\omega.$$

Proof. Only the equality is in need of proof. If ω is a 0-form, that is, a continuous function $f: W \to \mathbb{R}$, then

$$(ST)^*\omega = (S \circ T)^* f = f \circ (S \circ T) = (f \circ S) \circ T = (S^*f) \circ T = T^*(S^*f) = (T^*S^*)f.$$

Thus, the equality holds for 0-forms.

Let us denote points of U, V, W by x, y, z, respectively. Let $t_1, \ldots, t_m$ be the component functions of T and $s_1, \ldots, s_r$ be the component functions of S. We denote the component functions of ST by $u_1, \ldots, u_r$. If $\omega = dz_q$, then

$$S^*\omega = S^*(dz_q) = ds_q = \sum_{j=1}^{m}(D_j s_q)\, dy_j,$$

so that

$$T^*S^*\omega = T^*(S^*\omega) = \sum_{j=1}^{m}((D_j s_q) \circ T)\, dt_j = \sum_{j=1}^{m}((D_j s_q) \circ T)\sum_{i=1}^{n}(D_i t_j)\, dx_i$$

$$= \sum_{i=1}^{n}(D_i u_q)\, dx_i \qquad \text{by the chain rule}$$

$$= du_q = (ST)^*\omega.$$

If $\omega = \sum_{q=1}^{r} f_q\, dz_q$ is a 1-form in W, we have

$$(ST)^*\omega = \sum_{q=1}^{r}((ST)^*f_q)(ST)^*\, dz_q = \sum_{q=1}^{r}(T^*S^*f_q)(T^*S^*\, dz_q) = T^*(\sum_{q=1}^{r}(S^*f_q)(S^*\, dz_q))$$

$$= T^*(S^*\omega) = T^*S^*\omega.$$

The general case of the equality to be proved now follows from Remark 8-5.2(e).   □

The next proposition shows that exterior differentiation of forms and of induced forms have the expected relationship.

8-5.4. Proposition. *Let U be an open set in $\mathbb{R}^n$, V an open set in $\mathbb{R}^m$ and suppose $T: U \to V$ is a C^2 map. Then*

$$d(T^*\omega) = T^*(d\omega)$$

for any k-form ω in V of class C^1.

Proof. We use $y = (y_1,\ldots,y_m)$ for points of V and $x = (x_1,\ldots,x_n)$ for points of U. Let $t_1,\ldots,t_m$ be the component functions of T. If $\omega = f$ is a 0-form, then

$$T^*(d\omega) = T^*\left(\sum_{j=1}^{m}(D_j f)\,dy_j\right) = \sum_{j=1}^{m}T^*(D_j f)\,T^*(dy_j) = \sum_{j=1}^{m}((D_j f)\circ T)\,dt_j$$

$$= \sum_{j=1}^{m}((D_j f)\circ T)\sum_{i=1}^{n}(D_i t_j)\,dx_i = \sum_{i=1}^{n}\left(\sum_{j=1}^{m}((D_j f)\circ T)(D_i t_j)\right)dx_i$$

$$= \sum_{i=1}^{n}D_i(f\circ T)\,dx_i \qquad \text{by the chain rule}$$

$$= d(T^*f) = d(T^*\omega).$$

Thus, the result holds for 0-forms.

Now let $\omega = f\,dy_{i_1}\wedge dy_{i_2}\wedge\cdots\wedge dy_{i_k} = f\,dy_I$, where $I = \langle i_1,i_2,\ldots,i_k\rangle$ is an ascending k-index. Then by definition of T^* and Remark 8-5.2(f), we have

$$d(T^*\omega) = d(T^*f\,dt_{i_1}\wedge dt_{i_2}\wedge\cdots\wedge dt_{i_k})$$

$$= d(T^*f)\wedge dt_{i_1}\wedge dt_{i_2}\wedge\cdots\wedge dt_{i_k} + (-1)^0 T^*f\,d(dt_{i_1}\wedge dt_{i_2}\wedge\cdots\wedge dt_{i_k}),$$

where we have used (1) of Theorem 8-4.3 in the second step. Since the result has been shown above to hold for 0-forms, we have $d(T^*f) = T^*(df)$. Apply the identity (1) of Theorem 8-4.3 to $d(dt_{i_1}\wedge dt_{i_2}\wedge\cdots\wedge dt_{i_k})$ repeatedly $k-1$ times and use Corollary 8-4.4 k times, which we may, because T is of class C^2. Upon doing so, we find that $d(dt_{i_1}\wedge dt_{i_2}\wedge\cdots\wedge dt_{i_k}) = 0$. Therefore,

$$d(T^*\omega) = T^*(df)\wedge dt_{i_1}\wedge dt_{i_2}\wedge\cdots\wedge dt_{i_k} = T^*(df)\wedge T^*(dy_I)$$

$$= T^*(df\wedge dy_I) \qquad \text{by Remark 8-5.2(f)}$$

$$= T^*(d\omega).$$

Thus, the result holds for simple k-forms. Since d and T^* are both additive [see Theorem 8-4.3 and Remark 8-5.2(d)], it holds for all k-forms. $\square$

Problem Set 8-5

8-5.P1. Consider the mapping $T:\mathbb{R}\to\mathbb{R}^2$ defined by $T(x) = (x^2,x^3)$. If $\omega = y_1\,dy_2$ is a 1-form in $\mathbb{R}^2$, show that $T^*\omega = 3x^2\,dx$.

8-5.P2. Let $T:\mathbb{R}^2\to\mathbb{R}$ be defined by $T(x,y) = x - y$. Find $T^*(dx)$.

8-5.P3. Let $T:\mathbb{R}^2\to\mathbb{R}^2$ be given by $T(x_1,x_2) = (ax_1 + bx_2, cx_1 + ex_2)$ and let $\omega = dy_1\wedge dy_2$. Show that $T^*\omega = (ae - bc)\,dx_1\wedge dx_2$.

8-5.P4. Verify in $\mathbb{R}^n$ that $df_1\wedge\ldots\wedge df_n = \dfrac{\partial(f_1,\ldots,f_n)}{\partial(x_1,\ldots,x_n)}\,dx_1\wedge\ldots\wedge dx_n$ for any C^1 functions $f_1,\ldots,f_n$.

8-5.P5. In the notation used for Def. 8-5.1, show for any simple k-form $\omega = b_I \cdot dy_I$ in V with $k \leq n$ that

$$T^*\omega = (b_I \circ T) \sum_J \frac{\partial(t_{i_1}, \ldots, t_{i_k})}{\partial(x_{j_1}, \ldots, x_{j_k})}\, dx_J,$$

where $\langle i_1, \ldots, i_k \rangle = I$ and the summation extends over all ascending k-indices $J = \langle j_1, \ldots, j_k \rangle$ in $\{1, \ldots, n\}$. If $k > n$, then $T^*\omega = 0$. It is not assumed that I is ascending.

8-5.P6. Let ω be a k-form and Φ a k-surface in an open set $U \subseteq \mathbb{R}^m$. Let $\iota_k:[0,1]^k \to \mathbb{R}^k$ be the inclusion map, that is, $\iota_k(y) = y$. Show that $\int_\Phi \omega = \int_{\iota_k} \Phi^*\omega$.

8-5.P7. Let $T:\mathbb{R}^2 \to \mathbb{R}^3$ and $f:\mathbb{R}^3 \to \mathbb{R}^3$ be C^2 mappings and the component functions of f be f_1, f_2, f_3. Suppose ω is the 1-form defined by $\omega = f_1\, dx + f_2\, dy + f_3\, dz$. Prove that $T^*(d\omega) = \langle (\text{curl}\, f) \circ T, N \rangle$, where $N(u,v) \in \mathbb{R}^3$ has respective components

$$\frac{\partial(T_1, T_2)}{\partial(u,v)}, \quad \frac{\partial(T_2, T_3)}{\partial(u,v)}, \quad \frac{\partial(T_3, T_1)}{\partial(u,v)},$$

$\text{curl}\, f$ is as defined on p.296 and $\langle , \rangle$ denotes the inner product in $\mathbb{R}^3$.

8-6 Chains and Their Boundaries

The modern language of differential forms originated with É. Cartan but the general Stokes theorem was proposed by H. Poincaré as the formula:

$$\int_{\partial\Phi} \omega = \int_\Phi d\omega,$$

where Φ is a k-surface with 'boundary' $\partial\Phi$ in an open set $U \subseteq \mathbb{R}^n$, in which the $(k-1)$-form ω is defined.

George Stokes was the first to bring the classical result attributed to him into the public domain, but he did not claim credit, as he had come to know of it from Lord Kelvin.

The general Stokes theorem transforms an integral over a surface into another over the region enclosed by the surface, and includes the well-known theorems of Green, Gauss and Stokes, but with somewhat restrictive hypotheses.

First we verify the Stokes formula for the n-surface defined by the identity map of $[0,1]^n$ onto itself and then go to the general case with the help of induced forms.

The boundary of the 1-surface defined by the identity map of $[0,1]$ onto itself, intuitively speaking, is the 'sum' of two zero-dimensional surfaces that map onto $\{1\}$ and $\{0\}$, the latter taken with a negative sign, for reasons discussed in Section 8-1. For Green's theorem on $[0,1]^2$ in calculus, the boundary is supposed to be traversed 'anticlockwise', which means it is taken to consist of the four 1-surfaces given by

$$\Gamma_{10}(t) = (0,t), \quad \Gamma_{20}(t) = (t,0), \quad \Gamma_{11}(t) = (1,t) \quad \text{and} \quad \Gamma_{21}(t) = (t,1),$$

but Γ_{10} and Γ_{21} are to be traversed in the reverse direction in order that the anticlockwise orientation be maintained. Reversal of direction means that integrals computed using the above parametrisation are to be multiplied by -1. Equivalently, the parametrisation is to be reversed (replace t by $1-t$). Thus we may think of the anticlockwise boundary as $-\Gamma_{10} + \Gamma_{11} + \Gamma_{20} - \Gamma_{21}$, or

$$\sum_{i=1}^{2} (-1)^i (\Gamma_{i0} - \Gamma_{i1}).$$

To make this precise, we need to have a way of (i) forming sums of k-surfaces with 1 and -1 permitted as coefficients; (ii) setting up integrals of differential forms over such sums and (iii) generalising Γ_{i0} and Γ_{i1} to higher dimensions.

8-6.1. Definition. *Let U be an open subset of $\mathbb{R}^n$ and $\Phi_1, \Phi_2, \ldots, \Phi_r$ be k-surfaces. By a k-**chain** in U we mean a formal linear combination of k-surfaces*

$$c = \sum_{q=1}^{r} a_q \Phi_q,$$

where $\Phi_1, \Phi_2, \ldots, \Phi_r$ are k-surfaces and $a_1, a_2, \ldots, a_r$ are real numbers.

It may be emphasised that c is not a linear combination of the functions $\Phi_1, \Phi_2, \ldots, \Phi_r$ defined on $[0,1]^k$ but a 'formal' linear combination, meaning thereby a function defined on the set of all k-surfaces with respective values $a_1, a_2, \ldots, a_r$ on the surfaces $\Phi_1, \Phi_2, \ldots, \Phi_r$ and value 0 at every other k-surface.

The integral of a k-form ω over a chain $c = \sum_{q=1}^{r} a_q \Phi_q$ is defined by

$$\int_c \omega = \sum_{q=1}^{r} a_q \int_{\Phi_q} \omega.$$

We can add k-chains and multiply them by real constants as we do with any real valued functions defined on a set, which is the set of all k-surfaces in the present case. Thus, if $c = \sum_{q=1}^{r} a_q \Phi_q$ and $c' = \sum_{q'=1}^{r'} a_{q'} \Phi_{q'}$, then $c + c' = \sum_{q=1}^{r} a_q \Phi_q + \sum_{q'=1}^{r'} a_{q'} \Phi_{q'}$, where terms may be combined in the usual manner, and for any real number a, the chain ac is $\sum_{q=1}^{r} (aa_q) \Phi_q$. We give below some elementary properties of chains.

8-6.2. Proposition. *Let a be a real number and let ω, ω' be k-forms in an open set $U \subseteq \mathbb{R}^n$. Suppose c, c' are k-chains in U. Then*

$$\int_c (\omega + \omega') = \int_c \omega + \int_c \omega',$$

$$\int_{c + c'} \omega = \int_c \omega + \int_{c'} \omega,$$

$$\int_{ac} \omega = a \int_c \omega.$$

Proof. If $c = \sum_{q=1}^{r} a_q \Phi_q$, then

$$\int_c (\omega + \omega') = \sum_{q=1}^r a_q \int_{\Phi_q} (\omega + \omega') = \sum_{q=1}^r a_q \int_{\Phi_q} \omega + \sum_{q=1}^r a_q \int_{\Phi_q} \omega' = \int_c \omega + \int_c \omega'.$$

The remaining assertions can be proved analogously.   □

We proceed to define the boundary of a k-surface in $U \subseteq \mathbb{R}^n$. For each positive integer i, $1 < i \le k$, define mappings Γ_{i0} and Γ_{i1} as

$$\Gamma_{i0}(x_1, \ldots, x_{k-1}) = (x_1, \ldots, x_{i-1}, 0, x_i, \ldots, x_{k-1})$$

and

$$\Gamma_{i1}(x_1, \ldots, x_{k-1}) = (x_1, \ldots, x_{i-1}, 1, x_i, \ldots, x_{k-1}).$$

These functions map $[0,1]^{k-1}$ into various faces of $[0,1]^k$. For instance, if $k = 2$, then Γ_{21} takes $[0,1]$ onto the edge of $[0,1]^2$ between the vertex with coordinates $(0,1)$ and the vertex with coordinates $(1,1)$, whereas Γ_{20} takes $[0,1]$ onto the edge between the vertex $(0,0)$ and the vertex $(1,0)$. Actually, they map $\mathbb{R}^{k-1}$ into $\mathbb{R}^k$ and have derivatives of all orders. Therefore, if V is an open set in $\mathbb{R}^k$ containing $[0,1]^k$, then the inverse images $\Gamma_{i0}^{-1}(V)$ and $\Gamma_{i1}^{-1}(V)$ are open sets in $\mathbb{R}^{k-1}$ [see 2-6.P11], which Γ_{i0} and Γ_{i1} map into V. Consequently, if Φ is a k-surface, then the composed maps $\Phi \circ \Gamma_{i1}$ and $\Phi \circ \Gamma_{i0}$ are $(k-1)$-surfaces and hence the summation in Def. 8-6.4 below describes a $(k-1)$-chain. Besides, if Φ is of class C^2, then so are $\Phi \circ \Gamma_{i1}$ and $\Phi \circ \Gamma_{i0}$.

Strictly speaking, our notation for the maps Γ_{i0} and Γ_{i1} should indicate k as well, but we prefer not to complicate our symbols and instead take k as understood from the context. When we work with a composition such as $\Gamma_{i0} \circ \Gamma_{j0}$, it should be borne in mind that the value of k for Γ_{i0} is 1 higher than that for Γ_{j0}. So, the symbols Γ_{i0} and Γ_{j0} do not mean quite the same thing in the composition $\Gamma_{j0} \circ \Gamma_{i0}$ as they do in the composition $\Gamma_{i0} \circ \Gamma_{j0}$. This caveat applies to 8-6.P2–8-6.P4.

8-6.3. Remarks. (a) The component functions of Γ_{i0} and Γ_{i1} may be denoted by $(\Gamma_{i0})_j$ and $(\Gamma_{i1})_j$, $1 \le j \le k$. With this notation, we can describe Γ_{i0} by setting

$$(\Gamma_{i0})_j = \begin{cases} x_j & j < i \\ 0 & j = i \\ x_{j-1} & j > i \end{cases}$$

and analogously for Γ_{i1} with 1 replacing 0 when $j = i$.

(b) Since Γ_{i0} and Γ_{i1} map $[0,1]^{k-1}$ into $\mathbb{R}^k$, there are k Jacobians associated with each, depending on which $k-1$ component functions we are taking the Jacobian of:

$$\frac{\partial((\Gamma_{i0})_1, \ldots, (\Gamma_{i0})_{j-1}, (\Gamma_{i0})_{j+1}, \ldots, (\Gamma_{i0})_k)}{\partial(x_1, \ldots, x_{k-1})}, \qquad 1 \le j \le k,$$

and similarly for Γ_{i1}. However, since $(\Gamma_{i0})_i$ and $(\Gamma_{i1})_i$ are constant functions (0 and 1, respectively), the Jacobian vanishes unless $j = i$, in which case, the Jacobian is $\frac{\partial(x_1, \ldots x_{k-1})}{\partial(x_1, \ldots x_{k-1})} = 1$ everywhere.

8-6.4. Definition. *If* $\Phi:[0,1]^k \to U \subseteq \mathbb{R}^n$ *is a k-surface, $k > 1$, the* **boundary of Φ** *is defined to be the $(k-1)$-chain*

$$\partial\Phi = \sum_{i=1}^{k}(-1)^i(\Phi \circ \Gamma_{i0} - \Phi \circ \Gamma_{i1}).$$

and that of a 1-surface $\Phi:[0,1] \to U \subseteq \mathbb{R}^n$ *is the 0-chain* $\Phi(1) - \Phi(0)$. *Furthermore, the* **boundary of a k-chain** $c = \sum_{q=1}^{r} a_q \Phi_q$, *where $k > 0$, is defined to be*

$$\partial c = \sum_{q=1}^{r} a_q \partial\Phi_q.$$

One can informally think of $\partial\Phi$ as essentially the restriction of Φ to the faces of $[0,1]^k$, each of which is suitably 'reparametrised'.

If $k = 2$, then

$$\partial\Phi = \sum_{i=1}^{2}(-1)^i(\Phi \circ \Gamma_{i0} - \Phi \circ \Gamma_{i1}) = \Phi \circ \Gamma_{20} + \Phi \circ \Gamma_{11} - \Phi \circ \Gamma_{10} - \Phi \circ \Gamma_{21}.$$

The boundary being the sum of 1-surfaces with appropriate signs is a 1-chain.

8-6.5. Example. Let the 2-surface $\Phi:[0,1]^2 \to \mathbb{R}^2$ be given by

$$\Phi(r,\theta) = (r\cos(3\pi\theta), r\sin(3\pi\theta)).$$

Since $\Gamma_{10}(t) = (0,t)$, $\Gamma_{11}(t) = (1,t)$, $\Gamma_{20}(t) = (t,0)$, $\Gamma_{21}(t) = (t,1)$, we have

$\Phi \circ \Gamma_{10}(t) = (0,0)$; range is just the origin;

$\Phi \circ \Gamma_{11}(t) = (\cos(3\pi t), \sin(3\pi t))$; range is the circle of radius 1 about the origin, the subsets $[0,\frac{1}{3}]$ and $[\frac{2}{3},1]$ of $[0,1]$ both being mapped into the upper semicircle;

$\Phi \circ \Gamma_{20}(t) = (t,0)$; range is the segment between the origin and $(1,0)$;

$\Phi \circ \Gamma_{21}(t) = (-t,0)$; range is the segment between the origin and $(-1,0)$.

Note that the ranges of $\Phi \circ \Gamma_{10}$, $\Phi \circ \Gamma_{20}$ and $\Phi \circ \Gamma_{21}$ *contain interior points* of the range of Φ. Moreover, integrals over $\Phi \circ \Gamma_{20}$ and $\Phi \circ \Gamma_{21}$ need not cancel.

8-6.6. Remark. Consider the inclusion mapping $\iota_k:[0,1]^k \to \mathbb{R}^k$, that is, $\iota_k(y) = y$. Since it is the restriction of a C^2 map of $\mathbb{R}^k$ into itself, it provides a k-surface and the induced map ι_k^* is defined. Also, Γ_{i0} and Γ_{i1} can be considered as mapping $\mathbb{R}^{k-1}$ into $\mathbb{R}^k$, and moreover the compositions $\iota_k \circ \Gamma_{i0}$ and $\iota_k \circ \Gamma_{i1}$ reduce to Γ_{i0} and Γ_{i1}. Consequently,

$$\partial\iota_k = \sum_{i=1}^{k}(-1)^i(\Gamma_{i0} - \Gamma_{i1}).$$

Problem Set 8-6

8-6.P1. Show that $\Gamma_{i0}{}^*(dy_j) = \Gamma_{i1}{}^*(dy_j) = \begin{cases} dx_j & \text{if } j < i \\ 0 & \text{if } j = i \\ dx_{j-1} & \text{if } j > i. \end{cases}$

8-6.P2. If Φ is a k-surface, $k > 1$, prove that

$$\partial(\partial\Phi) = \sum_{i=1}^{k}(-1)^i \sum_{j=1}^{k-1}(-1)^j(\Phi\circ\Gamma_{i0}\circ\Gamma_{j0} - \Phi\circ\Gamma_{i0}\circ\Gamma_{j1} - \Phi\circ\Gamma_{i1}\circ\Gamma_{j0} + \Phi\circ\Gamma_{i1}\circ\Gamma_{j1}).$$

8-6.P3. Prove that, for any i,j, $1 \leq i \leq j \leq k-1$, we have

(i) $\Gamma_{i0}\circ\Gamma_{j0} = \Gamma_{j+1,0}\circ\Gamma_{i0}$,

(ii) $\Gamma_{i0}\circ\Gamma_{j1} = \Gamma_{j+1,1}\circ\Gamma_{i0}$,

(iii) $\Gamma_{i1}\circ\Gamma_{j0} = \Gamma_{j+1,0}\circ\Gamma_{i1}$,

(iv) $\Gamma_{i1}\circ\Gamma_{j1} = \Gamma_{j+1,1}\circ\Gamma_{i1}$.

8-6.P4. If Φ is a k-surface, $k > 0$, prove that

(i) $\sum_{i=1}^{k}\sum_{j=1}^{k-1}(-1)^{i+j}\Phi\circ\Gamma_{i0}\circ\Gamma_{j0} = 0$,

(ii) $\sum_{i=1}^{k}\sum_{j=1}^{k-1}(-1)^{i+j}(\Phi\circ\Gamma_{i0}\circ\Gamma_{j1} + \Phi\circ\Gamma_{i1}\circ\Gamma_{j1}) = 0$,

(iii) $\sum_{i=1}^{k}\sum_{j=1}^{k-1}(-1)^{i+j}\Phi\circ\Gamma_{i1}\circ\Gamma_{j1} = 0$.

8-6.P5. Let c be a k-chain, $k > 1$. Show that $\partial(\partial c) = 0$.

8-6.P6. State conditions under which the formula

$$\int_{\Phi}(f\,d\omega) = \int_{\partial\Phi}(f\omega) - \int_{\Phi}(df)\wedge\omega$$

holds, and show that it generalises the formula of integration by parts.

8-6.P7. For the 2-surfaces Φ of Example 8-2.2(e) and Example 8-2.2(h), describe the four maps $\Phi\circ\Gamma_{10}, \Phi\circ\Gamma_{11}, \Phi\circ\Gamma_{20}, \Phi\circ\Gamma_{21}$ on the domain $[0,1]$. Also, describe their ranges in the terminology of analytic geometry of $\mathbb{R}^2$. Which, if any, of the four maps has range contained in the the boundary of the range of Φ in the sense of Def. 2-4.12? (Answer this on the basis of a figure; precise proof not required.)

8-6.P8. Let ω be an n-form of class C^1 in an open set $U \subseteq \mathbb{R}^n$ and c an $(n+1)$-chain of class C^2. Show that $\int_{\partial c}\omega = 0$.

8-7 The General Stokes Theorem

We begin with the substitution form of the fundamental theorem of calculus in the language of differential forms.

8-7.1. Theorem. *Let F be a C^1 function on an open set $U \subseteq \mathbb{R}$ and Φ a 1-surface of class C^2 in U. Then $\int_\Phi dF = \int_{\partial\Phi} F$.*

Proof. $\int_\Phi dF = \int_{[0,1]} (F' \circ \Phi)\Phi' = \int_0^1 (F' \circ \Phi)(s)\Phi'(s)\,ds = F(\Phi(1)) - F(\Phi(0))$.

Also, $\partial\Phi = \Phi(1) - \Phi(0)$. (Here $\Phi(0)$ and $\Phi(1)$ denote 0-surfaces.) Consequently,

$$\int_{\partial\Phi} F = F(\Phi(1)) - F(\Phi(0)). \qquad \square$$

8-7.2. Example. Let $F = x$ be the 0-form in an open set U, where $[0,1] \subset U \subset \mathbb{R}^1$. Let $\Phi:[-1,1] \to U$ be given by $\Phi(x) = x^2$. Then

$$\int_\Phi dF = \int_{-1}^1 1 \cdot 2t\,dt = 0$$

and

$$\int_{\partial\Phi} F = F(\Phi(1)) - F(\Phi(-1)) = F(1) - F(1) = 0.$$

8-7.3. Theorem. Green's Theorem for Differential Forms. *Let U be an open subset of $\mathbb{R}^2$, ω a 1-form of class C^1 in U and c a 2-chain in U of class C^2. Then*

$$\int_{\partial c} \omega = \int_c d\omega.$$

Proof. We use y for points of $\mathbb{R}^2$ and x for points of $\mathbb{R}^1$. It is clear from the definitions of boundary and of integral over a chain that we need establish the equality only when c is a C^2 2-surface Φ.

Since ω is a 1-form, it can be represented as

$$\omega = f_2\,dy_1 + f_1\,dy_2.$$

Therefore,

$$d\omega = (D_1 f_1 - D_2 f_2)\,dy_1 \wedge dy_2,$$

where the negative sign comes from interchanging dy_2 with dy_1.

To begin with, suppose the 2-surface Φ is given by the inclusion map $\iota:[0,1]^2 \to \mathbb{R}^2$. Observe that $(D_i f_i) \circ \iota = (D_i f_i)$ for $i = 1,2$ and the Jacobian of ι is 1. Consequently,

$$\int_\iota d\omega = \int_\iota (D_1 f_1)\,dy_1 \wedge dy_2 - \int_\iota (D_2 f_2)\,dy_1 \wedge dy_2$$

$$= \int_{[0,1]^2} (D_1 f_1)\,dy_1\,dy_2 - \int_{[0,1]^2} (D_2 f_2)\,dy_1\,dy_2.$$

Using Fubini's theorem, we have

$$\int_\iota d\omega = \int_{[0,1]} dy_2 \int_{[0,1]} (D_1 f_1) \, dy_1 - \int_{[0,1]} dy_1 \int_{[0,1]} (D_2 f_2) \, dy_2.$$

Now,

$$\int_{[0,1]} (D_1 f_1) \, dy_1 = f_1(1, y_2) - f_1(0, y_2) = (f_1 \circ \Gamma_{11} - f_1 \circ \Gamma_{10})(y_2)$$

and

$$\int_{[0,1]} (D_2 f_2) \, dy_2 = f_2(y_1, 1) - f_2(y_1, 0) = (f_2 \circ \Gamma_{21} - f_2 \circ \Gamma_{20})(y_1).$$

Therefore, $\int_\iota d\omega$ is seen to be equal to

$$\int_{[0,1]} (f_1 \circ \Gamma_{11} - f_1 \circ \Gamma_{10})(y_2) \, dy_2 - \int_{[0,1]} (f_2 \circ \Gamma_{21} - f_2 \circ \Gamma_{20})(y_1) \, dy_1$$

$$= \int_{[0,1]} (f_1 \circ \Gamma_{11} - f_1 \circ \Gamma_{10})(x) \, dx - \int_{[0,1]} (f_2 \circ \Gamma_{21} - f_2 \circ \Gamma_{20})(x) \, dx \qquad (1)$$

by renaming the variables y_2 and y_1 as x.

We next compute the integral $\int_{\partial \iota} \omega$. Now, $\partial \iota = -\Gamma_{10} + \Gamma_{11} + \Gamma_{20} - \Gamma_{21}$; therefore,

$$\int_{\partial \iota} \omega = \int_{-\Gamma_{10}+\Gamma_{11}+\Gamma_{20}-\Gamma_{21}} \omega = -\int_{\Gamma_{10}} \omega + \int_{\Gamma_{11}} \omega + \int_{\Gamma_{20}} \omega - \int_{\Gamma_{21}} \omega. \qquad (2)$$

Now,

$$\int_{\Gamma_{i1}} \omega = \int_{\Gamma_{i1}} f_1 \, dy_2 + \int_{\Gamma_{i1}} f_2 \, dy_1$$

$$= \int_{[0,1]} f_1 \circ \Gamma_{i1} \frac{d\Gamma_{i1}}{dx} \, dx + \int_{[0,1]} f_2 \circ \Gamma_{i1} \frac{d\Gamma_{i1}}{dx} \, dx.$$

Together with Remark 8-6.3(b), this implies

$$\int_{\Gamma_{i1}} \omega = \int_{[0,1]} f_i \circ \Gamma_{i1} \, dx. \qquad (3)$$

Similarly,

$$\int_{\Gamma_{i0}} \omega = \int_{[0,1]} f_i \circ \Gamma_{i0} \, dx. \qquad (4)$$

Substituting (3) and (4) in (2) and using (1), we obtain the required result in the case when the 2-surface $c = \Phi$ is the inclusion map $\iota:[0,1]^2 \to \mathbb{R}^2$.

Suppose now that c is a 2-surface $\Phi:[0,1]^2 \to \mathbb{R}^2$. Then by the result of 8-5.P6,

$$\int_\Phi d\omega = \int_\iota \Phi^*(d\omega)$$

$$= \int_\iota d(\Phi^*\omega) \quad \text{by Proposition 8-5.4.}$$

Now, Φ is (can be extended to be) a C^2 map from an open subset of $\mathbb{R}^2$ to U and ω is a 1-form in U. Therefore $\Phi^*\omega$ is a 1-form in an open subset of $\mathbb{R}^2$. This permits us to use the special case that has already been proved. Together with Remark 8-6.6, this leads to

$$\int_\iota d(\Phi^*\omega) = \int_{\partial\iota} \Phi^*\omega = \int_{\sum_{i=1}^{2}(-1)^i(\Gamma_{i0}-\Gamma_{i1})} \Phi^*\omega = \sum_{i=1}^{2}(-1)^i(\int_{\Gamma_{i0}}\Phi^*\omega - \int_{\Gamma_{i1}}\Phi^*\omega)$$

$$= \sum_{i=1}^{2}(-1)^i\int_{\iota_{k-1}}(\Gamma_{i0}{}^*(\Phi^*\omega) - \Gamma_{i1}{}^*(\Phi^*\omega)) \quad \text{by 8-5.P6 again}$$

$$= \sum_{i=1}^{2}-1)^i\int_{\iota_{k-1}}((\Phi\circ\Gamma_{i0})^*(\omega) - (\Phi\circ\Gamma_{i1})^*(\omega)) \quad \text{by Proposition 8-5.3}$$

$$= \int_{\partial\Phi}\omega,$$

applying 8-5.P6 to each of the four integrals in the previous sum. As recorded at the beginning of this proof, it is sufficient to prove the theorem when the chain c is a 2-surface, as has now been done.   □

8-7.4. Example. Let Φ be the 2-surface of Example 8-3-2(h). Below we transform the integral $\int_{\partial\Phi} x\,dy$ via Theorem 8-7.3 and show that it does not yield the area (content in $\mathbb{R}^2$) of the range of Φ.

Theorem 8-7.3 transforms the integral into $\int_\Phi dx \wedge dy$. By definition, this means $\int_{[0,1]^2} J_\Phi\,dr\,d\theta$, where J_Φ means the Jacobian of Φ with respect to (r,θ). Computation leads to $J_\Phi = \frac{3\pi}{2}(3r-1)$. Therefore, the integral evaluates to $\frac{3\pi}{4}$, which differs from the area of the range of Φ, as the latter is π.

8-7.5. Stokes Theorem for Differential Forms. *Let U be an open subset of $\mathbb{R}^3$, ω be a 1-form in U of class C^1 and c a 2-chain in U of class C^2. Then*

$$\int_{\partial c}\omega = \int_c d\omega.$$

Proof. Let $\iota:[0,1]^2 \to \mathbb{R}^2$ be the inclusion map. Then by 8-5.P6 and Proposition 8-5.P4, we have

$$\int_\Phi d\omega = \int_\iota \Phi^*(d\omega) = \int_\iota d(\Phi^*\omega).$$

It follows from here by using Green's Theorem (Theorem 8-7.3) that

$$\int_\Phi d\omega = \int_{\partial\iota}\Phi^*\omega.$$

Now,

$$\int_{\partial\iota}\Phi^*\omega = \sum_{i=1}^{2}(-1)^i(\int_{\Gamma_{i0}}\Phi^*\omega - \int_{\Gamma_{i1}}\Phi^*\omega)$$

$$= \sum_{i=1}^{2}(-1)^i\int_\iota(\Gamma_{i0}{}^*(\Phi^*\omega) - \Gamma_{i1}{}^*(\Phi^*\omega)) \quad \text{by 8-5.P6 again}$$

$$= \sum_{i=1}^{2}(-1)^i\int_\iota((\Phi\circ\Gamma_{i0})^*(\omega) - (\Phi\circ\Gamma_{i1})^*(\omega)) \quad \text{by Proposition 8-5.3}$$

$$= \int_{\partial\Phi}\omega,$$

applying 8-5.P6 to each of the four integrals in the previous sum.   □

8-7.6. Example. Let Φ be the 2-surface described by the equations

$$x = \cos u, \quad y = (b + \sin u)\cos v, \quad z = (b + \sin u)\sin v,$$

where $0 \le u \le 2\pi$, $0 \le v \le 2\pi$ and $b \in \mathbb{R}$. (When $b > 1$, this parametrises a torus.)

Let $\omega = P\,dx + Q\,dy + R\,dz$ be a 1-form in $\mathbb{R}^3$ of class C^1. Then $d\omega =$

$\left(\dfrac{\partial R}{\partial y} - \dfrac{\partial Q}{\partial z} \right) dy \wedge dz + \left(\dfrac{\partial P}{\partial z} - \dfrac{\partial R}{\partial x} \right) dz \wedge dx + \left(\dfrac{\partial Q}{\partial x} - \dfrac{\partial P}{\partial y} \right) dx \wedge dy$. We shall prove

that $\int_{\partial\Phi} \omega = \int_{\Phi} d\omega$, where $\partial\Phi = \sum_{i=1}^{2} (-1)^i (\Phi \circ \Gamma_{i0} - \Phi \circ \Gamma_{i2\pi})$.

From the definition of integrals of forms, we get

$$\int_{\Phi} \left(\frac{\partial P}{\partial z} dz \wedge dx - \frac{\partial P}{\partial y} dx \wedge dy \right)$$

$$= \int_{[0,\,2\pi]^2} \left(\left(\frac{\partial P}{\partial z} \circ \Phi \right)(u,v) \frac{\partial(z,x)}{\partial(u,v)} - \left(\frac{\partial P}{\partial y} \circ \Phi \right)(u,v) \frac{\partial(x,y)}{\partial(u,v)} \right) du\,dv$$

$$= \int_{[0,\,2\pi]^2} \left(\left(\frac{\partial P}{\partial z} \circ \Phi \right)(u,v) \cos v - \left(\frac{\partial P}{\partial y} \circ \Phi \right)(u,v) \sin v \right) \sin u\,(b + \sin u)\,du\,dv . \quad (1)$$

Set $F(u,v) = P \circ \Phi$. Then

$$\frac{\partial F}{\partial v} = \left(\frac{\partial P}{\partial y} \circ \Phi \right)(u,v)\frac{\partial y}{\partial v} + \left(\frac{\partial P}{\partial z} \circ \Phi \right)(u,v)\frac{\partial z}{\partial v}$$

$$= -\sin v\,(b + \sin u)\left(\frac{\partial P}{\partial y} \circ \Phi \right)(u,v) + \cos v\,(b + \sin u)\left(\frac{\partial P}{\partial z} \circ \Phi \right)(u,v). \quad (2)$$

Substituting from (2) into (1), we find that the right side of (1) becomes

$$\int_{[0,\,2\pi]^2} \sin u \frac{\partial F}{\partial v}\,du\,dv = \int_{[0,\,2\pi]} \sin u\,(F(u,2\pi) - F(u,0))\,du = 0$$

since $F(u,2\pi) = (P \circ \Phi)(u,2\pi) = P(\cos u, b + \sin u, 0) = (P \circ \Phi)(u,0) = F(u,0)$. Therefore, it follows from (1) that

$$\int_{\Phi} \left(\frac{\partial P}{\partial z} dz \wedge dx - \frac{\partial P}{\partial y} dx \wedge dy \right) = 0. \quad (3)$$

On using the definition of integrals of forms, we get

$$\int_{\Phi} \left(\frac{\partial Q}{\partial x} dx \wedge dy - \frac{\partial Q}{\partial z} dy \wedge dz \right)$$

$$= \int_{[0,\,2\pi]^2} \left(\left(\frac{\partial Q}{\partial x} \circ \Phi \right)(u,v) \sin u \sin v - \left(\frac{\partial Q}{\partial z} \circ \Phi \right)(u,v) \cos u \right)(b + \sin u)\,du\,dv . \quad (4)$$

Observe that

$$\frac{\partial}{\partial u}(\sin v(b + \sin u)(Q\circ\Phi)(u,v))$$

$$= \sin v \cos u (Q\circ\Phi)(u,v) + \sin v(b + \sin u)\left\{(\frac{\partial Q}{\partial x}\circ\Phi)(u,v)(-\sin u)\right.$$

$$\left. + (\frac{\partial Q}{\partial y}\circ\Phi)(u,v)\cos u\cos v + (\frac{\partial Q}{\partial z}\circ\Phi)(u,v)\cos u\sin v\right\} \quad (5)$$

and

$$\frac{\partial}{\partial v}(\cos u\cos v\,(Q\circ\Phi)(u,v))$$

$$= (-\sin v)\cos u\,(Q\circ\Phi)(u,v) + \cos u\cos v\left\{(\frac{\partial Q}{\partial y}\circ\Phi)(u,v)(-\sin v)(b + \sin u)\right.$$

$$\left. + (\frac{\partial Q}{\partial z}\circ\Phi)(u,v)\cos v(b + \sin u)\right\}. \quad (6)$$

Substituting from (5) and (6) into the right side of (4), we get

$$-\int_{[0,2\pi]^2} \frac{\partial}{\partial u}(\sin v(b + \sin u)(Q\circ\Phi)(u,v)) + \frac{\partial}{\partial v}(\cos u\cos v\,(Q\circ\Phi)(u,v)). \quad (7)$$

Now,

$$\int_{[0,2\pi]} \frac{\partial}{\partial u}(\sin v(b + \sin u)(Q\circ\Phi)(u,v))\,du = 0$$

and

$$\int_{[0,2\pi]} \frac{\partial}{\partial v}(\cos u\cos v\,(Q\circ\Phi)(u,v))\,dv = 0.$$

Therefore

$$\int_\Phi\left(\frac{\partial Q}{\partial x}dx\wedge dy - \frac{\partial Q}{\partial z}dy\wedge dz\right) = 0.$$

Similarly,

$$\int_\Phi\left(\frac{\partial R}{\partial y}dy\wedge dz - \frac{\partial R}{\partial z}dz\wedge dx\right) = 0.$$

Together with (3), these two equalities show that $\int_\Phi d\omega = 0$. In order to prove the required equality, we must thus show that $\int_{\partial\Phi}\omega = 0$.
   Now,

$$\int_{\partial\Phi}P\,dx = -\int_{\Phi\circ\Gamma_{10}}P\,dx + \int_{\Phi\circ\Gamma_{12\pi}}P\,dx + \int_{\Phi\circ\Gamma_{20}}P\,dx - \int_{\Phi\circ\Gamma_{22\pi}}P\,dx. \quad (8)$$

Also, using the subscript 1 to denote first components, we have

$$\frac{d}{dt}((\Phi\circ\Gamma_{10})_1(t)) = \frac{\partial}{\partial t}\Phi_1(0,t) = \frac{\partial}{\partial t}\cos 0 = \frac{\partial}{\partial t}\cos 2\pi = \frac{\partial}{\partial t}\Phi_1(2\pi,t)$$

$$= \frac{\partial}{\partial t}((\Phi \circ \Gamma_{12\pi})_1(t)).$$

Therefore,

$$\int_{\Phi \circ \Gamma_{10}} P\,dx = \int_{[0,2\pi]} ((P \circ \Phi \circ \Gamma_{10})(t))\frac{\partial}{\partial t}((\Phi \circ \Gamma_{10})_1(t))\,dt = \int_{\Phi \circ \Gamma_{12\pi}} P\,dx.$$

Since $\frac{d}{dt}((\Phi \circ \Gamma_{20})_1(t)) = \frac{d}{dt}\,\Phi_1(t,0) = \frac{d}{dt}\,\Phi_1(t,2\pi) = \frac{d}{dt}((\Phi \circ \Gamma_{22\pi})_1(t)),$ therefore,

$$\int_{\Phi \circ \Gamma_{20}} P\,dx = \int_{[0,2\pi]} ((P \circ \Phi \circ \Gamma_{20})(t))\frac{d}{dt}((\Phi \circ \Gamma_{20})_1(t))\,dt = \int_{\Phi \circ \Gamma_{22\pi}} P\,dx.$$

Using (8), we therefore conclude that $\int_{\partial \Phi} P\,dx = 0$.

Similar arguments, all exploiting the fact that $\Phi(0,t) = \Phi(2\pi,t)$ and $\Phi(t,0) = \Phi(t,2\pi)$, lead to $\int_{\partial \Phi} Q\,dy = 0 = \int_{\partial \Phi} R\,dz$. Consequently, $\int_{\partial \Phi} \omega = 0$, as remained to be shown.

8-7.7. Theorem. Divergence Theorem for Differential Forms. *Let U be an open subset of $\mathbb{R}^3$, ω a 2-form of class C^1 in U and c a 3-chain in U of class C^2. Then*

$$\int_{\partial c} \omega = \int_c d\omega.$$

Proof. We use y for points of $\mathbb{R}^3$ and x for points of $\mathbb{R}^2$. It is clear from the definitions of boundary and of integral over a chain that we need establish the equality only when c is a C^2 3-surface Φ.

Since ω is a 2-form, it can be represented as

$$\omega = f_1\,dy_2 \wedge dy_3 + f_2\,dy_1 \wedge dy_3 + f_3\,dy_1 \wedge dy_2.$$

Therefore

$$d\omega = (D_1 f_1 - D_2 f_2 + D_3 f_3)\,dy_1 \wedge dy_2 \wedge dy_3,$$

where the negative sign comes from interchanging dy_2 with dy_1.

To begin with, suppose the 3-surface Φ is given by the inclusion map $\iota:[0,1]^3 \to \mathbb{R}^3$. Observe that $(D_i f_i)\circ\iota = (D_i f_i)$ for $i = 1,2,3$ and the Jacobian of ι is 1. Consequently,

$$\int_\iota d\omega = \int_\iota (D_1 f_1)\,dy_1 \wedge dy_2 \wedge dy_3 - \int_\iota (D_2 f_2)\,dy_1 \wedge dy_2 \wedge dy_3 + \int_\iota (D_3 f_3)\,dy_1 \wedge dy_2 \wedge dy_3$$

$$= \int_{[0,1]^3} (D_1 f_1)\,dy_1\,dy_2\,dy_3 - \int_{[0,1]^3} (D_2 f_2)\,dy_1\,dy_2\,dy_3 + \int_{[0,1]^3} (D_3 f_3)\,dy_1\,dy_2\,dy_3.$$

Using Fubini's theorem, we have

$$\int_\iota d\omega = \int_{[0,1]} dy_3 \int_{[0,1]} dy_2 \int_{[0,1]} (D_1 f_1)\,dy_1$$

$$- \int_{[0,1]} dy_1 \int_{[0,1]} dy_3 \int_{[0,1]} (D_2 f_2)\,dy_2$$

$$+ \int_{[0,1]} dy_1 \int_{[0,1]} dy_2 \int_{[0,1]} (D_3 f_3) \, dy_3 .$$

Now,

$$\int_{[0,1]} (D_1 f_1) \, dy_1 = f_1(1,y_2,y_3) - f_1(0,y_2,y_3) = (f_1 \circ \Gamma_{11} - f_1 \circ \Gamma_{10})(y_2,y_3),$$

$$\int_{[0,1]} (D_2 f_2) \, dy_2 = f_2(y_1,1,y_3) - f_2(y_1,0,y_3) = (f_2 \circ \Gamma_{21} - f_2 \circ \Gamma_{20})(y_1,y_3)$$

and

$$\int_{[0,1]} (D_3 f_3) \, dy_3 = f_2(y_1,y_2,1) - f_2(y_1,y_2,0) = (f_3 \circ \Gamma_{31} - f_3 \circ \Gamma_{30})(y_1,y_2).$$

Therefore, $\int_\iota d\omega$ is seen to be equal to

$$\int_{[0,1]} dy_2 \int_{[0,1]} dy_3 \, (f_1 \circ \Gamma_{11} - f_1 \circ \Gamma_{10})(y_2,y_3)$$

$$- \int_{[0,1]} dy_1 \int_{[0,1]} dy_3 \, (f_2 \circ \Gamma_{21} - f_2 \circ \Gamma_{20})(y_1,y_3)$$

$$+ \int_{[0,1]} dy_1 \int_{[0,1]} dy_2 \, (f_3 \circ \Gamma_{31} - f_3 \circ \Gamma_{30})(y_1,y_2)$$

$$= \int_{[0,1]} dx_1 \int_{[0,1]} dx_2 \, (f_1 \circ \Gamma_{11} - f_1 \circ \Gamma_{10})(x_1,x_2)$$

$$- \int_{[0,1]} dx_1 \int_{[0,1]} dx_2 \, (f_2 \circ \Gamma_{21} - f_2 \circ \Gamma_{20})(x_1,x_2)$$

$$+ \int_{[0,1]} dx_1 \int_{[0,1]} dx_2 \, (f_3 \circ \Gamma_{31} - f_3 \circ \Gamma_{30})(x_1,x_2) \tag{1}$$

by renaming each of the variable pairs $(y_2,y_3,)$, (y_1,y_3), (y_1,y_2) as (x_1,x_2).

We next compute the integral $\int_{\partial\iota} \omega$. Now,

$$\partial\iota = -\Gamma_{10} + \Gamma_{11} + \Gamma_{20} - \Gamma_{21} - \Gamma_{30} + \Gamma_{31} .$$

Therefore

$$\int_{\partial\iota} \omega = \int_{-\Gamma_{10}+\Gamma_{11}+\Gamma_{20}-\Gamma_{21}-\Gamma_{30}+\Gamma_{31}} \omega$$

$$= -\int_{\Gamma_{10}} \omega + \int_{\Gamma_{11}} \omega + \int_{\Gamma_{20}} \omega - \int_{\Gamma_{21}} \omega - \int_{\Gamma_{30}} \omega + \int_{\Gamma_{31}} \omega. \tag{2}$$

Now,

$$\int_{\Gamma_{il}} \omega = \int_{\Gamma_{il}} f_1 \, dy_2 \wedge dy_3 + \int_{\Gamma_{il}} f_2 \, dy_1 \wedge dy_3 + \int_{\Gamma_{il}} f_3 \, dy_1 \wedge dy_2$$

$$= \int_{[0,1]^2} f_1 \circ \Gamma_{il} \frac{\partial((\Gamma_{il})_2,(\Gamma_{il})_3)}{\partial(x_1,x_2)} \, dx_1 \, dx_2$$

$$+ \int_{[0,1]^2} f_2 \circ \Gamma_{il} \frac{\partial((\Gamma_{il})_1,(\Gamma_{il})_3)}{\partial(x_1,x_2)} \, dx_1 \, dx_2$$

$$+ \int_{[0,1]^2} f_3 \circ \Gamma_{il} \frac{\partial((\Gamma_{il})_1,(\Gamma_{il})_2)}{\partial(x_1,x_2)} \, dx_1 \, dx_2 .$$

Together with Remark 8-6.3(b), this implies

$$\int_{\Gamma_{i1}} \omega = \int_{[0,1]^2} f_i \circ \Gamma_{i1}\, dx_1\, dx_2 \ . \tag{3}$$

Similarly,

$$\int_{\Gamma_{i0}} \omega = \int_{[0,1]^2} f_i \circ \Gamma_{i0}\, dx_1\, dx_2 \ . \tag{4}$$

Substituting (3) and (4) in (2) and using (1), we obtain the required result in the case when the 3-surface $c = \Phi$ is the inclusion map $\iota{:}[0,1]^3 \to \mathbb{R}^3$.

Suppose now that c is a 3-surface $\Phi{:}[0,1]^3 \to \mathbb{R}^2$. Then by the result of 8-5.P6,

$$\int_\Phi d\omega = \int_\iota \Phi^*(d\omega) = \int_\iota d(\Phi^*\omega) \quad \text{by Proposition 8-5.4.}$$

Now, Φ is (can be extended to be) a C^2 map from an open subset of $\mathbb{R}^3$ to U and ω is a 2-form in U. Therefore, $\Phi^*\omega$ is a 2-form in an open subset of $\mathbb{R}^3$. This permits us to use the special case that has already been proved. Together with Remark 8-6.6, this leads to

$$\int_\iota d(\Phi^*\omega) = \int_{\partial\iota} \Phi^*\omega = \int_{\sum\limits_{i=1}^{3}(-1)^i(\Gamma_{i0}-\Gamma_{i1})} \Phi^*\omega \ = \sum_{i=1}^{3}(-1)^i\left(\int_{\Gamma_{i0}}\Phi^*\omega - \int_{\Gamma_{i1}}\Phi^*\omega\right)$$

$$= \sum_{i=1}^{3}(-1)^i\int_{\iota_{k-1}}(\Gamma_{i0}^*(\Phi^*\omega) - \Gamma_{i1}^*(\Phi^*\omega)) \quad \text{by 8-5.P6 again}$$

$$= \sum_{i=1}^{3}-1)^i\int_{\iota_{k-1}}((\Phi\circ\Gamma_{i0})^*(\omega) - (\Phi\circ\Gamma_{i1})^*(\omega)) \quad \text{by Proposition 8-5.3}$$

$$= \int_{\partial\Phi}\omega,$$

applying 8-5.P6 to each of the 6 integrals in the previous sum. As recorded at the beginning of this proof, it is sufficient to prove the theorem when the chain c is a 3-surface, as has now been done. $\square$

8-7.8. Example. The divergence theorem is used to evaluate

$$\int_{\partial\Phi} x\, dy \wedge dz + y\, dz \wedge dx + z\, dx \wedge dy \tag{1}$$

where Φ is described by

$$\Phi(t,u,v) = (t\cos u, (b + t\sin u)\cos v, (b + t\sin u)\sin v),$$

where $0 \le t \le a$, $0 \le u \le 2\pi$, $0 \le v \le 2\pi$ and $a, b \in \mathbb{R}$, and $0 < a < b$.

By the divergence theorem, the integral to be evaluated equals

$$\int_\Phi 3\, dx \wedge dy \wedge dz = 3\int_{[0,a]\times[0,2\pi]\times[0,2\pi]} J_\Phi\, dt\, du\, dv$$

$$= 3\int_{[0,a]\times[0,2\pi]\times[0,2\pi]} \det\begin{bmatrix} \cos u & -t\sin u & 0 \\ \sin u\cos v & t\cos u\cos v & -(b+t\sin u)\sin v \\ \sin u\sin v & t\cos u\sin v & (b+t\sin u)\cos v \end{bmatrix} dt\,du\,dv$$

$$= 3\int_{[0,a]\times[0,2\pi]\times[0,2\pi]} t(b+t\sin u)\,dt\,du\,dv = 6\pi\int_{[0,a]\times[0,2\pi]} t(b+t\sin u)\,dt\,du$$

$$= 6\pi^2 a^2 b.$$

We next evaluate (1) to verify the answer.

By definition,

$$d(x\,dy\wedge dz + y\,dz\wedge dx + z\,dx\wedge dy) = 3\,dx\wedge dy\wedge dz$$

and

$$\partial\Phi = \sum_{i=1}^{3}(-1)^i(\Phi\circ\Gamma_{i0} - \Phi\circ\Gamma_{i1}),$$

where Γ_{i0} and Γ_{i1} are understood as

$$\Gamma_{10}(u,v) = (0,u,v),\ \ \Gamma_{11}(u,v) = (a,u,v),\ \ 0\le u\le 2\pi,\ 0\le v\le 2\pi,$$

$$\Gamma_{20}(t,v) = (t,0,v),\ \ \Gamma_{21}(t,v) = (t,2\pi,v),\ \ 0\le t\le a,\ \ 0\le v\le 2\pi,$$

$$\Gamma_{30}(t,u) = (t,u,0),\ \ \Gamma_{31}(t,u) = (t,u,2\pi),\ \ 0\le t\le a,\ \ 0\le u\le 2\pi.$$

What we have to prove is the equality

$$\sum_{i=1}^{3}(-1)^i\int_{\Phi\circ\Gamma_{i0}} x\,dy\wedge dz + y\,dz\wedge dx + z\,dx\wedge dy$$

$$- \sum_{i=1}^{3}(-1)^i\int_{\Phi\circ\Gamma_{i1}} x\,dy\wedge dz + y\,dz\wedge dx + z\,dx\wedge dy$$

$$= \int_{\Phi} 3\,dx\wedge dy\wedge dz.$$

The six compositions $\Phi\circ\Gamma_{i0}$ and $\Phi\circ\Gamma_{i1}$ occurring in the boundary chain $\partial\Phi$ are

$$(\Phi\circ\Gamma_{10})(u,v) = \Phi(0,u,v) = (0,b\cos v,b\sin v),$$

$$(\Phi\circ\Gamma_{20})(t,v) = \Phi(t,0,v) = (t,b\cos v,b\sin v),$$

$$(\Phi\circ\Gamma_{30})(t,u) = \Phi(t,u,0) = (t\cos u,b+t\sin u,0),$$

$$(\Phi\circ\Gamma_{11})(u,v) = \Phi(a,u,v) = (a\cos u,(b+a\sin u)\cos v,(b+a\sin u)\sin v),$$

$$(\Phi\circ\Gamma_{21})(t,v) = \Phi(t,2\pi,v) = (t,b\cos v,b\sin v),$$

$$(\Phi\circ\Gamma_{31})(t,u) = \Phi(t,u,2\pi) = (t\cos u,b+t\sin u,0).$$

Using the above compositions, we first compute

$$\sum_{i=1}^{3}(-1)^i\int_{\Phi\circ\Gamma_{i0}} x\,dy\wedge dz - \sum_{i=1}^{3}(-1)^i\int_{\Phi\circ\Gamma_{i1}} x\,dy\wedge dz.$$

By Def. 8-2.4, we have

$$\int_{\Phi\circ\Gamma_{10}} x\, dy\wedge dz = \int_{[0,2\pi]^2} 0\, \frac{\partial(b\cos v, b\sin v)}{\partial(u,v)}\, du\, dv = 0. \tag{1}$$

Similarly,

$$\int_{\Phi\circ\Gamma_{20}} x\, dy\wedge dz = \int_{[0,a]\times[0,2\pi]} t\, \frac{\partial(b\cos v, b\sin v)}{\partial(t,v)}\, dt\, dv = 0 \tag{2}$$

because $\dfrac{\partial(b\cos v, b\sin v)}{\partial(t,v)} = 0.$

$$\int_{\Phi\circ\Gamma_{30}} x\, dy\wedge dz = \int_{[0,a]\times[0,2\pi]} t\cos u\, \frac{\partial(b+t\sin u, 0)}{\partial(t,u)}\, dt\, du = 0. \tag{3}$$

$$\int_{\Phi\circ\Gamma_{11}} x\, dy\wedge dz = \int_{[0,2\pi]^2} a\cos u\, \frac{\partial((b+a\sin u)\cos v, (b+a\sin u)\sin v)}{\partial(u,v)}\, du\, dv$$

$$= \int_{[0,2\pi]^2} (a\cos u)\cdot \det\begin{bmatrix} a\cos u\cos v & -(b+a\sin u)\sin v \\ a\cos u\sin v & (b+a\sin u)\cos v \end{bmatrix} du\, dv$$

$$= \int_{[0,2\pi]^2} (a^2\cos^2 u)(b+a\sin u)\, du\, dv = 2\pi^2 a^2 b. \tag{4}$$

$$\int_{\Phi\circ\Gamma_{21}} x\, dy\wedge dz = \int_{[0,a]\times[0,2\pi]} t\, \frac{\partial(b\cos v, b\sin v)}{\partial(t,v)}\, dt\, dv = 0 \tag{5}$$

because $\dfrac{\partial(b\cos v, b\sin v)}{\partial(t,v)} = 0.$

$$\int_{\Phi\circ\Gamma_{31}} x\, dy\wedge dz = \int_{[0,a]\times[0,2\pi]} (t\cos u)\, \frac{\partial(b+t\sin u, 0)}{\partial(t,u)}\, dt\, du = 0. \tag{6}$$

From (1)–(6), it follows that

$$\sum_{i=1}^{3} (-1)^i \int_{\Phi\circ\Gamma_{i0}} x\, dy\wedge dz - \sum_{i=1}^{3} (-1)^i \int_{\Phi\circ\Gamma_{i1}} x\, dy\wedge dz$$

$$= -(0 - 2\pi^2 a^2 b) + (0-0) - (0-0) = 2\pi^2 a^2 b. \tag{7}$$

Next we compute

$$\sum_{i=1}^{3} (-1)^i \int_{\Phi\circ\Gamma_{i0}} y\, dz\wedge dx - \sum_{i=1}^{3} (-1)^i \int_{\Phi\circ\Gamma_{i1}} y\, dz\wedge dx.$$

By Def. 8-2.4, we have

$$\int_{\Phi\circ\Gamma_{10}} y\, dz\wedge dx = \int_{[0,2\pi]^2} (b\cos v)\frac{\partial(b\sin v, 0)}{\partial(u,v)}\, du\, dv = 0. \tag{8}$$

Similarly,

$$\int_{\Phi\circ\Gamma_{20}} y\,dz\wedge dx = \int_{[0,a]\times[0,2\pi]} (b\cos v)\frac{\partial(b\sin v,t)}{\partial(t,v)}\,dt\,dv$$

$$= \int_{[0,a]\times[0,2\pi]} (b\cos v)\cdot\det\begin{bmatrix} 0 & b\cos v \\ 1 & 0 \end{bmatrix} dt\,dv$$

$$= \int_{[0,a]\times[0,2\pi]} (-b^2\cos^2 v)\,dt\,dv = -\pi ab^2. \tag{9}$$

$$\int_{\Phi\circ\Gamma_{30}} y\,dz\wedge dx = \int_{[0,a]\times[0,2\pi]} (b+t\sin u)\frac{\partial(0,t\cos u)}{\partial(u,v)}\,dt\,du = 0. \tag{10}$$

$$\int_{\Phi\circ\Gamma_{11}} y\,dz\wedge dx = \int_{[0,\,2\pi]^2} (b+a\sin u)\cos v\frac{\partial(0,t\cos u)}{\partial(t,u)}\,du\,dv$$

$$= \int_{[0,\,2\pi]^2} (b+a\sin u)\cos v\cdot\det\begin{bmatrix} a\cos u\sin v & (b+a\sin u)\cos v \\ -a\sin u & 0 \end{bmatrix} du\,dv$$

$$= \int_{[0,\,2\pi]^2} a(b+a\sin u)^2\sin u\cos^2 v\,du\,dv = 2\pi^2 a^2 b. \tag{11}$$

$$\int_{\Phi\circ\Gamma_{21}} y\,dz\wedge dx = \int_{[0,a]\times[0,2\pi]} (b\cos v)\frac{\partial(b\sin v,t)}{\partial(t,v)}\,dt\,dv = -\pi ab^2 \tag{12}$$

as in (2).

As in (3), we have

$$\int_{\Phi\circ\Gamma_{31}} y\,dz\wedge dx = \int_{[0,a]\times[0,2\pi]} (b+t\sin u)\frac{\partial(0,t\cos u)}{\partial(t,u)}\,dt\,du = 0. \tag{13}$$

From (8)–(13), it follows that

$$\sum_{i=1}^{3}(-1)^i\int_{\Phi\circ\Gamma_{i0}} y\,dz\wedge dx - \sum_{i=1}^{3}(-1)^i\int_{\Phi\circ\Gamma_{i1}} y\,dz\wedge dx$$

$$= -(0-2\pi^2 a^2 b)+(-\pi ab^2-(-\pi ab^2))-(0-0) = 2\pi^2 a^2 b. \tag{14}$$

Finally, we use the compositions to compute

$$\sum_{i=1}^{3}(-1)^i\int_{\Phi\circ\Gamma_{i0}} z\,dx\wedge dy - \sum_{i=1}^{3}(-1)^i\int_{\Phi\circ\Gamma_{i1}} z\,dx\wedge dy.$$

$$\int_{\Phi\circ\Gamma_{10}} z\,dx\wedge dy = \int_{[0,\,2\pi]^2} (b\sin v)\frac{\partial(0,b\cos v)}{\partial(u,v)}\,du\,dv = 0. \tag{15}$$

$$\int_{\Phi\circ\Gamma_{20}} z\,dx\wedge dy = \int_{[0,a]\times[0,2\pi]} (b\sin v)\,\frac{\partial(t,b\cos v)}{\partial(t,v)}\,dt\,dv$$

$$= \int_{[0,a]\times[0,2\pi]} (-b^2\sin^2 v)\,dt\,dv = \pi ab^2. \tag{16}$$

$$\int_{\Phi\circ\Gamma_{30}} z\,dx\wedge dy = 0. \tag{17}$$

$$\int_{\Phi\circ\Gamma_{11}} z\,dx\wedge dy = \int_{[0,\,2\pi]^2} (b + a\sin u)\sin v\frac{\partial(a\cos u,(b+a\sin u)\cos v)}{\partial(u,v)}\,du\,dv$$

$$= \int_{[0,\,2\pi]^2} (b + a\sin u)\sin v\cdot\det\begin{bmatrix} -a\sin u & 0 \\ a\cos u\cos v & -(b+a\sin u)\sin v \end{bmatrix}du\,dv$$

$$= \int_{[0,\,2\pi]^2} a(b + a\sin u)^2 \sin u\sin^2 v\,du\,dv = 2\pi^2 a^2 b. \tag{18}$$

$$\int_{\Phi\circ\Gamma_{21}} z\,dx\wedge dy = \int_{[0,a]\times[0,2\pi]} (b\sin v)\frac{\partial(t,b\cos v)}{\partial(t,v)}\,dt\,dv = \pi ab^2 \tag{19}$$

as in (16).

$$\int_{\Phi\circ\Gamma_{31}} z\,dx\wedge dy = 0. \tag{20}$$

From (15)–(20), it follows that

$$\sum_{i=1}^{3}(-1)^i\int_{\Phi\circ\Gamma_{i0}} z\,dx\wedge dy - \sum_{i=1}^{3}(-1)^i\int_{\Phi\circ\Gamma_{i1}} z\,dx\wedge dy$$

$$= -(0 - 2\pi^2 a^2 b) + (\pi ab^2 - (\pi ab^2)) - (0 - 0) = 2\pi^2 a^2 b. \tag{21}$$

From (7), (14) and (21), we find that

$$\sum_{i=1}^{3}(-1)^i\int_{\Phi\circ\Gamma_{i0}} x\,dy\wedge dz + y\,dz\wedge dx + z\,dx\wedge dy$$

$$- \sum_{i=1}^{3}(-1)^i\int_{\Phi\circ\Gamma_{i1}} x\,dy\wedge dz + y\,dz\wedge dx + z\,dx\wedge dy = 6\pi^2 a^2 b.$$

8-7.9. The General Stokes Theorem. *Let U be an open subset of $\mathbb{R}^n$, ω a $(k-1)$-form of class C^1 in U and c a k-chain of class C^2 in U. Then*

$$\int_{\partial c}\omega = \int_c d\omega.$$

Proof. We use y for points of $\mathbb{R}^n$ and x for points of $\mathbb{R}^{k-1}$.

It is clear from the definitions of boundary and of integral over a chain that we need establish the equality only when c is a C^2 k-surface.

The case when $n = k = 1$ easily handled. Indeed, ω is a 0-form, which is simply a C^1 function F on the open set $U \subseteq \mathbb{R}$, and the k-surface c is a 1-surface Φ, which is nothing but the restriction to $[0,1]$ of a C^2 function that maps an open set containing $[0,1]$ into U. Therefore, $\partial c = \partial \Phi = \Phi(1) - \Phi(0)$. (Here $\Phi(0)$ and $\Phi(1)$ denote 0-surfaces.) Consequently,

$$\int_{\partial c} \omega = F(\Phi(1)) - F(\Phi(0)).$$

Also,

$$\int_c d\omega = \int_{[0,1]} (F' \circ \Phi)\Phi' = \int_0^1 (F' \circ \Phi)(s)\Phi'(s)\,ds$$

and the required equality holds by virtue of the fundamental theorem of calculus. [See the substitution form of the FTC in Section 8-1; the latter does not assume either F' or Φ' to be continuous and is therefore more general than the present case under discussion.]

Now suppose that $n = k > 1$ and that the k-surface c is given by the inclusion map $\iota_k : [0,1]^k \to \mathbb{R}^k$, that is, $\iota_k(y) = y$.

Since $n = k > 1$, the $(k-1)$-form ω is an $(n-1)$-form and hence [see Remark 8-2.8] we can represent ω as

$$\omega = \sum_{i=1}^k f_i\,dy_1 \wedge \cdots \wedge dy_{i-1} \wedge dy_{i+1} \wedge \cdots \wedge dy_k.$$

Therefore,

$$d\omega = \sum_{i=1}^k df_i \wedge dy_1 \wedge \cdots \wedge dy_{i-1} \wedge dy_{i+1} \wedge \cdots \wedge dy_k$$
$$= \sum_{i=1}^k (-1)^{i-1}(D_i f_i)\,dy_1 \wedge \cdots \wedge dy_k,$$

where $(-1)^{i-1}$ comes from interchanging dy_i successively with $dy_1, \ldots, dy_{i-1}$. Recall that the chain c under consideration is the k-surface given by the inclusion map $\iota_k : [0,1]^k \to \mathbb{R}^k$. Moreover, $(D_i f_i) \circ \iota_k = (D_i f_i)$ and the Jacobian of ι_k is 1. Consequently,

$$\int_c d\omega = \sum_{i=1}^k (-1)^{i-1} \int_{\iota_k} (D_i f_i)\,dy_1 \wedge \cdots \wedge dy_k$$
$$= \sum_{i=1}^k (-1)^{i-1} \int_{[0,1]^k} (D_i f_i)\,dy_1 \cdots dy_k.$$

Using Fubini's theorem [see Remark 6-3.3], we have

$$\int_c d\omega = \sum_{i=1}^k (-1)^{i-1} \int_{[0,1]^{k-1}} dy_1 \cdots dy_{i-1}\,dy_{i+1} \cdots dy_k \int_{[0,1]} (D_i f_i)\,dy_i.$$

Now,

$$\int_{[0,1]} (D_i f_i)\,dy_i = f_i(y_1, \ldots, y_{i-1}, 1, y_{i+1}, \ldots, y_k) - f_i(y_1, \ldots, y_{i-1}, 0, y_{i+1}, \ldots, y_k)$$
$$= (f_i \circ \Gamma_{i1} - f_i \circ \Gamma_{i0})(y_1, \ldots, y_{i-1}, y_{i+1}, \ldots, y_k).$$

Therefore $\int_c d\omega$ is seen to be equal to

$$\sum_{i=1}^{k}(-1)^{i-1}\int_{[0,1]^{k-1}}(f_i\circ\Gamma_{i1}-f_i\circ\Gamma_{i0})(y_1,\ldots,y_{i-1},y_{i+1},\ldots,y_k)\,dy_1\cdots dy_{i-1}\,dy_{i+1}\cdots dy_k$$

$$=\sum_{i=1}^{k}(-1)^{i-1}\int_{[0,1]^{k-1}}(f_i\circ\Gamma_{i1}-f_i\circ\Gamma_{i0})\,dx_1\cdots dx_{k-1}, \tag{1}$$

by renaming the variables $y_1,\ldots,y_{i-1},y_{i+1},\ldots,y_k$ as $x_1,\ldots,x_{k-1}$. (One can consider the renaming as a transformation of variables with Jacobian 1.)

We next compute the integral $\int_{\partial c}\omega$. As noted in Remark 8-6.6 above, $\partial\iota_k$ $=\sum_{i=1}^{k}(-1)^i(\Gamma_{i0}-\Gamma_{i1})$; therefore

$$\int_{\partial c}\omega=\int_{\sum_{i=1}^{k}(-1)^i(\Gamma_{i0}-\Gamma_{i1})}\omega=\sum_{i=1}^{k}(-1)^i\int_{\Gamma_{i0}}\omega+\sum_{i=1}^{k}(-1)^{i+1}\int_{\Gamma_{i1}}\omega. \tag{2}$$

Now,

$$\int_{\Gamma_{i1}}\omega=\sum_{j=1}^{k}\int_{\Gamma_{i1}}f_j\,dy_1\wedge\cdots\wedge dy_{j-1}\wedge dy_{j+1}\wedge\cdots\wedge dy_k$$

$$=\sum_{j=1}^{k}\int_{[0,1]^{k-1}}f_j\circ\Gamma_{i1}\frac{\partial((\Gamma_{i1})_1,\ldots,(\Gamma_{i1})_{j-1},(\Gamma_{i1})_{j+1},\ldots,(\Gamma_{i1})_k)}{\partial(x_1,\ldots,x_{k-1})}\,dx_1\cdots dx_{k-1}.$$

Together with Remark 8-6.3(b), this implies

$$\int_{\Gamma_{i1}}\omega=\int_{[0,1]^{k-1}}f_i\circ\Gamma_{i1}\,dx_1\cdots dx_{k-1}. \tag{3}$$

Similarly,

$$\int_{\Gamma_{i0}}\omega=\int_{[0,1]^{k-1}}f_i\circ\Gamma_{i0}\,dx_1\cdots dx_{k-1}. \tag{4}$$

Substituting (3) and (4) in (2) and using (1), it follows that the theorem is valid for the special case when $n=k\geq 1$, ω is an $(n-1)$-form and $c=\iota_k=\iota_n$.

Suppose now that n may or may not be equal to k and that c is a k-surface $\Phi:[0,1]^k\to\mathbb{R}^n$. Then by the result of 8-5.P6,

$$\int_{\Phi}d\omega=\int_{\iota_k}\Phi^*(d\omega)=\int_{\iota_k}d(\Phi^*\omega)\quad\text{by Proposition 8-5.4.}$$

Now, Φ is (can be extended to be) a C^2 map from an open subset of $\mathbb{R}^k$ to U and ω is a $(k-1)$-form in U. Therefore $\Phi^*\omega$ is a $(k-1)$-form in an open subset of $\mathbb{R}^k$. This permits us to use the special case that has already been proved. Together with Remark 8-6.6, this leads to

$$\int_{\iota_k}d(\Phi^*\omega)=\int_{\partial\iota_k}\Phi^*\omega=\int_{\sum_{i=1}^{k}(-1)^i(\Gamma_{i0}-\Gamma_{i1})}\Phi^*\omega$$

$$=\sum_{i=1}^{k}(-1)^i(\int_{\Gamma_{i0}}\Phi^*\omega-\int_{\Gamma_{i1}}\Phi^*\omega)$$

$$=\sum_{i=1}^{k}(-1)^i\int_{\iota_{k-1}}(\Gamma_{i0}^*(\Phi^*\omega)-\Gamma_{i1}^*(\Phi^*\omega))\quad\text{by 8-5.P6 again}$$

$$= \sum_{i=1}^{k} (-1)^i \int_{\iota_{k-1}} ((\Phi \circ \Gamma_{i0})^*(\omega) - (\Phi \circ \Gamma_{i1})^*(\omega)) \quad \text{by Proposition 8-5.3}$$

$$= \int_{\partial \Phi} \omega,$$

applying 8-5.P6 to each of the $2k$ integrals in the previous sum. As recorded at the beginning of this proof, it is sufficient to prove the theorem when the chain c is a $(k-1)$-surface, as has now been done.   □

8-7.10. Example. We shall verify the equality in the general Stokes theorem when the differential form ω is

$$dx_2 \wedge dx_3 \wedge dx_4 + (x_1 + x_2) dx_1 \wedge dx_3 \wedge dx_4 + (x_2 - x_4) dx_1 \wedge dx_2 \wedge dx_4 + x_3 dx_1 \wedge dx_2 \wedge dx_3$$

in $\mathbb{R}^4$ and the chain c is the inclusion map $\iota : [0,1]^4 \to \mathbb{R}^4$.

From the given ω, we obtain

$$d\omega = (0 - 1 + 0 - 0) dx_1 \wedge dx_2 \wedge dx_3 \wedge dx_4 = -dx_1 \wedge dx_2 \wedge dx_3 \wedge dx_4.$$

Therefore,

$$\int_\iota d\omega = -\int_\iota dx_1 \wedge dx_2 \wedge dx_3 \wedge dx_4 = -\int_{[0,1]^4} dy_1 dy_2 dy_3 dy_4 = -1.$$

Now,

$$\Gamma_{10}(y_1, y_2, y_3) = (0, y_1, y_2, y_3) \text{ and } \Gamma_{11}(y_1, y_2, y_3) = (1, y_1, y_2, y_3).$$

So, $\int_{\Gamma_{10}} \omega = \int_{[0,1]^3} (1)(1) dy_1 dy_2 dy_3 = 1 = \int_{\Gamma_{11}} \omega$. Hence,

$$\int_{\Gamma_{10}} \omega - \int_{\Gamma_{11}} \omega = 0.$$

Also,

$$\Gamma_{20}(y_1, y_2, y_3) = (y_1, 0, y_2, y_3) \text{ and } \Gamma_{21}(y_1, y_2, y_3) = (y_1, 1, y_2, y_3).$$

So, $\int_{\Gamma_{20}} \omega = \int_{[0,1]^3} y_1 dy_1 dy_2 dy_3 = \frac{1}{2}$ and $\int_{\Gamma_{21}} \omega = \int_{[0,1]^3} (y_1 + 1) dy_1 dy_2 dy_3 = \frac{3}{2}$. Hence,

$$\int_{\Gamma_{10}} \omega - \int_{\Gamma_{11}} \omega = -1.$$

Next,

$$\Gamma_{30}(y_1, y_2, y_3) = (y_1, y_2, 0, y_3) \text{ and } \Gamma_{31}(y_1, y_2, y_3) = (y_1, y_2, 1, y_3).$$

So, $\int_{\Gamma_{30}} \omega = \int_{[0,1]^3} (y_1 - y_2) dy_1 dy_2 dy_3 = 0 = \int_{\Gamma_{31}} \omega$. Hence,

$$\int_{\Gamma_{30}} \omega - \int_{\Gamma_{31}} \omega = 0.$$

Finally,

$$\Gamma_{40}(y_1, y_2, y_3) = (y_1, y_2, y_3, 0) \text{ and } \Gamma_{41}(y_1, y_2, y_3) = (y_1, y_2, y_3, 1).$$

So, $\int_{\Gamma_{40}} \omega = \int_{[0,1]^3} y_3 dy_1 dy_2 dy_3 = \frac{1}{2} = \int_{\Gamma_{41}} \omega$. Hence,

$$\int_{\Gamma_{40}} \omega - \int_{\Gamma_{41}} \omega = 0.$$

Taking the alternating sum of $\int_{\Gamma_{i0}} \omega - \int_{\Gamma_{i1}} \omega$, $i = 1, 2, 3, 4$, we get $\int_{\partial \iota} \omega = -0 + (-1) - 0 + 0 = -1$, which is the same as the value obtained for $\int_\iota d\omega$.

Problem Set 8-7

8-7.P1. Use the divergence theorem to evaluate

$$\int_{\partial\Phi} F_1\, dy \wedge dz + F_2\, dz \wedge dx + F_3\, dx \wedge dy, \tag{1}$$

where

$$\Phi{:}[0,1] \times [0,2\pi] \times [0,1] \rightarrow \mathbb{R}^3$$

is given by

$$\Phi(r,\theta,z) = (r\cos\theta, r\sin\theta, z)$$

and

$$F(x,y,z) = (1 - (x^2 + y^2)^3, 1 - (x^2 + y^2)^3, x^2 z^2).$$

8-7.P2. Using Green's theorem, calculate the integral $\int_{\partial\iota} P\, dx + Q\, dy$, where $P = 5 - xy - y^2$ and $Q = 2xy - x^2$ and ι is the identity mapping of $[0,1]^2 = \{(x,y) : 0 \le x \le 1, 0 \le y \le 1\}$. Verify the answer by evaluating the integral directly.

8-7.P3. Use the Stokes theorem to evaluate the integral

$$\int_{\partial\Phi} y\, dx + z\, dy + x\, dz, \tag{1}$$

where $\Phi{:}(r,\theta) \rightarrow (r\cos\theta, r\sin\theta, \frac{r^2 \sin 2\theta}{2b})$, $0 \le r \le a$, $0 \le \theta \le 2\pi$ is a 2-surface in $\mathbb{R}^3$.

Verify your answer by actually evaluating the integral (1).

8-8 The Integral Formulas of Vector Analysis

We conclude this chapter with some discussion that helps exlain the connection between Stokes theorem (Theorem 8-7.5) and the divergence theorem (Theorem 8-7.7) of differential forms to the classical Stokes and divergence theorems of analysis by reducing integration of forms over 'parametrised' surfaces to integrals over domains in Euclidean spaces.

We begin with the following:

8-8.1. Definition. *Let U be an open set in $\mathbb{R}^n$. A **vector field** is a map $F{:}U \rightarrow \mathbb{R}^n$ of the open set U into $\mathbb{R}^n$. Since F associates a vector $F(v)$, $v \in U$, to each point of U, F is called a **vector field**. The map F is represented by coordinate functions, $F = (f_1, f_2, \ldots, f_n)$.*

Recall that F is continuous (respectively, differentiable) if each f_i is continuous (respectively, differentiable).

8-8.2. Example. Let U be the complement of the origin in $\mathbb{R}^2$. For $(x,y) \in U$, set

$$F(x,y) = \left(\frac{-x}{(x^2+y^2)^{\frac{1}{2}}}, \frac{y}{(x^2+y^2)^{\frac{1}{2}}} \right).$$

Then F Is a vector field which to each point $(x,y) \in U$ associates $\left(\frac{-x}{(x^2+y^2)^{1/2}}, \frac{y}{(x^2+y^2)^{1/2}} \right)$, having the same number of coordinates, namely, two, in this case.

Suppose f is a differentiable function on $U \subseteq \mathbb{R}$. Then $\mathbf{grad}f$ is the vector field

$$\text{grad}f = \left(\frac{\partial f}{\partial x_1}, \frac{\partial f}{\partial x_2}, \cdots, \frac{\partial f}{\partial x_n} \right). \tag{1}$$

With every vector field F defined in an open set $U \subseteq \mathbb{R}^n$ is associated a 1-form

$$\omega_F = f_1 dx_1 + f_2 dx_2 + \cdots + f_n dx_n. \tag{2}$$

We define the *divergence* of F to be the function $\mathbf{div}\,F{:}U{\to}\mathbb{R}$ given by

$$\text{div}\,F = \frac{\partial F_1}{\partial x_1} + \frac{\partial F_2}{\partial x_2} + \cdots + \frac{\partial F_n}{\partial x_n}. \tag{3}$$

The 1-form corresponding to the vector field $\text{grad}f$ is

$$\sum_{i=1}^{n} \frac{\partial f}{\partial x_i} dx_i. \tag{4}$$

The 1-form (4) is precisely the exterior differential of f.

The correspondence between forms and scalar and vector fields in $\mathbb{R}^3$ deserves special mention. The above said correspondence will be needed in the latter part of the chapter.

Let U be an open set in $\mathbb{R}^3$. A vector field on U is a continuous function $F{:}U{\to}\mathbb{R}^3$ with component functions F_1, F_2, F_3.

With every such F is associated a 1-form in U, namely, $\lambda_F = F_1 dx + F_2 dy + F_3 dz$ and a 2-form $\omega_F = F_1 dy \wedge dz + F_2 dz \wedge dx + F_3 dx \wedge dy$. Conversely, every 1-form λ in U is λ_F and every 2-form ω is ω_F for some vector field F on U.

Let f be a C^1 function defined on U. The gradient of f is the vector

$$\text{grad}\,f = \left(\frac{\partial f}{\partial x_1}, \frac{\partial f}{\partial x_2}, \frac{\partial f}{\partial x_3} \right).$$

Thus $\text{grad}\,f$ is a vector field on U. It is often denoted by ∇f.

Observe that $F = \operatorname{grad} f$ if and only if $\lambda_F = df$. Let $F = (F_1, F_2, F_3)$ be a C^1 vector field on U. Its *curl* is the vector field

$$\operatorname{curl} F = \nabla \times F = \left(\frac{\partial F_3}{\partial y} - \frac{\partial F_2}{\partial z}, \frac{\partial F_1}{\partial z} - \frac{\partial F_3}{\partial x}, \frac{\partial F_2}{\partial x} - \frac{\partial F_1}{\partial y} \right)$$

and its *divergence* is the function

$$\operatorname{div} F = \nabla \cdot F = \frac{\partial F_1}{\partial x} + \frac{\partial F_2}{\partial y} + \frac{\partial F_3}{\partial z}.$$

Since $\omega_F = F_1 dy \wedge dz + F_2 dz \wedge dx + F_3 dx \wedge dy$, when F is C^1, we have

$$d\omega_F = \left(\frac{\partial F_1}{\partial x} + \frac{\partial F_2}{\partial y} + \frac{\partial F_3}{\partial x} \right) dx \wedge dy \wedge dz = (\operatorname{div} F)\, dx \wedge dy \wedge dz$$

as all other terms vanish. Therefore, $\operatorname{div} F = 0$ if and only if $d\omega_F = 0$.

Furthermore, since

$$d(\lambda_F) = d(F_1 dx + F_2 dy + F_3 dz) = \left(\frac{\partial F_1}{\partial y} dy \wedge dx + \frac{\partial F_1}{\partial z} dz \wedge dx \right)$$

$$+ \left(\frac{\partial F_2}{\partial x} dx \wedge dy + \frac{\partial F_2}{\partial z} dz \wedge dy \right) + \left(\frac{\partial F_3}{\partial x} dx \wedge dz + \frac{\partial F_3}{\partial y} dy \wedge dz \right)$$

$$= \left(\frac{\partial F_3}{\partial y} - \frac{\partial F_2}{\partial z} \right) dy \wedge dz + \left(\frac{\partial F_1}{\partial z} - \frac{\partial F_3}{\partial x} \right) dz \wedge dx + \left(\frac{\partial F_2}{\partial x} - \frac{\partial F_1}{\partial y} \right) dx \wedge dy,$$

it follows that $\operatorname{curl} F = 0$ if and only if $d(\lambda_F) = 0$.

Example. Observe that $\operatorname{curl}(\operatorname{grad} f) = 0$, where $f : U \rightarrow \mathbb{R}$ is a C^2 function, and $\operatorname{div}(\operatorname{curl} F) = 0$ when F is a C^2 vector field.

If $F = \operatorname{grad} f$, then $\lambda_F = df$. Since $d(\lambda_F) = d(df) = 0$, it follows that $\operatorname{curl} F = 0$, that is, $\operatorname{curl}(\operatorname{grad} f) = 0$.

By definition of divergence and curl,

$$\operatorname{div}(\operatorname{curl} F) = \frac{\partial}{\partial x} \left(\frac{\partial F_3}{\partial y} - \frac{\partial F_2}{\partial z} \right) + \frac{\partial}{\partial y} \left(\frac{\partial F_1}{\partial z} - \frac{\partial F_3}{\partial x} \right) + \frac{\partial}{\partial z} \left(\frac{\partial F_2}{\partial x} - \frac{\partial F_1}{\partial y} \right) = 0,$$

since F is of class C^2 and Schwarz's theorem (Theorem 3-5.3) is therefore applicable.

Now, let $F = (f_1, f_2, \ldots, f_n)$ be a vector field in an open set $U \subseteq \mathbb{R}^n$. Consider the $(n-1)$-form corresponding to F:

$$\omega = \sum_{i=1}^{n} (-1)^{i-1} f_i\, dx_1 \wedge dx_2 \wedge \cdots \wedge dx_{i-1} \wedge dx_{i+1} \wedge \cdots \wedge dx_n. \tag{5}$$

Then $d\omega$ is the n-form

$$d\omega = \left(\sum_{i=1}^{n} \frac{\partial f_i}{\partial x_i} \right) dx_1 \wedge dx_2 \wedge \cdots \wedge dx_n$$

$$= (\operatorname{div} F) dx_1 \wedge dx_2 \wedge \cdots \wedge dx_n. \qquad (6)$$

See proof of Theorem 8-7.9 and (3) above.

8-8.3. Definition. *Let $I^n = [0,1]^n$ be the unit cuboid in $\mathbb{R}^n$ and let ι_n be the inclusion mapping with domain I^n into $\mathbb{R}^n$. Then ι_n is called a* **positively oriented surface** *and its boundary*

$$\partial \iota_n = \sum_{i=1}^{n} (-1)^i (\Gamma_{i0} - \Gamma_{i1}) \qquad (7)$$

is said to be the **positively oriented boundary**.

Now let Φ be an injective mapping of $[0,1]^n$ into $\mathbb{R}^n$ of class C^2 whose Jacobian is positive (at least in the interior of $[0,1]^n$). Let $\Omega = \Phi([0,1]^n)$. By Inverse Function Theorem 4-2.1, Ω is the closure of an open subset of $\mathbb{R}^n$. We define the positively oriented boundary of the set Ω to be the chain

$$\partial \Phi = \Phi(\partial [0,1]^n). \qquad (8)$$

We denote this $(n-1)$-chain by $\partial \Omega$.

8-8.4. Volume Element. In Examples 8-2.6(b), the value of the form $dx \wedge dy \wedge dz$ over the surface $\Phi : I^3 \to \mathbb{R}^3$ defined by

$$\Phi(r,\phi,\theta) = (ar\cos 2\pi\theta \sin \pi\phi, \ br\sin 2\pi\theta \sin \pi\phi, \ cr\cos \pi\phi), \qquad (r,\theta,\phi) \in [0,1]^3$$

turns out to be $\frac{4}{3} \pi abc$. This is the Jordan content of the ellipsoid $\frac{x^2}{a^2} + \frac{y^2}{b^2} + \frac{z^2}{c^2} \leq 1$.

Let $\Phi : I^n \to \mathbb{R}^n$ be an n-surface in $\mathbb{R}^n$. Assume that Φ is injective, continuously differentiable with positive Jacobian J_Φ. Let f be a continuous real-valued function on the range Ω of Φ. By Transformation Formula 7-4.4,

$$\int_{[0,1]^n} f(\Phi(u)) J_\Phi(u) \, du = \int_{\Phi([0,1]^n)} f(x) \, dx. \qquad (9)$$

It follows from the definition of the form $\omega = f dx_1 \wedge dx_2 \wedge \cdots \wedge dx_n$ and (9) above that

$$\int_{\Phi} \omega = \int_{[0,1]^n} f(\Phi(u)) J_\Phi(u) \, du = \int_{\Phi([0,1]^n)} f(x) \, dx. \qquad (10)$$

Consequently, when $f = 1$ everywhere, (10) becomes

$$\int_{\Phi} dx_1 \wedge dx_2 \wedge \cdots \wedge dx_n = \int_{[0,1]^n} dx. \qquad (11)$$

The above discussion leads to the following:

8-8.5. Definition. *The n-form*

$$dx_1 \wedge dx_2 \wedge \ldots \wedge dx_n \tag{12}$$

is called the **volume element** *in* $\mathbb{R}^n$. *It is often denoted by* dV_n (*the subscript is dropped when it is not necessary to specify the dimension.*)

Remark. Let $F = (f_1, f_2, \ldots, f_n)$ be a vector field in an open set $U \subseteq \mathbb{R}^n$. Define

$$\omega_F = f_1 \, dx_1 + f_2 \, dx_2 + \cdots + f_n \, dx_n$$

and let $\gamma: I \to \mathbb{R}$ be a 1-surface, $\gamma(u) = (\gamma_1(u), \gamma_2(u), \ldots, \gamma_n(u))$. Then the integral of ω_F can be written in the following way:

$$\int_\gamma \omega_F = \sum_{i=1}^{n} \int_0^1 f_i(\gamma(u)) \gamma'(u) \, du$$

$$= \int_0^1 F(\gamma(u)) \gamma'(u) \, du$$

$$= \int_0^1 F(\gamma(u)) \cdot t \, |\gamma'(u)| \, du, \tag{13}$$

where t denotes the unit vector in the direction of $\gamma'(u)$. We call $|\gamma'(u)| \, du$ the **element of arclength** along γ and denote it by the customary notation ds. The formula (13) can then be written in the form

$$\int_\gamma \omega_F = \int_\gamma (F \cdot t) \, ds. \tag{14}$$

The Surface Area. Let Φ be a 2-surface in an open set $U \subseteq \mathbb{R}^3$ of class C^1 with parameter domain $I^2 \subseteq \mathbb{R}^2$ given by

$$x_1 = \Phi_1(u,v), \quad x_2 = \Phi_2(u,v), \quad x_3 = \Phi_3(u,v), \quad (u,v) \in I^2. \tag{15}$$

Assume that Φ is an injective mapping of I^2 onto $\Phi(I^2)$. It is well known that the vector

$$\frac{\partial(\Phi_2, \Phi_3)}{\partial(u,v)} e_1 + \frac{\partial(\Phi_3, \Phi_1)}{\partial(u,v)} e_2 + \frac{\partial(\Phi_1, \Phi_2)}{\partial(u,v)} e_3 \tag{16}$$

represents a normal to the surface described by (15) and is denoted by $N(u,v)$. We denote by n the unit vector in the direction of N, that is,

$$n = \frac{N}{\|N\|}.$$

If Φ is a k-surface in $\mathbb{R}^k$ of class C^1 with parameter domain I^{k-1}, associate with $(u_1, u_2, \ldots, u_{k-1})$ the vector

$$N(u_1, u_2, \ldots, u_{k-1}) = \sum_{1 \leq i_1, i_2, \ldots, i_{k-1} \leq k} \frac{\partial(\Phi_{i_1}, \Phi_{i_2}, \ldots, \Phi_{i_{k-1}})}{\partial(u_1, u_2, \ldots, u_{k-1})} e_{i_k}, \tag{17}$$

where $e_1, e_2, \ldots, e_k$ denotes the standard basis in $\mathbb{R}^k$. The Jacobian in (17) corresponds to the equation

$$(x_1, x_2, \ldots, x_k) = \Phi(u_1, u_2, \ldots, u_{k-1}).$$

If f is a continuous function on $\Phi(I^{k-1})$, the area integral of f is defined to be

$$\int_\Phi f \, dA = \int_{I^{k-1}} f(\Phi(u_1, u_2, \ldots, u_{k-1})) \, |N(u_1, u_2, \ldots, u_{k-1})| \, du_1 \cdots du_{k-1}. \quad (18)$$

In particular, we obtain the area of Φ, namely,

$$A(\Phi) = \int_{I^{k-1}} \|N(u_1, u_2, \ldots, u_{k-1})\| \, du_1 \cdots du_{k-1}. \quad (19)$$

We next compute the area of the boundary $\sum_{i=1}^{k} (-1)^i (\Gamma_{i0} - \Gamma_{i1})$ of I^k using (19). Observe that this is a linear combination of $(k-1)$-surfaces in $\mathbb{R}^k$. They are, in fact, $2k$ in number. Fix $i = i_j$. Then the mapping $\Gamma_{i_j 0}$: $(u_1, u_2, \ldots, u_{k-1}) \to (u_1, u_2, \ldots, 0, \ldots, u_{k-1})$, where 0 occurs in the i_jth coordinate of the k-tuple on the right. A straightforward computation shows that the sum on the right side of (17) equals e_{i_j}. Hence the $(k-1)$-dimensional area of the i_jth face, using (19), equals 1. Hence the total area of the boundary is $2k$. Moreover, $N \cdot e_{i_p} = 0, p \neq j, p = 1, 2, \ldots, k$.

We next use the formula (19) to evaluate the surface area of of the sphere $\Phi: (u, v) \to (a\sin u \cos v, a\sin u \sin v, a\cos u)$, $a > 0$ and $0 \le u \le \pi$, $0 \le v \le 2\pi$. A straightforward computation shows

$$\frac{\partial(\Phi_1, \Phi_2)}{\partial(u, v)} = a^2 \sin u \cos u, \quad \frac{\partial(\Phi_2, \Phi_3)}{\partial(u, v)} = a^2 \sin^2 u \cos v, \quad \frac{\partial(\Phi_3, \Phi_1)}{\partial(u, v)} = a^2 \sin^2 u \sin v,$$

$$N = a^2 \sin^2 u \cos v \, e_1 + a^2 \sin^2 u \sin v \, e_2 + a^2 \sin u \cos u \, e_3,$$

$$\|N\| = a^2 |\sin u|,$$

$$A(\Phi) = a^2 \int_0^\pi \int_0^{2\pi} \sin u \, du \, dv = 4\pi a^2.$$

8-8.6. Stokes Theorem. *Let $F = (f_1, f_2, f_3)$ be a vector field of class C^1 defined in an open set $U \subseteq \mathbb{R}^3$ and Φ be an injective transformation of I^2 into U of class C^2 with positive Jacobian. Then*

$$\int_\Phi (\text{curl} \, F) \cdot n \, dA = \int_{\partial \Phi} (F \cdot t) \, ds.$$

Proof. Let $\omega = f_1 \, dx_1 + f_2 \, dx_2 + f_2 \, dx_3$ be the 1-form associated with the vector field F. Then

$$d\omega = \left(\frac{\partial f_3}{\partial x_2} - \frac{\partial f_2}{\partial x_3} \right) dx_2 \wedge dx_3 + \left(\frac{\partial f_1}{\partial x_3} - \frac{\partial f_3}{\partial x_1} \right) dx_3 \wedge dx_1 + \left(\frac{\partial f_2}{\partial x_1} - \frac{\partial f_1}{\partial x_2} \right) dx_1 \wedge dx_2,$$

$$\int_\Phi d\omega = \int_{I^2} \left[\left(\left(\frac{\partial f_3}{\partial x_2} - \frac{\partial f_2}{\partial x_3} \right) \circ \Phi \right) \frac{\partial(\Phi_2,\Phi_3)}{\partial(u,v)} + \left(\left(\frac{\partial f_1}{\partial x_3} - \frac{\partial f_3}{\partial x_1} \right) \circ \Phi \right) \frac{\partial(\Phi_3,\Phi_1)}{\partial(u,v)} \right.$$

$$\left. + \left(\left(\frac{\partial f_2}{\partial x_1} - \frac{\partial f_1}{\partial x_2} \right) \circ \Phi \right) \frac{\partial(\Phi_1,\Phi_2)}{\partial(u,v)} \right] dudv$$

$$= \int_{I^2} ((\operatorname{curl} F) \circ \Phi) \cdot N(u,v) \, dudv$$

$$= \int_{I^2} ((\operatorname{curl} F) \circ \Phi) \cdot n(u,v) \, \| N(u,v) \| \, dudv$$

$$= \int_{I^2} ((\operatorname{curl} F) \circ \Phi) \cdot n \, dA$$

$$= \int_\Phi (\operatorname{curl} F) \cdot n \, dA, \tag{20}$$

using (18). Also,

$$\int_{\partial\Phi} \omega = \int_{\partial\Phi} (F \cdot t) \, ds, \tag{21}$$

using (15).

The proof is completed on using Theorem 8-7.5 , (20) and (21).   □

The connection between the divergence theorem for differential forms (Theorem 8-7.7) and the classical form of the theorem is the following.

8-8.7. The Divergence Theorem. *Let* $F = (f_1, f_2, f_3)$ *be a vector field of class* C^1 *in an open set* $U \subseteq \mathbb{R}^3$ *and* Φ *be a 3-surface in U that is injective on* $[0,1]^3$ *and is of class* C^2 *with positive Jacobian. If* $\Omega = \Phi([0,1]^3)$ *and* $\partial\Omega = \Phi(\partial I_n)$, *then*

$$\int_\Omega \operatorname{div} F \, dV = \int_{\partial\Omega} F \cdot n \, dA.$$

Proof. Let

$$\omega = f_1 \, dx_2 \wedge dx_3 + f_2 \, dx_3 \wedge dx_1 + f_3 \, dx_1 \wedge dx_2$$

be the 2-form associated with the vector field F. Then

$$d\omega = \left(\frac{\partial f_1}{\partial x_1} + \frac{\partial f_2}{\partial x_2} + \frac{\partial f_3}{\partial x_3} \right) dx_1 \wedge dx_2 \wedge dx_3$$

$$= (\operatorname{div} F) \, dx_1 \wedge dx_2 \wedge dx_3 \, .$$

By Transformation Formula 7-4.4,

$$\int_\Phi d\omega = \int_\Phi \operatorname{div} F \, dx_1 \wedge dx_2 \wedge dx_3$$

$$= \int_{[0,1]^3} \mathrm{div}\,(F(\Phi(u)) J_\Phi(u)\,du_1\,du_2\,du_3$$

$$= \int_\Omega (\mathrm{div}\,F)\,dx_1\,dx_2\,dx_3, \tag{22}$$

using (9).

Also,

$$\int_{\partial\Phi} \omega = \int_{\Phi(\sum_{i=1}^3 (-1)^i (\Gamma_{i0} - \Gamma_{i1}))} \omega$$

$$= \sum_{i=1}^3 \int_{[0,1]^3} f_i(\Phi(u_1, u_2, u_3)\, \frac{\partial(\Phi_{i_1}, \Phi_{i_2})}{\partial(u_1, u_2)}\, du_1\,du_2\,du_3$$

$$= \int_{\partial\Omega} F{\cdot}n\,dA, \tag{23}$$

using (11).

In view of Theorem 8-7.7, (22) and (23), it follows that

$$\int_\Omega (\mathrm{div}\,F)\,dx_1\,dx_2\,dx_3 = \int_{\partial\Omega} F{\cdot}n\,dA. \qquad \square$$

The proof of the following generalisation to $\mathbb{R}^n$ of Theorem 8-8.7 is no different from that of the above said theorem; it is therefore not included.

8-8.8. The Divergence Theorem (Generalisation). *Let $F = (f_1, f_2, \dots, f_n)$ be a vector field of class C^1 in an open set $U \subseteq \mathbb{R}^n$ and Φ be an n-surface in U that is injective on I^n and is of class C^2 with positive Jacobian. If $\Omega = \Phi(I^n)$ and $\partial\Omega = \Phi(\partial\iota_n)$, then*

$$\int_\Omega \mathrm{div}\,F\,dV = \int_{\partial\Omega} F{\cdot}n\,dA.$$

9

Solutions

Problem Set 2-2

2-2.P1. (a) $\|-x\|_1 = \sum_{k=1}^{n} |-x_k| = \sum_{k=1}^{n} |x_k| = \|x\|_1 \geq 0$. Also, $\|x\|_1 = 0$ means $\sum_{k=1}^{n} |x_k| = 0$. But each term in the sum $\sum_{k=1}^{n} |x_k|$ is nonnegative. Therefore, $\sum_{k=1}^{n} |x_k| = 0$ if and only if each $x_k = 0$, or equivalently, $x = 0$.

(b) $\|\alpha x\|_1 = \sum_{k=1}^{n} |\alpha x_k| = |\alpha|(\sum_{k=1}^{n} |x_k|) = |\alpha|\,\|x\|_1$.

(d) $\|x + y\|_1 = \sum_{k=1}^{n} |x_k + y_k| \leq \sum_{k=1}^{n} |x_k| + \sum_{k=1}^{n} |y_k| = \|x\|_1 + \|y\|_1$.

(e) $\|x - z\|_1 = \|(x - y) + (y - z)\|_1 \leq \|x - y\|_1 + \|y - z\|_1$ by part (d).

(f) $\|x\|_1 = \|(x - y) + y\|_1 \leq \|x - y\|_1 + \|y\|_1$ by part (d). Therefore,

$$\|x\|_1 - \|y\|_1 \leq \|x - y\|_1.$$

By an analogous argument, $\|y\|_1 - \|x\|_1 \leq \|y - x\|_1$. But what has been proved in part (a) shows that $\|y - x\|_1 = \|x - y\|_1$. Therefore,

$$\|y\|_1 - \|x\|_1 \leq \|x - y\|_1.$$

The two inequalities displayed above together yield $|\|x\|_1 - \|y\|_1| \leq \|x - y\|_1$.

2-2.P2. Let y be the n-vector with every component equal to 1 and z be the one with $z_k = |x_k|$. Then $\|x\|_1 = \sum_{k=1}^{n} |x_k| = y \cdot z \leq \|y\|_2 \|z\|_2 = n^{1/2}\|x\|_2$.

2-2.P3. $\|x + y\|_2^2 = (x+y) \cdot (x+y) = \sum_{k=1}^{n} (x+y)^2 = \sum_{k=1}^{n} x_k^2 + \sum_{k=1}^{n} 2x_k y_k + \sum_{k=1}^{n} y_k^2 = \|x\|_2^2 + \|y\|_2^2$, because $\sum_{k=1}^{n} x_k y_k = x \cdot y$ is given to be 0.

2-2.P4. Write $a^2 bc$ as $a^{3/2}b^{1/2}a^{1/2}b^{1/2}c$ and so on. Let x denote the 4-vector $(a^{3/2}b^{1/2}, b^{3/2}c^{1/2}, c^{3/2}a^{1/2})$ and y denote the 4-vector $(a^{1/2}b^{1/2}c, b^{1/2}c^{1/2}a, c^{1/2}a^{1/2}b)$. Then $x \cdot y = abc(a + b + c)$ and $\|x\|\|y\| = (a^3 b + b^3 c + c^3 a)^{1/2}[abc(a + b + c)]^{1/2}$. The Cauchy–Schwarz inequality yields the required result.

2-2.P5. $|\|x_p\| - \|x\|| \leq \|x_p - x\|$.

2-2.P6. (a) The inequality is valid if $a_1 = a_2 = \cdots = a_n = 0$; so suppose at least one $a_i > 0$. Let $b_k = a_k/(\sum_{j=1}^{n} a_j^p)^{1/p}$ for $1 \leq k \leq n$. Observe that $0 \leq b_k \leq 1$, so that $b_k^q \leq$

S. Shirali, H.L. Vasudeva, *Multivariable Analysis*,
DOI 10.1007/978-0-85729-192-9_9, © Springer-Verlag London Limited 2011

b_k^p. Hence $(\sum_{k=1}^{n} a_k{}^q)/[(\sum_{j=1}^{n} a_j{}^p)^{1/p}]^q = \sum_{k=1}^{n} [a_k{}^q/[(\sum_{j=1}^{n} a_j{}^p)^{1/p}]^q] = \sum_{k=1}^{n} b_k{}^q \leq \sum_{k=1}^{n} b_k{}^p = \sum_{k=1}^{n} [a_k{}^p/[(\sum_{j=1}^{n} a_j{}^p)^{1/p}]^p] = 1$. So, $\sum_{k=1}^{n} a_k{}^q \leq [(\sum_{j=1}^{n} a_j{}^p)^{1/p}]^q$.

(b) $\max_{1 \leq j \leq n} |x_j| \leq (\sum_{j=1}^{n} |x_j|^p)^{1/p} \leq n^{1/p} \max_{1 \leq j \leq n} |x_j|$. Therefore $\|x\|_\infty \leq (\sum_{j=1}^{n} |x_j|^p)^{1/p} \leq n^{1/p} \|x\|_\infty$. But $\lim_{p \to \infty} n^{1/p} = 1$. Therefore $\lim_{p \to \infty} \|x\|_p$ exists and equals $\|x\|_\infty$.

Problem Set 2-3

2-3.P1. $A(x) = A(x \cdot 1) = xA(1) = ax$. So, $A(x + y) = a(x + y) = ax + ay = A(x) + A(y)$.

2-3.P2. $A(cx) = A(cx_1, cx_2) = ((c^3x_1{}^3 + c^3x_2{}^3)^{1/3}, 0) = (c(x_1{}^3 + x_2{}^3)^{1/3}, 0) = c((x_1{}^3 + x_2{}^3)^{1/3}, 0) = cA(x)$. But when $x = (1,0)$ and $y = (0,1)$, we have $A(x + y) = (2^{1/3}, 0)$, $A(x) = A(y) = (1,0)$.

2-3.P3. Straightforward computation.

2-3.P4. A straightforward computation confirms that $f_1(x,y)^2 + 4f_2(x,y)^2 = 1$. So the range is included in $\{(u,v) \in \mathbb{R}^2 : u^2 + 4v^2 = 1\}$, an ellipse. The reverse inclusion can be proved: Take any $(u,v) \in \mathbb{R}^2$ with $u^2 + 4v^2 = 1$. Then $-1 \leq u \leq 1$. If $u = 1$, then $v = 0$ and $(x,y) = (1,0)$ satisfies $f_1(x,y) = 1 = u$, $f_2(x,y) = 0 = v$, so that $f(x,y) = (u,v)$; if $u = -1$, then again $v = 0$ and $(x,y) = (0,1)$ satisfies $f(x,y) = (u,v)$. Suppose $-1 < u < 1$; then $v \neq 0$ and hence $v \gtrless 0$. Take $y = 1$ and $x = \pm\sqrt{[(1 + u)/(1 - u)]}$ according as $v \gtrless 0$. One can then verify that $f(x,y) = (u,v)$. Thus the range is precisely $\{(u,v) \in \mathbb{R}^2 : u^2 + 4v^2 = 1\}$.

2-3.P5. Since f' is continuous, it maintains the same sign in some open interval I containing x_0. Therefore f is injective on I and has an inverse g when restricted to I. Then ϕ is injective on $I \times \mathbb{R}$: Consider distinct (x_1, y_1), $(x_2, y_2) \in I \times \mathbb{R}$. If $x_1 \neq x_2$, then since both are in I and f is injective on I, we have $f(x_1) \neq f(x_2)$ and $\phi(x_1, y_1), \phi(x_2, y_2)$ differ in the first component. Suppose $x_1 = x_2$; then $y_1 \neq y_2$ while $x_1 f(x_1) = x_2 f(x_2)$, so that $-y_1 + x_1 f(x_1) \neq -y_2 + x_2 f(x_2)$, which means $\phi(x_1, y_1), \phi(x_2, y_2)$ differ in the second component. Thus ϕ is injective and hence invertible on $I \times \mathbb{R}$. If $x \in I$, we have $u = f(x) \Leftrightarrow x = g(u) \Rightarrow xf(x) = ug(u)$. Also, $v = -y + xf(x) \Leftrightarrow y = -v + xf(x) = -v + ug(u)$.

2-3.P6. $B(x_1 + x_2) = A(x_1 + x_2, 0) = A((x_1, 0) + (x_2, 0)) = A(x_1, 0) + A(x_2, 0) = B(x_1) + B(x_2)$ and $B(cx) = A(cx, 0) = A(c(x, 0)) = cA(x, 0) = cB(x)$. Similarly for C.

2-3.P7. $C((x_1,y_1) + (x_2,y_2)) = C(x_1 + x_2, y_1 + y_2) = A(x_1 + x_2) + B(y_1 + y_2) = A(x_1) + B(y_1) + A(x_2) + B(y_2) = C(x_1,y_1) + C(x_2,y_2); \quad C(c(x,y)) = C(cx,cy) = A(cx) + B(cy) = cA(x) + cB(y) = c(A(x) + B(y)) = cC(x,y).$

2-3.P8. Direct computation shows A to be linear. The required matrix is

$$\begin{bmatrix} 0 & 0 & 0 & 0 & 0 \\ 0 & 0 & 0 & 0 & 0 \\ 0 & 0 & 1 & 0 & 0 \\ 0 & 0 & 0 & 1 & 0 \\ 0 & 0 & 0 & 0 & 1 \end{bmatrix}.$$

2-3.P9. The function $f:\mathbb{R}^2 \to \mathbb{R}^2$ is $f(xu_1 + yu_2) = (e^x \cos y)v_1 + (e^x \sin y)v_2$. The single equation is $f(xu_1 + yu_2) = pv_1 + qv_2$. In coordinate language, $f(x,y) = (p,q)$, where $f(x,y)$ is defined as $(e^x \cos y, e^x \sin y)$.

2-3.P10. $x_2 + x_3 = 20$, $-3x_2 - 4x_3 = 2000$. (a) Yes, Yes. (b) $x_1 = -5\beta - 6\gamma + 9$.

2-3.P11. The three equations $x \cdot z_i = 0$ for $i = 1,2,3$ are $x_1 + 3x_2 + 2x_3 - x_4 = 0$, $3x_1 + 10x_2 + 4x_3 = 0$ and $4x_1 + 13x_2 + 7x_3 + 4x_4 = 0$. Eliminating x_1 from the second and third equations, we get $x_2 - 2x_3 + 3x_4 = 0$ and $x_2 - x_3 + 8x_4 = 0$. Eliminating x_2 from the last equation, we get $x_3 + 5x_4 = 0$. We can now choose $x_4 = 1$, which leads to $x_3 = -5$, $x_2 = -13$, $x_1 = 50$. So one possibility is $x = (50, -13, -5, 1)$.

2-3.P14. If $a = 0$, then the points (t,t,t) and $(t,-t,t)$ serve the purpose, provided that $0 < t < \frac{1}{3}$ and $3t^2 < \delta^2$. If $a = 1$, then $(1,t,t)$ and $(1,-t,-t)$ serve the purpose, provided that $0 < 2t^2 < \delta^2$. Suppose $0 \neq a \neq 1$. Choose t so that $0 < 2|t| < |a - 1|$ and $2t^2 < \delta^2$. Then the points (a,t,t) and $(a,t,-t)$ both lie in the δ-ball centred at $(a,0,0)$; also, $a + 2t - 1$ has the same sign as $a - 1$ and hence $F(a,t,t) = at^2(a + 2t - 1)$ and $F(a,t,-t) = -at^2(a - 1)$ have opposite signs.

2-3.P15. Choose t such that $0 < |t| < |b|,|c|$ and $3t^2 < \delta^2$. Then $(\pm t, b + t, c + t)$ both lie in the required δ-ball. Also, $|b + \frac{1}{2}t| \geq |b| - \frac{1}{2}|t| = |b| - |t| + \frac{1}{2}|t| > \frac{1}{2}|t|$ and hence $(b + \frac{1}{2}t)^2 > \frac{1}{4}t^2$, which implies $b(b + t) > 0$, i.e., $b + t$ has the same sign as b. Similarly, $c + t$ has the same sign as c. It follows that $(b + t)(c + t)$ has the same sign as bc. Therefore $F(t, b + t, c + t) = 3t^2(b + t)(c + t)$ has the same sign as bc, whereas $F(-t, b + t, c + t) = -t^2(b + t)(c + t)$ has sign opposite to that of bc.

2-3.P16. $\|A(1,0)\| = \|(1 + 0,0)\| = \|(1,0)\| = \sqrt{(1^2 + 0^2)} = 1$; $\|A(0,1)\| = \|(0 + 1,0)\| = 1$; $\|A(3/5,4/5)\| = \|(7/5,0)\| = 7/5$; $\|A(12/13,5/13)\| = \|(17/13,0)\| = 17/13$; and $\|A(1/\sqrt{2}, 1/\sqrt{2})\| = \|(1/\sqrt{2} + 1/\sqrt{2},0)\| = \sqrt{(2 + 0)} = \sqrt{2}$. So the largest one is $\|A(1/\sqrt{2},1/\sqrt{2})\|$. Lastly, $x_1^2 + x_2^2 \le 1 \Rightarrow (x_1 + x_2)^2 + 0^2 = x_1^2 + x_2^2 + 2x_1x_2 \le (x_1^2 + x_2^2) + (x_1^2 + x_2^2) \le 2$.

2-3.P17. 5, 2, 5/2, 50/13. (a) $\|A(x_1,x_2)\|^2 = (x_1 + x_2)^2 + (2x_1 - x_2)^2 = 5x_1^2 + 2x_2^2 - 2x_1x_2$. (b) Since $|2x_1x_2| \le x_1^2 + x_2^2$, it follows from (a) that $\|A(x_1,x_2)\|^2 \le 5x_1^2 + 2x_2^2 + x_1^2 + x_2^2 \le 6(x_1^2 + x_2^2) \le 6$ when $\|x\| \le 1$. (c) Using (b), we have sup $\{\|Ax\| : \|x\| \le 1\} \le \sqrt{6}$.

2-3.P18. $A(x_1,x_2)^2 = (x_1 + x_2)^2 + (x_1 - x_2)^2 = 2(x_1^2 + x_2^2)$. So sup $\{\|Ax\| : \|x\| \le 1\} \le \sqrt{2}$. When $x = (x_1,x_2) = (1/\sqrt{2},1/\sqrt{2})$, we have $\|x\| = 1$ and $\|Ax\| = \|(\sqrt{2},0)\| = \sqrt{2}$. Hence sup $\{\|Ax\| : \|x\| \le 1\} = \sqrt{2}$.

2-3.P19. When $x \ne 0$, the vector $v = (a/2\|x\|)x$ lies in the ball V and $Av = (a/2\|x\|)Ax$.

Problem Set 2-4

2-4.P1. Let A and B be closed. Then by definition, their complements A^c and B^c are open. It has been shown (in the paragraph following Def.2-4.2) that an intersection of two open sets is open. Therefore $A^c \cap B^c$ is open and hence its complement $(A^c \cap B^c)^c$ is closed. But $(A^c \cap B^c)^c = A \cup B$. A similar argument applies to the intersection of any family of closed sets, because it has been noted that a union of any family of open sets is open.

2-4.P2. Denote the ball by B. Let $e_1 = (1,0,0,\dots,0)$, so that $\|e_1\| = 1$, and consider $y = a + re_1$. Then $\|y - a\| = r$ and hence y does not belong to B. It is nevertheless a closure point of the ball. To see why, let B_1 be an open ball about y with radius α, say. Take $\beta = \min\{\frac{1}{2}\alpha, r\}$. Then the point $z = y - \beta e_1 = a + (r - \beta)e_1$ is in B because $\|z - a\| = \|(r - \beta)e_1\| = r - \beta < r$; it is also in B_1 because $\|z - y\| = \beta < \alpha$. Thus any ball B_1 about y contains a point of B, which means y is a closure point of B. But y does not belong to B, as noted earlier. Since we have found a closure point of B that does not belong to it, it follows that B is not closed.

2-4.P3. $\|x\| = \|(x - u) + u\| \le \|x - u\| + \|u\| \le M + \|u\|$.

2-4.P4. Since F is closed, its complement F^c is open. Since an intersection of two open sets is open, the set $U \cap F^c$ is open. But $U \cap F^c = U \setminus F$. The argument that $F \setminus U$ is closed is similar.

2-4.P5. Let $A \subseteq \mathbb{R}^n$. Since $A^\circ \subseteq A$ (as is immediate from Def.2-4.7) and is an open set by Proposition 2-4.9, it is a subset of the union of all open sets contained in A. For the reverse inclusion, consider any a belonging to the union in question. Then $a \in B$, where $B \subseteq A$ and is open. Since B is open, a is an interior point of it. Since $B \subseteq A$, it follows from Def.2-4.7 that a is an interior point of A as well, thereby proving the reverse inclusion.

As noted immediately after Def.2-4.10, it is trivial that $A \subseteq \overline{A}$; moreover, $\overline{A}$ is closed by Proposition 2-4.11. Therefore it contains the intersection of all closed sets containing A. For the reverse inclusion, consider any $a \in \overline{A}$ and any closed set $C \supseteq A$. By Def.2-4.10, any ball about a contains a point of A; but then that point must belong to C, because $C \supseteq A$. It follows that a is a closure point of C and hence belongs to C, considering that C is closed. Thus it has been shown that every point of $\overline{A}$ belongs to every closed set containing A, whereby the reverse inclusion has been found to hold.

2-4.P6. Suppose x is a closure point. For each p, the ball $B(x,\frac{1}{p})$ contains some $x_p \in X$. The sequence $\{x_p\}_{p\geq 1}$ then converges to x. Conversely, suppose a sequence $\{x_p\}_{p\geq 1}$ in X converges to x. Consider any ball $B(x,\varepsilon)$ about x. Some $p \in \mathbb{N}$ satisfies $\|x_p - x\| < \varepsilon$. For any such p, we have $x_p \in B(x,\varepsilon)$ as well as $x_p \in X$. This means $B(x,\varepsilon)$ contains a point of X. It follows that x is a closure point.

2-4.P7. Suppose $\{x_p\}_{p\geq 1}$ is a convergent sequence in $\{x \in \mathbb{R}^2 : \|x\| = 1\}$. That is, $\|x_p\| = 1$ for every p and $x_p \to x$ for some $x \in \mathbb{R}^n$. If we show that $\|x\| = 1$, it will follow by Proposition 2-4.5 that $\{x \in \mathbb{R}^2 : \|x\| = 1\}$ is closed. From the inequality $\|x\| = \|x_p - (x_p - x)\| \leq \|x_p\| + \|x_p - x\| = 1 + \|x_p - x\|$, we obtain $\|x\| \leq 1$ upon taking the limit as $p \to \infty$. Similarly, from the inequality $\|x\| = \|x_p - (x_p - x)\| \geq \|x_p\| - \|x_p - x\| = 1 - \|x_p - x\|$, we obtain $\|x\| \geq 1$ by taking the limit as $p \to \infty$. So, $\|x\| = 1$.

2-4.P8. Consider any ball B about $(-1,0)$ with radius ε. Let $0 < r < \varepsilon$. Then the point $(-1,r)$, which obviously belongs to E, also belongs to B by virtue of the inequality $\|(-1,r)-(-1,0)\| = \|(0,r)\| = r < \varepsilon$. On the other hand, the point $(-1-r,0)$ obviously does not belong to E but does belong to B by virtue of the inequality $\|(-1-r,0)-(-1,0)\| = \|(-r,0)\| = r < \varepsilon$. Thus any ball about $(-1,0)$ contains a point of E as well as a point of its complement. As noted immediately after Def.2-4.12, this means $(-1,0)$ belongs to the boundary of E.

To show that $(1,0)$ is an interior point of $F = \{(x_1,x_2) \in \mathbb{R}^2 : 0 \leq x_1 \leq 2\}$, let $\delta = 1$. Since $\|(x_1,x_2)-(1,0)\| = \|(x_1-1,x_2)\|$, then $\|(x_1,x_2)-(1,0)\| < \delta \Rightarrow |x_1-1| < 1$ as well as $|x_2| < 1 \Rightarrow 0 < x_1 < 2 \Rightarrow (x_1,x_2) \in F$. This means $\delta = 1$ has the property that the δ-ball about (x_1,x_2) is contained in F.

2-4.P9. It is immediate from the definitions of closure and boundary that $\overline{F} \supseteq F \cup \partial F$. Also, $F \supseteq F^\circ$ by definition of interior. So, $F \cup \partial F \supseteq F^\circ \cup \partial F$. It remains to note why $F^\circ \cup \partial F \supseteq \overline{F}$. But this too is trivial from the definition of boundary.

Problem Set 2-5

2-5.P1. Let $K_1, \ldots, K_m$ be compact subsets of $\mathbb{R}^n$ and $\{x_p\}_{p \geq 1}$ be a sequence in their union K. For some j, $1 \leq j \leq m$, there must be infinitely many p such that $x_p \in K_j$. Therefore some subsequence has every term in K_j. Since K_j is compact, it follows that the subsequence has a subsequence converging to a limit in K_j, which must then belong to the union K. But this subsequence is itself a subsequence of the sequence that we started with. The latter is therefore seen to have a subsequence converging to a limit in K.

2-5.P2. Consider a closed ball $B = \{x \in \mathbb{R}^n : \|x - u\| \leq r\}$. Let $a, b \in B$ and $0 \leq \lambda \leq 1$. Then $\|a - u\| \leq r$ and $\|b - u\| \leq r$ and therefore

$$\|\lambda a + (1-\lambda)b - u\| = \|\lambda a + (1-\lambda)b - \lambda u - (1-\lambda)u\| \leq \lambda\|a - u\| + (1-\lambda)\|b - u\|$$

$$\leq \lambda r + (1-\lambda)r = r,$$

so that $\lambda a + (1-\lambda)b \in B$. A similar argument works for an open ball $\{x \in \mathbb{R}^n : \|x - u\| < r\}$.

2-5.P3. Suppose $\{x_p\}_{p \geq 1}$ does not converge to x. Then there exists $\varepsilon > 0$ such that some subsequence $\{x_{p_q}\}_{q \geq 1}$ satisfies $\|x_{p_q} - x\| \geq \varepsilon$ for all q. Since K is compact, $\{x_{p_q}\}_{q \geq 1}$ has a convergent subsequence, which we shall denote by $\{\xi_r\}_{r \geq 1}$. Then $\{\xi_r\}_{r \geq 1}$ is a convergent subsequence of $\{x_p\}_{p \geq 1}$ and also satisfies the inequality $\|\xi_r - x\| \geq \varepsilon$ for all r. But according to the hypothesis, this subsequence must converge to x, in contradiction with the inequality. To show that compactness of K cannot be dropped, consider the sequence $1, 0, 2, 0, 3, 0, 4, \ldots$ in $\mathbb{R}$. Any convergent subsequence must converge to 0, but the sequence itself does not.

2-5.P4. The set is not bounded and therefore not compact in view of Theorem 2-5.7.

2-5.P5. Let $\mathcal{U}$ be an open cover of $X = \{x\} \cup \{x_p : p \in \mathbb{N}\}$. By definition of cover, some set $U \in \mathcal{U}$ has to contain x. Since U is open, there is some $\varepsilon > 0$ such that the ε-ball B about x is a subset of U. Now convergence of $\{x_p\}_{p \geq 1}$ to x means that there is some $N \in \mathbb{N}$ such that $p \geq N \Rightarrow x_p \in B \Rightarrow x_p \in U$. Once again by definition of cover, for each $p < N$, some set $U_p \in \mathcal{U}$ has to contain x_p. Hence the finite sub-family $\{U, U_1,\ldots,U_{p-1}\}$ of $\mathcal{U}$ covers X. We have shown that any open cover of X contains a finite subcover, which is what it means for X to be compact.

2-5.P6. Consider the disjoint open sets $U = \{(x,y) \in \mathbb{R}^2 : x < 0\}$ and $U = \{(x,y) \in \mathbb{R}^2 : x > 0\}$. Since $(x,y) \in A \Rightarrow x^2 \geq 1 + y^2 > 0$, we have $A \subseteq U \cup V$. Consequently, $A = (U \cap A) \cup (V \cap A)$ and $(U \cap A) \cap (V \cap A) \subseteq U \cap V = \varnothing$. It remains only to see why $U \cap A \neq \varnothing \neq V \cap A$. This follows from the fact that $(-2,0) \in U \cap A$ and $(2,0) \in V \cap A$.

Problem Set 2-6

2-6.P1. Denote $\lim_{t \to x} f(t)$ by L. We must show $\|L - y\| \leq K$. Consider any $\varepsilon > 0$. By definition of limit, $\exists \delta' > 0$ such that $\|f(t) - L\| < \varepsilon$ whenever $0 < \|t - x\| < \delta'$ and $t \in A$. Since x is a limit point of A, there exists $\xi \in A$ such that $0 < \|\xi - x\| <$ min $\{\delta, \delta'\}$. This ξ must satisfy $\|f(\xi) - L\| < \varepsilon$ as well as $\|f(\xi) - y\| \leq K$. It follows that $\|L - y\| \leq K$.

2-6.P2. (a) Denote $\lim_{t \to x} f(t)$ by L. For any $\varepsilon > 0$, there exists $\delta_1 > 0$ such that $0 < \|u - a\| < \delta_1$, $0 < \|v - b\| < \delta_1 \Rightarrow \|f(u,v) - L\| < \varepsilon/2$. It follows by 2-6.P1 that $0 < \|u - a\| < $ min $\{\delta_1, \mu\} \Rightarrow \| \lim_{v \to b} f(u,v) - L\| \leq \varepsilon/2 < \varepsilon$. Since $\varepsilon > 0$ is arbitrary, it follows that $\lim_{u \to a} [\lim_{v \to b} f(u,v)]$ exists and equals L. Proceeding in an analogous manner, one can show that if there exists a positive number ν such that $\lim_{u \to a} f(u,v)$ exists whenever $\|v - b\| < \nu$, then the $\lim_{v \to b} [\lim_{u \to a} f(u,v)]$ exists and is equal to $\lim_{t \to x} f(t)$.

(b) Here $n = m = k = 1$. Now, $\lim_{y \to 0} f(x,y) = 0$ whether x is 0 or not; for $x = 0$, this is trivial because $f(0,y) = 0$ for all y, and for $x \neq 0$, $\lim_{y \to 0} f(x,y) = \lim_{y \to 0} y \frac{\sin(xy)}{xy} = 0$. Thus we have $\lim_{y \to 0} f(x,y) = 0$ for $|x| < \mu$, where μ is any positive number. Consequently, $\lim_{x \to 0} \lim_{y \to 0} f(x,y) = 0$. Now, $\lim_{x \to 0} f(x,y) = \lim_{x \to 0} y \frac{\sin(xy)}{xy} = y$; it follows that

$\lim_{y\to 0}\ \lim_{x\to 0} f(x,y) = 0$ as well. Moreover, $\lim_{(x,y)\to(0,0)}\ \frac{\sin(xy)}{x} =$
$\lim_{(x,y)\to(0,0)}\ y\frac{\sin(xy)}{xy} = 0.$

2-6.P3. As seen in Example 2-6.6(a), the function satisfies the inequality $|f(x,y)|$ $= |x\sin(1/y)| \le |x|$, which shows that the limit as $(x,y)\to(0,0)$ is 0. The same inequality shows that $\lim_{x\to 0} f(x,y) = 0$ for $y \ne 0$, which implies $\lim_{y\to 0}[\lim_{x\to 0} f(x,y)] =$ 0. However, $\lim_{y\to 0} f(x,y)$ does not exist unless $x = 0$. So the domain of $\lim_{y\to 0} f(x,y)$ consists of a single point and consequently hs no limit point. Thus, there is no such thing as $\lim_{x\to 0}[\lim_{y\to 0} f(x,y)]$.

2-6.P4. $\lim_{x\to 0} f(x,0) = 1 = -\lim_{y\to 0} f(0,y).$

2-6.P5. We have $f(x+h,y+k) - f(x,y) = (x+h)^2(y+k) - x^2y$
$$= x^2k + 2xyh + 2xhk + yh^2 + h^2k.$$
Let $0 < \delta < 1$. Then
$\sqrt{(h^2+k^2)} < \delta \Rightarrow |h|,|k|,|hk|,h^2 < \delta < 1$
$\qquad\qquad \Rightarrow |x^2k + 2xyh + 2xhk + yh^2 + h^2k| < \delta(x^2 + 2|xy| + 2|x| + |y| + 1).$
To ensure that this is less than ε, choose
$$\delta < \min\{1, 1/(x^2 + 2|xy| + 2|x| + |y| + 1)\}.$$

2-6.P6. At any point $(x,y) \ne (0,0)$, the function is continuous because it is a quotient of continuous functions and the denominator does not vanish at the point in question. Continuity at $(0,0)$ follows from the inequality $|f(x,y)| \le |x|[x^2/(x^2+y^2)] \le |x|$.

2-6.P7. First assume the function $f: S\to\mathbb{R}^m$ is continuous at $x \in S$ and let $\{s_p\}$ be a sequence in S converging to x. We shall show that $\{f(s_p)\}$ converges to $f(x)$. Consider any $\varepsilon > 0$. By continuity of f at x, some $\delta > 0$ satisfies $\|s-x\| < \delta \Rightarrow$ $\|f(s)-f(x)\| < \varepsilon$. Since $\lim_{p\to\infty} s_p = x$, there exists some p_0 such that $p \ge p_0 \Rightarrow$ $\|s_p - x\| < \delta$. Therefore $p \ge p_0 \Rightarrow \|f(s_p)-f(x)\| < \varepsilon$. Such an integer p_0 has been shown to exist for an arbitrary $\varepsilon > 0$. Therefore $\lim_{p\to\infty} f(s_p) = f(x)$.

For the converse, assume that every sequence $\{s_p\}$ in S converging to x satisfies $\lim_{p\to\infty} f(s_p) = f(x)$. We shall argue why f must be continuous at x. Suppose, if possible, that f is not continuous at x. Then there must exist $\varepsilon > 0$ for which no δ > 0 can fulfill the requirement that $\|s-x\| < \delta \Rightarrow \|f(s)-f(x)\| < \varepsilon$. That is to say, whatever the number $\delta > 0$ may be, there exists some s such that $\|s-x\| < \delta$ but $\|f(s)-f(x)\| \ge \varepsilon$. For every $p \in \mathbb{N}$, the number $\delta = \frac{1}{p}$ is positive and so, there exists s_p such that $\|s_p - x\| < \frac{1}{p}$ but $\|f(s_p)-f(x)\| \ge \varepsilon$. These inequalities show that the sequence $\{s_p\}$ converges to x but the sequence $\{f(s_p)\}$ does not converge to $f(x)$. This contradicts the assumption that *every* sequence $\{s_p\}$ in S converging to x satisfies $\lim_{p\to\infty} f(s_p) = f(x)$. Therefore the supposition that f is not continuous at x must be incorrect.

2-6.P8. (a) Since $\big|\,|x_1| - |\xi_1|\,\big| \le |x_1 - \xi_1|$, we have

$$\|f(x_1,x_2) - f(\xi_1,\xi_2)\| = \|(|x_1|,x_2) - (|\xi_1|,\xi_2)\| \le \|(x_1,x_2) - (\xi_1,\xi_2)\|.$$

This shows that f is continuous (take $\delta = \varepsilon$).

(b) It was seen in Problem 2-4.P8 that $u = (-1,0)$ is a boundary point of E. Now, $f(u) = f(-1,0) = (1,0)$ and $f(E) = \{(x_1,x_2) \in \mathbb{R}^2 : 0 \le x_1 \le 2\}$. It was also seen in Problem 2-4.P8 that $(1,0)$ is an interior point of $\{(x_1,x_2) \in \mathbb{R}^2 : 0 \le x_1 \le 2\}$.

(c) Let $u \in U \cap \partial E$ and consider any ε-ball B_2 about $f(u)$. Then for some $\delta > 0$, the δ-ball B_1 about u has the property that $f(B_1) \subseteq B_2$. Since u is a boundary point of E, the ball B_1 contains a point $x \in E$ as well as a point $y \in E^c$. Therefore $f(x) \in f(B_1) \cap f(E)$. Since $f(B_1) \subseteq B_2$, it follows that B_2 contains the point $f(x) \in f(E)$. Similarly, it contains the point $f(y) \in f(E^c)$. Since f is injective, one can show that $f(y) \in f(E)^c$ as follows: If not, then $f(y) \in f(E)$ and hence $f(y) = f(z)$ for some $z \in E$. Since f is injective, this implies $y = z$. But this is a contradiction, because $z \in E$ and $y \in E^c$. Thus, B_2 has been shown to contain the point $f(x) \in f(E)$ as well as the point $f(y) \in f(E)^c$. Therefore, $f(u) \in \partial(f(E))$.

2-6.P9. Whichever norm we may use,

$$\|f(x) - f(\xi)\| \ge |f_j(x) - f_j(\xi)| \text{ for each } j$$

and, by Proposition 2-2.6,

$$\|f(x) - f(\xi)\| \le \|f(x) - f(\xi)\|_1 = \sum_{j=1}^{n} |f_j(x) - f_j(\xi)|.$$

2-6.P10. $\|f(x_1,\ldots,x_n) - f(\xi_1,\ldots,\xi_n)\| = |g(x_1) - g(\xi_1)|.$

2-6.P11. To prove the sufficiency part, suppose f is continuous and consider any $s \in f^{-1}(V)$, where $V \subseteq \mathbb{R}^m$ is open. Since $f(s)$ is an interior point of V, there exists $\varepsilon > 0$ such that the ε-ball about $f(s)$ is a subset of V, which means $\|y - f(s)\| < \varepsilon \Rightarrow y \in V$. By continuity at s, there exists some $\delta > 0$ such that, for any $x \in S$,

$$\|x - s\| < \delta \;\Rightarrow\; \|f(x) - f(s)\| < \varepsilon \;\Rightarrow\; f(x) \in V \;\Rightarrow\; x \in f^{-1}(V).$$

Thus, the intersection of S with the δ-ball about s is a subset of $f^{-1}(V)$. Now let U be the union of all such balls, one for each $s \in f^{-1}(V)$. Then U is open (being a union of open sets) and contains the centres of all the balls in the union, so that $f^{-1}(V) \subseteq U$. Moreover, $f^{-1}(V) \subseteq S$ by definition of inverse image. So,

$$f^{-1}(V) \subseteq S \cap U.$$

But since the intersection of S with any ball in the union is a subset of $f^{-1}(V)$, we also have

$$S \cap U \subseteq f^{-1}(V).$$

Thus $f^{-1}(V) = S \cap U$. This proves the sufficiency part.

To prove the necessity part, suppose that, for any open set $V \subseteq \mathbb{R}^m$, the inverse image $f^{-1}(V) = \{x \in S : f(x) \in V\}$ is the intersection of S with some open set $U \subseteq \mathbb{R}^n$. Let s be any point of S and $\varepsilon > 0$. Choose V to be the ε-ball about $f(s)$. Then

$$f^{-1}(V) = \{x \in S : f(x) \in V\} = \{x \in S : \|f(x) - f(s)\| < \varepsilon\} = S \cap U$$

for some open set $U \subseteq \mathbb{R}^n$. Since V has been chosen as the ε-ball about $f(s)$, we have $f(s) \in V$, i.e., $s \in f^{-1}(V)$, and therefore $s \in S \cap U$. But U is open and therefore some $\delta > 0$ satisfies

$$x \in S, \|x - s\| < \delta \Rightarrow x \in U.$$

Therefore

$$x \in S, \|x - s\| < \delta \Rightarrow x \in S \cap U = f^{-1}(V) \Rightarrow f(x) \in V.$$

But V was chosen to be the ε-ball about $f(s)$. Therefore

$$x \in S, \|x - s\| < \delta \Rightarrow \|f(x) - f(s)\| < \varepsilon.$$

Since such a positive δ has been shown to exist for every $\varepsilon > 0$, we see that f is continuous at s.

2-6.P12. Let $\{y_p\}$ be a sequence in $f(K)$. Then there is a sequence $\{x_p\}$ in K such that $y_p = f(x_p) \ \forall \ p \in \mathbb{N}$. Since K is compact, $\{x_p\}$ has a convergent subsequence $\{x_{p(k)}\}$ with limit $x \in K$, in view of Theorem 2-5.7 and Theorem 2-5.2. Since f is continuous at x, it follows that $\{f(x_{p(k)})\}$ converges to $f(x)$ [see 2-6.P7]. This means that the subsequence $\{y_{p(k)}\}$ of $\{y_p\}$ is convergent with limit in $f(K)$. Thus any sequence in $f(K)$ contains a convergent subsequence with limit in $f(K)$. By Theorem 2-5.7 and Theorem 2-5.2, $f(K)$ is compact.

2-6.P13. Taking $\varepsilon = 1$ in the definition of uniform continuity, we find that there exists some $\delta > 0$ such that

$$\|f(\xi) - f(x)\| < 1 \text{ as long as } \|\xi - x\| < \delta, \text{ where } \xi \in X, \ x \in X. \tag{1}$$

By Proposition 2-5.3, there exists a finite family $\mathcal{U}$ of δ-balls B, all centred at points of X, such that $\mathcal{U}$ covers X. Since $\mathcal{U}$ covers X, the family $\{f(B \cap X) : B \in \mathcal{U}\}$ covers $f(X)$. In view of (1), each set $f(B \cap X)$ is contained in a 2-ball in $\mathbb{R}^m$. Therefore $f(X)$ is covered by the finite family of sets $f(B \cap X)$, each of which is contained in a 2-ball. Using the facts that (a) any ball is bounded (b) the union of a finite family of bounded sets is bounded and (c) a subset of a bounded set is bounded, we conclude that $f(X)$ is bounded. This means that f is bounded.

The result of the next problem can be obtained as a consequence of the following: (1) A continuous image of a connected set is connected. (2) A connected subset of $\mathbb{R}$ is an interval. But here we ask for a direct proof. The result will be needed for 6-4.P7.

2-6.P14. Suppose not. Then there exist $\alpha, \beta, \gamma \in \mathbb{R}$ such that $\alpha, \beta \in f(X)$, $\alpha < \gamma < \beta$, but $\gamma \notin f(X)$. The sets $(-\infty, \gamma) \subseteq \mathbb{R}$ and $(\gamma, \infty) \subseteq \mathbb{R}$ are open. Since their union consists of all real numbers except $\gamma \notin f(X)$, we have

$$f^{-1}((-\infty, \gamma)) \cup f^{-1}((\gamma, \infty)) = X. \tag{1}$$

Also,

$$f^{-1}((-\infty, \gamma)) \cap f^{-1}((\gamma, \infty)) = \varnothing \tag{2}$$

because $(-\infty, \gamma) \cap (\gamma, \infty) = \varnothing$. By continuity of f and 2-6.P11, there exist open sets $U \subseteq \mathbb{R}^n$ and $V \subseteq \mathbb{R}^n$ such that

$$f^{-1}((-\infty, \gamma)) = X \cap U \text{ and } f^{-1}((\gamma, \infty)) = X \cap V.$$

From (1) and (2), it follows that

$$(X \cap U) \cup (X \cap V) = X \text{ and } (X \cap U) \cap (X \cap V) = \varnothing.$$

Since the intersections $(-\infty, \gamma) \cap f(X)$, $(\gamma, \infty) \cap f(X)$ are nonempty (considering they contain α and β, respectively), the same is true of their inverse images $f^{-1}((-\infty, \gamma) \cap f(X)) = f^{-1}((-\infty, \gamma)) = X \cap U$ and $f^{-1}((\gamma, \infty) \cap f(X)) = f^{-1}((\gamma, \infty)) = X \cap V$. The existence of such open sets U, V contradicts the connectedness of X.

2-6.P15. Since $I = [a, b] \times [c, d]$ is compact, f is uniformly continuous on I. Since $|g|$ is Riemann integrable, it has an upper bound M, say. Given any $\varepsilon > 0$, there exists $\delta > 0$ such that for every pair of points $z = (x, y)$, $z' = (x', y')$ in I such that $\|z - z'\| < \delta$, we have $|f(x, y) - f(x', y')| < \varepsilon/M$. Now consider any y, y' such that $|y - y'| < \delta$ and any $x \in [a, b]$. Setting $z = (x, y)$ and $z' = (x, y')$, we have $\|z - z'\| < \delta$ and therefore $|f(x, y) - f(x, y')| < \varepsilon/M$ and hence $|g(x)f(x, y) - g(x)f(x, y')| < \varepsilon$. But this means $|y - y'| < \delta \Rightarrow |g(x)f(x, y) - g(x)f(x, y')| < \varepsilon$ for all $x \in [a, b]$. Hence

$$|y - y'| < \delta \Rightarrow |F(y) - F(y')| \leq \int_a^b |g(x)f(x, y) - g(x)f(x, y')| \, dx \leq \varepsilon(b - a).$$

2-6.P16. Suppose, if possible, that λ_1 and λ_2 are both limits at x and that $\lambda_1 \neq \lambda_2$. Then $\varepsilon = \|\lambda_1 - \lambda_2\| > 0$. By Def. 2-6.5, there exist $\delta_1 > 0$ and $\delta_2 > 0$ such that

and

$$\|f(x + h) - \lambda_1\| < \frac{\varepsilon}{2} \quad \text{whenever} \quad 0 < \|h\| < \delta_1 \text{ and } x + h \in A$$

$$\|f(x + h) - \lambda_2\| < \frac{\varepsilon}{2} \quad \text{whenever} \quad 0 < \|h\| < \delta_2 \text{ and } x + h \in A.$$

Choose h such that $0 < \|h\| < \min\{\delta_1, \delta_2\}$ and $x + h \in A$. Such an h exists because x is a limit point of A. Then the inequalities $\|f(x + h) - \lambda_1\| < \frac{\varepsilon}{2}$ and $\|f(x + h) - \lambda_1\| < \frac{\varepsilon}{2}$ both hold and hence

$$\varepsilon = \|\lambda_1 - \lambda_2\| \le \|f(x+h) - \lambda_1\| + \|f(x+h) - \lambda_2\| < \tfrac{\varepsilon}{2} + \tfrac{\varepsilon}{2} = \varepsilon,$$

a contradiction.

2-6.P17. Let $x, y \in \mathbb{R}^n$ and and $\varepsilon > 0$. There exists $s \in S$ such that $\|x - s\| < d(x,S) + \varepsilon$. Now, $d(y,S) \le \|y - s\| \le \|y - x\| + \|x - s\| < \|y - x\| + d(x,S) + \varepsilon$. It follows that $d(y,S) - d(x,S) < \|y - x\| + \varepsilon$. By interchanging x and y in this argument, we get $d(x,S) - d(y,S) < \|y - x\| + \varepsilon$. Therefore $|d(x,S) - d(y,S)| < \|y - x\| + \varepsilon$. Since this holds, for every $\varepsilon > 0$, we conclude that $|d(x,S) - d(y,S)| \le \|y - x\|$. Uniform continuity is now immediate.

2-6.P18. (a) Straightforward computation.

(b) Observe that $\Phi_1^2 + \Phi_2^2 = (1 - \alpha(2u - 1)\sin \pi v)^2$ and $(1 - \alpha(2u - 1)\sin \pi v)$ is always positive in view of the hypothesis that $0 < \alpha < 1$.

Consider $(u,v) \ne (u',v')$ such that $\Phi(u,v) = \Phi(u',v')$.

Suppose $u = u'$. Then $v \ne v'$ and the equality $\Phi_3(u,v) = \Phi_3(u',v')$ leads to $\cos \pi v = \cos \pi v'$ unless $u = u' = \tfrac{1}{2}$. But πv and $\pi v'$ both belong to the interval $[0,\pi]$ and $\cos$ is injective on this interval. So we find that $u = u' = \tfrac{1}{2}$. Therefore $2u - 1 = 2u' - 1 = 0$. Consequently, $\Phi(u,v) = (\cos 2\pi v, \sin 2\pi v, 0)$ and similarly for $\Phi(u',v')$. It follows that $\cos 2\pi v = \cos 2\pi v'$ and $\sin 2\pi v = \sin 2\pi v'$. But $2\pi v$ and $2\pi v'$ both belong to the interval $[0, 2\pi]$ and are distinct. So, one among $2\pi v$ and $2\pi v'$ must be 0 and the other must be 2π. This means one among v, v' is 0 and the other is 1.

Now suppose that $u \ne u'$. First we rule out the possibility that either u or u' is $\tfrac{1}{2}$. If $u = \tfrac{1}{2}$, then the observation recorded at the beginning and the fact that

$$\Phi_1(u,v)^2 + \Phi_2(u,v)^2 = \Phi_1(u',v')^2 + \Phi_2(u',v')^2$$

together lead to

$$(1 - \alpha(2(\tfrac{1}{2}) - 1)\sin \pi v) = (1 - \alpha(2u' - 1)\sin \pi v'),$$

which is to say,

$$(1 - \alpha(2u' - 1)\sin \pi v') = 1, \text{ or } \alpha(2u' - 1)\sin \pi v' = 0.$$

But also

$$\alpha(2u' - 1)\cos \pi v' = \Phi_3(u',v') = \Phi_3(u,v) = \alpha(2(\tfrac{1}{2}) - 1)\cos \pi v = 0.$$

Hence $\alpha^2(2u' - 1)^2 = 0$ It follows that $u' = \tfrac{1}{2}$, which is a contradiction because $u' \ne u$. We conclude that $u \ne \tfrac{1}{2}$; a similar argument shows that $u' \ne \tfrac{1}{2}$.

Having ruled out the possibility that either u or u' is $\tfrac{1}{2}$, we again use the observation recorded at the beginning to arrive at the equality

$$\alpha(2u - 1)\sin \pi v = \alpha(2u' - 1)\sin \pi v'.$$

Since $\Phi_3(u',v') = \Phi_3(u,v)$, we also have

$$\alpha(2u - 1)\cos \pi v = \alpha(2u' - 1)\cos \pi v'.$$

The foregoing two equalities show that $(2u-1)^2 = (2u'-1)^2$. Since $u \neq u'$, we obtain from here that $2u-1 = -(2u'-1)$, which yields $u = 1-u'$, as desired. Since neither u nor u' is $\frac{1}{2}$, we further obtain $\sin \pi v = -\sin \pi v'$ and $\cos \pi v = -\cos \pi v'$. But πv and $\pi v'$ both belong to the interval $[0,\pi]$ and sin is nonnegative on this interval. It follows that $\sin \pi v = -\sin \pi v' = 0$. Therefore $\cos \pi v = -\cos \pi v' = \pm 1$. Once again using the fact that πv and $\pi v'$ both belong to the interval $[0,\pi]$, we conclude that one among πv and $\pi v'$ must be 0 and the other must be π. This means one among v, v' is 0 and the other is 1.

Problem Set 2-7

2-7.P1. By the result of 2-3.P18, $\|A\| = \sqrt{2}$.

2-7.P2. From 2-7.P1, $\|A\| = \sqrt{2}$, and from 2-3.P16, $\|B\| = \sqrt{2}$. Now, $(BA)(x_1,x_2) = (2x_1,0)$. Therefore $\|(BA)(x_1,x_2)\|^2 = (2x_1)^2 + 0^2 = 4x_1^2 \leq 4(x_1^2 + x_2^2) = 4\|(x_1,x_2)\|^2$, which implies that $\|BA\| \leq 2 = (\sqrt{2})(\sqrt{2}) = \|B\| \cdot \|A\|$, as required.

2-7.P3. $\|A\| = \sqrt{2} = \|B\|$, so that $\|B\|\,\|A\| = 2$. But $\|BA\| = 0$, because $BA = O$.

2-7.P4. No matter which norms are being used, $\|(0,b)\| \leq \|(a,b)\|$, which is to say, $\|A(a,b)\| \leq \|(a,b)\|$. Therefore $\|A\| \leq 1$. Since $\|A(0,b)\| = \|(0,b)\|$, we have $\|A\| = 1$.

2-7.P5. Modify the argument of the theorem above.

2-7.P6. $\|B^{-1}\| = \|AA^{-1}B^{-1} - BA^{-1}B^{-1} + BA^{-1}B^{-1}\| = \|BA^{-1}B^{-1} + (A-B)A^{-1}B^{-1}\| \leq \|BA^{-1}B^{-1}\| + \|(A-B)A^{-1}\|\|B^{-1}\|$. Therefore $\|B^{-1}\|(1 - \|(B-A)A^{-1}\|) \leq \|BA^{-1}B^{-1}\|$. The required inequality now follows because $1 - \|(B-A)A^{-1}\|$ is positive.

2-7.P7. Let $\{e_k : 1 \leq k \leq n\}$ be *the* standard basis in $\mathbb{R}^n$ and $y_k = Ae_k$ for $1 \leq k \leq n$. Set $y = (y_1,y_2,\ldots,y_n)$. For $x \in \mathbb{R}^n$, we have $x = \sum_{k=1}^{n} x_k e_k$ and $Ax = \sum_{k=1}^{n} x_k Ae_k = \sum_{k=1}^{n} x_k y_k = x \cdot y$.

Problem Set 2-8

2-8.P1. $\sum_{n=1}^{\infty} f(m,n) = \frac{1}{2^m}$ and consequently, $\sum_{m=1}^{\infty}(\sum_{n=1}^{\infty} f(m,n)) = \sum_{m=1}^{\infty} \frac{1}{2^m} = 1$. Also, $\sum_{m=1}^{\infty} f(m,n) = (-1)^{n-1}$ and hence $\sum_{n=1}^{\infty}(\sum_{m=1}^{\infty} f(m,n))$ is undefined. Since $s(2n+1,2n+1) \to 1$ and $s(2n,2n) \to 0$ as $n \to \infty$, the series is not convergent.

2-8.P2. $\sum_{n=1}^{\infty} f(1,n) = \infty$ and $\sum_{n=1}^{\infty} f(2,n) = -\infty$; $\sum_{m=1}^{\infty} f(m,1) = \infty$ and $\sum_{m=1}^{\infty} f(m,2) = -\infty$.

2-8.P3. The terms of the finite sum $s(p,p)$ correspond to the points with integral coordinates in the square with vertices $(1,1)$, $(p,1)$, (p,p) and $(1,p)$. The number of points on the line segments passing through $(n,1)$ and $(1,n)$, both terminating at (n,n), is $2n-1$, where $1 \le n \le p$. The sum of the squares of the coordinates of any point on either of the segments is greater than n^2 and consequently, $s(p,p) \le \sum_{n=1}^{p}(2n-1)n^{-\alpha} = \sum_{n=1}^{p} 2n^{1-\alpha} - \sum_{n=1}^{p} n^{-\alpha}$. These are partial sums of convergent series and hence, Proposition 2-8.10 shows that that the double series $\Sigma_{m,n}(m^2 + n^2)^{-\alpha/2}$ is convergent.

Problem Set 3-2

3-2.P1. Show that $(x+h)^3 + (y+k) - (x^3 + y)$ can be expressed as $(3x^2)h + k + [\sqrt{(h^2 + k^2)}] \cdot u(h,k)$, where $u(h,k) = (3xh^2 + h^3)/\sqrt{(h^2 + k^2)}$ for $(h,k) \ne (0,0)$. Then use the fact that $|h| \le \sqrt{(h^2 + k^2)}$ to show that $|u(h,k)| \le (|3x| + |h|)\sqrt{(h^2 + k^2)}$.

3-2.P3. If $h = 0$, then $[\phi(th,tk) - \phi(0,0)]/t = 0$ and derivative in the direction (h,k) is 0. Suppose $h \ne 0$. Then $[\phi(th,tk) - \phi(0,0)]/t = tk^3/h \to 0$ as $t \to 0$. Thus, the derivative in every direction is 0. Since $f(0,0) = 0$ and $f(y^3,y) = 1$ for $y \ne 0$, it follows that f is not continuous at $(0,0)$.

3-2.P4. Since $f(x) - f(0) = \|x\|$, for any $\varepsilon > 0$, choosing $\delta = \varepsilon$ ensures that $0 < \|x\| < \delta \Rightarrow |f(x) - f(0)| < \varepsilon$. Thus f is continuous. If f were to be differentiable, then by Remark 3-2.5(b), it would have a derivative at 0 in the direction of e_1. Therefore $[f(te_1) - f(0)]/t$ would have a limit as $t \to 0$. But $[f(te_1) - f(0)]/t = \frac{|t|}{t}$, and it is well known that this does not have a limit as $t \to 0$.

3-2.P5. Denoting the 'increment' vector by (h,k) we have

$$f(x+h,y+k) - f(x,y) = (2xh + h^2, yh + xk + hk, k) = (2xh, yh + xk, k) + (h^2, hk, 0)$$
$$= A(h,k) + (h^2, hk, 0),$$

where $A:\mathbb{R}^2 \to \mathbb{R}^3$ is the map for which $A(h,k) = (2xh, yh + xk, k)$. Now, $(h^2, hk, 0) = \|(h,k)\| u(h,k)$, where $u(h,k) = (h^2, hk, 0)/\sqrt{(h^2 + k^2)}$. Since A is linear with matrix as stated in the problem and since $u(h,k) \to 0$ as $(h,k) \to 0$, the required conclusion follows.

3-2.P6. Denote the 'increment' vector by (h,k). The expression we must deal with is $((x + h)^2 + (y + k))^{10} - (x^2 + y)^{10}$. Since it would be cumbersome to expand the 10th powers, we prefer to start with

$$(z + s)^{10} - z^{10} = 10z^9 s + \sum_{j=2}^{10} C(10, j) z^{10-j} s^j,$$

where $C(10, j)$ is the usual binomial coefficient $10!/j!(10-j)!$ Denoting the sum $\sum_{j=2}^{10} C(10, j) z^{10-j} s^{j-1}$ by $u(s)$, we have

$$(z + s)^{10} - z^{10} = 10 z^9 s + su(s), \text{ where } u(0) = 0 \text{ and } u(s) \to 0 \text{ as } s \to 0.$$

It follows from here by taking

$$z = (x^2 + y) \text{ and } s = (x + h)^2 + (y + k) - (x^2 + y) = 2hx + h^2 + k$$

that $((x + h)^2 + (y + k))^{10} - (x^2 + y)^{10}$

$$= 10(x^2 + y)^9 (2hx + h^2 + k) + (2hx + h^2 + k) \cdot u(2hx + h^2 + k)$$

$$= 20x(x^2 + y)^9 h + 10(x^2 + y)^9 k + 10h^2(x^2 + y)^9 + (2hx + h^2 + k) \cdot u(2hx + h^2 + k).$$

Now,

$$|10h^2(x^2 + y)^9| \le (10(x^2 + y)^9 \cdot \sqrt{(h^2 + k^2)}) \cdot \sqrt{(h^2 + k^2)} \quad \text{and}$$

$10(x^2 + y)^9 \cdot \sqrt{(h^2 + k^2)}$ approaches 0 as (h, k) approaches $(0, 0)$.

Also, by the Cauchy–Schwarz inequality,

$$|2hx + h^2 + k| \le |h \cdot (2x + h) + k \cdot (1)| \le \sqrt{(h^2 + k^2)} \cdot \sqrt{[(2x + h)^2 + 1^2]}$$

and $(2x + h)^2 + 1^2$ is bounded as (h, k) approaches $(0, 0)$. But this also implies that $2hx + h^2 + k \to 0$ as $(h, k) \to (0, 0)$ and hence that $u(2hx + h^2 + k)$ does the same. It follows that, if we take $v(h, k)$ to be the quotient,

$$[10h^2(x^2 + y)^9 + (2hx + h^2 + k) \cdot u(2hx + h^2 + k)] / \sqrt{(h^2 + k^2)} ,$$

when $(h, k) \ne (0, 0)$, then $v(h, k) \to 0$ as $(h, k) \to (0, 0)$ and

$$((x + h)^2 + (y + k))^{10} - (x^2 + y)^{10} = [20x(x^2 + y)^9 \quad 10(x^2 + y)^9] \begin{bmatrix} h \\ k \end{bmatrix}$$
$$+ \|(h, k)\| v(h, k).$$

Thus, $(x^2 + y)^{10}$ is a differentiable function of (x, y) with derivative given by the 1×2 matrix

$$[20x(x^2 + y)^9 \quad 10(x^2 + y)^9].$$

3-2.P7. We have

$$(x + h)^{20}(y + k)^{10} - x^{20} y^{10} = [(x + h)^2(y + k)]^{10} - (x^2 y)^{10}.$$

Letting $s = (x + h)^2(y + k) - x^2 y$ in this, we obtain

$$(x + h)^{20}(y + k)^{10} - x^{20} y^{10} = (x^2 y + s)^{10} - (x^2 y)^{10}$$
$$= 10(x^2 y)^9 s + |s| \cdot v(s), \text{ where } v(s) \to 0 \text{ as } s \to 0,$$

according to what is given. From Example 3-2.3(b),

$$s = A(h,k) + \|(h,k)\| u(h,k),$$

where $A:\mathbb{R}^2 \to \mathbb{R}$ is the linear map for which $A(h,k) = 2xyh + x^2k$ and

$$u(h,k) = \frac{2xhk + yh^2 + kh^2}{\sqrt{h^2 + k^2}} \quad \text{when } (h,k) \neq (0,0).$$

It follows that

$$(x+h)^{20}(y+k)^{10} - x^{20}y^{10} = 10(x^2y)^9[A(h,k) + \|(h,k)\| u(h,k)] + |s|\cdot v(s),$$

so that

$$(x+h)^{20}(y+k)^{10} - x^{20}y^{10} = 10(x^2y)^9 A(h,k) + 10(x^2y)^9\cdot\|(h,k)\| u(h,k) + |s|\cdot v(s)$$

and also that

$$|s| \leq \|A\|\cdot\|(h,k)\| + \|(h,k)\|\cdot|u(h,k)|,$$

which implies that

$$\frac{|s|}{\|(h,k)\|} \quad \text{is bounded as } \|(h,k)\| \to 0.$$

This and the fact that $s \to 0$ as $\|(h,k)\| \to 0$ together have the consequence that

$$w(h,k) = 10(x^2y)^9\cdot u(h,k) + \frac{|s|}{\|(h,k)\|}\cdot v(s) \to 0 \quad \text{as } \|(h,k)\| \to 0.$$

Since the above equality for $(x+h)^{20}(y+k)^{10} - x^{20}y^{10}$ can be recast as

$$(x+h)^{20}(y+k)^{10} - x^{20}y^{10} = 10(x^2y)^9 A(h,k) + \|(h,k)\| w(h,k),$$

we have obtained the derivative in question as $10(x^2y)^9 A(h,k)$, which is to say, the required derivative is the linear map $B:\mathbb{R}^2 \to \mathbb{R}$ such that

$$B(h,k) = 10(x^2y)^9(2xyh + x^2k) = 20x^{19}y^{10}h + 10x^{20}y^9k.$$

It has the 1×2 matrix $[20x^{19}y^{10} \quad 10x^{20}y^9]$.

3-2.P8. Since $\frac{f(x+th)-f(x)}{t} = -\frac{f(x+(-t)(-h))-f(x)}{-t}$, it is immediate from Def. 3-2.4 that $D_h f(x) = -D_{-h} f(x)$. Therefore the derivatives in the directions h and $-h$ cannot both be positive. The function $f(x_1,x_2) = x_1$ has derivative 1 in the direction $h = e_1$ at every point $(x_1,x_2) \in \mathbb{R}^2$.

3-2.P9. Since each f_k is differentiable in (a,b), we have $f_k(\alpha + h_k) - f_k(\alpha) = f_k'(\alpha)h_k + |h_k| u_k(h_k)$, where $u_k(h_k) \to 0$ as $h_k \to 0$. Let $x = (x_1,\dots,x_n) \in E$ and $h = (h_1,\dots,h_n) \in \mathbb{R}^n$. Then

$$f(x+h) - f(x) = \sum_{k=1}^{n} [f_k(x_k + h_k) - f_k(x_k)] = \sum_{k=1}^{n} f_k'(x_k)h_k + \sum_{k=1}^{n} |h_k| u_k(h_k),$$

so that

$$|f(x + h) - f(x) - \sum_{k=1}^{n} f_k{}'(x_k)h_k| \le \|h\| \sum_{k=1}^{n} \frac{|h_k|}{\|h\|} u_k(h_k)|$$

and $|\sum_{k=1}^{n} u_k(h_k)| \to 0$ as $h \to 0$ because $u_k(h_k) \to 0$ as $h_k \to 0$.

3-2.P10. By definition of $a_k(x)$, we have $f_k(x + h) - f_k(x) = a_k(x)h_k + h_k E_k(h)$, where $E_k(h) \to 0$ as $h \to 0$. So, $f(x + h) - f(x) = \sum_{k=1}^{n} [f_k(x + h) - f_k(x)] = \sum_{k=1}^{n} a_k(x)h_k + \sum_{k=1}^{n} h_k E_k(h)$, so that

$$|f(x + h) - f(x) - \sum_{k=1}^{n} a_k(x)h_k| = |\sum_{k=1}^{n} h_k E_k(h)| \le \|h\| |\sum_{k=1}^{n} E_k(h)|$$

and $|\sum_{k=1}^{n} E_k(h)| \to 0$ as $h \to 0$.

3-2.P11. $F(c + h) - F(c) = g(c + h) \cdot f(c + h) - g(c) \cdot f(c)$

$$= g(c + h) \cdot \{f(c + h) - f(c)\}$$

$$= g(c + h) \cdot \{f'(c)(h) + \|h\| u(h)\}$$

$$= g(c + h) \cdot \{f'(c)(h)\} + \|h\| g(c + h) \cdot u(h)$$

$$= g(c) \cdot \{f'(c)(h)\} + [g(c + h) - g(c)] \cdot \{f'(c)(h)\}$$
$$+ \|h\| g(c + h) \cdot u(h).$$

Now,

$$|[g(c + h) - g(c)] \cdot \{f'(c)(h)\} + \|h\| g(c + h) \cdot u(h)|$$
$$\le \|h\| [\|g(c + h) - g(c)\| \|f'(c)\| + \|g(c + h)\| \|u(h)\|].$$

Since g is continuous at c, it follows that $g(c + h)$ is bounded as $h \to 0$. Therefore both terms in the bracket on the right side tend to 0 as $h \to 0$.

3-2.P12. $D_1 f(0,0) = \lim_{t \to 0} \frac{f(t,0) - f(0,0)}{t} = \lim_{t \to 0} \frac{t}{t} = 1$ and similarly, $D_2 f(0,0) = 1$. If $h = (a_1, a_2)$, where $a_1 a_2 \ne 0$, is any other direction, then $D_h f(0,0) = \lim_{t \to 0} \frac{f(ta_1, ta_2) - f(0,0)}{t} = \lim_{t \to 0} \frac{1}{t}$, which does not exist. It follows from Remark 3-2.5(b) that f is not differentiable.

Problem Set 3-3

3-3.P1. (a) As shown in 3-2.P3, the derivative in every direction is 0, so that the zero linear map serves as the Gateaux derivative.
(b) Consider any $h \in \mathbb{R}^n$. Then

$$\lim [f(x + th) - f(x)]/t = Ah \text{ as } t \to 0. \tag{1}$$

For k belonging to some ball K centred at $0 \in \mathbb{R}^n$,

$$g(f(x) + k)) - g(f(x)) = g'(f(x))k + \| k \| v(k), \tag{2}$$

where $v(k) \to 0$ as $k \to 0$. Since f is continuous at x, then for all sufficiently small t, we have $f(x + th) - f(x) \in K$. Then we may take $k = f(x + th) - f(x)$ in (2). But if we choose k in this manner, then we have $f(x) + k = f(x + th)$. Therefore by (2) and the linearity of $g'(f(x))$,

$$\frac{g(f(x + th)) - g(f(x))}{t} = g'(f(x))\frac{k}{t} + \frac{\| k \|}{t} v(k), \tag{3}$$

where, by (1), $k/t \to Ah$ and hence $\| k \|/t$ is bounded as $t \to 0$. Since $k \to 0$ as $t \to 0$, it follows from (3) and (2) that

$$\lim \frac{g(f(x + th)) - g(f(x))}{t} = g'(f(x))Ah \text{ as } t \to 0.$$

3-3.P2. Since f is continuous, x is an interior point of the domain of $g \circ f$. Consider any $k \in \mathbb{R}^m$. Then

$$\lim [g(f(x) + tk) - g(f(x))]/t = Gk \quad \text{as } t \to 0. \tag{1}$$

For sufficiently small t and any $h \in \mathbb{R}^n$, we have

$$(g \circ f)(x + th) = g(Ax + t(Ah) + b) = g(f(x) + t(Ah))$$

and hence $[(g \circ f)(x + th) - (g \circ f)(x)]/t = [g(f(x) + t(Ah)) - g(f(x))]/t$. By (1), this has limit $G(Ah)$ as $t \to 0$.

3-3.P3. Modify the proof of Corollary 3-3.4, keeping in mind that the chain rule is not available for Gateaux derivatives. So, fish for something else that will serve the purpose.

3-3.P4. (a) Since f is real valued, each $G(e_j)$ is a real number. For a general $h \in \mathbb{R}^n$, the linearity of G leads to $G(h) = G(h_1, \ldots, h_n) = G(\sum_{j=1}^n h_j e_j) = \sum_{j=1}^n h_j G(e_j)$. Since $G \neq 0$, at least one among $G(e_j)$ must be nonzero. So $h'_j = G(e_j)$ describes a nonzero element of $\mathbb{R}^n$. The Cauchy–Schwarz inequality yields $|G(h_1, \ldots, h_n)| \leq \sqrt{(\sum_{j=1}^{n} h_j^2)} \sqrt{\sum_{j=1}^{n} G(e_j)^2} = \|h\|_2 \sqrt{\sum_{j=1}^{n} G(e_j)^2}$, with the rider that equality holds here only if there is some real number α such that $h = \alpha h'$. The inequality shows that $\|G\| \leq \sqrt{\sum_{j=1}^{n} G(e_j)^2}$. Since equality indeed holds when $h = h'/\|h'\|_2$ while $\|h\|_2 = 1$, we find that $\|G\| = \sqrt{\sum_{j=1}^{n} G(e_j)^2}$. Hence, the equality $\|G\| = |G(h)|$ holds with $\|h\|_2 = 1$ if and only if $|G(h_1, \ldots, h_n)| = \sqrt{(\sum_{j=1}^{n} h_j^2)} \sqrt{\sum_{j=1}^{n} G(e_j)^2} = \|h\|_2 \sqrt{\sum_{j=1}^{n} G(e_j)^2}$ with $\|h\|_2 = 1$. According to the rider, this is only possible when h also satisfies $h = \alpha h'$ where α is some real number. Such an α must satisfy $|\alpha| = \|h\|_2/\|h'\|_2 = 1/\|h'\|_2$. There are exactly two such real numbers α, namely, $\pm 1/\|h'\|_2$. Therefore $h = \pm h'/\|h'\|_2$ are the only two $h \in \mathbb{R}^n$ satisfying $\|G\| = |G(h)|$ and $\|h\|_2 = 1$.

(b) Proceeding as in (a), we have $G(h) = \sum_{j=1}^{n} h_j G(e_j)$. Therefore $|G(h)| \leq (\sum_{j=1}^{n} |h_j|) \cdot \max\{|G(e_j)| : 1 \leq j \leq n\} = \|h\|_1 \cdot \max\{|G(e_j)| : 1 \leq j \leq n\}$, so that $\|G\| \leq \max\{|G(e_j)| : 1 \leq j \leq n\} = |G(e_p)|$, say. Note that p need not be unique. Select $h \in \mathbb{R}^n$ such that $h_p = 1$ and $h_i = 0$ for $i \neq p$. Then $\|h\|_1 = 1$ and $|G(h)| = |G(e_p)| = \max\{|G(e_j)| : 1 \leq j \leq n\}$, which shows that $\|G\| = \max\{|G(e_j)| : 1 \leq j \leq n\}$ and also that $|G(h)| = \|G\|$. Since $|G(-h)| = \|G\|$ and $\|-h\|_1 = 1$, we get two elements of $\mathbb{R}^n$ of the required kind. When p is not unique, we can get at least two more.

3-3.P5. Denote $F(b) - F(a)$ by p, and let $\phi:[a, b] \to \mathbb{R}$ be defined by

$$\phi(t) = p \cdot F(t).$$

Also, ϕ is the composition of the maps $t \to F(t)$ and $x \to p \cdot x$ in that order. The first one has derivative $h \to F'(t)h$ and the second one has derivative $h \to p \cdot h$ [Remark 3-2.2(d)]. Their norms are, respectively, $\|F'(t)\|$ and $\|p\|$ [see Example 2-7.3(c)]. By the chain rule, the derivative $\phi'(t)$ exists and equals the composition of the linear maps

$$h \to F'(t)h \quad \text{and} \quad h \to p \cdot h,$$

in that order. Using the property that $\|ST\| \leq \|S\| \|T\|$ for any linear maps S and T for which the product ST is defined, we find that

$$\|\phi'(t)\| \leq \|p\| \cdot \|F'(t)\| \quad \forall \, t \in (a, b).$$

However, by the mean value theorem,

$$|\phi(b) - \phi(a)| = |\phi'(c)| \cdot |b - a| \quad \text{for some } c \in (a, b)$$
$$= \|\phi'(c)\| \cdot (b - a).$$

Moreover, $\phi(b) = p \cdot F(b)$ and $\phi(a) = p \cdot F(a)$. Hence

$$|p \cdot (F(b) - F(a))| \leq \|p\| \|F'(c)\| \cdot (b - a),$$

i.e., $\|p\|^2 \leq (b - a)\|p\| \|F'(c)\|$ (because $p = F(b) - F(a)$).

So, $\|p\| \leq (b - a)\|F'(c)\|$.

Since $p = F(b) - F(a)$, this is the same as $\|F(b) - F(a)\| \leq (b - a)\|F'(c)\|$.

3-3.P6. (a) Suppose $x = \Sigma x_j e_j$ and $y = \Sigma y_j e_j$, where $a_j \leq x_j \leq b_j$ and $a_j \leq y_j \leq b_j$ for $1 \leq j \leq n$, and let $0 \leq t \leq 1$. Then $tx + (1 - t)y = \Sigma (tx_j + (1 - t)y_j)e_j$ and $a = ta + (1 - t)a \leq tx_j + (1 - t)y_j \leq tb + (1 - t)b = b$.
(b) Let $x = \Sigma x_k e_k$. Since x and $x + h$ both belong to the cuboid, then for $1 \leq k \leq n$, we have (i) $a_k \leq x_k \leq b_k$ and (ii) $a_k \leq x_k + h_k \leq b_k$. When $0 \leq t \leq 1$, we therefore have $a_j \leq x_j + th_j \leq b_j$. This, together with the first $j - 1$ inequalities in (ii) and the last $n - j$ inequalities in (i) leads to the required conclusion.
(c) Yes: If $\sqrt{(h_1^2 + \ldots + h_n^2)}$ is less than the radius of the ball (or equal), then so is $\sqrt{(h_1^2 + \ldots + t^2 h_j^2)}$.

3-3.P7. $\phi'(t)$ is the limit as $h \to 0$ of the quotient $\Phi(h) = \frac{f(x+(t+h)\mu e) - f(x+t\mu e)}{h}$ and the derivative of f at $x + t(\mu e)$ in the direction of e is the limit as $h \to 0$ of $F(h) = \frac{f(x+t\mu e + he) - f(x+t\mu e)}{h}$. The relation is $\Phi(h) = \mu F(\mu h)$ for $\mu \neq 0$. This implies (i) and that (ii) $\phi'(t)$ is μ times $(D_e f)(x + t(\mu e))$, the directional derivative in question. For (iii), we note that

$$f(x + \mu e) - f(x) = \phi(1) - \phi(0) = \phi'(\theta) = \mu \cdot (D_e f)(x + \theta(\mu e)), 0 < \theta < 1.$$

3-3.P8. The map $x \to f_1(x)u_1$ from E to $\mathbb{R}^m$ is the composition $A \circ f_1$, where $A:\mathbb{R} \to \mathbb{R}^m$ is the linear map given by $Az = zu_1$. By the chain rule, its derivative at x_0 is the composition $A \circ f_1'(x_0)$. This composition maps $h \in \mathbb{R}^n$ into $A(f_1'(x_0)(h)) = [f_1'(x_0)(h)]u_1$. If $f_2:E \to \mathbb{R}$ is also differentiable at $x_0 \in E$, then the derivative of the map $x \to f_2(x)u_2$ from E to $\mathbb{R}^m$ is given by $h \to [f_2'(x_0)(h)]u_2$. Therefore $\phi'(x_0)(h) = [f_1'(x_0)(h)]u_1 + [f_2'(x_0)(h)]u_2$.

3-3.P9. (a) $f(2\pi) - f(0) = (0,0)$ has norm 0 but $f'(\theta) = (-\sin\theta, \cos\theta)$ has norm 1.
(b) Apply the one variable mean value theorem to $\phi(t) = [f(a + t(b - a))] \cdot c$ on $[0,1]$, noting that $\phi'(t) = \lim_{h \to 0} \left[\frac{f(a+(t+h)(b-a)) - f(a+t(b-a))}{h} c \right] = [(D_{b-a}f)(a + t(b - a)] \cdot c$, because $\lim_{u \to v} u \cdot c = v \cdot c$.
(c) The function satisfies $f(b) - f(a) = (0,0)$; since the function has a linear derivative represented by the matrix $[-\sin t \quad \cos t]$, it follows by Remark 3-2.5(b) that it has a directional derivative given by $[(D_{b-a}f)(a + \theta(b - a)] =$

$2\pi[-\sin 2\pi\theta, \cos 2\pi\theta]$. The required θ must satisfy $-c_1 \sin 2\pi\theta + c_2 \cos 2\pi\theta = 0$. If $c_1 = 0$, then $\theta = \frac{1}{4}$ or $\frac{3}{4}$. If $c_2 = 0$, then $\theta = \frac{1}{2}$. If $c_2/c_1 > 0$, then there are two possibilities for θ: $\frac{1}{2\pi}\arctan(c_2/c_1)$ and $\frac{1}{2} + \frac{1}{2\pi}\arctan(c_2/c_1)$. If $c_2/c_1 < 0$, then again there are two possibilities for θ: $\frac{1}{2} + \frac{1}{2\pi}\arctan(c_2/c_1)$ and $1 + \frac{1}{2\pi}\arctan(c_2/c_1)$.

(d) $f(1) - f(0) = (0,0)$ and $f'(\theta) = (1 - 2\theta, 1 - 3\theta^2)$, which cannot be $(0,0)$ for any value of θ. Now $c_1(1 - 2\theta) + c_2(1 - 3\theta^2) = -3c_2\theta^2 - 2c_1\theta + (c_1 + c_2)$ and $(f(1) - f(0))\cdot(c_1, c_2) = 0$. If $c_2 = 0$, then these can be equal only if $\theta = \frac{1}{2}$. However, if $c_2 \neq 0$, then these can be equal only if θ is one among the numbers $-[c_1 \pm \sqrt{(c_1^2 + 3c_2^2 + 3c_1c_2)}]/3c_2$. It remains to check when both of them lie in $(0,1)$. To do so, we express them as $-\frac{1}{3}[u \pm \sqrt{(u^2 + 3u + 3)}]$, where $u = c_1/c_2$. By applying usual differentiation techniques to the functions $-\frac{1}{3}[u \pm \sqrt{(u^2 + 3u + 3)}]$, one can show that $-[c_1 + \sqrt{(c_1^2 + 3c_2^2 + 3c_1c_2)}]/3c_2$ lies in $(0,1) \Leftrightarrow c_1/c_2 < -1$ and that $-[c_1 - \sqrt{(c_1^2 + 3c_2^2 + 3c_1c_2)}]/3c_2$ lies in $(0,1) \Leftrightarrow c_1/c_2 > -2$. In case $-2 < c_1/c_2 < -1$, of course both lie in $(0,1)$. Thus θ is nonunique $\Leftrightarrow c_2 \neq 0$ and $-2 < c_1/c_2 < -1$.

3-3.P10. (a) Denote the centre of the ball B by a and consider any $b \in B$. By Problem 3-3.P9(b), we have $(f(b) - f(a))\cdot c = 0$ for every $c \in \mathbb{R}^m$; in particular, for $c = f(b) - f(a)$. Therefore $f(b) = f(a)$. Since b is an arbitrary point of B here, it follows that f is constant on B.

(b) For any $a, b \in B$ such that $0 \neq b - a = pu$ for some $p \in \mathbb{R}$, we have $f(b) = f(a)$. This follows by using the result of Problem 3-3.P9(b) and noting that $D_{b-a}f = pD_u f$.

3-3.P11. Let s_0 be any point of S. By Corollary 3-3.4, f is constant in any open ball contained in S. Therefore $S_1 = \{s \in S : f(s) = f(s_0)\}$ is an open subset of S. However, by continuity of f, the set $S_2 = \{s \in S : f(s) \neq f(s_0)\}$ is also an open subset of S. Since S_1 and S_2 are disjoint open subsets of the connected set S with union equal to S, one among the two must be empty. Now S_1 cannot be empty because it contains s_0. Therefore S_2 must be empty, which means $S_1 = S$, i.e., $f(s) = f(s_0) \ \forall \ s \in S$.

3-3.P12. Apply 3-3.P7(iii) with (x,t) as x, t as μ and $(1,-1)$ as e. Then

$$f(x + t, 0) - f(x, t) = t\cdot[D_{(1,-1)}f((x,t) + \theta(t,-t))] = t\cdot[D_{(1,-1)}f(x + \theta t, (1 - \theta)t)]$$

$$= t\cdot[\tfrac{\partial f}{\partial x}(x + \theta t, (1 - \theta)t)(1) + \tfrac{\partial f}{\partial t}(x + \theta t, (1 - \theta)t)(-1)] = 0, \text{ because } \tfrac{\partial f}{\partial x}$$

$$= \tfrac{\partial f}{\partial t} \text{ by hypothesis.}$$

Therefore $f(x, t) = f(x + t, 0) > 0$ for all (x, t).

3-3.P13. Applying 3-3.P9(b) with $m = 1$, we get for some $\theta \in (0,1)$,

$$f(y) - f(x) = (D_{y-x}f)(x + \theta(y-x)) = \sum_{i=1}^{n} (y_i - x_i)\frac{\partial f}{\partial x_i}(z), \text{ where } z = x + \theta(y-x).$$

Therefore   by   the   Cauchy–Schwarz   inequality, $|f(y) - f(x)| \leq$

$[\sum_{i=1}^{n}(y_i - x_i)^2]^{1/2}[\sum_{i=1}^{n}(\frac{\partial f}{\partial x_i}(z))^2]^{1/2}$. On letting $y \to 0$ and using (ii) and (iii), we get

$|f(0) - f(x)| \leq \varepsilon[\sum_{i=1}^{n} x_i^2]^{1/2}$ for sufficiently small $\sum_{i=1}^{n} x_i^2$. This means f is differentia-

ble at 0 (with derivative zero).

3-3.P14. It suffices to consider only ϕ. Continuity at points $(x,y) \neq (0,0)$ is clear. The observation that

$$\frac{x^4 + 4x^2y^2 - y^4}{(x^2 + y^2)^2} = 1 + 2\frac{y^2}{(x^2 + y^2)}\frac{x^2 - y^2}{(x^2 + y^2)} \qquad \text{so that } |\phi(x,y)| \leq 3|y|$$

proves continuity at (0,0). For any $(h,k) \neq (0,0)$ and any $t \neq 0$, we have

$$\frac{\phi(th, tk) - \phi(0,0)}{t} = k\frac{h^4 + 4h^2k^2 - k^4}{(h^2 + k^2)^2},$$

which shows that the right side here is the derivative in the direction (h,k). Hence the partial derivatives $D_1\phi(0,0)$ and $D_2\phi(0,0)$ are obtained from here by setting $(h,k) = (1,0)$ and $(h,k) = (0,1)$, respectively. Thus $D_1\phi(0,0) = 0$ and $D_2\phi(0,0) = -1$. If ϕ were to be differentiable at (0,0), we would have

$$\phi(h,k) - \phi(0,0) = (-1)k + (h^2 + k^2)^{1/2}u(h,k),$$

where $u(h,k) \to 0$ as $(h,k) \to (0,0)$. This would imply that

$$u(h,k) = (h^2 + k^2)^{-1/2}[k\frac{h^4 + 4h^2k^2 - k^4}{(h^2 + k^2)^2} + k] = 2(h^2 + k^2)^{-1/2}kh^2\frac{h^2 + 3k^2}{(h^2 + k^2)^2},$$

which does not approach 0 as $(h,k) \to (0,0)$, because its value when $k = h$ is $2^{1/2}h/|h|$. This would be a contradiction. The directional derivative is not linear in (h,k); so no Gateaux derivative.

3-3.P15. Suppose $a_1 \neq 0$. Then $\frac{f(ta_1, ta_2) - f(0,0)}{t} = \frac{t^3a_1a_2^2}{t^3(a_1^2 + t^2a_2^4)} = \frac{a_1a_2^2}{a_1^2 + t^2a_2^4}$, which

has limit $\frac{a_2^2}{a_1}$ as $t \to 0$. If $a_1 = 0$, then $\frac{f(ta_1, ta_2) - f(0,0)}{t} = 0$. The value of the func-

tion at any point of the parabola $x = y^2$ other than (0,0) is $\frac{1}{2}$, whereas its value at

(0,0) is 0, i.e., $f(y^2, y) - f(0,0) = \frac{1}{2}$ for all $y \neq 0$. The directional derivative is not

linear in (a_1, a_2); so no Gateaux derivative.

Problem Set 3-4

3-4.P1. $f(x,y) = (f_1(x,y), f_2(x,y), f_3(x,y))$, where $f_1(x,y) = x^2$, $f_2(x,y) = xy$ and $f_3(x,y) = y$. The 3×2 matrix having rows $[\partial f_i/\partial x \quad \partial f_i/\partial y]$, where $i = 1,2,3$, is the one given in the problem. Since each partial derivative is continuous, the function f is differentiable [Theorem 3-4.4] and hence the derivative is given [Theorem 3-4.2] by the aforementioned matrix.

3-4.P2. Since $f(x,0) = 0 = f(0,y)$ for all x,y, both partial derivatives exist at $(0,0)$ and are 0. To prove discontinuity at $(0,0)$, we shall work with the norm $\|(u,v)\| = \sqrt{(u^2 + v^2)}$. For any $\delta > 0$, the point $(x,y) = (\delta/2, \delta/2)$ satisfies $\|(x,y) - (0,0)\| < \delta$ but $|f(x,y) - 0| = 1/2$, whereby f is seen to be discontinuous at $(0,0)$. If $h \neq 0 \neq k$, then $[f(th,tk) - f(0,0)]/t = hk/t(h^2 + k^2)$, which has no limit as $t \to 0$.

3-4.P3. Since $f(x,0) = 0 = f(0,y)$, both partial derivatives exist and are 0. Since $|x|,|y| \leq (x^2 + y^2)^{1/2}$, we have $|f(x,y)| \leq (x^2 + y^2)/(x^2 + y^2)^{1/2} = (x^2 + y^2)^{1/2}$. Therefore f is continuous at $(0,0)$. However, there is no derivative in any direction (h,k) for which $h \neq 0 \neq k$, because $[f(th,tk) - f(0,0)]/t = thk/|t|(h^2 + k^2)^{1/2}$, which has no limit as $t \to 0$.

3-4.P4. The partial derivatives $\partial x/\partial p$, $\partial x/\partial q$, $\partial y/\partial p$, $\partial y/\partial q$ are all continuous and the map is therefore differentiable. The Jacobian is the determinant of

$$\begin{bmatrix} \cos q & -p\sin q \\ \sin q & p\cos q \end{bmatrix},$$

which is p.

3-4.P5. $\partial x/\partial p = -\sin p \cosh q$, $\partial x/\partial q = \cos p \sinh q$, $\partial y/\partial p = \cos p \sinh q$, $\partial y/\partial q = \sin p \cosh q$. Upon computing the relevant determinant, we find that the Jacobian is $-\sin^2 p - \sinh^2 q$. This vanishes $\Leftrightarrow (p,q) = (k\pi, 0)$, where $k \in \mathbb{Z}$.

3-4.P6. Proceed as in the proof of Theorem 3-4.4. Statement (1) is now available only for $2 \leq j \leq n$. So, split $g(x + h) - g(x)$ as $[g(x + z_n) - g(x + z_1)] + [g(x + z_1) - g(x + z_0)] = \sum_{j=2}^{n}(g(x + z_j) - g(x + z_{j-1})) + [g(x + h_1 e_1) - g(x)]$ and handle the summation by using (1) as in the theorem. For $g(x + h_1 e_1) - g(x)$, invoke the definition of $D_1 g$.

3-4.P7. As in Theorem 3-4.4, $g(x + h) - g(x) = \sum_{j=1}^{n} h_j (D_j g)(x + z_{j-1} + \theta_j h_j e_j)$ for any $x \in E$. Now boundedness of partial derivatives yields the desired continuity.

3-4.P8. Here $f_1(x,y) = \sin x \cos y$, $f_2(x,y) = x + y$, $f_3(x,y) = x^2 - y$. Therefore the partial derivatives are

$D_1 f_1 = \cos x \cos y$, $D_2 f_1 = -\sin x \sin y$, $D_1 f_2 = 1$, $D_2 f_2 = 1$, $D_1 f_3 = 2x$, $D_2 f_3 = -1$.

The Jacobian matrix is therefore

$$\begin{bmatrix} \cos x \cos y & -\sin x \sin y \\ 1 & 1 \\ 2x & -1 \end{bmatrix}.$$

3-4.P9. As in Example 3-2.6, the derivative at $(0,0)$ in any direction (h,k) is $h^3/(h^2 + k^2)$, which is not linear in (h,k). Therefore there is no Gateaux derivative at $(0,0)$. But partial derivatives are 1 and 0. Since $[f(t,0) - f(0,0)]/t = 1$, then $(D_1 f)(0,0) = 1$; since $f(0,y) = 0 \; \forall \; y$, then $(D_2 f)(0,0) = 0$. For $(x,y) \neq (0,0)$,

$$(D_1 f)(x,y) = \frac{3x^2(x^2 + y^2) - x^3(2x)}{(x^2 + y^2)^2} = \frac{x^4 + 3x^2 y^2}{(x^2 + y^2)^2} = \frac{(x^2 + y^2)^2 + x^2 y^2 - y^4}{(x^2 + y^2)^2}$$

$$= 1 + \frac{y^2(x^2 - y^2)}{(x^2 + y^2)^2}$$

and $\quad (D_2 f)(x,y) = -\frac{2x^3 y}{(x^2 + y^2)^2} = -\frac{2xy}{x^2 + y^2} \frac{x^2}{x^2 + y^2}.$

Therefore $|(D_1 f)(x,y)| \leq 2$ and $|(D_2 f)(x,y)| \leq 1$. Thus the partial derivatives are bounded.

3-4.P10. For $(x, y) \neq (0,0)$, we have $(D_1 f)(x,y) = 2xy^2 \ln(x^2 + y^2) + 2x^3 y^2/(x^2 + y^2)$ and similarly for $(D_2 f)(x,y)$. Also, $(D_1 f)(0,0) = (D_2 f)(0,0) = 0$. (It may be recalled from 3-4.P9 that mere existence of partial derivatives does not ensure that there is a Gateaux derivative.) Since the partial derivatives we have obtained are continuous, the function is differentiable everywhere by Theorem 3-4.4. Therefore the linear derivative also provides the Gateaux derivative. By Theorem 3-4.2, the matrix representation, which must be 1×2, has entries $(D_1 f)(x,y)$, $(D_2 f)(x,y)$, respectively.

3-4.P11. (a) $(D_1 f)(x,y) = (1 + y^2)/(1 + x^2 + y^2 + x^2 y^2)$ and $(D_2 f)(x,y) = -(1 + x^2)/(1 + x^2 + y^2 + x^2 y^2)$. Since both are continuous, f is differentiable everywhere by Theorem 3-4.4 and thus $f(x,y) = f(0,0) + x(D_1 f)(0,0) + y(D_2 f)(0,0) + \|(x,y)\| u(x,y)$, where $u(x,y) \to 0$ as $(x,y) \to 0$. Now $f(0,0) = 0$, $(D_1 f)(0,0) = 1$ and

$(D_2f)(0,0) = -1$. So, $f(x,y) = x - y + \|(x,y)\| u(x,y)$. Since $u(x,y) \to 0$ as $(x,y) \to 0$, the required approximation is $x - y$.

(b) Proceed similarly, noting that $f(3,\tfrac{1}{2}) = \tfrac{\pi}{4}$ and that $(D_1f)(3,\tfrac{1}{2}) = \tfrac{1}{10}$, $(D_2f)(3,\tfrac{1}{2}) = -\tfrac{4}{5}$. The answer is $\tfrac{1}{10}(x-3) - \tfrac{4}{5}(y-\tfrac{1}{2})$.

3-4.P12. By the chain rule,

$$g'(t) = [(D_1f)(ty_1 + (1-t)x_1, y_2)]\cdot(y_1 - x_1) + [(D_2f)(x_1, ty_2 + (1-t)x_2)]\cdot(y_2 - x_2).$$

Now, $f(y_1,y_2) - f(x_1,x_2) = g(1) - g(0) = g'(\theta)$, where $0 < \theta < 1$, using the mean value theorem. So,

$$f(y_1,y_2) - f(x_1,x_2) = [(D_1f)(\theta y_1 + (1-\theta)x_1, y_2)]\cdot(y_1 - x_1)$$
$$+ [(D_2f)(x_1, \theta y_2 + (1-\theta)x_2)]\cdot(y_2 - x_2).$$

Take $z_i = \theta y_i + (1-\theta)x_i$.

3-4.P13. Follow the argument of part (b) of Theorem 3-4.9, keeping in mind that x is now not an arbitrary point of S.

3-4.P14. Direct computation will yield the result. However, it is easier when determinants are used. First, $J(y,z) = \det \begin{bmatrix} y_u & y_v \\ z_u & z_v \end{bmatrix}$ and similarly for $J(z,x)$ and $J(x,y)$. Second,

$$\det \begin{bmatrix} x_u & x_u & x_v \\ y_u & y_u & y_v \\ z_u & z_u & z_v \end{bmatrix} = \det \begin{bmatrix} x_v & x_u & x_v \\ y_v & y_u & y_v \\ z_v & z_u & z_v \end{bmatrix} = 0.$$

Expanding these by the first column obviously leads to the required equalities.

3-4.P15. (a) Since the cofactor A_{ik} is a function of only those entries not in the ith row (or kth column), we have $\dfrac{\partial}{\partial a_{ij}} A_{ik} = 0$ for any i,j,k. It follows that $\dfrac{\partial}{\partial a_{ij}}(a_{ik}A_{ik}) = 0$ if $k \neq j$ and A_{ij} if $k = j$. The equality $\det A = \Sigma_j a_{ij}A_{ij}$ now implies that $\dfrac{\partial}{\partial a_{ij}} \det A = A_{ij}$ for any i,j.

(b) By the chain rule, $\dfrac{d}{dx} \det A = \Sigma_{ij}\left[(\dfrac{\partial}{\partial a_{ij}} \det A)a_{ij}'\right] = \Sigma_{ij}A_{ij}a_{ij}'$. (*)

From this and the fact that $A_{ij} = (A_B^i)_{ij}$, we further obtain $\dfrac{d}{dx} \det A = \Sigma_{ij}(A_B^i)_{ij}a_{ij}' = \Sigma_i[\Sigma_j(A_B^i)_{ij}a_{ij}] = \Sigma_i \det A_B^i$.

(c) Since $A_{ij} = c_{ji} \det A$, it follows from (*) of part (b), that $\frac{d}{dx} \det A = $ det$A \sum_{ij} a_{ij}' c_{ji}$. This says that $\frac{d}{dx}(\ln(\det A)) = \sum_{ij} a_{ij}' c_{ji}$.

3-4.P16. z is homogeneous of degree 0. By Euler's Theorem 3-4.9(a), the answer is 0.

3-4.P17. If $y_0 \in [c,d]$, then

$$|F(y_0 + h) - F(y_0) - h\left(\int_a^b \frac{\partial f}{\partial y}(x,y_0)\, dx\right)|$$

$$= |\int_a^b (f(x,y_0 + h) - f(x,y_0) - h\frac{\partial f}{\partial y}(x,y_0))\, dx|$$

$$\leq \int_a^b |f(x,y_0 + h) - f(x,y_0) - h\frac{\partial f}{\partial y}(x,y_0)|\, dx$$

$$\leq |h|\int_a^b |\frac{\partial f}{\partial y}(x,y_0 + \theta h) - \frac{\partial f}{\partial y}(x,y_0)|\, dx,$$

where $0 < \theta < 1$. Since $\frac{\partial f}{\partial y}$ is continuous on the compact domain $[a,b]\times[c,d]$, we know that $\frac{\partial f}{\partial y}(x,y_0 + h)$ converges uniformly to $\frac{\partial f}{\partial y}(x,y_0)$ as $h\to 0$. Therefore the last mentioned integral tends to 0 as $h\to 0$. Using this limit in the inequality proved shows $F'(y)$ to be as claimed. If $[c,d]$ is replaced by (c,∞), then we can apply the previous case to $[c',d]$ for every $c' > c$ and every $d > c'$ to conclude that the result continues to hold after the replacement. Similarly, it holds if $[c,d]$ is replaced by $\mathbb{R}$.

3-4.P18. First consider any $u \in [0,\frac{1}{2}]$. We have $0 \leq -\ln(1 - u) = u + \frac{u^2}{2} + \frac{u^3}{3} + \cdots = u(1 + \frac{u}{2} + \frac{u^2}{3} + \cdots) \leq u(1 + u + u^2 + \cdots) = \frac{u}{1-u}$ and also $1 - u \geq \frac{1}{2}$. Therefore $0 \leq u \leq \frac{1}{2} \Rightarrow 0 \leq -\ln(1 - u) \leq 2u$. So, $\alpha \geq \sqrt{2} \Rightarrow 0 \leq \frac{\sin^2\phi}{\alpha^2} \leq \frac{1}{2} \Rightarrow 0 \leq -\ln(1 - \frac{\sin^2\phi}{\alpha^2}) \leq 2\frac{\sin^2\phi}{\alpha^2} \leq \frac{2}{\alpha^2}$ for all ϕ. Thus $0 \leq -\ln(\alpha^2 - \sin^2\phi) + 2\ln\alpha \leq \frac{2}{\alpha^2}$ for $\alpha \geq \sqrt{2}$ and all ϕ. Hence

$$0 \leq -\int_0^{\pi/2} \ln(\alpha^2 - \sin^2\phi)\, d\phi + \pi\ln\alpha \leq \frac{\pi}{\alpha^2} \quad \text{for } \alpha \geq \sqrt{2}. \quad (1)$$

Now, by Leibnitz's formula [see last part of 3-4.P17], when $\alpha > 1$, we have

$\frac{d}{d\alpha} \int_0^{\pi/2} \ln(\alpha^2 - \sin^2\phi)\, d\phi = \int_0^{\pi/2} [\frac{\partial}{\partial\alpha} \ln(\alpha^2 - \sin^2\phi)]\, d\phi = \int_0^{\pi/2} \frac{2\alpha}{\alpha^2 - \sin^2\phi}\, d\phi =$

$\int_0^{\pi/2} \frac{4\alpha}{2\alpha^2 - 1 + \cos 2\phi}\, d\phi = \int_0^{\pi} \frac{2\alpha}{2\alpha^2 - 1 + \cos\phi}\, d\phi = \frac{\pi}{\sqrt{\alpha^2 - 1}}$ in view of the general equality

$\int_0^{\pi} \frac{d\theta}{a + b\cos\theta} = \frac{\pi}{\sqrt{a^2 - b^2}}$ when $a > b > 0$.

But $\frac{d}{d\alpha} \pi \ln(\alpha + \sqrt{(\alpha^2 - 1)}) = \frac{\pi}{\sqrt{\alpha^2 - 1}}$ and therefore,

$\int_0^{\pi/2} \ln(\alpha^2 - \sin^2\phi)\, d\phi = \pi \ln(\alpha + \sqrt{(\alpha^2 - 1)}) + c$

$\qquad = \pi \ln\alpha + \pi \ln[1 + \sqrt{(1 - \frac{1}{\alpha^2})}] + c$ for some constant c and all $\alpha > 1$.   (2)

Since $\ln[1 + \sqrt{(1 - \frac{1}{\alpha^2})}] \to \ln 2$ as $\alpha \to \infty$, the equality (2) shows that

$\qquad [\int_0^{\pi/2} \ln(\alpha^2 - \sin^2\phi)\, d\phi - \pi\ln\alpha - \pi\ln 2 - c] \to 0$ as $\alpha \to \infty$.

On the other hand, (1) implies that

$\qquad [\int_0^{\pi/2} \ln(\alpha^2 - \sin^2\phi)\, d\phi - \pi\ln\alpha] \to 0$ as $\alpha \to \infty$.

Thus, $c = -\pi\ln 2$. Together with (2), this leads to the required equality.

3-4.P19. By Leibnitz's formula [see last part of 3-4.P17], $u'(x) = \int_0^{\pi} \sin\phi \sin(n\phi - x\sin\phi)\, d\phi$ and $u''(x) = -\int_0^{\pi} \sin^2\phi \cos(n\phi - x\sin\phi)\, d\phi$. Therefore

$x^2 u'' + xu + (x^2 - n^2)u$

$\qquad = -x^2 \int_0^{\pi} \sin^2\phi \cos(n\phi - x\sin\phi)\, d\phi + x \int_0^{\pi} \sin\phi \sin(n\phi - x\sin\phi)\, d\phi$

$\qquad \quad + (x^2 - n^2) \int_0^{\pi} \cos(n\phi - x\sin\phi)\, d\phi$

$\qquad = \int_0^{\pi} [(x^2\cos^2\phi - n^2)\cos(n\phi - x\sin\phi) + x\sin\phi \sin(n\phi - x\sin\phi)]\, d\phi$

$\qquad = [-(n + x\cos\phi)\sin(n\phi - x\sin\phi)]_0^{\pi} = 0$.

3-4.P20. In view of the given continuity of f, α, β and $D_2 f$, the function $\Phi(s,t) = \int_s^t f(x,y)\, dx$ has partial derivatives $\frac{\partial\Phi}{\partial s} = -f(s,y)$, $\frac{\partial\Phi}{\partial t} = f(t,y)$ by the fundamental theorem of calculus, and $\frac{\partial\Phi}{\partial y} = \int_s^t \frac{\partial f}{\partial y}(x,y)\, dx$ by Leibnitz's formula [see 3-4.P17]. The required equality for F' now follows by the chain rule. Continuity of the derivative follows from the given continuity of f, α', β' and $D_2 f$.

3-4.P21. For $n = 1$, we have $F_1(x) = \int_0^x f(t)\,dt$ and hence $F_1'(x) = f(x)$. By Problem 3-4.P20, we find for $n > 1$, that $F_n'(x) = \frac{1}{(n-1)!}(x-x)^{n-1}f(x) + \frac{1}{(n-2)!}\int_0^x (x-t)^{n-2}f(t)\,dt = \frac{1}{(n-2)!}\int_0^x (x-t)^{n-2}f(t)\,dt$. It therefore follows by induction that $F_n^{(n)}(x) = f(x) \ \forall n \in \mathbb{N}$.

3-4.P22. To show that $\Gamma(t) \notin D$, it is sufficient to prove that $\Gamma_1(t)^2 + \Gamma_2(t)^2 > 1$. By (b), $\Gamma_1(0)^2 + \Gamma_2(0)^2 = 1$. Therefore, if we can establish that $\frac{d}{dt}(\Gamma_1(t)^2 + \Gamma_2(t)^2) > 0$ when $t = 0$, the required conclusion will follow. Since this derivative is nonnegative in view of (a) and (b), all we need to establish is that it is nonzero.

Define $a = \Gamma_1'(0)$, $b = \Gamma_2'(0)$, $\alpha = \gamma_1'(\theta_0) = -\sin\theta_0$, $\beta = \gamma_2'(\theta_0) = \cos\theta_0$. Then hypotheses (c) and (d) yield

$$\alpha^2 + \beta^2 = 1, \quad a^2 + b^2 > 0, \quad a\alpha + b\beta = 0, \tag{1}$$

where we have availed ourselves of the fact that $\gamma_3'(\theta_0) = 0$ while using (d). When $t = 0$,

$$
\begin{aligned}
\tfrac{1}{2}\frac{d}{dt}(\Gamma_1(t)^2 + \Gamma_2(t)^2) &= \Gamma_1(0)\,\Gamma_1'(0) + \Gamma_2(0)\,\Gamma_2'(0) \\
&= \cos\theta_0\,\Gamma_1'(0) + \sin\theta_0\,\Gamma_2'(0) \qquad \text{by (b)} \\
&= a\beta - b\alpha.
\end{aligned}
$$

Because of (1), this cannot be 0. As noted at the end of the first paragraph, this is all we need to establish.

3-4.P23. Recall that Φ is given by

$$\Phi_1(u,v) = (1-\alpha(2u-1)\sin\pi v)\cos 2\pi v, \quad \Phi_2(u,v) = (1-\alpha(2u-1)\sin\pi v)\sin 2\pi v,$$
$$\Phi_3(u,v) = \alpha(2u-1)\cos\pi v.$$

where $0 < \alpha < 1$.

(a) is obvious, because $\Gamma(u_0) = (u_0, 0) = \gamma(0)$.

(b) $(\Phi \circ \Gamma)(s) = (1, 0, \alpha(2s-1))$. This yields $(\Phi \circ \Gamma)'(s) = (0, 0, 2\alpha)$ for all $s \in [0,1]$.

(c) We note the following simple computational facts:

    (i) replacing ξ by $1-\xi$ reverses the signs of $2\xi-1$, $\cos\pi\xi$, $\sin 2\pi\xi$ but preserves $\sin\pi\xi$ and $\cos 2\pi\xi$;

    (ii) replacing ξ by $-\xi$ reverses the signs of $\sin\pi\xi$, $\sin 2\pi\xi$ but preserves $\cos\pi\xi$ and $\cos 2\pi\xi$.

Using these, it is an easy computation that, for all $t \in [-1,1]$,

$$(\Phi \circ \gamma)_1(t) = (\Phi_1 \circ \gamma)(t) = (1 + \alpha(2u_0 - 1)\sin \pi t)\cos 2\pi t \,,$$
$$(\Phi \circ \gamma)_2(t) = (\Phi_2 \circ \gamma)(t) = -(1 + \alpha(2u_0 - 1)\sin \pi t)\sin 2\pi t \,,$$
$$(\Phi \circ \gamma)_3(t) = (\Phi_3 \circ \gamma)(t) = \alpha(2u_0 - 1)\cos \pi t \,.$$

This shows that $\Phi \circ \gamma$ is continuously differentiable on $[0,1]$.

(d) It follows from (b) that $(\Phi \circ \Gamma)'(u_0) = (0,0,2\alpha) \neq 0$. Therefore the required conclusion will follow if we can show that $(\Phi \circ \gamma)_3'(0) = 0$ but $(\Phi \circ \gamma)_2'(0) \neq 0$. From the computation of $(\Phi \circ \gamma)_3(t)$ in (c), we find that its derivative is

$$(\Phi \circ \gamma)_3'(t) = -\pi\alpha(2u_0 - 1)\sin \pi t \,,$$

so that $(\Phi \circ \gamma)_3'(0) = 0$. From the computation of $(\Phi \circ \gamma)_2(t)$ in (c), we find that its derivative is

$$(\Phi \circ \gamma)_2'(t) = -2\pi(1 + \alpha(2u_0 - 1)\sin \pi t)\cos 2\pi t - \pi\alpha(2u_0 - 1)\cos \pi t \sin 2\pi t \,.$$

For $t = 0$, this leads to
$$(\Phi \circ \gamma)_2'(0) = -2\pi \neq 0.$$
This establishes (d).

The interpretation about the graph of $\Phi \circ \Gamma$ lying on the 'edge' of M is that the points with $0 < t < 1$ cannot lie on the edge.

Problem Set 3-5

3-5.P1. No, because θ depends on h and there is no telling how it will behave as $h \to 0$.

3-5.P2. Since $f(h,0) = 0$, we have $D_1 f(0,0) = 0$. By an elementary computation, for $(x,y) \neq (0,0)$, we have $D_1 f(x,y) = 2xy^4/(x^2 + y^2)^2$. Therefore $D_1 f(0,k) = 0$ when $k \neq 0$. This shows that $D_{2\,1} f(0,0) = 0$. Similarly, $D_{1\,2} f(0,0) = 0$. An elementary computation also shows that, for $(x,y) \neq (0,0)$, we have $D_{2\,1} f(x,y) = [2xy/(x^2 + y^2)]^3$, which is not continuous at $(0,0)$.

3-5.P3. Since $f(h,0) = 0$, we have $D_1 f(0,0) = 0$. By an elementary computation, for $(x,y) \neq (0,0)$, we have $D_1 f(x,y) = 2xy^4/(x^2 + y^2)^2$. The partial derivatives of $D_1 f(x,y)$ at $(0,0)$ are both 0. If $D_1 f$ were differentiable at $(0,0)$, we would have

$$D_1 f(h,k) = D_1 f(h,k) - D_1 f(0,0) = hD_{1\,1} f(0,0) + kD_{2\,1} f(0,0) + (h^2 + k^2)^{1/2} u(h,k)$$
$$= (h^2 + k^2)^{1/2} u(h,k),$$

where $u(h,k)\to 0$ as $(h^2 + k^2)^{1/2}\to 0$, which amounts to $D_1 f(h,k)/(h^2 + k^2)^{1/2}\to 0$, i.e., $2hk^4/(h^2 + k^2)^{5/2}\to 0$ as $(h^2 + k^2)^{1/2}\to 0$. But this is false, because when $k = \lambda h$, we have $2hk^4/(h^2 + k^2)^{5/2} = 2\lambda^4/(1 + \lambda^2)^{5/2}$.

3-5.P4. $D_1 F = f'(x + g(y))$, $D_2 F = f'(x + g(y))g'(y)$, $D_{11} F = f''(x + g(y))$, $D_{12} F = f''(x + g(y))g'(y)$. So $(D_1 F)(D_{12} F) = f'(x + g(y))\, f''(x + g(y))g'(y) = (D_2 F)(D_{11} F)$.

3-5.P5. Let $f(t) = F(tx, ty)$. Then on the one hand, $f'(t) = x(D_1 F)(tx, ty) + y(D_2 F)(tx, ty)$, so that

$$f''(t) = x^2(D_{11} F)(tx, ty) + xy(D_{12} F)(tx, ty) + xy(D_{21} F)(tx, ty) + y^2(D_{22} F)(tx, ty)$$
$$= x^2(D_{11} F)(tx, ty) + 2xy(D_{12} F)(tx, ty) + y^2(D_{22} F)(tx, ty)$$

by Young's theorem (Theorem 3-5.4). On the other hand, $f(t) = t^p F(x,y)$, which leads to $f'(t) = pt^{p-1}F(x,y)$, so that $f''(t) = p(p-1)t^{p-2}F(x,y)$. So, $x^2(D_{11} F)(tx, ty) + 2xy(D_{12} F)(tx, ty) + y^2(D_{22} F)(tx, ty) = p(p-1)t^{p-2}F(x,y)$. Setting $t = 1$, we get the desired equality. We remark that one can do this with higher derivatives too and obtain an equality that can be expressed in self-explanatory notation as $(xD_1 + yD_2)^m F(x,y) = p(p-1)\cdots(p-m+1)F(x,y)$.

3-5.P6. Proceed as in the proof of Schwarz's theorem (Theorem 3-5.3) up to (2). But now $D_{21}f$ is also continuous and the analogue of (2) for $\psi(x) = f(x, b+h) - f(x,b)$ also holds. As in the proof of Young's theorem (Theorem 3-5.4), $\psi(a+h) - \psi(a) = \phi(b+h) - \phi(b)$. So, when $h = k$, the right sides of (2) and its analogue are equal. Cancel h^2 and use the given continuity.

3-5.P7. For $h \neq 0$,

$$f(h,y) - f(0,y) = h^2 \arctan\tfrac{y}{h} - y^2 \arctan\tfrac{h}{y} = h[h\arctan\tfrac{y}{h} - y(\tfrac{y}{h}\arctan\tfrac{h}{y})].$$

Therefore $D_1 f(0,y) = -y$ (including the case when $y = 0$). It follows that $D_{21}f(0,0) = -1$. A similar argument shows that $D_{12}f(0,0) = 1$.

3-5.P8. Write $D_1 f$ as f_1 and so on. By Euler's theorem (Theorem 3-4.9(a)), we have (1) $xf_1 + yf_2 + zf_3 = nf$. Upon differentiating with respect to x,y,z, we get (2) $xf_{11} + yf_{12} + zf_{13} = (n-1)f_1$ and two more equations (3) and (4). The reader would do well to write them out! It may be noted that the first order partial derivatives appear on the right sides with the factor $(n-1)$, but appear on the left side in (1) without the factor. We have four linear equations for three 'unknowns' x,y,z. Applying Cramer's rule to (2)–(4), we get $z = (n-1)B/H$, where B is the determinant of a certain matrix, which we shall denote by $[B]$. Moreover, the numerator of z when using Cramer's rule with (2),(3),(1) in that order works out to be $(n-1)A$. Also, the coefficient matrix is the transpose of $[B]$. Since it is given that $z \neq 0$, we know $B \neq 0$. Therefore $z = (n-1)A/B$. By multiplying the two solutions obtained for z, we get the required equality.

3-5.P9. Proceed as in 3-5.P8, but since nothing is guaranteed to be nonzero, Cramer's Rule only yields $Hz = (n-1)B$ and $Bz = (n-1)A$. Multiply to get $B^2 z(n-1) = AHz(n-1)$. First suppose $z \neq 0$. Then the required equality is immediate if $n-1 \neq 0$; but $n-1 = 0 \Rightarrow Hz = 0 = Bz$, $\Rightarrow H = 0 = B$ (as $z \neq 0$) $\Rightarrow B^2 = AH$, which establishes the required equality for all $z \neq 0$. For the rest, take the limit as $z \to 0$.

3-5.P10. Differentiate the equalities of 3-4.P14 with respect to v and u, respectively, and subtract.

3-5.P11. Use the chain rule twice successively in $F(x,y) = 0$, to get

$$F_x + F_y y' = 0 = F_{xx} + 2F_{xy} y' + F_{yy}(y')^2 + F_y y''.$$

Eliminate y': $F_y^3 y'' = -F_{xx} F_y^2 + 2F_{xy} F_x F_y - F_{yy} F_x^2$, which equals the given determinant.

3-5.P12. Observe that, since $f(s,t)$ is a continuous function of t for each s, it follows by the fundamental theorem of calculus (FTC) that $\int_c^y f(s,t)\, dt$ is a differentiable function of y with a continuous derivative $f(s,y)$ at every $(s,y) \in [a,b] \times [c,d]$. Therefore, by Leibnitz's formula [see 3-4.P17],

$$D_2 F(x,y) = \int_a^x \left[\frac{\partial}{\partial y} \left(\int_c^y f(s,t)\, dt \right) \right] ds = \int_a^x f(s,y)\, ds.$$

One further application of the FTC leads to $D_{1\,2} F(x,y) = f(x,y)$. Now, since f must be uniformly continuous, $\int_c^y f(s,t)\, dt$ is a continuous function of s for each y. Therefore by the FTC, $D_1 F(x,y) = \int_c^y f(x,t)\, dt$. Yet another application of the FTC leads to $D_{2\,1} F(x,y) = f(x,y)$.

3-5.P13. Proceed as in Corollary 3-5.6 after showing for each k that $|f_k(x+h) - f_k(x) - \sum_{j=1}^n h_j \cdot D_j f_k(x)| \leq Ln^{\frac{1}{2}} \|h\|_2$. To arrive at this, set up ϕ as in the proof of Theorem 3-5.5 and note that it satisfies

$$|\phi(1) - \phi(0) - \phi'(0)| = |\phi'(\theta) - \phi'(0)| = |\sum_{j=1}^n h_j \cdot D_j f_k(x + \theta h) - \sum_{j=1}^n h_j \cdot D_j f_k(x)|$$

$$\leq \sum_{j=1}^n |h_j| \, |D_j f_k(x + \theta h) - D_j f_k(x)| \leq L\|h\|_2 \sum_{j=1}^n |h_j| \leq L\|h\|_2^2 n^{\frac{1}{2}}$$

by Cauchy–Schwarz. But

$$|\phi(1) - \phi(0) - \phi'(0)| = |f_k(x+h) - f_k(x) - \sum_{j=1}^n h_j \cdot D_j f_k(x)|.$$

Problem Set 4-1

4-1.P1. Let $X = (0,1) \subseteq \mathbb{R}$ and $Tx = \frac{1}{2}x$. Then T is a contraction map without any fixed point. For (i), $[x \in \mathbb{R}, x = x^2] \Leftrightarrow x = 0$ or 1. So the fixed points are 0 and 1. For (ii), there are no fixed points unless $\alpha = 0$, in which case every $x \in \mathbb{R}$ is a fixed point. For (iii), the domain and range are disjoint and there is no question of fixed points.

4-1.P2. Let x_0 be the unique fixed point of T^3. Then $T^3(Tx_0) = T(T^3(x_0)) = Tx_0$. Therefore Tx_0 is a fixed point of T^3, so that $Tx_0 = x_0$. Thus x_0 is a fixed point of T as well. If T were to have another fixed point, then so would T^3. The reader will see that the result is true for any power of T.

4-1.P3. (a) If x_0 and y_0 are both fixed points of T, then $\|Tx_0 - Ty_0\| = \|x_0 - y_0\|$, which contradicts the hypothesis unless $x_0 = y_0$.
(b) The hypothesis here immediately implies that of part (a).

4-1.P4. (a) $\|Tx - Ty\| = |(x + 1/x) - (y + 1/y)| = |(x - y) + (1/x - 1/y)| = |(x - y) - (x - y)/_{xy}| = |x - y||1 - 1/_{xy}| < |x - y| = \|x - y\|$, and $x + 1/_x \neq x$ for all x.
(b) $f'(x) = 1 - e^x(1 + e^x)^{-2}$ and $0 < 1 - e^x(1 + e^x)^{-2} < 1$ everywhere. If z were to be a fixed point, we would have $z = f(z) = z + (1 + e^z)^{-1}$, which implies $(1 + e^z)^{-1} = 0$, a contradiction.

4-1.P5. Consider the real-valued map $g : X \to \mathbb{R}$ defined by $g(x) = \|x - Tx\|$. Then

$$|g(x) - g(y)| = |\,\|x - Tx\| - \|y - Ty\|\,|$$
$$\leq |\,\|x - Tx\| - \|y - Tx\|\,| + |\,\|y - Tx\| - \|y - Ty\|\,|$$
$$\leq \|x - y\| + \|Tx - Ty\| \leq 2\|x - y\|.$$

Therefore g is continuous. Since X is compact, g attains its minimum at some $z \in X$. If $z \neq T(z)$, then $g(T(z)) = \|Tz - T(Tz)\| < \|z - Tz\| = g(z)$, contradicting the minimality of $g(z)$. Therefore $z = T(z)$, i.e., z is a fixed point of T. If w were to be another fixed point, then we would have $\|Tz - Tw\| = \|z - w\|$ although $z \neq w$, a contradiction.

4-1.P6. $\frac{d}{dx} Tx = 1 - 7x^6/500$. Therefore, when $x \in [1,2]$ the double inequality $0 < \frac{d}{dx} Tx < 1$ must hold. From the first part of this double inequality, it follows that $x - (x^7 - 6)/500$ describes an increasing function on $[1,2]$. Besides, $T(1) = 1 +$

5/500 > 1 and $T(2) = 2 - (128 - 6)/500 = 439/250 < 2$. Therefore T maps $[1,2]$ into itself. Now max $(\frac{d}{dx}Tx)$ exists because $\frac{d}{dx}Tx$ is a continuous function on the closed bounded interval $[1,2]$. The second part of the above double inequality implies that $|\frac{d}{dx}Tx| < 1$. By Proposition 4-1.5, it follows that T is a contraction and hence by Contraction Principle 4-1.6, the given sequence converges to the unique fixed point of T, which is easily seen to be $\sqrt[3]{6}$.

4-1.P7. $T^2 x = 1$ for every $x \in [0,3]$ and therefore T^n is a contraction when $n = 2$.

4-1.P8. By (i), $\alpha_n \geq 0$ for all n. By (ii), there exists a positive $k < 1$ and a positive integer $m > N$ such that $0 \leq \alpha_m < k$. From (i), it now follows that

$$\|T^m x - T^m y\| \leq k\|x-y\| \text{ for all } x, y \in X.$$

That is, T^m is a contraction. Now apply Corollary 4-1.8.

4-1.P9. Note that $x \in \mathbb{R}$ is a fixed point of f if and only if x is a root of the polynomial $g(x) = x^3 - 3x + 1$.
(a) By direct evaluation, we find $g(-2) < 0 < g(-1)$, $g(0) > 0 > g(1)$ and $g(1) < 0 < g(2)$. So, there is a root u between -2 and -1, a root v between 0 and 1 and a root w between 1 and 2. Then $u < v < w$. Since g is a polynomial of degree 3, there are no further roots. It follows that these are the only three fixed points of f. This establishes (a). We remark for later reference that $g > 0$ on (u,v) and $g < 0$ on (v,w).
(b) Since the map $t \to t^3$ is strictly increasing on $\mathbb{R}$, therefore f is strictly increasing on $[u,w]$. But $f(u) = u$ and $f(w) = w$. So, f maps the open interval (u,w) onto itself.
(c) If $x = v$, then $f^n(v) = v$ for all n. Suppose $u < x < v$. Then on the one hand, $f(x) < f(v) = v$ and on the other hand, $x^3 - 3x + 1 = g(x) > 0$ [see remark above at the end of (a)], so that $\frac{1}{3}(1 + x^3) > x$, i.e., $f(x) > x$. Thus $u < x < v \Rightarrow u < x < f(x) < v$. Therefore $f^n(x)$ is an increasing sequence in (u,v) and thus has a limit in $(u,v]$. Its limit is easily seen to be a fixed point of f and must therefore be equal to v. A similar argument when $v < x < w$ shows that $f^n(x)$ is a decreasing sequence in (v,w) with limit v.
(d) Since $w > 1$ (as seen in (a)), there exist x_1 and x_2 in (u,w) such that both are greater than 1. By the mean value theorem, $|f(x_1) - f(x_2)| = |x_1 - x_2||f'(t)|$, where t lies between x_1 and x_2; in particular $t > 1$. But $f'(t) = t^2 > 1$. So, $|f(x_1) - f(x_2)| > |x_1 - x_2|$.

4-1.P10. (a) $f(x) = x \Leftrightarrow g(x) = x^5 + x - 1 = 0$. Now $g(\frac{3}{4}) = \frac{3}{4}((\frac{3}{4})^4 + 1) - 1 =$ $\frac{1011}{1024} - 1 < 0 < g(1)$ and $g'(x) = 5x^4 + 1 > 0$ everywhere. So, g has a unique root in $\mathbb{R}$, which lies in $[\frac{3}{4}, 1]$, i.e., f has a unique fixed point $\alpha \in \mathbb{R}$, which lies in $[\frac{3}{4}, 1]$. (b) $f'(x) = -4x^3/(x^4 + 1)^2 < 0$ when $x > 0$. This and the fact that $f(y) > 0 \; \forall \; y \in \mathbb{R}$ lead to the required conclusion.

(c) Since $0 < \frac{3}{4} < \alpha$ by (a), we have $f(\frac{3}{4}) > \alpha > \frac{3}{4}$ by (b). First we show that f is a self map of the interval $[\frac{3}{4}, f(\frac{3}{4})]$. Consider any x such that $\frac{3}{4} \le x \le f(\frac{3}{4})$. It follows from (b) and (i) that $f(\frac{3}{4}) \ge f(x) \ge f^2(\frac{3}{4}) > \frac{3}{4}$. So, f is a self map of $[\frac{3}{4}, f(\frac{3}{4})]$. Next, we show that $|f'(x)| \le |f'(\frac{16}{21})|$ on $[\frac{3}{4}, f(\frac{3}{4})]$. To this end, $\frac{d}{dx}|f'(x)| = \frac{d}{dx}[4x^3/(x^4 + 1)^2] = 4x^2(-5x^4 + 3)/(x^4 + 1)^3$, which is positive when $x^4 < \frac{3}{5}$. Now, $(\frac{16}{21})^4 < (\frac{16}{21})^2 = \frac{256}{441} < \frac{3}{5}$. Therefore $|f'(x)|$ is increasing on $[\frac{3}{4}, \frac{16}{21}]$. Hence $|f'(x)| \le |f'(\frac{16}{21})|$ on $[\frac{3}{4}, \frac{16}{21}]$. But $\frac{3}{4} < f(\frac{3}{4}) = \frac{256}{337} < \frac{256}{336} = \frac{16}{21}$. Therefore the interval in question, namely $[\frac{3}{4}, f(\frac{3}{4})]$, is a subinterval of $[\frac{3}{4}, \frac{16}{21}]$. Consequently, the inequality $|f'(x)| \le |f'(\frac{16}{21})|$ holds on $[\frac{3}{4}, f(\frac{3}{4})]$ as well. From (i) and Proposition 4-1.5, it now follows that f is a contraction. Note: The contraction f approximates α via the contraction principle, but Newton's method applied to the polynomial g works much faster.

4-1.P11. Consider a sequence $\{r_p\}$ in $(0,1)$ such that $r_p \to 1$. For each p, the map $T_p(x) = r_p(Tx)$ is a contraction in S and therefore has a fixed point x_p, i.e., $x_p \in S$ and $r_p(Tx_p) = x_p$. Since S is compact, some subsequence $\{x_{p(k)}\}$ converges to a limit $x \in S$. Since T must be continuous (in view of the inequality it satisfies), the equality $r_{p(k)}(Tx_{p(k)}) = x_{p(k)} \; \forall \; k$ immediately shows that x is a fixed point of T. The map T given by $Tx = x$ satisfies the condition; the reader may observe that *every* point is a fixed point.

4-1.P12. Define $f:X \to \mathbb{R}$ by $f(x) = \|Tx - x\| \ge 0$. Then $\|f(x) - f(\xi)\| \le (1 + c)$ $\|x - \xi\|$, where c is the contraction constant. This inequality shows that f is continuous. Moreover,

$$f(Tx) = \|T(Tx) - Tx\| \le c\|Tx - x\| = cf(x) \text{ for all } x \in X.$$

In the case when X is closed as well as bounded, it is compact [Theorem 2-5.7] and we can deduce that $f(x_0) = 0$ for some $x_0 \in X$ by Theorem 2-6.13. This x_0 is then a fixed point of T, considering that $\|Tx_0 - x_0\| = f(x_0) = 0$.

Now suppose X is not bounded (but is closed). Take any $z \in X$ and consider the set $Y = \{x \in X : f(x) \le f(z)\}$. Since f is continuous, Y is closed. For $y \in Y$, we have

$$\|z - y\| \le \|Tz - z\| + \|Tz - Ty\| + \|Ty - y\| \le 2f(z) + c\|z - y\|.$$

Hence

$$\|z - y\| \le 2f(z)/(1 - c).$$

Consequently, Y is not only closed but also bounded. Besides, T maps Y into Y because

$$y \in Y \Rightarrow f(Ty) \le cf(y) \le cf(z) \le f(z) \Rightarrow Ty \in Y.$$

Therefore T has a fixed point in Y and hence also in X.

Finally, since

$$\|x - x_0\| = \|x - Tx_0\| \le \|Tx - x\| + \|Tx - Tx_0\| \le f(x) + c\|x - x_0\|,$$

we have $\|x - x_0\| \le f(x)/(1 - c)$. It follows that

$$\|x_p - x_0\| \le f(x_p)/(1 - c) = f(T^{p-1}x_1)/(1 - c) \le c^{p-1}f(x_1)/(1 - c),$$

which shows that the sequence $\{x_p\}_{p \ge 1}$ converges to x_0.

Problem Set 4-2

4-2.P1. For both g and g_1, the interval $(3,5)$ may be taken as the open set; $g_1(y) = -y^{1/2}$. No, because any open set containing 0 must also contain negative numbers, which cannot be squares of any real numbers.

4-2.P2. Since $\frac{dy}{dx} = (1 + x)e^x > 0$ on $(-1, \infty)$, the given function is injective when restricted to this interval and has range $(-e^{-1}, \infty)$, which is an open subset U of $\mathbb{R}$ containing e. Therefore there is a continuous inverse g with domain U. These will serve the purpose; so will any subset V of U that contains e and is open, taken with the restriction of g to V. Corresponding question for $(-2, -2/e^2)$: The point $(-2, -2/e^2)$ lies on the graph of $y = xe^x$. Find an open set containing $y = -2/e^2$ such that there is a continuous function $x = g(y)$ defined on it, for which $x = g(y) \Rightarrow y = xe^x$ and $g(-2/e^2) = -2$. The answer is that $\frac{dy}{dx} < 0$ on $(-\infty, -\frac{3}{2})$ and the given function when restricted to this interval is injective and has range $U =$

$(-\frac{3}{2}e^{-3/2}, 0)$. Take g to be the inverse defined on U. Note that we know the existence of the required map though we cannot compute it explicitly.

4-2.P3. The first part is the same as Example 4-2.2(d). As $f(x,y) = f(x, y + 2\pi)$, f is not injective.

4-2.P4. The inverse function theorem says only that the function is *locally* invertible. Since it is not injective [as seen in 4-2.P3], it is certainly not invertible.

4-2.P5. Let $U \subseteq E$ be open and $b \in f(U)$. Then $b = f(a)$ for some $a \in U$. Since $f'(a)$ is invertible, the inverse function theorem yields open sets $U_1 \subseteq U$ and $V \subseteq \mathbb{R}^n$ such that $a \in U_1$ and $f(U_1) = V$. But then $f(a) \in V \subseteq f(U)$. Since $b = f(a)$ and V is open, this means $f(U)$ is open.

4-2.P6. Let b be any point of V. Then $b = f(a)$ for some $a \in U$. By hypothesis, $f'(a)$ is invertible and hence the inverse function theorem yields open sets $U_1 \subseteq U$ and $V_1 \subseteq \mathbb{R}^n$ such that $a \in U_1$ and $f(U_1) = V_1$ and f has a differentiable local inverse on V_1. But then $b = f(a) \in V_1$ and the local inverse is therefore differentiable at b. However, g has to agree with the local inverse and must therefore also be differentiable at b.

4-2.P7. Let K denote the closure of $f(V)$. It is trivial to show that $f(\overline{V}) \subseteq K$. To prove the reverse inclusion, consider any $y \in K$. Then there exists a sequence $\{x_n\}$ in V such that $f(x_n) \to y$. Since $\overline{V}$ is compact, $x_n \to x \in \overline{V}$ when $\{x_n\}$ is replaced by a suitable subsequence. From the continuity of f, it follows that $f(x_n) \to f(x)$, so that $y = f(x)$. Since $x \in \overline{V}$, we have $y \in f(\overline{V})$. So $K \subseteq f(\overline{V})$, and hence $K = f(\overline{V})$. Finally, we note that the set $f(V)$, of which K is the closure, is an open set, as proved in 4-2.P5.

4-2.P8. $[f(h) - f(0)]/h = 1 + 2h\sin(1/h) \to 1$ as $h \to 0$. Also

$$f'(x) = 1 + 4x\sin(1/x) - 2\cos(1/x) \text{ for } x \neq 0.$$

Now consider any $(-\delta, \delta)$. When n is any natural number greater than $1/2\pi\delta$, the number $x = 1/2\pi n$ belongs to $(-\delta, \delta)$ and satisfies $f'(x) = -1 < 0$, while the number $x = 1/(2n+1)\pi$ also belongs to $(-\delta, \delta)$ but satisfies $f'(x) = 3$. Thus f' takes positive as well as negative values in $(-\delta, \delta)$, and consequently, f cannot be injective on $(-\delta, \delta)$.

4-2.P9. (a) $\Phi(s,x) = \Phi(s',x') \Rightarrow (\phi(s) + \psi(x), x) = (\phi(s') + \psi(x'), x') \Rightarrow [x = x', \phi(s) + \psi(x) = \phi(s') + \psi(x')] \Rightarrow [\psi(x) = \psi(x'), \phi(s) + \psi(x) = \phi(s') + \psi(x')] \Rightarrow \phi(s) = \phi(s') \Rightarrow s = s'$. (b) Let $(s',x') \in \mathbb{R}^n \times X$. Then $s' - \psi(x') = \phi(s)$ for some s, so that $(s',x') = (\phi(s) + \psi(x'), x') = \Phi(s,x')$.

4-2.P10. Since all partial derivatives are continuous in open set containing $(0,0,0)$, f is continuously differentiable. The determinant of the Jacobian matrix at $(0,0,0)$ is found to be 20, which is nonzero. By the inverse function theorem, f has a continuously differentiable local inverse at (meaning, on some open set containing) the point $(0,0,0)$.

4-2.P11. Sufficient conditions are that f and g be continuously differentiable mappings of all pairs (u,v) belonging to some open set containing (u_0,v_0), where $x_0 = f(u_0,v_0)$ and $y_0 = g(u_0,v_0)$ and $\partial(f,g)/\partial(u,v) \neq 0$ at (u_0,v_0). Differentiating the equalities $u = F(x,y)$, $v = G(x,y)$ with respect to u and v, we get

$$1 = \frac{\partial F}{\partial x}\frac{\partial f}{\partial u} + \frac{\partial F}{\partial y}\frac{\partial g}{\partial u}, \qquad 0 = \frac{\partial F}{\partial x}\frac{\partial f}{\partial v} + \frac{\partial F}{\partial y}\frac{\partial g}{\partial v},$$

$$0 = \frac{\partial G}{\partial x}\frac{\partial f}{\partial u} + \frac{\partial G}{\partial y}\frac{\partial g}{\partial u}, \qquad 1 = \frac{\partial G}{\partial x}\frac{\partial f}{\partial v} + \frac{\partial G}{\partial y}\frac{\partial g}{\partial v}.$$

We get the required equalities by solving these four equations. (Remark: Actually these equations merely state that the Jacobian matrix of a map and of its inverse are inverse matrices. Therefore, if one knows how to express each entry of the inverse of a matrix in terms of the 'cofactors' from the latter, one can easily generalise the result of this problem to higher dimensions.)

4-2.P12. Since $f(-x,-y) = f(x,y)$ and $(-x,-y) \neq (x,y)$ whenever $(x,y) \neq (0,0)$, therefore f always maps *at least* two points of U into the same point. To show that precisely two points of U are mapped into the same point, consider $(x,y),(u,v) \in U$ that are mapped into the same point. Then we have the two equations $x^2 - y^2 = u^2 - v^2$ and $2xy = 2uv$. Squaring the first and using the second, we get $x^2 + y^2 = u^2 + v^2$, which, upon being combined with the first, leads to $x^2 = u^2$, $y^2 = v^2$. We may suppose $x \neq 0$. If $y = 0$, then the preceding equations imply that $v = 0$ and $x = \pm u$, so that either $(x,y) = (u,v)$ or $(x,y) = (-u,-v)$. If $y \neq 0$, then the second of the original two equations shows that u,v have the same or opposite signs according as x,y do. It follows again that either $(x,y) = (u,v)$ or $(x,y) = (-u,-v)$. This shows that the mapping is 'two-to-one'. To show that the mapping is surjective, we note that the equations $s = x^2 - y^2$, $t = 2xy$ can be solved for (x^2,y^2) by elementary methods by first obtaining $s^2 = (x^2 + y^2)^2 - t^2$ and then $x^2 + y^2 = \sqrt{(s^2 + t^2)}$, leading to $x^2 = \frac{1}{2}(\sqrt{(s^2 + t^2)} + s)$ and $y^2 = \frac{1}{2}(\sqrt{(s^2 + t^2)} - s)$. These equations show that $(s,t) \neq (0,0) \Rightarrow (x,y) \neq (0,0)$. This demonstrates surjectivity but falls short of obtaining an explicit differentiable local inverse, because we have not shown that the sign ambiguity can be settled on some open set in such a manner as to ensure differentiability. Rather than doing this, we use the inverse function theorem. The linear derivative Df at any (x,y) has matrix with first row $[2x \quad -2y]$ and second row $[2y \quad 2x]$; it can be seen to be invertible for any $(x,y) \neq (0,0)$, for instance via its determinant, which is $4(x^2 + y^2)$.

4-2.P13. Since the three functions mentioned are continuously differentiable and $x^2 + y^2$ does not vanish on U, the component functions f_1 and f_2 are continuously differentiable. Therefore, so is f. If f' is invertible at some $(a,b) \in U$, then by the inverse function theorem, f must be injective on some open set containing (a,b). However, any open set containing (a,b) contains points $(\lambda a, \lambda b) \neq (a,b)$ and $f(\lambda a, \lambda b) = f(a,b)$. This contradicts the injectivity of f on the open set.

4-2.P14. What is to be proved is that the map $\phi: U \to \mathbb{R}^n$ defined by $\phi(x) = \left(\frac{f_1(x)}{h(x)}, \ldots, \frac{f_n(x)}{h(x)} \right)$ has a noninvertible linear derivative everywhere on U.

Since $f_1, f_1, \ldots, f_n$ and h are continuously differentiable and h vanishes nowhere on U, the component functions of ϕ are continuously differentiable and hence so is ϕ. Therefore by Theorem 2-7.15, if ϕ' is invertible at some $(a,b) \in U$, it is invertible on an open set containing (a,b) and we may assume that $(a,b) \neq (0,0)$. Now, by the inverse function theorem, ϕ must be injective on some open set containing (a,b). However, any open set containing (a,b) contains points $(\lambda a, \lambda b) \neq (a,b)$ and, at the same time, by the homogeneity hypothesis, $\phi(\lambda a, \lambda b) = \phi(a,b)$. This contradicts the injectivity of f on the open set. To deduce 4-2.P13, take $n = 2$, $U = \{(x,y) \in \mathbb{R}^2 : (x,y) \neq (0,0)\}$, $f_1(x,y) = x^2 - y^2$, $f_2(x,y) = xy$ and $h(x,y) = x^2 + y^2$.

4-2.P15. Since u, v, w are continuously differentiable and homogeneous of degree 0, the same argument as in 4-2.P14 applies. If $f_1, f_2, \ldots, f_n$ are continuously differentiable and homogeneous of degree 0 on an open set in $\mathbb{R}^n$, then their Jacobian is zero everywhere on that set.

4-2.P16. (a) Obvious that $r > 0$. Since the range of $\cos^{-1}$ is $[0,\pi]$, we have $\theta \in [0,\pi]$ when $y \geq 0$. But when $y < 0$, we have $x/(x^2 + y^2)^{1/2} \neq -1$, so that $\cos^{-1}(x/(x^2 + y^2)^{1/2}) \neq \pi$ and hence $\theta \neq -\pi$.

(b) First, note that $\sin(\cos^{-1} u) = \sqrt{(1 - u^2)} \ \forall \ u \in [-1,1]$. Therefore when $y < 0$, we have $\sin\theta = -\sin[\cos^{-1}(x/(x^2 + y^2)^{1/2})] = -|y|/(x^2 + y^2)^{1/2} = -(-y)/(x^2 + y^2)^{1/2} = y/r$, and when $y \geq 0$, we have $\sin\theta = \sin[\cos^{-1}(x/(x^2 + y^2)^{1/2})] = |y|/(x^2 + y^2)^{1/2} = y/(x^2 + y^2)^{1/2} = y/r$. Also, $\cos\theta = \cos(-\theta) = x/(x^2 + y^2)^{1/2} = x/r$, whether $y < 0$ or ≥ 0.

(c) Suppose $(r,\theta) \in (0,\infty) \times (-\pi,\pi]$ and $x = r\cos\theta$, $y = r\sin\theta$. By definition of g, the first component of $g(x,y)$ is $(x^2 + y^2)^{1/2} = (r^2\cos^2\theta + r^2\sin^2\theta)^{1/2} = r$, remembering that $r \in (0,\infty)$. If $y = r\sin\theta \geq 0$, then $\sin\theta \geq 0$ and so $\theta \notin (-\pi,0)$; hence $\theta \in [0,\pi]$. Moreover, the second component of $g(x,y)$ is $\cos^{-1}(x/(x^2 + y^2)^{1/2}) = \cos^{-1}(\cos\theta) = \theta$, considering that $\theta \in [0,\pi]$. If $y = r\sin\theta < 0$, then $\sin\theta < 0$ and so $\theta \notin [0,\pi]$; hence $\theta \in (-\pi,0)$. Moreover, the second component of $g(x,y)$ is

$-\cos^{-1}(x/(x^2+y^2)^{1/2}) = -\cos^{-1}(\cos\theta) = \theta$, considering that $\theta \in (-\pi, 0)$. In either case, $g(x,y) = (r,\theta)$.

(d) For the sequence $(x_n, y_n) = (-1, \frac{1}{n})$, we have $\theta_n \to \cos^{-1}(-1) = \pi$, but for $(x_n, y_n) = (-1, -\frac{1}{n})$, we have $\theta_n \to -\cos^{-1}(-1) = \pi$. So $\lim g(x,y)$ as $(x,y) \to (-1,0)$ does not exist.

4-2.P17. $f'(a) = 2e$. So, $\phi_y(x) = x + \frac{1}{2e}(y - xe^x)$; also, $x_2(y) = 1 + \frac{1}{2e}(y - e)$ and

$$x_3(y) = 1 + \frac{1}{2e}(y - e) + \frac{1}{2e}[y - (1 + \frac{1}{2e}(y - e)) \cdot \exp(1 + \frac{1}{2e}(y - e))].$$

Only x_1 and x_2 are partial sums of the Taylor series of f^{-1} at $y = e$. The point in question is $f(a) = e$.

4-2.P18. Clearly f' is continuously differentiable everywhere and its derivative at $(0,0)$ is the identity map. Moreover, it maps $(0,0)$ into itself. By the inverse function theorem, it has a local inverse at $(0,0)$. By the argument in the proof of the theorem, an approximating sequence for the local inverse, valid in some ball centred at $(0,0)$, is generated by the contraction map

$$\phi_{(u,v)}(x,y) = (x,y) + ((u,v) - f(x,y)) = (u - y^2, v - x^3).$$

Thus, if the approximating sequence starts with $(x_1, y_1) = (0,0)$, then the second term is

$$(x_2, y_2) = (x_1, y_1) + ((u,v) - f(x_1, y_1)) = (0,0) + ((u,v) - (0,0)) = (u,v)$$

and the third term is

$$(x_3, y_3) = (x_2, y_2) + ((u,v) - f(x_2, y_2)) = (x_2, y_2) + ((u,v) - (x_2 + y_2^2, x_2^3 + y_2))$$
$$= (u,v) + ((u,v) - (u + v^2, u^3 + v)) = (u - v^2, v - u^3).$$

Problem Set 4-3

4-3.P1. No; the relevant Jacobian has value 0.

4-3.P2. Let $f_1(x,y,z,u) = 3x + y - z - u^3$, $f_2(x,y,z,u) = x - y + 2z + u$, $f_3(x,y,z,u) = 2x + 2y - 3z + 2u$. Then $\partial(f_1, f_2, f_3)/\partial(x,y,u) = -12 - 12u^2$, which can never be 0. If there were to exist a solution for x, y, z valid on some interval (in which u varies), then the fact that $(2x + 2y - 3z) = (3x + y - z) - (x - y + 2z)$ would imply that $-2u = u^3 + u$ on that interval, which is plainly impossible.

4-3.P3. Clearly, $f(1,0,0,1) = (0,0)$. Denote the component functions of f by f_1 and f_2. Then

$$f_1(x,y) = f_1(x_1,x_2,y_1,y_2) = x_1 y_2 + x_2 y_1 - 1$$

and

$$f_2(x,y) = f_2(x_1,x_2,y_1,y_2) = x_1 x_2 - y_1 y_2 .$$

The linear derivative $D_y f$ at $(1,0,01)$ is given by the matrix

$$\begin{bmatrix} \partial f_1/\partial y_1 & \partial f_1/\partial y_2 \\ \partial f_2/\partial y_1 & \partial f_2/\partial y_2 \end{bmatrix} = \begin{bmatrix} 0 & 1 \\ -1 & 0 \end{bmatrix},$$

which is invertible. Therefore the theorem shows that y is a function of x near $(1,0)$. Also, the linear derivative $D_x f$ at $(1,0,01)$ is given by the matrix

$$\begin{bmatrix} \partial f_1/\partial x_1 & \partial f_1/\partial x_2 \\ \partial f_2/\partial x_1 & \partial f_2/\partial x_2 \end{bmatrix} = \begin{bmatrix} 1 & 0 \\ 0 & 1 \end{bmatrix}.$$

Therefore the required linear derivative at the point $(1,0,01)$ has matrix

$$-\begin{bmatrix} 0 & 1 \\ -1 & 0 \end{bmatrix}^{-1} \begin{bmatrix} 1 & 0 \\ 0 & 1 \end{bmatrix} = \begin{bmatrix} 0 & 1 \\ -1 & 0 \end{bmatrix}.$$

4-3.P4. We have $f(x(y,z),y,z) = 0 \ \forall \ y, z$. Using the chain rule to differentiate with respect to y, we get $\frac{\partial f}{\partial x}(x(y,z),y,z) \cdot \frac{\partial x}{\partial y}(y,z) + \frac{\partial f}{\partial y}(x(y,z),y,z) = 0$. Similarly, from the identity $f(x,y(z,x),z) = 0 \ \forall \ z, x$, we get $\frac{\partial f}{\partial y}(x,y(z,x),z) \cdot \frac{\partial y}{\partial z}(z,x) + \frac{\partial f}{\partial z}(x,y(z,x),z) = 0$. Finally, $\frac{\partial f}{\partial z}(x,y,z(x,y)) \cdot \frac{\partial z}{\partial x}(x,y) + \frac{\partial f}{\partial x}(x,y,z(x,y)) = 0$. If $f(x,y,z) = 0$, then $(x,y,z) = (x(y,z),y,z) = (x,y(z,x),z) = (x,y,z(x,y))$. Therefore we can write the three equations as $\frac{\partial f}{\partial x}\frac{\partial x}{\partial y} + \frac{\partial f}{\partial y} = 0$, $\frac{\partial f}{\partial y}\frac{\partial y}{\partial z} + \frac{\partial f}{\partial z} = 0$, $\frac{\partial f}{\partial z}\frac{\partial z}{\partial x} + \frac{\partial f}{\partial x}$ $= 0$. If $\frac{\partial f}{\partial x} = 0$, it follows from the first equation that $\frac{\partial f}{\partial y} = 0$ and then from the second that $\frac{\partial f}{\partial z} = 0$ as well. Thus, if one is nonzero, so are the other two. Furthermore, from the first equation, we have $\frac{\partial f}{\partial y} = -\frac{\partial f}{\partial x}\frac{\partial x}{\partial y}$; substituting this in the second, we get $-\frac{\partial f}{\partial x}\frac{\partial x}{\partial y}\frac{\partial y}{\partial z} + \frac{\partial f}{\partial z} = 0$. By substituting this in the third, we get $\frac{\partial f}{\partial x}\frac{\partial x}{\partial y}\frac{\partial y}{\partial z}\frac{\partial z}{\partial x} + \frac{\partial f}{\partial x} = 0$, or $\frac{\partial f}{\partial x}(\frac{\partial x}{\partial y}\frac{\partial y}{\partial z}\frac{\partial z}{\partial x} + 1) = 0$. The desired equality is now immediate.

4-3.P5. The product equals 1, not -1, because $\frac{\partial x}{\partial y} = -y/\sqrt{(1 - y^2 - z^2 - u^2)} = -1 = \frac{\partial y}{\partial z} = \frac{\partial z}{\partial u} = \frac{\partial u}{\partial x}$.

4-3.P6. By definition, $h = f \circ \phi$, where $\phi(x_1, \ldots, x_n) = (g_1(x_1), g_2(x_2), \ldots, g_n(x_n))$. By the chain rule and the fact that the determinant of a product of matrices is the product of their determinants, we have $J_h(x) = J_f(\phi(x)) \cdot J_\phi(x_1, \ldots, x_n)$. Now, the Jacobian matrix of ϕ has diagonal entries $g_i'(x_i)$, $1 \le i \le n$, and all other entries are 0. Therefore its determinant is $J_\phi(x_1, \ldots, x_n) = g_1'(x_1) \cdot g_2'(x_2) \cdot \cdots \cdot g_n'(x_n)$.

4-3.P7. f maps $\mathbb{R}^1 \times \mathbb{R}^2$ into $\mathbb{R}^1 = \mathbb{R}$ and has a continuous linear derivative everywhere because all its three partial derivatives

$$D_1 f(x, y_1, y_2) = 2xy_1 + e^x, \qquad D_2 f(x, y_1, y_2) = x^2, \qquad D_3 f(x, y_1, y_2) = 1$$

are continuous. Also, $f(0, 1, -1) = 0 + 1 - 1 = 0$ and $D_x f(0, 1, -1) = D_1 f(0, 1, -1) = 1 \ne 0$. By Implicit Function Theorem 4-3.2, there exists a differentiable function g on an open set containing $(1, -1)$ in $\mathbb{R}^2$ such that $g(1, -1) = 0$ and $f(g(y_1, y_2), y_1, y_2) = 0$ on that open set. Moreover, its linear derivative $[(D_1 g)(1, -1) \quad (D_2 g)(1, -1)]$ is $-A_1^{-1} A_2$, where

$$A_1 = D_x f(0, 1, -1) = 1$$

and

$$A_2 = D_{(y_1, y_2)} f(0, 1, -1) = [D_2 f(0, 1, -1) \quad D_3 f(0, 1, -1)] = [0 \quad 1].$$

Thus $[(D_1 g)(1, -1) \quad (D_2 g)(1, -1)] = [0 \quad -1]$. This means $(D_1 g)(1, -1) = 0$ and $(D_2 g)(1, -1) = -1$.

4-3.P8. Let $F : \mathbb{R}^2 \to \mathbb{R}$ be defined by $F(x, t) = f(x) - tg(x)$. Then $F(0, 0) = 0$ and $(D_x F)(0, 0) = f'(0) - 0 \cdot g'(0) \ne 0$. It follows from the implicit function theorem that the function $x = x(t)$ of the required kind exists on a suitable interval $(-\delta, \delta)$ and that $x'(0) = -(D_t F)(0, 0)/(D_x F)(0, 0)$. Since $(D_t F)(0, 0) = -g(0)$, we have $x'(0) = g(0)/f'(0)$. When $g(0) = 0$, we take $x(t) = 0$ on $\mathbb{R}$.

Problem Set 4-4

4-4.P1. (i) $W = (-1/\sqrt{2}, 1/\sqrt{2})$. (ii) With $W = (-1, -1/\sqrt{2}) \cup (-1/\sqrt{2}, 1/\sqrt{2}) \cup (1/\sqrt{2}, 1)$, take $g(y)$ to be $\sqrt{(1 - y^2)}$ on the second interval and either $\sqrt{(1 - y^2)}$ or $-\sqrt{(1 - y^2)}$ on the other two (in any combination); all satisfy $g(0) = 1$. (iii) Same as part (ii); the only solution satisfying $g(y) > 0$ is $g(y) = \sqrt{(1 - y^2)} \; \forall \, y \in W$.

4-4.P2. There would be no purpose to this problem if it merely called for the two proofs to be written one after the other. The uniqueness established *during the proof* of (a) of Theorem 4-3.2 (just before defining g) can be used in the proof of (b) of Theorem 4-4.1 to show instantly that G agrees with G_1 on B_1. So the construction of the map G_2 becomes unnecessary. Since that uniqueness was not recorded as one of the conclusions of Theorem 4-3.2, it was unavailable while proving Theorem 4-4.1 separately.

4-4.P3. (a) $f(x,y) = y^3 - |x|^{1/2}$; $g(x) = |x|^{1/6}$.
(b) $f(x,y) = y^2 - x^2$; $g_1(x) = x$ and $g_2(x) = -x$ are both differentiable solutions. Any solution g must satisfy $|g(x)| = |x|$. If it is also differentiable, then the identity $g(x)^2 = x^2$ leads to $g(x)g'(x) = x$ so that $|g'(x)| = 1$. Since a derivative has the intermediate value property, we must have either $g'(x) = 1$ everywhere or $g'(x) = -1$ everywhere. So, g_1 and g_2 are the only possibilities.
(c) $f(x,y) = y^2 - x^4$; $g_1(x) = x^2$, $g_2(x) = -g_1(x)$, $g_3(x) = x^2$ for $x \geq 0$ and $-x^2$ for $x < 0$, $g_4(x) = -g_3(x)$.

4-4.P4. Modify the proof of Theorem 4-4.4 as follows: The function $y \to F(x,y)$ has a positive derivative at $y = b$. Therefore there exists a positive $\eta_1 < \eta$ such that the function is negative at $b - \eta_1$ and positive at $b + \eta_1$. Now apply the intermediate value theorem on the interval $[b - \eta_1, b + \eta_1]$. The proof of differentiability carries over without any modification.

4-4.P5. The proof of Theorem 4-4.1 carries over almost verbatim.

Problem Set 5-1

5-1.P1. $\phi(x,y) = xy$, $f(x,y) = y - x^2$, so that $\Phi(x) = x^3$. The Lagrange equations $y - 2\lambda x = 0$, $x + \lambda = 0$ taken with constraint $y - x^2 = 0$ have solution $\lambda = x = y = 0$. However, for every $\delta > 0$, the points $(-\delta, \delta^2), (\delta, \delta^2)$, both satisfy the constraint and yet $\phi(-\delta, \delta^2) < \phi(0,0) < \phi(\delta, \delta^2)$.

5-1.P2. (a) Consider any $y_1 \in W$ such that $(x_1, y_1) \in T$. Suppose, if possible, that $g(y_1) \neq x_1$. Define g_1 on W to agree with g except that $g_1(y_1) = x_1$. Then $g_1 \neq g$ but $(g_1(y), y) \in T$ whenever $y \in W$ (even when $y = y_1$, because $(x_1, y_1) \in T$). This contradicts the uniqueness of g.
(b) The given local minimum at b means that there exists an open ball $W_1 \subseteq W$ such that $b \in W_1$ and $\phi(g(y), y) \geq \phi(g(b), b) = \phi(a, b)$ whenever $y \in W_1$. The set $U_1 = \{(x,y) \in S : y \in W_1\}$ is open and contains (a, b) because $b \in W_1$. Consider any $(x,y) \in T \cap U_1$. We have $(x,y) \in T$ and $y \in W_1 \subseteq W$, so that $x = g(y)$ by (a). Therefore $\phi(x,y) = \phi(g(y), y)$. Since $y \in W_1$, this implies $\phi(g(y), y) \geq \phi(a, b)$. Thus $\phi(x,y)$

$\geq \phi(a,b)$ whenever $(x,y) \in T \cap U_1$. Since U_1 is an open set containing (a,b), there exists an open ball $B \subseteq U_1$ such that $(a,b) \in B$. Now, $(x,y) \in T \cap B \Rightarrow (x,y) \in T \cap U_1 \Rightarrow \phi(x,y) \geq \phi(a,b)$.

5-1.P3. The Jacobian matrix of the $\mathbb{R}^2$-valued constraint function has rows $[1 \; -1 \; 1]$ and $[2 \; 1 \; 4]$. The first two columns provide an invertible matrix, so that the invertibility condition holds on the entire constraint set. Denoting the point (a,b) of Theorem 5-1.3 by (x,y,z) and $\lambda_1, \ldots, \lambda_m$, by λ, μ, equations (1) there become $2x + \lambda + 2\mu = 0$, $2y - \lambda + \mu = 0$, $2z + \lambda + 4\mu = 0$. These lead to $x = -(\lambda + 2\mu)/2$, $y = (\lambda - \mu)/2$, $z = -(\lambda + 4\mu)/2$. Substituting in the constraints, we get $-3\lambda - 5\mu = 4$ and $-5\lambda - 21\mu = 32$. Therefore $\lambda = 2$ and $\mu = -2$ and hence $(x,y,z) = (1,2,3)$. Since this solution of the Lagrange equations is unique, the minimum must occur at $(1,2,3)$. The minimum value of $x^2 + y^2 + z^2$ is 14.

5-1.P4. (a) The distance of any point (x,y) from the circle $x^2 + y^2 = 1$ is $(x^2 + y^2)^{1/2} - 1$, which is therefore our objective function. The constraint is $x + y = 4$. The Jacobian matrix of the constraint function is $[1 \; 1]$. Both entries are non-zero everywhere and therefore the invertibility condition holds on the entire constraint set. The Lagrange equations are $x(x^2 + y^2)^{-1/2} + \lambda = y(x^2 + y^2)^{-1/2} + \lambda = 0$ and the constraint is $x + y = 4$. The only solution is $(x,y) = (2,2)$. If the minimum exists, then it must occur at $(2,2)$.
(b) It is sufficient to minimise double the square of the distance of a point (x,y) from the line $x + y = 4$, which is $(x + y - 4)^2$. The constraint is $x^2 + y^2 = 1$. Since the constraint set is compact, we expect a minimum as well as a maximum. The Jacobian matrix of the constraint function is $[2x \; 2y]$, the entries of which cannot both vanish at the same point of the constraint set. The Lagrange equations are $2(x + y - 4) + 2\lambda x = 0 = 2(x + y - 4) + 2\lambda y$ and the constraint. There are precisely two solutions: $(x,y) = (1/\sqrt{2}, 1/\sqrt{2})$ and $(-1/\sqrt{2}, -1/\sqrt{2})$. By evaluating $(x + y - 4)^2$ at the two points, we find that $(1/\sqrt{2}, 1/\sqrt{2})$ is a point of minimum and the other one is a maximum.

5-1.P5. The objective function is $\phi(x,y,u,v) = (x-u)^2 + (y-v)^2$ and the constraint functions are $u + v - 4$ and $x^2 + y^2 - 1$. Observe that the constraint equations rule out the possibility that $x - u = 0 = y - v$, because this would imply $x + y = u + v = 4$, whereas

$$x^2 + y^2 = 1 \Rightarrow |x|, |y| \leq 1 \Rightarrow x + y \leq 2 \Rightarrow x + y \neq 4.$$

Now, the Lagrangian is $L = (x-u)^2 + (y-v)^2 + \lambda(x^2 + y^2 - 1) + \mu(u + v - 4)$. Using subscripts to denote partial differentiation,

$$L_x = 2(x-u) + 2\lambda x, \; L_y = 2(y-v) + 2\lambda y, \; L_u = -2(x-u) + \mu$$
$$\text{and } L_v = -2(y-v) + \mu.$$

For these to be zero, we must have $x - u = y - v$ and hence $\lambda x = \lambda y$. But $\lambda \neq 0$, because, as already observed, $x - u \neq 0$. Therefore $x = y$, which implies $u = v$.

This leads to the two solutions $u = v = 2$ and $x = y = \pm 1/\sqrt{2}$ with $\lambda = \pm 2\sqrt{2} - 1$ and $\mu = \pm\sqrt{2} - 4$. The Jacobian matrix of the $\mathbb{R}^2$- valued constraint function is

$$\begin{bmatrix} 0 & 0 & 1 & 1 \\ 2x & 2y & 0 & 0 \end{bmatrix}.$$

On the constraint set, one cannot have $x = 0 = y$; therefore the 2×2 matrices formed by the first and third columns and by the second and third columns cannot both fail to be invertible at the same point of the constraint set. Thus the invertibility condition holds everywhere on the constraint set and there are no points of extremum other than the solutions obtained.

5-1.P6. Since the constraints define a compact set and the function $x^2 + y^2 + z^2$ is continuous, an absolute maximum and absolute minimum must exist. The Jacobian matrix of the $\mathbb{R}^2$- valued constraint function has rows $\begin{bmatrix} \frac{x}{2} & \frac{2y}{5} & \frac{2z}{25} \end{bmatrix}$ and $\begin{bmatrix} 1 & 1 & -1 \end{bmatrix}$. These cannot be proportional at any point of the constraint set. It therefore follows that the absolute extrema occur among the solutions to the Lagrange equations.

The partial derivatives of

$$L = x^2 + y^2 + z^2 + \lambda(\tfrac{x^2}{4} + \tfrac{y^2}{5} + \tfrac{z^2}{25} - 1) + \mu(z - x - y)$$

are

$$\partial L/\partial x = 2x + \tfrac{1}{2}\lambda x - \mu, \ \partial L/\partial y = 2y + \tfrac{2}{5}\lambda y - \mu, \ \partial L/\partial z = 2z + \tfrac{2}{25}\lambda z + \mu.$$

For these to be zero, we must have $x = \frac{2\mu}{\lambda+4}$, $y = \frac{5\mu}{2\lambda+10}$ and $z = -\frac{25\mu}{2\lambda+50}$. Substituting in $z = x + y$, we get $\frac{25\mu}{2\lambda+50} + \frac{5\mu}{2\lambda+10} + \frac{2\mu}{\lambda+4} = 0$. Now $\mu \neq 0$, because otherwise the first of the given constraints will not be satisfied. So,

$$25(\lambda + 4)(2\lambda + 10) + 2(2\lambda + 50)(2\lambda + 10) + 5(2\lambda + 50)(\lambda + 4) = 0.$$

This is a quadratic in λ with roots $\lambda = -10$ and $\lambda = -75/17$. Corresponding to the root $\lambda = -10$, we have $x = -\mu/3$, $y = -\mu/2$ and $z = -5\mu/6$. Substituting this in the quadratic constraint, we get $\mu = \pm 6(5/19)^{1/2}$. The corresponding points are $(x,y,z) = (\pm 2(5/19)^{1/2}, \pm 3(5/19)^{1/2}, \pm 5(5/19)^{1/2})$. Both yield $x^2 + y^2 + z^2 = 10$. Similarly, corresponding to the root $\lambda = -75/17$, we get $\mu = \pm 140/17(646)^{1/2}$ and $(x,y,z) = (\pm 40/(646)^{1/2}, \mp 35/(646)^{1/2}, \pm 5/(646)^{1/2})$. Both yield $x^2 + y^2 + z^2 = 2850/646 < 10$. It follows that these two points give the absolute minimum and the other two the absolute maximum.

5-1.P7. (a) The Jacobian matrix of the constraint function is $[\, 6x-2y-2 \quad -2x+4y-6 \,]$, the entries of which can both vanish only when $x = 1$ and $y = 2$. Thus the set $S = \{(x,y,z) \in \mathbb{R}^3 : 6x-2y-2 = 0 = -2x+4y-6 \}$ is the same as $\{(1,2,z) : z \in \mathbb{R}\}$. Observe that this set S can be verified to be contained in the constraint set. Since the invertibility condition fails on the set S, extrema that lie in it may not be detected by the Lagrange multiplier method. First we seek extrema outside S. The Lagrange equations are

$$2x+\lambda(6x-2y-2)= 0, \quad 2y+\lambda(-2x+4y-6) = 0, \quad 2(z-1)= 0.$$

If $6x-2y-2 = 0$, then $x = 0$ and $y = -1$, which violates the constraint. Therefore $6x-2y-2 \neq 0$, and similarly, $-2x+4y-6 \neq 0$. So $2x/(6x-2y-2) = 2y/(-2x+4y-6)$ and hence $x^2 +xy-y^2 +3x-y = 0$. Combining this with the constraint, we obtain $5x^2 +4x-8y+7 = 0$, so that $y = (5x^2 +4x+7)/8$. When we substitute for y from here in the constraint, the resulting equation is

$$5(x-1)^2 \, [5(x+1)^2 + 16] = 0,$$

showing that there is no solution with $x \neq 1$. Thus there are no extrema outside S. For extrema in S, we recall that the set is defined by $6x-2y-2 = 0$ and $-2x+4y-6 = 0$. We may use these two as constraints and apply the Lagrange method, taking advantage of the fact that the Jacobian matrix of the two constraint functions is invertible everywhere. Alternatively, we may note that, when $x = 1$, $y = 2$, we have $x^2 + y^2 +(z-1)^2 = 5+(z-1)^2$, which is obviously minimum when $z = 1$, whereby the absolute minimum is seen to be at $(1,2,1)$.

(b) Upon writing the constraint as $2y^2 -(2x+6)y+(3x^2 -2x+7) = 0$ (a quadratic equation for y), we get $y = \frac{1}{2}[\, x+3\pm\sqrt{-5(x-1)^2} \,]$, so that the constraint set is $\{(1,2,z) : z \in \mathbb{R}\}$, which is the same set as what we called S above. (So the search for an extremum outside S was doomed before it began!) Then we proceed as in the last sentence of the preceding paragraph. There is no need for differentiation at any stage.

5-1.P8. (a) $|f(r,\theta)| \leq r^2$. (b) Given a ball $|r| < a < 1$ centred at the origin, choose $\theta = a/2\pi$, so that the point $(r,\theta) = (a/2,\theta)$ satisfies $|r| < a$ as well as $r/\theta = \pi$. Then (r,θ) is in the given ball and $f(r,\theta) = -a^2/4 < 0$. Therefore f does not have a local minimum at the origin. (c) According to our choice of polar coordinates, a line through the origin is represented by $\theta =$ constant (same on both sides of the origin) $= \alpha$, say. If $\alpha = 0$ (the line is the x-axis), then $f(r,\theta) = f(r,0) = r^2$, which shows that the restriction of f to the line $\theta = \alpha$ has a local strict minimum at the origin when $\alpha = 0$. Now suppose $\alpha \neq 0$. Then for $0 < |r| < |\alpha|\frac{\pi}{2}$, we have $\cos(r/\alpha) > 0$ and hence $f(r,\alpha) > 0$. Therefore the restriction of f to the line $\theta = \alpha$ has a local strict minimum at the origin in this case too.

5-1.P9. (a) Since $4x^4y^2 \le (x^4 + y^2)^2$, we have $0 \le \frac{4x^6y^2}{(x^4+y^2)^2} \le x^2$. Since $\lim x^2 = 0$ as

$(x,y) \rightarrow (0,0)$, the same is true of $\frac{4x^6y^2}{(x^4+y^2)^2}$. Also, $\lim (x^2 + y^2 - 2x^2y) = 0$. Hence

$\lim f(x,y) = 0$ as $(x,y) \rightarrow (0,0)$. Thus f is continuous at $(0,0)$.

(b) Clearly, $g_\theta(0) = f(0,0) = 0$. Also,

$$\frac{g_\theta(t) - g_\theta(0)}{t} = \frac{1}{t}[t^2 - 2t^3 \cos^2\theta \sin\theta - \frac{4t^4 \cos^6\theta \sin^2\theta}{(t^2 \cos^4\theta + \sin^2\theta)^2}]$$

$$= t - 2t^2 \cos^2\theta \sin\theta - \frac{4t^3 \cos^6\theta \sin^2\theta}{(t^2 \cos^4\theta + \sin^2\theta)^2}.$$

Therefore $g_\theta'(0) = 0$ (when $\theta = 0$ as well as otherwise). Besides, for $t \ne 0$,

$$g_\theta'(t) = 2t - 6t^3 \cos^2\theta \sin\theta - \frac{16t^3 \cos^6\theta \sin^2\theta}{(t^2 \cos^4\theta + \sin^2\theta)^2} + \frac{16t^5 \cos^{10}\theta \sin^2\theta}{(t^2 \cos^4\theta + \sin^2\theta)^3}.$$

Therefore

$$\frac{g_\theta'(t) - g_\theta'(0)}{t} = 2 - 6t^2 \cos^2\theta \sin\theta - \frac{16t^2 \cos^6\theta \sin^2\theta}{(t^2 \cos^4\theta + \sin^2\theta)^2}$$

$$+ \frac{16t^4 \cos^{10}\theta \sin^2\theta}{(t^2 \cos^4\theta + \sin^2\theta)^3},$$

so that $g_\theta''(0) = 2$ (when $\theta = 0$ as well as otherwise).

(c) $f(x,x^2) = -x^4$, which shows that f takes negative values arbitrarily close to $(0,0)$.

5-1.P10. (a) $|f(x,y)| \le x^2 + y^2$. (b) Along the y-axis ($x = 0$), the function becomes $\phi(y) = f(0,y) = y^2 \cos(2y^2/\pi)$. For $0 < |y| < \pi/2$, we therefore have $\phi(y) > 0$; also $\phi(0) = 0$. So ϕ has a local strict minimum at $y = 0$. Along the line $y = kx$, the function becomes

$$\psi(x) = f(x, kx) = x^2(1 + k^2)\cos\left[x^2 \frac{1+k^2}{\arctan k}\right] \text{ when } k \ne 0 \text{ and } x^2 \text{ when } k = 0.$$

When $k \ne 0$, we have $\psi(x) > 0$ for $0 < |x| < [\frac{\pi}{2}|\arctan k|/(1 + k^2)]^{1/2}$ while $\psi(0) = 0$, so that ψ has a local strict minimum at $x = 0$. The case when $k = 0$ is trivial. (c) Consider any open set $x^2 + y^2 < \delta^2 < \pi$ containing $(0,0)$. Take (x,y) with $x^2 = \delta^2/4[1 + \tan^2(\delta^2/4\pi)]$ and $y = x\tan(\delta^2/4\pi) \ne 0$. Then

$$x^2 + y^2 = \delta^2/4 < \delta^2 \quad \text{while} \quad f(x,y) = (\delta^2/4)\cos[(\delta^2/4)/(\delta^2/4\pi)] = -(\delta^2/4) < 0.$$

Problem Set 5-2

5-2.P1. $f_x = 6x^2 - 6x$ and $f_y = 6y^2 + 6y$. These vanish simultaneously only at $(0,0), (0,-1), (1,0)$ and $(1,-1)$. So, extrema can occur only at these points, though not necessarily at all of them. Now $f_{xx} = 12x - 6$, $f_{xy} = 0$ and $f_{yy} = 12y + 6$. Therefore $f_{xy}^2 > f_{xx}f_{yy}$ at $(0,0)$ and $(1,-1)$. It follows by (c) of Theorem 5-2.3 that there is no extremum at $(0,0)$ and $(1,-1)$. Also, $f_{xy}^2 < f_{xx}f_{yy}$ at $(1,0)$ and $(0,-1)$. Since $f_{xx} > 0$ at $(1,0)$, it follows from (a) of Theorem 5-2.3 that there is a local strict minimum at $(1,0)$; since $f_{xx} < 0$ at $(0,-1)$, it follows from (b) of Theorem 5-2.3 that there is a local strict maximum at $(0,-1)$.

5-2.P2. (a) The first partial derivatives are

$$D_1 f = 4x_1^3 - 4x_2x_3, \qquad D_2 f = 4x_2^3 - 4x_3x_1, \qquad D_3 f = 4x_3^3 - 4x_1x_2.$$

For all three to vanish, we must have $x_1^4 = x_2^4 = x_3^4 = x_1x_2x_3 \geq 0$, from which it follows that $|x_1| = |x_2| = |x_3|$ and further that $|x_1|^3 = |x_2||x_3|$, so that $|x_1| = |x_2| = |x_3| = 0$ or 1. Although there are nine such points in $\mathbb{R}^3$, four of them fail to satisfy $x_1x_2x_3 \geq 0$. At the remaining five points

$$P_0 = (0,0,0), \quad P_1 = (1,1,1), \quad P_2 = (1,-1,-1), \quad P_3 = (-1,1,-1), \quad P_4 = (-1,-1,1),$$

all three partial derivatives vanish. So a local extremum, if any, must occur at one of these five points. In order to apply the theorem, we compute second partial derivatives:

$$D_{1\,1}f = 12x_1^2, \qquad D_{2\,1}f = -4x_3, \qquad D_{3\,1}f = -4x_2$$
$$D_{1\,2}f = -4x_3, \qquad D_{2\,2}f = 12x_2^2, \qquad D_{3\,2}f = -4x_1$$
$$D_{1\,3}f = -4x_2, \qquad D_{2\,3}f = -4x_1, \qquad D_{3\,3}f = 12x_3^2.$$

At P_0, we find that all values are 0 and consequently, Q vanishes everywhere. So, Theorem 5-2.1 tells us nothing about this point! At P_2, we find that

$$D_{1\,1}f = 12, \qquad D_{2\,1}f = 4, \qquad D_{3\,1}f = 4$$
$$D_{1\,2}f = 4, \qquad D_{2\,2}f = 12, \qquad D_{3\,2}f = -4$$
$$D_{1\,3}f = 4, \qquad D_{2\,3}f = -4, \qquad D_{3\,3}f = 12.$$

Hence Q is given by

$$Q(h_1, h_2, h_3) = 4[3(h_1^2 + h_2^2 + h_3^2) + 2(-h_2h_3 + h_3h_1 + h_1h_2)].$$

Upon recasting this as

$$Q(h_1, h_2, h_3) = 4[(h_2 - h_3)^2 + (h_3 + h_1)^2 + (h_1 + h_2)^2 + (h_1^2 + h_2^2 + h_3^2)],$$

we see that Q is positive definite. From the theorem we can now conclude that f has a strict local minimum at P_2. Similar computations show that f has a strict local minimum at P_3, P_4 and also at P_1.

Regarding $P_0 = (0,0,0)$, when $x_1 = x_2 = x_3 = a$, say, we have

$$x_1{}^4 + x_2{}^4 + x_3{}^4 - 4x_1x_2x_3 = a^3(3a - 4),$$

which is negative for $0 < a < \frac{4}{3}$ and positive for $a < 0$. Therefore there is no extremum at P_0.

(b) $\quad x_1{}^4 + x_2{}^4 + x_3{}^4 - 4x_1x_2x_3 = -1 + \frac{1}{3}\Big[(x_1{}^2 - 1)^2 + (x_2{}^2 - 1)^2 + (x_3{}^2 - 1)^2$

$$+ 2\big((x_2x_3 - x_1)^2 + (x_3x_1 - x_2)^2 + (x_1x_2 - x_3)^2 \big)$$

$$+ (x_2{}^2 - x_3{}^2)^2 + (x_3{}^2 - x_1{}^2)^2 + (x_1{}^2 - x_2{}^2)^2 \Big].$$

The above identity shows that, at the points $P_1, \ldots, P_4$, the function has not only local minima but also *absolute* minima.

(c) In the above answer to part (b), drop the initial constant and also the factor $\frac{1}{3}$; then drop the last three squares. The partial derivatives of the resulting function $g(x_1, x_2, x_3) = (x_1{}^2 + x_2{}^2 + x_3{}^2)^2 - 12x_1x_2x_3 + 3$ vanish at the same five points as those of f and there are strict minima at precisely the same four points.

(d) Proceed as in part (a). The Hessian form at P_2 turns out to be

$$Q(h_1, h_2, h_3) = 10[(h_2 - h_3)^2 + (h_3 + h_1)^2 + (h_1 + h_2)^2 + 7(h_1{}^2 + h_2{}^2 + h_3{}^2)].$$

5-2.P3. Proceeding as in the example illustrated with $a > b > c > 0$, we find that $(\pm 1, 0, 0)$, $(0, \pm 1, 0)$, $(0, 0, \pm 1)$ and $(0,0,0)$ are the only points where all three partial derivatives vanish. The hypothesis $a > b > 0 > c$ implies that $f(0, t, 0) > f(0,0,0) = 0 > f(0,0,t)$ for all t. Therefore there is no extremum at $(0,0,0)$. There is a maximum at $(\pm 1, 0, 0)$ but no extremum at $(0, \pm 1, 0)$ for the same reasons as in the illustrated example. At $(0, 0, \pm 1)$ however, the hypothesis $a > b > 0 > c$ implies that the entries of the Hessian matrix are all positive and hence there is a minimum.

5-2.P4. The first partial derivatives are

$$D_x F = yz(2x + y + z - 1), \qquad D_y F = zx(x + 2y + z - 1),$$

$$D_z F = xy(x + y + 2z - 1).$$

At any point where some two coordinates are 0, all first partial derivatives vanish. Although there are three kinds depending on which two coordinates are 0, we need consider only one kind. These points are listed below under (A). We proceed to find points at which all three partial derivatives vanish and only one coordinate is 0. It is sufficient to consider only the possibility $x = 0$, $y \neq 0 \neq z$. In this event, vanishing of the partial derivatives is equivalent to $y + z - 1 = 0$.

These points are listed under (B). Lastly, we consider the case when all three coordinates are nonzero. When this is so, vanishing of the partial derivatives is equivalent to

$$2x + y + z - 1 = 0, \qquad x + 2y + z - 1 = 0, \qquad x + y + 2z - 1 = 0.$$

It is easily seen that the unique solution of this system is $(\frac{1}{4}, \frac{1}{4}, \frac{1}{4})$. Thus we have to consider the following three categories of points that are candidates for an extremum:

(A) x arbitrary with $y = z = 0$;

(B) $x = 0$, $y \neq 0 \neq z$ with $y + z - 1 = 0$; note that this implies $y \neq 1 \neq z$;

(C) $(\frac{1}{4}, \frac{1}{4}, \frac{1}{4})$.

(From 2-3.P14 and 2-3.P15, we know that the points in (A) and (B) are not extrema. Here we shall try to demonstrate the fact by applying Theorem 5-2.1.)

The second partial derivatives are

$$D_{xx}F = 2yz \quad D_{yx}F = 2zx + 2yz + z^2 - z \quad D_{zx}F = 2xy + 2yz + y^2 - y$$
$$D_{yy}F = 2zx \quad D_{zy}F = 2xy + 2zx + x^2 - x \quad D_{zz}F = 2xy.$$

At a point in category (A), the Hessian matrix becomes

$$(x^2 - x)\begin{bmatrix} 0 & 0 & 0 \\ 0 & 0 & 1 \\ 0 & 1 & 0 \end{bmatrix}.$$

The associated quadratic form is $Q(a,b,c) = 2(x^2 - x)bc$. For $0 \neq x \neq 1$, this can take positive as well as negative values, and therefore by Theorem 5-2.1, there is no extremum at the point in question. If $x = 0$ or 1, the Hessian form is identically zero and the theorem fails. We are forced to argue as in 2-3.P14 why there is no extremum at $(0,0,0)$ and $(1,0,0)$. We can then conclude that there is no extremum at any point listed in (A).

At a point in category (B), the Hessian matrix becomes

$$-(y^2 - y)\begin{bmatrix} 2 & 1 & 1 \\ 1 & 0 & 0 \\ 1 & 0 & 0 \end{bmatrix}.$$

The associated quadratic form is $Q(a,b,c) = -2(y^2 - y)(a(a + b + c))$. Considering that $0 \neq y \neq 1$, we know Q can take positive as well as negative values, and therefore by Theorem 5-2.1, there is no extremum at the point in question. Thus there is no extremum at any point listed under (B).

At the only point in category (C), the Hessian matrix becomes

$$\frac{1}{16}\begin{bmatrix} 2 & 1 & 1 \\ 1 & 2 & 1 \\ 1 & 1 & 2 \end{bmatrix}.$$

The associated quadratic form is

$$Q(a,b,c) = \tfrac{1}{16}(2a^2 + 2b^2 + 2c^2 + 2bc + 2ca + 2ab)$$

$$= \tfrac{1}{16}[(b+c)^2 + (c+a)^2 + (a+b)^2].$$

This is seen to be positive definite, so that by Theorem 5-2.1, there is a local minimum at $(\tfrac{1}{4},\tfrac{1}{4},\tfrac{1}{4})$.

To summarise, there is a local minimum at $(\tfrac{1}{4},\tfrac{1}{4},\tfrac{1}{4})$ and no local extremum anywhere else.

(Note: The reader is invited to prove by using the arithmetic mean-geometric mean inequality and *single variable methods* that $xyz(x + y + z - 1) \geq -\tfrac{1}{256}$ for all $x \geq 0, y \geq 0, z \geq 0$.)

5-2.P5. $Q(a,b,c) = a^2 + 13b^2 + 4c^2 - 10bc - 2ca + 4ab = (a + 2b - c)^2 + (3b - c)^2 + 2c^2$. For this to be zero, we must have $a + 2b - c = 3b - c = c = 0$, which implies $(a,b,c) = (0,0,0)$. So Q is positive definite.

5-2.P6. $Q(a,b,c) = a^2 + 5b^2 + 10c^2 - 10bc - 2ca - 2ab = (a - b - c)^2 + (2b - 3c)^2 \geq 0$. This shows that Q is positive semidefinite. Since $Q(5,3,2) = 0$, it is not positive definite.

5-2.P7. α is represented by the 3×2 matrix having rows $[1 \quad 0], [0 \quad 1], [1 \quad 1]$. Therefore $Q \circ \alpha$ is the quadratic form in $\mathbb{R}^2$ associated with the product

$$\begin{bmatrix} 1 & 0 & 1 \\ 0 & 1 & 1 \end{bmatrix}\begin{bmatrix} A & B & -B \\ B & A & -B \\ -B & -B & A \end{bmatrix}\begin{bmatrix} 1 & 0 \\ 0 & 1 \\ 1 & 1 \end{bmatrix}.$$

This product turns out to have rows

$$[2(A - B) \quad A - B] \quad \text{and} \quad [A - B \quad 2(A - B)].$$

The associated quadratic form $Q \circ \alpha$ is therefore $(Q \circ \alpha)(a,b) = 2(A - B)[(a^2 + b^2) + ab]$. Since $(a^2 + b^2) + ab = (a + \tfrac{1}{2}b)^2 + \tfrac{3}{4}b^2$, the form is positive or negative definite according as $A > B$ or $A < B$.

5-2.P8. $Q \circ \alpha$ is associated with the product

$$\begin{bmatrix} 1 & 0 & 1 \\ 0 & 1 & 1 \end{bmatrix} \begin{bmatrix} 1 & -1 & -1 \\ -1 & 5 & -5 \\ -1 & -5 & 10 \end{bmatrix} \begin{bmatrix} 1 & 0 \\ 0 & 1 \\ 1 & 1 \end{bmatrix} = \begin{bmatrix} 9 & 3 \\ 3 & 5 \end{bmatrix}.$$

So $(Q \circ \alpha)(a,b) = 9a^2 + 6ab + 5b^2 = (3a+b)^2 + 4b^2 \geq 0$. This can be zero only if $(a,b) = (0,0)$. Thus $Q \circ \alpha$ is positive definite.

5-2.P9. As discussed in Example 5-2.10(b), one need only check whether

$$F''(a) < F'(a) \frac{f_{11}(a,a,a) - f_{12}(a,a,a)}{f_1(a,a,a)} \quad \text{or} \quad F''(a) > F'(a) \frac{f_{11}(a,a,a) - f_{12}(a,a,a)}{f_1(a,a,a)},$$

where subscripts indicate partial differentiation, $F(x) = \tan x$, $f(x,y,z) = y^5 z + z^5 x + x^5 y - 3(\pi/4)^6$ and $a = \pi/4$. Elementary computations show that

$$F'(a) = 2, \quad F''(a) = 4, \quad f_1(a,a,a) = 6(\pi/4)^5,$$
$$f_{12}(a,a,a) = 5(\pi/4)^4 \quad \text{and} \quad f_{11}(a,a,a) = 20(\pi/4)^4.$$

So the question is whether $4 < 2(20(\pi/4)^4 - 5(\pi/4)^4)/6(\pi/4)^5$, which simplifies to whether $\pi < 5$, which it is. Therefore $(\pi/4, \pi/4, \pi/4)$ is a point of constrained local strict maximum.

5-2.P10. Proceed as in the previous problem to get

$$F'(a) = \sqrt{3}/2, \quad F''(a) = -1/2, \quad f_1(a,a,a) = \pi/3, \quad f_{11}(a,a,a) = 0, \quad \text{and} \quad f_{12}(a,a,a) = 1.$$

Then $F'(a)(f_{11}(a,a,a) - f_{12}(a,a,a))/f_1(a,a,a) = -3\sqrt{3}/2\pi$, which is less than $F''(a) = -1/2$, because $\pi < 3\sqrt{3}$. So, minimum.

5-2.P11. Proceed as in the previous problem to get

$$F'(a) = 2, \quad F''(a) = 4, \quad f_1(a,a,a) = \pi/2, \quad f_{11}(a,a,a) = 0, \quad \text{and} \quad f_{12}(a,a,a) = 1.$$

Then $F'(a)(f_{11}(a,a,a) - f_{12}(a,a,a))/f_1(a,a,a) = -4/\pi$, which is less than $F''(a) = 4$. So, minimum.

5-2.P12. Yes. Let $u' = (h,k)$ with $h \in \mathbb{R}^n$ and $k \in \mathbb{R}^m$. Using equalities established in the proof of the theorem, we argue thus: $f'(a,b)(u) = A_1 h + A_2 k$; hence $h = -A_1^{-1} A_2 k = g'(b)k$; so $u' = (g'(b)k, k) = G'(b)(k)$. Therefore $H_\Phi(h) = H(G'(b)(k)) = H(u') > 0$. Similarly for u''. Apply Theorem 5-2.1.

5-2.P13. One partial derivative of the constraint function never vanishes; so the invertibility condition holds everywhere on the constraint set. The Lagrangian is $L = x^2 + y^2 + z^2 + \lambda(z - xy - 2)$. We shall use subscripts to denote partial differentiation. We have $L_x = 2x - \lambda y$, $L_y = 2y - \lambda x$, $L_z = 2z + \lambda$. This confirms that $(x,y,z) = (0,0,2)$, $\lambda = -4$ is one solution. With these values of x, y, z, λ, we have

$$L_{xx} = 2 \qquad\qquad L_{yx} = 4 \qquad\qquad L_{zx} = 0$$
$$L_{yy} = 2 \qquad\qquad L_{zy} = 0$$
$$L_{zz} = 2.$$

Therefore the Hessian form of the Lagrangian is $H(a,b,c)$ = $2a^2 + 2b^2 + 2c^2 + 8ab$. The linear derivative of the constraint function $f(x,y,z)$ = $z - xy - 2$ at $(0,0,2)$ has the matrix $[0 \quad 0 \quad 1]$. The condition $(f'(0,0,2))(a,b,c)$ = 0 thus becomes $c = 0$. For such (a,b,c), the Hessian form of the Lagrangian is $H(a,b,0) = 2a^2 + 2b^2 + 8ab = 2[(a+b)^2 + 2ab]$. This is negative when $a = -b = 1$ and positive when $a = b = 1$. Therefore there is no extremum at $(0,0,2)$ by 5-2.P12.

5-2.P14. One partial derivative of the constraint function never vanishes; so the invertibility condition holds everywhere on the constraint set. The Lagrangian is $L = x^2 + y^2 + z^2 + u^2 + \lambda(u - xyz - 2)$. We shall use subscripts to denote partial differentiation. We have $L_x = 2x - \lambda yz$, $L_y = 2y - \lambda zx$, $L_z = 2z - \lambda xy$, $L_u = 2u + \lambda$. This confirms that $(x,y,z,u) = (0,0,0,2)$, $\lambda = -4$ is one solution. With these values of x,y,z,u,λ, we have

$L_{xx} = 2$	$L_{yx} = 0$	$L_{zx} = 0$	$L_{ux} = 0$
	$L_{yy} = 2$	$L_{zy} = 0$	$L_{uy} = 0$
		$L_{zz} = 2$	$L_{uz} = 0$
			$L_{uu} = 2$.

The Hessian matrix is thus seen to be twice the identity matrix. It follows without further ado that the quadratic form of interest to us is positive definite. Therefore the given point is a (local strict) minimum.

5-2.P15. Conversion leads to $x^3 + y^3 + z^3 + (1 + (yz + zx + xy))^3$, the second derivatives of which are an unpleasant prospect. We prefer not to convert.

One partial derivative of the constraint function never vanishes; so the invertibility condition holds everywhere on the constraint set. The Lagrangian is $L = x^3 + y^3 + z^3 + u^3 + \lambda(u - (yz + zx + xy) - 1)$. Denoting partial derivatives by subscripts, we have $L_x = 3x^2 - \lambda(y+z)$, $L_y = 3y^2 - \lambda(z+x)$, $L_z = 3z^2 - \lambda(x+y)$, $L_u = 3u^2 + \lambda$. This confirms that $(x,y,z,u) = (0,0,0,1)$, $\lambda = -3$ is one solution. With these values of x,y,z,u,λ,

$L_{xx} = 0$	$L_{yx} = 3$	$L_{zx} = 3$	$L_{ux} = 0$
	$L_{yy} = 0$	$L_{zy} = 3$	$L_{uy} = 0$
		$L_{zz} = 0$	$L_{uz} = 0$
			$L_{uu} = 6$.

Therefore the Hessian form of the Lagrangian is $H(a,b,c,d)$ = $6[(bc + ca + ab) + d^2]$. The linear derivative of the constraint function $f(x,y,z,u)$ = $u - (yz + zx + xy) - 1$ at $(0,0,0,1)$ has the matrix $[0 \quad 0 \quad 0 \quad 1]$. The condition $(f'(0,0,0,1))(a,b,c,d) = 0$ thus becomes $d = 0$. For such (a,b,c,d), the Hessian form of the Lagrangian is $H(a,b,c,0) = 6(bc + ca + ab)$, which is positive when $(a,b,c,d) = (1,1,0,0)$ and negative when $(a,b,c,d) = (1,-1,0,0)$. By the result of 5-2.P12, there is no extremum at $(0,0,0,1)$.

5-2.P16. One partial derivative of the constraint function never vanishes; so the invertibility condition holds everywhere on the constraint set. The Lagrangian is $L = F(x) + F(y) + F(z) + \lambda(z - g(x)g(y) - C)$. Denoting partial derivatives by subscripts, we have $L_x = F'(x) - \lambda g'(x)g(y)$, $L_y = F'(y) - \lambda g(x)g'(y)$, $L_z = F'(z) + \lambda$. This confirms that $(x,y,z) = (0, 0, C)$, $\lambda = -F'(C)$ is one solution. For these values of x, y, z, λ,

$$L_{xx} = F''(0) \qquad\qquad L_{yx} = F'(C)g'(0)^2 \qquad\qquad L_{zx} = 0$$
$$L_{yy} = F''(0) \qquad\qquad L_{zy} = 0$$
$$L_{zz} = F''(C).$$

The Hessian form of the Lagrangian is

$$H(a,b,c) = F''(0)a^2 + F''(0)b^2 + F''(C)c^2 + 2F'(C)g'(0)^2ab.$$

The linear derivative of the constraint function $f(x,y,z) = z - g(x)g(y) - C$ at $(0,0,C)$ has the matrix $[0 \quad 0 \quad 1]$. The condition $(f'(0,0,C))(a,b,c) = 0$ thus becomes $c = 0$. For such (a,b,c), the Hessian form of the Lagrangian is $H(a,b,0) = F''(0)a^2 + F''(0)b^2 + 2F'(C)g'(0)^2ab$. It follows upon completing squares that

$$H(a,b,0) = F''(0)\left(a + \tfrac{F'(C)g'(0)^2}{F''(0)}b\right)^2 + \tfrac{1}{F''(0)}\left(F''(0)^2 - F'(C)^2g'(0)^4\right)b^2.$$

Hence $H(a,b,0)$ retains the same sign for all nonzero $(a,b) \in \mathbb{R}^2$ if and only if $F''(0)^2 - F'(C)^2g'(0)^4 > 0$, or equivalently, $|F''(0)| > |F'(C)| \cdot g'(0)^2$. Therefore this inequality is a sufficient condition that there be an extremum at $(0,0,C)$. Furthermore, when the condition is satisfied, the sign of $H(a,b,0)$ is negative if $F''(0) < 0$. Thus the required further condition (for a maximum) is that $F''(0) < 0$.

5-2.P17. One partial derivative of the constraint function is nonzero at $(0,0,1)$; so the invertibility condition holds at the point. The Lagrangian is $L = (x^2 + y^2 + z^2) + \lambda(ze^z - xy(x^2 + y^2) - e)$. Denoting partial derivatives by subscripts, we have $L_x = 2x - 3\lambda x^2y - \lambda y^3$, $L_y = 2y - 3\lambda xy^2 - \lambda x^3$, $L_z = 2z + \lambda(z + 1)e^z$. This confirms that $(x,y,z) = (0,0,1)$, $\lambda = -1/e$ is one solution. For these values of x, y, z, λ, we have

$$L_{xx} = 2 \qquad\qquad L_{yx} = 0 \qquad\qquad L_{zx} = 0$$
$$L_{yy} = 2 \qquad\qquad L_{zy} = 0$$
$$L_{zz} = -1.$$

Therefore the Hessian form H of the Lagrangian is $H(a,b,c) = 2a^2 + 2b^2 - c^2$. The linear derivative of the constraint function $f(x,y,z) = ze^z - xy(x^2 + y^2) - e$ at $(0,0,1)$ has the matrix $[0 \quad 0 \quad 2e]$. The condition $(f'(0,0,1))(a,b,c) = 0$ thus becomes $c = 0$. For such (a,b,c), the Hessian form of the Lagrangian is $H(a,b,0) = 2a^2 + 2b^2$, which is positive definite. By Theorem 5-2.9, we conclude that the point in question is a local extremum (strict minimum).

5-2.P18. One partial derivative of the constraint function is nonzero at $(0,0,1)$; so the invertibility condition holds at the point. The Lagrangian is $L = xyz + \lambda(ze^z - xy(x^2 + y^2) - e)$. Denoting partial derivatives by subscripts, we have $L_x = yz - 3\lambda x^2 y - \lambda y^3, L_y = zx - 3\lambda xy^2 - \lambda x^3, L_z = xy + \lambda(z + 1)e^z$. This confirms that $(x,y,z) = (0, 0, 1)$, $\lambda = 0$ is one solution. For these values of x, y, z, λ, we have

$L_{xx} = 0$	$L_{yx} = 1$	$L_{zx} = 0$
	$L_{yy} = 0$	$L_{zy} = 0$
		$L_{zz} = 0.$

Therefore the Hessian form H of the Lagrangian is $H(a,b,c) = 2ab$. The linear derivative of the constraint function $f(x,y,z) = ze^z - xy(x^2 + y^2) - e$ at $(0,0,1)$ has the matrix $[0 \quad 0 \quad 2e]$. The condition $(f'(0,0,1))(a,b,c) = 0$ thus becomes $c = 0$. For such (a,b,c), the Hessian form of the Lagrangian is $H(a,b,0) = 2ab$, which takes positive as well negative values. By 5-2.P12, we conclude that the point in question is not a local extremum.

5-2.P19. It was already checked that the invertibility condition holds on the whole constraint set. The Hessian matrix turns out to be

$$\begin{bmatrix} 2+2\lambda & 0 & -2 & 0 \\ 0 & 2+2\lambda & 0 & -2 \\ -2 & 0 & 2 & 0 \\ 0 & -2 & 0 & 2 \end{bmatrix}.$$

The linear derivative of the $\mathbb{R}^2$-valued constraint function $f(x,y,u,v) = (x^2 + y^2 - 1, u + v - 4)$ at $(\pm 1/\sqrt{2}, \pm 1/\sqrt{2}, 2, 2)$ has the matrix

$$\begin{bmatrix} \pm\sqrt{2} & \pm\sqrt{2} & 0 & 0 \\ 0 & 0 & 1 & 1 \end{bmatrix}.$$

An element $(a,b,c,d) \in \mathbb{R}^4$ is mapped into zero by this matrix if and only if $a + b = 0 = c + d$. For such (a,b,c,d), the Hessian form of the Lagrangian is $H(a,-a,c,-c) = 4[(1 + \lambda)a^2 - 2ac + c^2]$. This is positive definite when $\lambda = 2\sqrt{2} - 1$ and can take positive as well as negative values when $\lambda = -2\sqrt{2} - 1$. Therefore $(1/\sqrt{2}, 1/\sqrt{2}, 2, 2)$ is a point of local strict minimum but $(-1/\sqrt{2}, -1/\sqrt{2}, 2, 2)$ is not an extremum.

5-2.P20. Since $b^2 > a^2$, the constraint cannot hold with $\sqrt{(x^2 + y^2)} = 0$. The linear derivative of the constraint function is

$$\left[2x(1 - \frac{b}{\sqrt{x^2+y^2}}) \qquad 2y(1 - \frac{b}{\sqrt{x^2+y^2}}) \qquad 2z\right].$$

If $z = 0$ at any point of the constraint set, then the constraint equation implies $(\sqrt{(x^2 + y^2)} - b)^2 = a^2 > 0$, so that $1 - b/\sqrt{(x^2 + y^2)} \neq 0$. It follows that, for the above linear derivative to be zero at some point of the constraint set, we must have $x = y = 0$, which is impossible (as already observed). Thus the invertibility condition is satisfied. The Lagrangian is

$$L(x,y,z) = x + \lambda(x^2 + y^2 + z^2 - 2b\sqrt{(x^2 + y^2)} + (b^2 - a^2))$$

and its partial derivatives are

$$L_x = 1 + 2\lambda x(1 - \frac{b}{\sqrt{x^2+y^2}}) = 1 + \frac{2\lambda x}{\sqrt{x^2+y^2}} (\sqrt{(x^2 + y^2)} - b),$$

$$L_y = 2\lambda y(1 - \frac{b}{\sqrt{x^2+y^2}}) = \frac{2\lambda y}{\sqrt{x^2+y^2}} (\sqrt{(x^2 + y^2)} - b),$$

$$L_z = 2\lambda z.$$

If λ, x or $\sqrt{(x^2 + y^2)} - b$ were to be 0, then we would have $L_x = 1$. Therefore if $L_x = 0$, then λ, x and $\sqrt{(x^2 + y^2)} - b$ must all be nonzero; hence, using the expressions for L_y and L_z, we see that $L_x = L_y = L_z = 0$ implies $y = z = 0$. Using this in the constraint we get $|x| - b = \pm a$ and from $L_x = 0$, we get the following four solutions of the Lagrange and constraint equations:

$$
\begin{array}{lll}
x = b + a, & y = z = 0, & \lambda = -1/2a, \\
x = b - a, & y = z = 0, & \lambda = 1/2a, \\
x = -(b + a), & y = z = 0, & \lambda = 1/2a, \\
x = -(b - a), & y = z = 0, & \lambda = -1/2a.
\end{array}
$$

When $y = z = 0$, we find that the linear derivative of the constraint function is

$$\left[2x(1 - b/|x|) \qquad 0 \qquad 0\right].$$

This maps into zero only those elements of $\mathbb{R}^3$ that are of the form $(0, \beta, \gamma)$. Also, $L_{xx} = L_{zz} = 2\lambda$, $L_{yy} = 2\lambda(1 - b/|x|)$ and $L_{yz} = L_{zx} = L_{xy} = 0$. At $(0, \beta, \gamma)$, the Hessian form of the Lagrangian therefore works out to be $2\lambda[(1 - b/|x|)\beta^2 + \gamma^2]$. This is positive definite if $(1 - b/|x|) > 0$ and $\lambda > 0$. Therefore there is a local minimum when $x = -(b + a)$, $y = z = 0$. Similarly, there is a local maximum when $x = b + a$, $y = z = 0$. The other two solutions correspond neither to a maximum nor a minimum.

5-2.P21. Let x_i be the length of the tangent from the ith vertex to the circle, $1 \leq i \leq 6$. We have to minimise Σx_i subject to $\Sigma \arctan x_i = \pi$. The six entries of the matrix of the linear derivative of the constraint function are $1/(1+x_i^2)$, all of which are nonzero. The invertibility condition is therefore fulfilled everywhere. The six partial derivatives of the Lagrangian L are $\partial L/\partial x_i = 1 + \lambda[1/(1+x_i^2)]$, and for all of these to be zero, we have to have $x_i = \sqrt{(-1-\lambda)}$, $1 \le i \le 6$. The constraint now implies that $-1-\lambda = \tan^2\frac{\pi}{6} = \frac{1}{3}$, so that $\lambda = -\frac{4}{3}$. It follows that $x_i = \frac{1}{\sqrt{3}}$. Now we have the unique solution of the Lagrange equations and the constraint. It remains to check the sufficient condition for a minimum. Since the second partial derivatives of the Lagrangian are $L_{x_i x_i} = -2\lambda x_i/(1+x_i^2)^2 = \sqrt{3}/2 > 0$ and $L_{x_i x_j} = 0$ when $i \ne j$, the Hessian form of the Lagrangian is positive definite (on all of $\mathbb{R}^6$). This verifies the sufficiency. Therefore, the minimum area is $\frac{6}{\sqrt{3}} = 2\sqrt{3}$.

Problem Set 6-1

6-1.P1. Suppose, if possible, that $[a_k, b_k] \ne [c_k, d_k]$ for some $k \ne p$. Then one of these intervals contains a number t that the other one does not. Let $t \in [a_k, b_k]$ but $t \notin [c_k, d_k]$. Then the point x for which $x_i = a_i$ for $i \ne k$ but $x_k = t$ belongs to the face of I in question but does not belong to J at all.

6-1.P2. If $p \ne q$, then the points x and x' with $x_i = x'_i = a_i$ for $i \ne q$ and $x_q = a_q$, $x'_q = b_q$ both belong to the first mentioned face but cannot both belong to the second, because $x_q \ne x'_q$ in view of the stipulation [see Def. 6-1.1] that $a_q < b_q$. Thus $p = q$. The point x with each $x_i = a_i$ belongs to the first mentioned face and therefore to the second, which means $x_p = c_p$, and hence $a_p = c_p$. The proof that $[a_i, b_i] = [c_i, d_i]$ for $i \ne p$ is as in Problem 6-1.P1. Now, either $b_p \le d_p$, in which case $I \subseteq J$, or $d_p \le b_p$, in which case $J \subseteq I$. For the last part, let $x_i = (a_i + b_i)/2$ for $i \ne p$ and $x_p = a_p + \frac{1}{2}\min\{b_p - a_p, d_p - a_p\}$. Then $x \in \mathbb{R}^n$ is an interior point of I as well as J.

6-1.P3. Consider dimension 1 first: If $[a, b]$ and $[c, d]$, where $a < b$ and $c < d$, have no common interior point and their union is an interval, we shall prove that either $b = c$ or $a = d$. We know $a < \frac{a+b}{2} < b$ and $c < \frac{c+d}{2} < d$. So, $a = c \Rightarrow a < \frac{a+d}{2} < d \Rightarrow a < \min\{\frac{a+b}{2}, \frac{a+d}{2}\} < \min\{b, d\} \Rightarrow \min\{\frac{a+b}{2}, \frac{a+d}{2}\} \in (a, b) \cap (a, d) = (a, b) \cap (c, d) = \emptyset$, a contradiction. Hence $a \ne c$. We may suppose $a < c$ (the contrary case being analogous). We shall argue that $b = c$. Since $c < \frac{c+d}{2} < d$, we have $b > c \Rightarrow a < c < \frac{c+b}{2} < b \Rightarrow c < \min\{\frac{c+d}{2}, \frac{c+b}{2}\} < \min\{b, d\} \Rightarrow \min\{\frac{c+d}{2}, \frac{c+b}{2}\} \in (c, b) \cap (c, d) \subseteq (a, b) \cap (c, d) = \emptyset$, a contradiction. Hence $b \le c$. But $b < c \Rightarrow b < \frac{c+b}{2} < c \Rightarrow \frac{c+b}{2} \notin [a, b] \cup [c, d]$, a contradiction because

$[a,b]\cup[c,d]$ is an interval containing b as well as c. Therefore $b = c$ and the one-dimensional case is established.

For dimension $n > 1$, we first observe that if $[a,b] \subseteq [c,d]$, where $a < b$, $c < d$, then the two intervals have an interior point in common. Now, let the n-dimensional cuboids in question be $I = I_1 \times \cdots \times I_n$ and $J = J_1 \times \cdots \times J_n$ and suppose $I \cup J = K_1 \times \cdots \times K_n$, where the I_i, J_i, K_i are all closed bounded intervals, i.e., one-dimensional cuboids. Then $I_i \cup J_i = \{z_i : z \in I \cup J\} = K_i$ by Remark 6-1.2(b). Thus each $I_i \cup J_i$ is a closed bounded interval K_i and the one-dimensional case applies. In particular, I_i and J_i must have a point in common. For some p ($1 \le p \le n$), neither I_p nor J_p contains the other, because otherwise, by the observation at the beginning of this paragraph, each pair I_i, J_i would have a common interior point x_i and $x = (x_1, \dots, x_n)$ would be a common interior point of I and J. We claim that $I_i = J_i$ for $i \ne p$. For, suppose if possible that $I_q \ne J_q$, where $q \ne p$. Then one of them contains a point not in the other. To be specific, suppose $x_q \in I_q$ but $x_q \notin J_q$. Since neither I_p nor J_p contains the other, there exists $x_p \in J_p$ such that $x_p \notin I_p$. For $i \ne p, q$ there exists some $x_i \in I_i \cup J_i$, as already noted. The point $x = (x_1, \dots, x_n)$ then belongs to $(I_1 \cup J_1) \times \cdots \times (I_n \cup J_n)$, which is the same as $K_1 \times \cdots \times K_n$, i.e., $I \cup J$. But $x \notin I = I_1 \times \cdots \times I_n$ because $x_p \notin I_p$ and also $x \notin J = J_1 \times \cdots \times J_n$ because $x_q \notin J_q$. This contradiction proves our claim that $I_i = J_i$ for $i \ne p$. This has the consequence that I_p and J_p do not have a common interior point and therefore it follows from the one-dimensional case (established above) that the right endpoint of one among I_p and j_p is equal to the left endpoint of the other. This and the fact that $I_i = J_i$ for $i \ne p$ quickly lead to required conclusion.

6-1.P4. By 2-4.P6, an element of $\mathbb{R}^n$ belongs to the closure of a subset of $\mathbb{R}^n$ if and only if some sequence lying in the subset converges to that element. Let $x \in \overline{I}$. Then some sequence $\{x^{(k)}\}_{k \in \mathbb{N}}$ in I converges to x. This means that the sequence $\{x^{(k)}_i\}_{k \in \mathbb{N}}$ of ith components converges to the ith component x_i of the limit x. Since each $x^{(k)}$ belongs to I, we have $a_i < x^{(k)}_i < b_i$ for $1 \le i \le n$. Taking limits, we have $a_i \le x_i \le b_i$ for $1 \le i \le n$. Thus $x \in J$. This proves $\overline{I} \subseteq J$. For the reverse inclusion, suppose $x \in J$. Then $a_i \le x_i \le b_i$ for $1 \le i \le n$. Therefore for each i, there exists a sequence $\{t_{i,k}\}_{k \in \mathbb{N}}$ in (a_i, b_i) converging to x_i. The sequence $\{x^{(k)}\}_{k \in \mathbb{N}}$ in $\mathbb{R}^n$ for which $x^{(k)}_i = t_{i,k}$ then lies in I and converges to x. Thus $x \in \overline{I}$.

6-1.P5. Let $J = [\alpha_1, \beta_1] \times \cdots \times [\alpha_n, \beta_n]$. By Remark 6-1.2(b), α_i is the minimum among the ith coordinates of the points of J. Since $J = \cup \mathcal{F}$, it follows that α_i is also the minimum among the left endpoints of the ith edges of the cuboids belonging to $\mathcal{F}$. In particular, α_i is a point of the partition P_i. Similarly, β_i is also a point of the partition P_i. It follows by Proposition 6-1.8 that there exists a paving $Q_1, \dots, Q_n$ of J such that the family $\mathcal{F}_1$ of all the cuboids formed by this paving is the unique subfamily of the family of all the cuboids formed by the given paving $P_1, \ldots, P_n$ satisfying $\cup \mathcal{F}_1 = J$. It follows that $\mathcal{F} = \mathcal{F}_1$. This completes the proof the existence of the kind of paving claimed. The last part about volumes is now merely a consequence of Remark 6-1.7(d).

Problem Set 6-2

6-2.P1. There are only two cuboids: $[0,1] \times [0,1]$ and $[0,1] \times [1,2]$. For the first one, m is $0 + 0 = 0$ and M is $1 + 1 = 2$; for the second one, m is $0 + 1 = 1$ and M is $1 + 2 = 3$.

6-2.P2. On the cuboid $K_{j,k} = \left[\frac{j-1}{n}, \frac{j}{n} \right] \times \left[\frac{k-1}{n}, \frac{k}{n} \right]$, we have $m_{j,k} = \inf \{ f(x) : x \in K_{j,k} \} = \frac{j-1}{n} + 2 \frac{k-1}{n}$ and $\mathrm{vol}(K_{j,k}) = \frac{2}{n^2}$. Therefore $L(f,P) = \sum_{j=1}^{n} \sum_{k=1}^{n} (\frac{j-1}{n} + 2 \frac{k-1}{n}) \frac{2}{n^2} = \frac{3}{n} (n-1)$.

6-2.P3. On each cuboid K formed by the given paving, $M_K - m_K = \frac{4}{n}$ and the total volume of all the cuboids is $(1)(2)(1)(4) = 8$. So, $U(f,P) - L(f,P) = (\frac{4}{n}) 8 = \frac{32}{n}$.

6-2.P4. It is still true (as before) that each cuboid K in $\mathcal{B}$ has at least one edge K_p with exactly one endpoint in J_p. So the rest of the proof of Proposition 6-2.11 goes through as before. The subfamilies need not be the same as before, because a cuboid having one edge K_p with exactly one endpoint in J_p and also one edge K_q ($q \neq p$, of course) with both endpoints outside J_q now belongs to $\mathcal{O}$ instead of $\mathcal{B}$.

6-2.P5. First suppose g is integrable. The function $f\chi$ is 0 outside J while it agrees with f on J and hence with g. It is therefore bounded, agrees with g on J and is 0 outside I. By Proposition 6-2.11, it is integrable and has integral $\int_J g$. Next, suppose $f\chi$ is integrable. By Proposition 6-2.7, the restriction of $f\chi$ to J is integrable; but the restriction is equal to g. By Proposition 6-2.11, $\int_J g = \int_I (f\chi)$.

6-2.P6. Let $\varepsilon > 0$. Then $f(x,y) \geq \varepsilon \Rightarrow x, y \in \mathbb{Q}$ and $x = \frac{p}{q}$, where the smallest such positive q satisfies $\frac{1}{q} \geq \varepsilon$. Since $x \in [0,1]$, we must have $0 \leq p \leq q$. There are only finitely many such positive integers q; let N be the largest. For each q, $1 \leq q \leq N$, there are $q + 1$ integers p such that $0 \leq p \leq q$. So, there are $\sum_{1 \leq q \leq N} (q+1)$ or fewer, say M, rational numbers $x_1, \ldots, x_M$ such that $f(x_j, y) \geq \varepsilon$. Let $\delta > 0$ be less than ε and also less than half the smallest distance between any two x_j. Then the numbers $x_j \pm \delta$, after excluding $0 - \delta$ and $1 + \delta$, form a partition of $[0,1]$. Together with the partition $a = y_0 < y_1 = b$ of $[a,b]$, it provides a paving P of the domain of f. Let $\mathcal{F}$ be the family of those cuboids formed by P that have $[0,\delta]$ or $[x_j - \delta,\ x_j + \delta]$ or $[1 - \delta, 1]$ as an edge, and $\mathcal{G}$ consist of the remaining cuboids. By Remark 6-2.9, the total volume of the cuboids in $\mathcal{F}$ cannot exceed $6\delta(b - a)$. Therefore, in view of the fact that $\sup f = 1$ and $\inf f = 0$, we obtain the inequality $\Sigma_{K \in \mathcal{F}} (M_K - m_K) \mathrm{vol}(K) \leq 6\delta(b - a)$. If $(x,y) \in K \in \mathcal{G}$, then $x \neq x_j \ \forall\, j$ and hence $f(x,y) < \varepsilon$, so that $M_K - m_K \leq \varepsilon$. It follows that $\Sigma_{K \in \mathcal{G}} (M_K - m_K) \mathrm{vol}(K) \leq \varepsilon \cdot \mathrm{vol}([0,1] \times [a,b]) = \varepsilon(b - a)$. Therefore $U(f,P) - L(f,P) \leq 6\delta(b - a) + \varepsilon(b - a) < 7\varepsilon(b - a)$, because $\delta < \varepsilon$. Since every cuboid contains points with irrational coordinates, every lower sum is 0 and hence the integral is 0.

6-2.P7. (a) For each cuboid K formed by any paving of I, w have $M_K \geq 0$ and $m_K \geq 0$ (usual notation). Therefore $U(f,P) = \Sigma_K M_K \mathrm{vol}(K) \geq 0$ and $L(f,P) = \Sigma_K m_K \mathrm{vol}(K) \geq 0$.

(b) First show that if $M = \sup \{f(s) : s \in X\}$, $m = \inf \{f(s) : s \in X\}$, where X is any subset of the domain of the bounded function f, then $M - m = \sup \{|f(s) - f(t)| : s,t \in X\}$. By definition of M and m, we have $f(s) - f(t) \leq M - m$ as well as $f(t) - f(s) \leq M - m \ \forall\, s,t \in X$, so that $|f(s) - f(t)| \leq M - m \ \forall\, s,t \in X$. To prove the reverse inequality, consider any $\varepsilon > 0$. There exist $s,t \in X$ such that $f(s) > M - \frac{\varepsilon}{2}$ and $f(t) < m + \frac{\varepsilon}{2}$. This implies $f(s) - f(t) > M - m - \varepsilon$. It follows from this inequality that $\sup \{|f(s) - f(t)| : s,t \in X\} \geq M - m - \varepsilon$. Since this is true for every positive ε, it further follows that $\sup \{|f(s) - f(t)| : s,t \in X\} \geq M - m$.

Next, let $\mathcal{F}$ be the family of cuboids formed by a paving P of I and let $m_K(f) = \inf\{f(x) : x \in K\}$, $M_K(f) = \sup\{f(x) : x \in K\}$, with corresponding meanings for $m_K(|f|)$, $M_K(|f|)$. According to what has been proved above, for any s and t belonging to K, we have $||f(s)| - |f(t)|| \leq |f(s) - f(t)| \leq M_K(f) - m_K(f)$, so that $\sup \{||f(s)| - |f(t)|| : s,t \in K\} \leq M_K(f) - m_K(f)$. Again using what has been proved above, we conclude that $M_K(|f|) - m_K(|f|) \leq M_K(f) - m_K(f)$. Upon multiplying by $\mathrm{vol}(K)$ and taking the sum over $K \in \mathcal{F}$, it follows that

$$U(|f|,P) - L(|f|,P) \leq U(f,P) - L(f,P)$$

for any paving P and any bounded function f. By Proposition 6-2.5, it now follows that, if f is integrable, then so is $|f|$. To prove the inequality, note that $-|f(x)| \leq f(x) \leq |f(x)| \ \forall\, x \in K$. Since f and $|f|$ are both integrable, the required inequality follows.

6-2.P8. Without loss of generality, we may assume that $a_i = 0, b_i = 1$ for $1 \leq i \leq n$. Subdivide the domain $I = [0,1] \times \cdots \times [0,1]$ into m^n congruent cuboids. Then

$$U(f,P) - L(f,P) = \frac{1}{m^n} \sum_{i_1,i_2,\dots,i_n=1}^{m} [f(\tfrac{i_1}{m}, \tfrac{i_2}{m}, \dots, \tfrac{i_n}{m}) - f(\tfrac{i_1-1}{m}, \tfrac{i_2-1}{m}, \dots, \tfrac{i_n-1}{m})]$$

$$= \frac{1}{m^n} \left[\sum_{i_1,i_2,\dots,i_n=1}^{m} f(\tfrac{i_1}{m}, \tfrac{i_2}{m}, \dots, \tfrac{i_n}{m}) - \sum_{i_1,i_2,\dots,i_n=1}^{m} f(\tfrac{i_1-1}{m}, \tfrac{i_2-1}{m}, \dots, \tfrac{i_n-1}{m}) \right]$$

$$= \frac{1}{m^n} \left[\sum_{i_1,i_2,\dots,i_n=1}^{m} f(\tfrac{i_1}{m}, \tfrac{i_2}{m}, \dots, \tfrac{i_n}{m}) - \sum_{j_1,j_2,\dots,j_n=1}^{m} f(\tfrac{j_1-1}{m}, \tfrac{j_2-1}{m}, \dots, \tfrac{j_n-1}{m}) \right].$$

Now, any term in the first summation that has $i_k < m$ for every k cancels with a unique term in the second summation, namely, the one with $j_k = i_k + 1$ for every k. Note that the latter term has $j_k > 1$ for every k. Similarly, any term in the second summation that has $j_k > 1$ for every k cancels with a unique term in the first summation, namely, the one with $i_k = j_k - 1$ for every k. Note that the latter term has $i_k < m$ for every k. After all the cancellations, the only surviving terms from the first (respectively, second) summation are those with $i_k = m$ (respectively, $j_k = 1$) for some k. For a given k, the number of terms in the first summation with $i_k = m$ is m^{n-1}; considering that there are n possibilities for k, the total number of such terms is at most nm^{n-1} (actually fewer because of double count). Similarly for the second summation. It follows that

$$0 \leq U(f,P) - L(f,P) \leq \frac{2nm^{n-1}M}{m^n} = \frac{2nM}{m} \to 0 \quad \text{as } m \to \infty,$$

where $M = \sup |f|$ over $[0,1] \times \cdots \times [0,1]$.

6-2.P9. Being integrable, each of f and g must be bounded. Denote by B a common upper bound for their absolute values. Given any $\varepsilon > 0$, there exists a paving P of I such that $U(f,P) - L(f,P) < \varepsilon/2B$ as well as $U(g,P) - L(g,P) < \varepsilon/2B$. Next, let $\mathcal{F}$ be the family of cuboids formed by P and, for each $K \in \mathcal{F}$, let $m_K(f) = \inf \{f(x) : x \in K\}$, $M_K(f) = \sup \{f(x) : x \in K\}$, with corresponding meanings for $m_K(g)$, $M_K(g)$ and for $m_K(fg)$, $M_K(fg)$. Then for any $s, t \in K$, we have

$$|f(s)g(s) - f(t)g(t)| = |f(s)g(s) - f(t)g(s) + f(t)g(s) - f(t)g(t)|$$
$$\leq |g(s)||f(s) - f(t)| + |f(t)||g(s) - g(t)|$$
$$\leq B(M_K(f) - m_K(f)) + B(M_K(g) - m_K(g)),$$

so that

$$M_K(fg) - m_K(fg) = \sup \{|f(s)g(s) - f(t)g(t)| : s, t \in K\}$$
$$\leq B(M_K(f) - m_K(f)) + B(M_K(g) - m_K(g)).$$

On multiplying by vol(K) and taking the sum over all $K \in \mathcal{F}$, we get

$$U(fg,P) - L(fg,P) \leq B[U(f,P) - L(f,P) + U(g,P) - L(g,P)] < B(\varepsilon/2B + \varepsilon/2B) = \varepsilon,$$

which implies fg is integrable.

6-2.P10. Let $\varepsilon > 0$. There exists a paving P of $I \subseteq \mathbb{R}^m$ such that $U(f,P) - L(f,P) = \Sigma_K(M_K - m_K)\mathrm{vol}(K) < \varepsilon$, taken over all the cuboids K formed by P, with M_K and m_K having the usual meanings. Now $I \times J$ has a paving P' such that the cuboids formed by it are precisely $K \times J$; also, $\mathrm{vol}(K \times J) = \mathrm{vol}(K)\mathrm{vol}(J)$ and, by definition of ϕ its sup over $K \times J$ is M_K and inf is m_K. It follows that $U(\phi,P') - L(\phi,P') = \Sigma_K(M_K - m_K)\mathrm{vol}(K \times J) = \Sigma_K(M_K - m_K)\mathrm{vol}(K)\mathrm{vol}(J) < \varepsilon \mathrm{vol}(J)$. This shows ϕ to be integrable over $I \times J$.

6-2.P11. First we assume that both f_1 and f_2 are nonnegative. Let $P_1 : a_1 = x_0 < x_1 < \dots < x_n = b_1$ be a partition of I_1 and $P_2 : a_2 = y_0 < y_1 < \dots < y_m = b_2$ be a partition of I_2. Then $P = P_1 \times P_2$ is a partition of I into mn subcuboids (subrectangles) K_{ij}. Set

$$M_i = \sup\{f_1(x) : x \in [x_{i-1}, x_i]\}, \ M'_i = \sup\{f_2(y) : y \in [y_{i-1}, y_i]\},$$

$$m_i = \inf\{f_1(x) : x \in [x_{i-1}, x_i]\}, \ m'_i = \inf\{f_2(y) : y \in [y_{i-1}, y_i]\},$$

$$M_{ij} = \sup\{f_1(x)f_2(y) : x \in [x_{i-1}, x_i], \ y \in [y_{j-1}, y_j]\},$$

$$m_{ij} = \inf\{f_1(x)f_2(y) : x \in [x_{i-1}, x_i], \ y \in [y_{j-1}, y_j]\}.$$

Since both f_1 and f_2 have been assumed nonnegative, we have $m_j m'_j = m_{ij}$ and $M_i M'_j = M_{ij}$ and hence

$$m_j m'_j = m_{ij} \leq f_1(x)f_2(y) \leq M_{ij} = M_i M'_j.$$

It follows that

$$m_j m'_j (x_i - x_{i-1})(y_j - y_{j-1}) \leq m_{ij}(x_i - x_{i-1})(y_j - y_{j-1}) \leq \underline{\int}_{K_{ij}} f_1 f_2$$

$$\leq \overline{\int}_{K_{ij}} f_1 f_2 \leq M_{ij}(x_i - x_{i-1})(y_j - y_{j-1}) \leq M_i M'_j (x_i - x_{i-1})(y_j - y_{j-1}).$$

Summing over i and j and using Remark 6-4.6(g), we get $L(f_1,P_1)L(f_2,P_2) \leq \underline{\int}_K f_1 f_2 \leq \overline{\int}_K f_1 f_2 \leq U(f_1,P_1)U(f_2,P_2)$. In view of the fact that, first, this holds for all partitions P_1, P_2, second, f_1, f_2 are integrable, and third, $L(f_1,P_1)$, $L(f_2,P_2)$, $U(f_1,P_1)$, $U(f_2,P_2) \geq 0$, it follows that

$$(\textstyle\int_{I_1} f_1(x)\,dx)(\int_{I_2} f_2(y)\,dy) \leq \underline{\int}_K f_1 f_2 \leq \overline{\int}_K f_1 f_2 \leq (\int_{I_1} f_1(x)\,dx)(\int_{I_2} f_2(y)\,dy).$$

To remove the assumption that f_1 and f_2 are nonnegative, let $g_k = \max\{f_k, 0\}$ and $h_k = \min\{f_k, 0\}$, $k = 1,2$. Then g_1, g_2, h_1, h_2 are all integrable and $\int_{I_k} f_k = \int_{I_k} g_k - \int_{I_k} h_k$, $k = 1,2$. A simple computation now leads to the required equality.

Problem Set 6-3

6-3.P1. Fix any $x \in I$. If $x \notin \mathbb{Q}$, then $f(x,y) = 0 \ \forall \ y \in [a,b]$ and therefore $\bar{\int}_J f(x,y)\,dy = 0$. Suppose $x \in \mathbb{Q}$. Then $x = p/q$ with some minimal positive q and $f(x,y) = 1/q$ if $y \in \mathbb{Q}$ while $f(x,y) = 0$ if $y \notin \mathbb{Q}$. Therefore $\bar{\int}_J f(x,y)\,dy = (b-a)/q$. Thus $\bar{\int}_J f(x,y)\,dy$ is the product of the constant $(b-a)$ with the Thomae function, which has integral 0. Therefore $\int_I dx \, \bar{\int}_J f(x,y)\,dy = 0$.

6-3.P2. No. For each fixed $x \in [0,3]$ other than $x = \sqrt{3}$, $f(x,y)$ takes only two values: 3 when $y \in \mathbb{Q}$ but x^2 when $y \notin \mathbb{Q}$. Therefore, f is the Dirichlet function, which is not integrable. Thus $\int_{[0,1]} f(x,y)\,dy$ does not exist except when $x = \sqrt{3}$.

6-3.P3. Consider any $x \in [0,3]$. If $x^2 = 3$, then $f(x,y) = 3 \ \forall \ y \in [0,1]$. So $\int_{[0,1]} f(x,y)\,dy = 3$. If $x^2 \neq 3$, then $f(x,y)$ takes only two values: 3 when $y \in \mathbb{Q}$ but x^2 when $y \notin \mathbb{Q}$. Therefore when $x^2 < 3$, $\underline{\int}_{[0,1]} f(x,y)\,dy = x^2$ and $\bar{\int}_{[0,1]} f(x,y)\,dy = 3$. Similarly, when $x^2 > 3$, $\underline{\int}_{[0,1]} f(x,y)\,dy = 3$ and $\bar{\int}_{[0,1]} f(x,y)\,dy = x^2$. This means $\underline{\int}_{[0,1]} f(x,y)\,dy = x^2$ when $x < \sqrt{3}$ but 3 when $x > \sqrt{3}$. For $\bar{\int}_{[0,1]} f(x,y)\,dy$, it is the other way around. Therefore,

$$\int_{[0,3]} dx \underline{\int}_{[0,1]} f(x,y)\,dy = \underline{\int}_{[0,\sqrt{3}]} x^2 \, dx + \underline{\int}_{[\sqrt{3},3]} 3 \, dx = 9 - 2\sqrt{3}$$
$$\text{and } \int_{[0,3]} dx \bar{\int}_{[0,1]} f(x,y)\,dy = 9 + 2\sqrt{3}.$$

Now consider $y \in [0,1]$. If $y \in \mathbb{Q}$, then $f(x,y) = 3 \ \forall \ x \in [0,3]$ and so, $\underline{\int}_{[0,3]} f(x,y)\,dx = \bar{\int}_{[0,3]} f(x,y)\,dx = 9$. But if $y \notin \mathbb{Q}$, then $f(x,y) = x^2 \ \forall \ x \in [0,3]$ and so, $\underline{\int}_{[0,3]} f(x,y)\,dx = \bar{\int}_{[0,3]} f(x,y)\,dx = 9$. Therefore

$$\int_{[0,1]} dy \underline{\int}_{[0,3]} f(x,y)\,dx = \int_{[0,1]} dy \bar{\int}_{[0,3]} f(x,y)\,dx = 9.$$

6-3.P4. If $y \in \mathbb{Q}$ then $f(x,y) = \cos x$. So $\int_{[0,\pi]} f(x,y)\,dx = \int_{[0,\pi]} \cos x \, dx = 0$. If $y \notin \mathbb{Q}$ then $f(x,y) = 0$. So $\int_{[0,\pi]} f(x,y)\,dx = 0$. Hence, for all y, $\int_{[0,\pi]} f(x,y)\,dx = 0$, so that $\int_{[0,1]} dy \int_{[0,\pi]} f(x,y)\,dx = 0$. For $x = \frac{\pi}{2}$, $f(x,y) = 0$ for all y. So $\int_{[0,1]} f(\frac{\pi}{2},y)\,dy = 0$. For $x \neq \frac{\pi}{2}$, $f(x,y) = \cos x$ if $y \in \mathbb{Q}$ and 0 if $y \notin \mathbb{Q}$. So, when $x \neq \frac{\pi}{2}$, $\int_{[0,1]} f(x,y)\,dy$ does not exist. Now let P be a paving of $[0, \frac{\pi}{4}] \times [0,1]$ into subcuboids $K_1, \dots, K_m$. Then $U(f,P) - L(f,P) = \Sigma_{1 \le i \le m} (\sup_{K_i} f - \inf_{K_i} f) \text{vol}(K_i) \ge \Sigma_{1 \le i \le m} (\frac{1}{\sqrt{2}} - 0) \text{vol}(K_i) = (\frac{1}{\sqrt{2}})(\frac{\pi}{4})$. The result now follows by Proposition 6-2.7.

6-3.P5. $\int_{[0,1]} f(x,y)\,dx = \int_{[0,y]} (1/y^2)\,dx + \int_{[y,1]} (-1/x^2)\,dx = 1/y + (1 - 1/y) = 1$. So, $\int_{[0,1]} dy \int_{[0,1]} f(x,y)\,dx = 1$. $\int_{[0,1]} f(x,y)\,dy = \int_{[0,x]} (-1/x^2)\,dy + \int_{[x,1]} (1/y^2)\,dy = -1/x - (1 - 1/x) = -1$. So, $\int_{[0,1]} dx \int_{[0,1]} f(x,y)\,dy = -1$.

6-3.P6. Suppose there exists a point (x_0, y_0) at which $\frac{\partial^2 f}{\partial x \partial y} - \frac{\partial^2 f}{\partial y \partial x} \neq 0$. Without loss of generality, we may assume that it is positive. In view of the continuity of the partial derivatives, (x_0, y_0) lies in the interior of a closed rectangle $R = [a,b] \times [c,d]$ on which $\frac{\partial^2 f}{\partial x \partial y} - \frac{\partial^2 f}{\partial y \partial x} > 0$. So $\int_R \frac{\partial^2 f}{\partial x \partial y} - \frac{\partial^2 f}{\partial y \partial x} > 0$. On the other hand, by Fubini's theorem,

$$\int_R \frac{\partial^2 f}{\partial x \partial y} - \frac{\partial^2 f}{\partial y \partial x} = \int_c^d dy \int_a^b \frac{\partial^2 f}{\partial x \partial y} dx - \int_a^b dx \int_c^d \frac{\partial^2 f}{\partial y \partial x} dy = 0,$$

since each equals $f(b,d) - f(b,c) - f(a,d) + f(a,c)$.

6-3.P7. At each point $(x,y) \in \mathbb{R}^2$, at most one term in the series is different from zero and therefore no convergence problem arises in the definition of the function. Also, f vanishes outside $(0,1] \times (0,1]$, is unbounded on every open set containing $(0,0)$ and is continuous except at $(0,0)$.

We begin by simplifying the description of f. In order to do so, consider any $(x,y) \in (0,1] \times (0,1]$. There exists a unique $j \in \mathbb{N}$ such that $x \in (2^{-j}, 2^{1-j}]$. If $j > 1$, then we have

$$f(x,y) = \sum_{i=1}^{\infty} [\phi_i(x) - \phi_{i+1}(x)] \phi_i(y)$$
$$= [\phi_{j-1}(x) - \phi_j(x)] \phi_{j-1}(y) + [\phi_j(x) - \phi_{j+1}(x)] \phi_j(y)$$
$$= -\phi_j(x) \phi_{j-1}(y) + \phi_j(x) \phi_j(y) = \phi_j(x)[-\phi_{j-1}(y) + \phi_j(y)].$$

However, if $j = 1$, then $f(x,y) = \sum_{i=1}^{\infty} [\phi_i(x) - \phi_{i+1}(x)] \phi_i(y) = \phi_1(x) \phi_1(y)$. Thus f has the simplified description:

$$f(x,y) = \begin{cases} \phi_j(x)[-\phi_{j-1}(y) + \phi_j(y)] & \text{if } \frac{1}{2^j} < x \le \frac{1}{2^{j-1}} \text{ with } j > 1 \\ \phi_1(x) \phi_1(y) & \text{if } \frac{1}{2^j} < x \le \frac{1}{2^{j-1}} \text{ with } j = 1. \end{cases}$$

This simplified description shows that, when $x \in (2^{-j}, 2^{1-j}]$ with $j > 1$ (so that $x \notin (2^{-j}, 2^{1-j}]$ with $j = 1$),

$$\int_0^1 f(x,y) dy = \int_0^1 \phi_j(x)[-\phi_{j-1}(y) + \phi_j(y)] dy$$
$$= \phi_j(x) \int_0^1 [-\phi_{j-1}(y) + \phi_j(y)] dy = \phi_j(x)[-1 + 1]$$
$$= 0 = \phi_1(x), \text{ considering that } x \notin (2^{-j}, 2^{1-j}] \text{ with } j = 1.$$

But when $x \in (2^{-j}, 2^{1-j}]$ with $j = 1$,

$$\int_0^1 f(x,y) dy = \int_0^1 \phi_1(x) \phi_1(y) dy = \phi_1(x).$$

Thus $\int_0^1 f(x,y)\,dy = \phi_1(x)$ for all $x \in (0,1]$. This implies $\int_0^1 dx \int_0^1 f(x,y)\,dy = \int_0^1 \phi_1(x)\,dx = 1$.

Before computing the integral iterated in the reverse order, note that

$$\int_{1/2^j}^{1/2^{j-1}} \phi_j(x)[-\phi_{j-1}(y) + \phi_j(y)]\,dx = [-\phi_{j-1}(y) + \phi_j(y)] \text{ for } j > 1$$

and $\qquad \int_{1/2^j}^{1/2^{j-1}} \phi_1(x)\phi_1(y)\,dx = \phi_1(y) \text{ for } j = 1$.

It follows from the simplified description of f derived above that

$$\int_0^1 f(x,y)\,dx = \sum_{j=1}^{\infty} \int_{1/2^j}^{1/2^{j-1}} f(x,y)\,dx = \int_{1/2}^1 f(x,y)\,dx + \sum_{j=2}^{\infty} \int_{1/2^j}^{1/2^{j-1}} f(x,y)\,dx$$

$$= \phi_1(y) + \sum_{j=2}^{\infty} [-\phi_{j-1}(y) + \phi_j(y)],$$

in view of what was noted in the preceding paragraph. Now, for any integer $m > 1$,

$$\phi_1(y) + \sum_{j=2}^{m} [-\phi_{j-1}(y) + \phi_j(y)] = \phi_m(y).$$

Therefore $\int_0^1 f(x,y)\,dx = \lim_{m\to\infty} \phi_m(y)$. But this limit is zero when $0 < y \le 1$, because $m > 1 - \frac{\ln y}{\ln 2} \Rightarrow 1/2^{m-1} < y \Rightarrow \phi_m(y) = 0$. Thus $\int_0^1 f(x,y)\,dx = 0$ for $0 < y \le 1$, and consequently, $\int_0^1 dy \int_0^1 f(x,y)\,dx = 0$.

Problem Set 6-4

6-4.P1. If $E° = \varnothing$, then $\partial E = \overline{E}$ and $c(\partial E) = c(\overline{E}) \ge c(E) > 0$, contradicting Proposition 6-4.14.

6-4.P2. $\overline{E}$ has content by Proposition 6-4.14 because $\partial\overline{E} \subseteq \partial E$. Since $c(E \cap \partial E) \le c(\partial E) = 0$ and $\overline{E} = E \cup \partial E$, it follows by Remark 6-4.6(d) that $c(\overline{E}) = c(E)$.

6-4.P3. If E is a cuboid, then $\alpha(E)$ is also a cuboid; moreover, its edges have the same length as those of E. Therefore they have the same volume and thus same content too. Since α is continuous on an open set containing E and is an injective map, it follows by 2-6.P8(c) that $\alpha(\partial E) \subseteq \partial(\alpha(E))$. But α has a continuous inverse on the open set $\mathbb{R}^n$ containing $\alpha(E)$ and therefore $\alpha(\partial E) = \partial(\alpha(E))$. Now suppose E has content. By Proposition 6-4.14, $c(\partial E) = 0$. Using this with Proposition 6-4.10 and what has just been proved about the behaviour of α on cuboids that $c(\partial(\alpha(E))) = 0$. By Proposition 6-4.14, $\alpha(E)$ has content. Finally, applying Proposition 6-4.10, for any $\varepsilon > 0$, we first obtain finitely many cuboids covering E and having total volume less than $c(E) + \varepsilon$; their images under α then provide finitely many cuboids covering $\alpha(E)$ and having the same total volume, which is less than $c(E) + \varepsilon$. By Remark 6-4.6(b), it follows that $c(\alpha(E)) \leq c(E) + \varepsilon$. Therefore $c(\alpha(E)) \leq c(E)$. Arguing the same way with α^{-1} (i.e., replacing s by $-s$), we find that $c(E) \leq c(\alpha(E))$.

6-4.P4. Let I be a cuboid such that $E \subseteq I$. By definition of integrability of $f: E \to \mathbb{R}$, the extension $f_I : I \to \mathbb{R}$ is integrable and $\int_E f = \int_I f_I$. Hence by 6-2.P7, $|f_I|$ is integrable and $|\int_E f| = |\int_I f_I| \leq \int_I |f_I|$. But $|f_I| = |f|_I$. Therefore, again by definition, $|f|$ is integrable and $|\int_E f| \leq \int_I |f_I| = \int_I |f|_I = \int_E |f|$.

6-4.P5. Let $m = \inf\{f(x) : x \in A\}$ and $M = \sup\{f(x) : x \in A\}$; let I be any cuboid containing A. Then $m \leq f_I(x) \leq M$ on I and $0 = m \cdot c(A) = \int_I m\chi_A \leq \underline{\int}_I f_I \leq \overline{\int}_I f_I \leq \int_I M\chi_A = M \cdot c(A) = 0$. Therefore $\int_I f_I$ exists and is zero, i.e., $\int_A f$ exists and is zero.

6-4.P6. We shall use the easily proven inequality $\underline{\int}_I (f + g) \geq \underline{\int}_I f + \underline{\int}_I g$ and its analogue for upper integrals.

Let $m = \inf\{f(x) : x \in E\}$ and $M = \sup\{f(x) : x \in E\}$; let I be any cuboid containing E. For any subset G of E, denote by f_G the *extension* to I of the restriction of f to G. Then $m\chi_{X_k} \leq f_{X_k} \leq M\chi_{X_k}$ and $f_E = f_{E \setminus X_k} + f_{X_k}$. It follows from the foregoing inequality that $m \cdot c(X_k) \leq \underline{\int}_I f_{X_k} \leq \overline{\int}_I f_{X_k} \leq M \cdot c(X_k)$ and further that $\lim \underline{\int}_I f_{X_k} = \lim \overline{\int}_I f_{X_k} = 0$; also, it follows from the equality that

$$\underline{\int}_I f_{X_k} + \int_I f_{E \setminus X_k} = \underline{\int}_I f_{X_k} + \underline{\int}_I f_{E \setminus X_k} \leq \underline{\int}_I f_E \leq \overline{\int}_I f_E \leq \overline{\int}_I f_{X_k} + \overline{\int}_I f_{E \setminus X_k}$$
$$= \overline{\int}_I f_{X_k} + \int_I f_{E \setminus X_k}.$$

The result follows from here upon using the limits established just earlier.

6-4.P7. If $c(E) = 0$, then $\int_E f = 0$ and therefore any $\mu \in [m, M]$ serves the purpose. When $c(E) \neq 0$, we have $m \cdot c(E) \leq \int_E f \leq M \cdot c(E)$ and hence $m \leq \frac{1}{c(E)} \int_E f \leq M$. Take $\mu = \frac{1}{c(E)} \int_E f$. If E is closed, then it is also compact (being bounded), so that m and M are in the range of f. If E is also connected then, the range of f must be an interval [see 2-6.P14] and must therefore be $[m, M]$. This implies the existence of the required ξ.

6-4.P8. Observe that $f(x)^m \leq M^m$ for all $x \in E$. So, $\int_E f^m \leq M^m c(E)$, which implies $\limsup_{m \to \infty} (\int_E f^m)^{1/m} \leq M$. Now, let $\varepsilon > 0$. There exists $x_0 \in E$ such that $f(x_0) > M - \varepsilon$.

Any open cuboid I containing any point of E, in particular x_0, must contain a point interior to E and hence must also contain an open cuboid $K \subseteq E$. Therefore $c(E \cap I) \geq c(K \cap I) = c(K) > 0$. By continuity of f, there exists an open cuboid I such that $f(x) > M - \varepsilon$ on I. Hence $\int_E f^m \geq \int_{E \cap I} f^m \geq (M - \varepsilon)^m c(E \cap I)$. Since $c(E \cap I) > 0$, we have $\lim_{m \to \infty} c(E \cap I)^{1/m} = 1$ and therefore $\liminf_{m \to \infty} (\int_E f^m)^{1/m} \geq M - \varepsilon$. Since this holds for every $\varepsilon > 0$, we have $\liminf_{m \to \infty} (\int_E f^m)^{1/m} \geq M$. The reverse inequality has already been proved. To see why the condition $E = \overline{E}^\circ$ cannot be omitted, take $E = [0,1] \cup \{5\}$ and $f(x) = 1$ on $[0,1]$ and $f(5) = 2$.

6-4.P9. Let I be a cuboid containing E. If f_I, g_I and $(fg)_I$ are the functions on I obtained by setting f, g and fg equal to zero outside E, then surely $(fg)_I$ equals the product $(f_I)(g_I)$. In view of Def. 6-4.3, we need prove only that $(f_I)(g_I)$ is integrable. This follows from 6-2.P9.

6-4.P10. Consider any $x, y \in K$. Observe that J_x is the disjoint union of $J_x \cap J_y$ and $J_x \backslash J_y$ and similarly for J_y. Therefore

$$|F(x) - F(y)| = |\int_{J_x} f - \int_{J_y} f| = |\int_{J_x \backslash J_y} f - \int_{J_y \backslash J_x} f| \leq |\int_{J_x \backslash J_y} f| + |\int_{J_y \backslash J_x} f|$$

$$= \int_{(J_x \backslash J_y) \cup (J_y \backslash J_x)} |f|.$$

If we set $u = (u_1, \ldots, u_n)$ and $v = (v_1, \ldots, v_n)$, where $u_i = \min\{x_i, y_i\}$ and $v_i = \max\{x_i, y_i\}$, it is easy to check that $(J_x \backslash J_y) \cup (J_y \backslash J_x) \subseteq J_v \backslash J_u$. Therefore $\int_{(J_x \backslash J_y) \cup (J_y \backslash J_x)} |f| \leq \int_{J_v \backslash J_u} |f| \leq M \cdot c(J_v \backslash J_u)$, where $M = \sup |f|$. Since $u_i \leq v_i$ for each i, we have $J_u \subseteq J_v$ and therefore $c(J_v \backslash J_u) = c(J_v) - c(J_u)$. Our choice of u and v implies that $c(J_v) - c(J_u)$ can be made arbitrarily small by taking x and y sufficiently close.

6-4.P11. The function is defined on the cuboid $[0,1] \times [0,1]$ except at the origin. Set it equal to 0 at the origin; the extended function is discontinuous only at the origin and is also bounded. By Theorem 6-4.15, it is integrable and hence so is the given function.

Problem Set 6-5

6-5.P1. Since $b_1 - a_1 > b_2 - a_2$, we have $a_1 + b_2 < b_1 + a_2$. It follows that, when $x_2 \in [a_2, b_2]$, we have

$$a_1 + a_2 \leq a_1 + x_2 \leq a_1 + b_2 < b_1 + a_2 \leq b_1 + x_2. \tag{1}$$

(a) From the last three inequalities in (1), we get

$$[a_1 + x_2, b_1 + x_2] = [a_1 + x_2, a_1 + b_2] \cup [a_1 + b_2, b_1 + a_2] \cup [b_1 + a_2, b_1 + x_2].$$

It is now immediate that $S = B \cup C \cup D$.

(b) From the first three inequalities in (1), we get

$$[a_1 + a_2, b_1 + a_2] = [a_1 + a_2, a_1 + x_2] \cup [a_1 + x_2, a_1 + b_2] \cup [a_1 + b_2, b_1 + a_2].$$

It is now immediate that $J = A \cup B \cup C$.

(c) By Proposition 6-5.1, the boundaries of A, B, C, D all have content zero. Therefore by Proposition 6-4.14, all the four sets have content. Since $A \cap B = \{(x_1, \dots, x_n) \in \mathbb{R}^n : (x_2, \dots, x_n) \in [a_2, b_2] \times \cdots \times [a_n, b_n], \; x_1 = a_1 + x_2\}$, it follows by Proposition 6-5.1 that it has content zero. A similar argument applies to $B \cap C$ and $C \cap D$.

(d) In view of the above, Remark 6-4.6(d) shows that S has content and that, firstly, $c(S) = c(B) + c(C) + c(D)$ and secondly, that $c(J) = c(A) + c(B) + c(C)$. But $c(J) = \text{vol}(J) = (b_1 - a_1) \cdots (b_n - a_n)$. Therefore, we need only prove that $c(D) = c(A)$. This results from applying 6-4.P3 with $s = a_1 - b_1$ and $p = 1$.

(e) Take $[a_1, b_1] = [0,1]$, $[a_2, b_2] = [0,2]$. Then $(x_1, \dots, x_n) \in B$ if $x_1 = 2$ and $x_2 = 0$; but $(x_1, \dots, x_n) \notin S$.

If $b_1 - a_1 \le b_2 - a_2$, let N be an integer large enough so that $(b_2 - a_2)/N < b_1 - a_1$. Now partition $[a_2, b_2]$ into N equal subintervals $I_1, \dots, I_N$, so that S is the union of N sets $S_1, \dots, S_N$ obtained by replacing $[a_2, b_2]$ in its definition by $I_1, \dots, I_N$. Each of these sets satisfies the hypothesis that has now been dropped. So, it is covered by part (d) and therefore has content $(b_1 - a_1) \cdots (b_n - a_n)/N$. Moreover, the intersection of the union of $S_1, \dots, S_k$ with S_{k+1} ($1 \le k < N$) has content zero. Hence, $c(S) = \Sigma_k c(S_k) = (b_1 - a_1) \cdots (b_n - a_n)$.

6-5.P2. By Theorems 6-3.2, 6-4.13 and 6-5.1

$$\int_T D_{2\,1}f = \int_0^a dx \int_0^{b(1-x/a)} D_{2\,1}f(x,y)\,dy = \int_0^a [D_1 f(x, b(1 - x/a)) - D_1 f(x,0)]\,dx$$

$$= \int_0^a [D_1 f(x, b(1 - x/a))\,dx + f(0,0) - f(a,0)$$

$$= aD_1 f(x_0, b(1 - x_0/a)) + f(0,0) - f(a,0),$$

where $0 \le x_0 \le a$, by the mean value theorem for integrals. Observe that $x_0/a + b(1 - x_0/a)/b = 1$. So, the point (x_0, y_0), where $y_0 = b(1 - x_0/a)$, lies on the line segment joining $(a, 0)$ to $(0, b)$.

6-5.P3. The given set E is the same as

$$\{(x,y) \in \mathbb{R}^2 : x \ge 0, \; y \ge 0, \; (x,y) \ne (0,0), \; 0 \le x < (A\varepsilon)^{1/2}, \; x[(1 - \varepsilon)/\varepsilon]^{1/2}$$

$$< y < (A - x^2)^{1/2}\}.$$

Since $\qquad 0 \leq x < (A\varepsilon)^{\frac{1}{2}} \Rightarrow x[(1-\varepsilon)/\varepsilon]^{\frac{1}{2}} < (A-x^2)^{\frac{1}{2}}$

and $\quad x = (A\varepsilon)^{\frac{1}{2}} \Rightarrow x[(1-\varepsilon)/\varepsilon]^{\frac{1}{2}} = (A-x^2)^{\frac{1}{2}}$,

the boundary is the union of the four (nondisjoint) sets

$$\{(0,y) \in \mathbb{R}^2 : 0 \leq y \leq A^{\frac{1}{2}}\}, \qquad\qquad \{((A\varepsilon)^{\frac{1}{2}}, [A(1-\varepsilon)]^{\frac{1}{2}})\},$$
$$\{(x,y) \in \mathbb{R}^2 : 0 \leq x \leq (A\varepsilon)^{\frac{1}{2}}, \, y = x[(1-\varepsilon)/\varepsilon]^{\frac{1}{2}}\}$$

and

$$\{(x,y) \in \mathbb{R}^2 : 0 \leq x \leq (A\varepsilon)^{\frac{1}{2}}, \, y = (A-x^2)^{\frac{1}{2}}\}.$$

The first two are easily seen to be subsets of rectangles of arbitrarily small content (area in the present context) and the latter two have content zero in view of Proposition 6-5.1. Therefore the given set has content. Also, it is a subset of the rectangle $[0, (A\varepsilon)^{\frac{1}{2}}] \times [0, A^{\frac{1}{2}}]$, which has content $A\varepsilon^{\frac{1}{2}}$.

Problem Set 7-1

7-1.P1. (a) $3,5$ because $1 < 2e/n \leq 2 \Leftrightarrow 3 \leq n \leq 5$. (b) $3,4$ because $2 < \sqrt{99}/n \leq 4$ $\Leftrightarrow 3 \leq n \leq 4$.

7-1.P2. $4,5$. A partition with n subintervals will do if and only if $1 < \frac{2e}{n} \leq \frac{3}{2}$, i.e., $\frac{4e}{3} \leq n < 2e$. Since $\frac{5}{2} < e < 3$, this implies $\frac{10}{3} < n < 6$. Therefore integers other than 4 and 5 are ruled out. One can check that $n = 4$ and $n = 5$ both fulfil $1 < \frac{2e}{n} \leq \frac{3}{2}$.

7-1.P3. As in the proof of Proposition 7-1.3, choose N such that $1 + 1/N < \frac{4}{3}$, e.g., $N = 4$. Then $\left[\frac{9}{6}(4)\right] = 6$ and $\left[\frac{11}{6}(4)\right] = 7$. Therefore the triplet (nonunique) may be taken as $4,6,7$. To check that this triplet works, we note that $\frac{4}{3}\frac{6}{4} = 2$, $\frac{9}{6}$ < 2 and $\frac{11}{7} < 2$.

7-1.P4. The inequality $l \leq L < \mu l$ will hold if and only if either $a \leq b/n < \mu a$ or $b/n \leq a < \mu\, b/n$, depending on which among a and b/n is bigger. These are respectively equivalent to $b/a\mu < n \leq b/a$ and $b/a \leq n < \mu b/a$. Therefore the inequality $l \leq L < \mu l$ will hold if and only if either $b/a\mu < n \leq b/a$ or $b/a \leq n < \mu b/a$. In other words if and only if $b/a\mu < n < \mu b/a$. Therefore it is possible to choose n as required if and only if the interval $(b/a\mu, \mu\, b/a)$ contains an integer.

(a) If $b/a < \mu$, then $b/a\mu < 1$ and also $\mu b/a > b/a > 1$. Therefore the interval in question contains the integer 1. Suppose $b/a \geq \mu$. Since $\mu > \sqrt{2}$, we have $\mu^2 - 1 > 1$ and hence $\mu/(\mu^2 - 1) < \mu \leq b/a$. It follows that $1 < (b/a)(\mu^2 - 1)/\mu = (b/a)(\mu - 1/\mu)$

= length of the interval ($b/_{a\mu}$, $\mu b/_a$). Therefore the interval in question contains an integer in this case as well.

(b) If $1 < \mu \leq \sqrt{2}$, take $a = 1$ and $b = \mu$. Then the interval ($b/_{a\mu}$, $\mu b/_a$) is $(1, \mu^2) \subseteq (1,2)$, which contains no integer.

Problem Set 7-2

7-2.P1. Let $W = (0,\infty)$, $\alpha(x) = 1/x$ and $F = (0,1)$. Then $\alpha(F) = (1,\infty)$, which is not even bounded. Note that $\overline{F}$ is not a subset of W and therefore Proposition 7-2.4 is not contradicted.

7-2.P2. Modify the proof as follows: Consider an arbitrary $\eta > 0$. Since α' is continuous, $\overline{E}$ is covered by open balls contained in the open set V (as in the proof) such that, on each ball, $\|\alpha'\|$ varies by no more than $\eta/2$. Replace V by the union of these balls; then $\sup\{\|\alpha'(x)\| : x \in \overline{V}\} \leq M_0^{1/n} + \eta$. Take an arbitrary $\mu > 1$ and, instead of Proposition 7-2-1, use Proposition 7-1.3 to ensure that (diam $K)^n < \mu^n \mathrm{vol}(K)$. Then the total volume of the family $\{\alpha(K)' : K \in \mathcal{H}\}$ of closed cuboids (as in the proof) is less than $(\mu(M_0^{1/n} + \eta))^n [c(F) + \varepsilon/(2M)^n]$. Since $\varepsilon > 0$, $\eta > 0$ and $\mu > 1$ are all arbitrary, the total volume no greater than $M_0 c(F)$.

7-2.P3. First injectivity. Consider (u_1,v_1), $(u_2,v_2) \in E$ and $(u_1,v_1) \neq (u_2,v_2)$. Denote $\alpha(u_1,v_1)$ by (x_1,y_1), and $\alpha(u_2,v_2)$ by (x_2,y_2). We claim $(x_1,y_1) \neq (x_2,y_2)$. The definition of α shows that $x_1^2 + y_1^2 = u_1^2$ and $x_2^2 + y_2^2 = u_2^2$. Therefore if $u_1 \neq u_2$, we have $x_1^2 + y_1^2 \neq x_2^2 + y_2^2$, so that $(x_1,y_1) \neq (x_2,y_2)$. If $u_1 = u_2$, then $v_1 \neq v_2$. Since by definition of E, we have $0 < v_1 < 2\pi$ as well as $0 < v_2 < 2\pi$, therefore either $\cos v_1 \neq \cos v_2$ or $\sin v_1 \neq \sin v_2$. Since $u_1 = u_2 \neq 0$, it follows that either $u_1 \cos v_1 \neq u_2 \cos v_2$ or $u_1 \sin v_1 \neq u_2 \sin v_2$, i.e., $(x_1,y_1) \neq (x_2,y_2)$. Thus α is injective.

For surjectivity, consider any $(x,y) \in G$. We must find $(u,v) \in E$ such that $\alpha(u,v) = (x,y)$, i.e., $x = u\cos v$ and $y = u\sin v$. Set

$$u = (x^2 + y^2)^{1/2} \tag{1}$$

and

$$v = \begin{cases} \cos^{-1} \frac{x}{u} & \text{if } y > 0 \\ 2\pi - \cos^{-1} \frac{x}{u} & \text{if } y \leq 0 \end{cases}. \tag{2}$$

The given definition of G ensures that $0 < u < A$. To show that $(u,v) \in E$, we need only show that $0 < v < 2\pi$. Since the range of $\cos^{-1}$ is $[0,\pi]$, it is immediate that $0 \leq v \leq \pi$ when $y > 0$ and that $\pi \leq v \leq 2\pi$ when $y \leq 0$. In particular, $v = 0 \Rightarrow y$

> 0 and $v = 2\pi \Rightarrow y \leq 0$. We argue why $v \neq 0$. Suppose if possible that $v = 0$. Then $y > 0$. This implies on the one hand that $x < u$ by (1) and on the other hand by (2) that $v = \cos^{-1}(\frac{x}{u})$, which further implies $\cos^{-1}(\frac{x}{u}) = 0$, so that $x = u$. This contradition shows that v cannot be 0. Now suppose, if possible, that $v = 2\pi$. Then $y \leq 0$. By (2), this implies that $v = 2\pi - \cos^{-1}(\frac{x}{u})$, which further implies $\cos^{-1}(\frac{x}{u}) = 0$, so that $x = u > 0$, and on using (1), we have $y = 0$. But by the given definition of G, we have $x > 0 \Rightarrow y \neq 0$. Thus v cannot be 2π either. So, $0 < v < 2\pi$. It remains to show that $\alpha(u,v) = (x,y)$, i.e., $u\cos v = x$ and $u\sin v = y$.

We use the two consequence of (2) that firstly, $\cos v = \frac{x}{u}$ regardless of whether $y > 0$ or $y \leq 0$, and secondly, $\sin v = \sqrt{(1 - (\frac{x}{u})^2)}$ for $y > 0$ and $-\sqrt{(1 - (\frac{x}{u})^2)}$ for $y \leq 0$. It follows from the former consequence that $u\cos v = u \cdot \frac{x}{u} = x$ regardless of whether $y > 0$ or $y \leq 0$. And it follows from the latter consequence that, for $y > 0$, we have $u\sin v = u\sqrt{(1 - (\frac{x}{u})^2)} = \sqrt{y^2} = y$, while for $y \leq 0$, we have $u\sin v = -u\sqrt{(1 - (\frac{x}{u})^2)} = -\sqrt{y^2} = y$.

Problem Set 7-4

7-4.P1. Since W_1 has content, its boundary ∂W_1 has content zero by Proposition 6-4.14. It follows by Remark 6-4.6(d) that the set $F = \{x \in W_1 : \det \alpha'(x) = 0\} \cup \partial W_1$ has content zero. Therefore, for each integer k, there exist [Proposition 6-4.10] finitely many closed cuboids which cover F and have total volume less than $1/k$. Denote their union by S_k. Then S_k is closed, $F \subseteq S_k$ and $c(S_k) < 1/k$. Each of the aforementioned cuboids is contained in an open cuboid with volume not more than twice as much [Remark 6-1.2(iv)]. Denote the union of these open cuboids by T_k. Then $T_k \supseteq S_k$ and $c(T_k) < 2/k$. Also, $E \backslash T_k$ has content by Remark 6-4.6(h). Note that $\overline{W_1 \cap T_k^c} = W_1 \cap T_k^c$ because T_k contains ∂W_1. Now,

$$\overline{E \backslash T_k} = \overline{E \cap T_k^c} \subseteq \overline{E} \cap \overline{T_k^c} = \overline{E} \cap T_k^c \subseteq \overline{W_1} \cap T_k^c = W_1 \cap T_k^c \subseteq W_1 \cap S_k^c = W_1 \backslash S_k$$

and the invertibility of α on W_1 yields

$$\alpha(E \backslash T_k) = \alpha(E \backslash (E \cap T_k)) = \alpha(E) \backslash \alpha(E \cap T_k).$$

Furthermore, the set $\alpha(E \backslash T_k)$ has content in view of Proposition 7-2.4 (take W_1 as the set 'W' there). Therefore the integrals

$$\int_{\alpha(E) \backslash \alpha(E \cap T_k)} f \quad \text{and} \quad \int_{E \backslash T_k} (f \circ \alpha)|\det \alpha'|$$

both exist. Since α is invertible and α' injective on the open set $W_1 \backslash S_k$ containing $\overline{E \backslash T_k}$, it follows from Theorem 7-4.4 that the two integrals are also equal. Moreover, since all the sets $E \cap T_k$ are contained in the single set E, whose closure is contained in the open set W on which α is continuously differentiable, it follows that $c(\alpha(E \cap T_k)) \to 0$ as $k \to \infty$ by Proposition 7-2.2. The required conclusion now follows from what was established in Problem 6-4.P6.

7-4.P2. Since f is continuous, the integrals all exist, while the second, third and fourth integrals are equal by Fubini's theorem [see Remark 6-3.3]. So, we need only prove that the first and second are equal. To this end, let $W = \mathbb{R}^2$ and $W_1 = \{(r,\theta) \in \mathbb{R}^2 : 0 < r < B, \ 0 < \theta < 2\pi\} = (0,B) \times (0,2\pi)$, where $B > A$. Then the transformation $\alpha : W \to \mathbb{R}^2$ defined by $\alpha(r,\theta) = (r\cos\theta, r\sin\theta)$ is continuously differentiable on W. On the open subset W_1, it is invertible with $\det \alpha' = r > 0$ everywhere. Besides, W_1 has content and $E \subseteq \overline{W}_1 = [0,B] \times [0,2\pi]$. Finally, $\alpha(E) = D$ and $(f \circ \alpha)(r,\theta) = f(r\cos\theta, r\sin\theta)$. The required equality therefore follows from 7-4.P1. (A similar argument justifies the usual procedure of evaluating the integral over the part of the disc in the first quadrant, i.e., $\{(x,y) \in \mathbb{R}^2 : 0 \le x^2 + y^2 \le A^2, \ x \ge 0, \ y \ge 0\}$, via polar coordinates in the above manner, the integration with respect to θ being taken over $[0, \frac{\pi}{2}]$.)

7-4.P3. Let $(x,y) = \alpha(u,v) = (u, v - u)$. Then α is injective and continuously differentiable on $\mathbb{R}^2$, with linear derivative $\alpha'(u,v)$ given by the matrix having first row $[1 \ \ 0]$ and second row $[-1 \ \ 1]$. If we take $f(x,y) = \tan^{-1}(x+y)$, then $[(f \circ \alpha)|\det\alpha'|](u,v) = \tan^{-1}v$. Using our definition of α, the set E is most easily 'described in terms of u and v' by the inequalities $u \ge 0$, $v - u \ge 0$, $v \le 1$, which is the same as, $u \ge 0$, $u \le v \le 1$. As a figure would immediately suggest and a little manipulation will easily confirm, it is also the same as $0 \le v \le 1$, $0 \le u \le v$. Such a description of E in terms of u and v simply means $E = \alpha(F)$, where

$$F = \{(u,v) \in \mathbb{R}^2 : 0 \le u \le 1, \ u \le v \le 1\} = \{(u,v) \in \mathbb{R}^2 : 0 \le v \le 1, \ 0 \le u \le v\}.$$

By the transformation formula (Theorem 7-4.4), the required integral is

$$\int_F [(f \circ \alpha)|\det\alpha'|](u,v)\,du\,dv = \int_F \tan^{-1}v\,du\,dv.$$

By Fubini's theorem [see Remark 6-3.3] this can be evaluated as either

$$\int_0^1 (\int_u^1 \tan^{-1}v\,dv)\,du \qquad \text{or} \qquad \int_0^1 (\int_0^v \tan^{-1}v\,du)\,dv.$$

The latter is easier to evaluate and leads to the answer $\frac{\pi}{4} - \frac{1}{2}$.

7-4.P4. By Proposition 7-4.12, we may choose the balloon according to our convenience, which means we take $E_m = \{(x,y) \in \mathbb{R}^2 : \frac{1}{m} \le x^2 + y^2 \le 1\}$ By the Proposition 7-4.15, $\int_{E_m} f = \int_{E_m} f\chi_{E_m} = \int_E f\chi_{E_m}$

$$= \int_0^{2\pi} d\theta \int_0^1 f(r\cos\theta, r\sin\theta) \cdot \chi_{E_m}(r\cos\theta, r\sin\theta) r\,dr$$

$$= \int_0^{2\pi} d\theta \int_{1/m}^1 f(r\cos\theta, r\sin\theta)\, r\, dr$$

$$= \int_0^{2\pi} d\theta \int_{1/m}^1 \frac{r}{r^{2\alpha}}\, dr = \begin{cases} \frac{2\pi}{2-2\alpha}\left(1 - \frac{1}{m^{2-2\alpha}}\right) < \infty \text{ for each } m & \text{if } \alpha \neq 1 \\[2mm] 2\pi(1 + \ln m) < \infty \text{ for each } m & \text{if } \alpha = 1 \end{cases}.$$

Therefore $\lim_{m\to\infty} \int_{E_m} f$ is finite (in fact, $\frac{2\pi}{2-2\alpha}$) if $\alpha < 1$ and does not exist otherwise.

7-4.P5. It is well known that: (1) $\lim_{A\to\infty} \int_0^A \frac{\sin x}{x}\, dx$ exists in $\mathbb{R}$; (2) $\int_{2k\pi}^{(2k+1)\pi} \frac{\sin x}{x}\, dx \geq \frac{2}{(2k+1)\pi}$ for $k \in \mathbb{N}$; (3) $\sum_{k=\mu}^{\nu} \frac{1}{k} \geq \ln(\nu + 1) - \ln\mu$ for $\mu < \nu$. Consider the function $f(x) = \frac{\sin x}{x}$. The sequence $\{E_m\}$, where $E_m = [0, m]$, is a balloon for $[0, \infty)$ and we know from (1) that $\lim_{m\to\infty} \int_{E_m} f$ is finite. Now let $\{F_m\}$ be the sequence of sets where

$$F_m = [0, (2m-1)\pi] \cup \bigcup_{k=m}^{m^2} [2k\pi, (2k+1)\pi]$$

$$= [0, (2m-1)\pi] \cup [2m\pi, (2m+1)\pi] \cup \bigcup_{k=m+1}^{m^2} [2k\pi, (2k+1)\pi].$$

Each F_m is a union of $m^2 - m + 2$ disjoint closed intervals and therefore has content. Since $[0, (2m-1)\pi] \cup [2m\pi, (2m+1)\pi] \subseteq [0, (2m+1)\pi]$, therefore $F_m \subseteq F_{m+1}$. Also, $\bigcup_{m=1}^{\infty} F_m \supseteq \bigcup_{m=1}^{\infty} [0, (2m-1)\pi] = [0, \infty)$. Thus $\{F_m\}$ is a balloon for $[0, \infty)$, and moreover,

$$\int_{F_m} f = \int_0^{(2m-1)\pi} \frac{\sin x}{x}\, dx + \sum_{k=m}^{m^2} \int_{2k\pi}^{(2k+1)\pi} \frac{\sin x}{x}\, dx.$$

In view of (2), we have

$$\sum_{k=m}^{m^2} \int_{2k\pi}^{(2k+1)\pi} \frac{\sin x}{x}\, dx \geq \sum_{k=m}^{m^2} \frac{2}{(2k+1)\pi} \geq \sum_{k=m}^{m^2} \frac{2}{(2k+2)\pi} = \frac{1}{\pi} \sum_{k=m}^{m^2} \frac{1}{k+1} = \frac{1}{\pi} \sum_{k=m+1}^{m^2+1} \frac{1}{k}.$$

By using (3) and the above representation of $\int_{F_m} f$ as a sum, we deduce from this inequality that

$$\int_{F_m} f \geq \int_0^{(2m-1)\pi} \frac{\sin x}{x}\, dx + \frac{1}{\pi}[\ln(m^2 + 2) - \ln(m + 1)].$$

The second term on the right tends to ∞ with m while the first has a finite limit by (1). It follows that $\int_{F_m} f$ tends to ∞ with m.

7-4.P6. Since $T(0,0) = (2,-1)$, we have $a_1 = 2$ and $a_2 = -1$. Furthermore, since $T(1,0) = (5,0)$, we have $2 + b_1 = 5$ and $-1 + b_2 = 0$ and hence $b_1 = 3$ and $b_2 = 1$. Similar considerations lead to $c_1 = 1$ and $c_2 = -1$. So,

$$T(x,y) = (u,v) = (2 + 3x + y, -1 + x - y).$$

Now $\frac{\partial}{\partial x}(2 + 3x + y) = 3$, $\frac{\partial}{\partial y}(2 + 3x + y) = 1$, $\frac{\partial}{\partial x}(-1 + x - y) = 1$, $\frac{\partial}{\partial y}(-1 + x - y) = -1$. Therefore the Jacobian is -4. Since the equations

$$u = 2 + 3x + y, \qquad v = -1 + x - y$$

lead to $x = \frac{1}{4}(u + v - 1)$ and $y = \frac{1}{4}(u - 3v - 5)$, the mapping T is bijective. By the transformation formula,

$$\int_D \exp(2u - v)\, du\, dv = \int_{[0,1] \times [0,1]} \exp(2(2 + 3x + y) - (-1 + x - y))(4)\, dx\, dy$$
$$= \int_{[0,1] \times [0,1]} \exp(5 + 5x + 3y)(4)\, dx\, dy = 4e^5 \int_{[0,1]} e^{5x}\, dx \int_{[0,1]} e^{3y}\, dy$$
$$= \tfrac{4}{15} e^5 (e^5 - 1)(e^3 - 1).$$

Problem Set 8-2

8-2.P1. Φ is a 1-surface and $\Phi(u) = (\Phi_1(u), \ldots, \Phi_n(u))$.

$$\int_\Phi \omega = \int_\Phi \sum_{i=1}^n f_i\, dx_i = \int_{[0,1]} \sum_{i=1}^n f_i(\Phi(u)) \frac{d\Phi_i}{du}\, du.$$

If $\Phi(u) = (u, u^2, u^3)$ and $\omega = dx + dz$, then the above integral equals

$$\int_{[0,1]} \frac{d\Phi_1}{du}\, du + \int_{[0,1]} \frac{d\Phi_3}{du}\, du = \int_{[0,1]} du + \int_{[0,1]} 3u^2\, du = 2.$$

8-2.P2. $\int_{[0,1]^2} [f_1(\Phi_1(u), \Phi_2(u), \Phi_3(u)) \frac{\partial(\Phi_2, \Phi_3)}{\partial(u_1, u_2)} + f_2(\Phi_1(u), \Phi_2(u), \Phi_3(u)) \frac{\partial(\Phi_3, \Phi_1)}{\partial(u_1, u_2)} + f_3(\Phi_1(u), \Phi_2(u), \Phi_3(u)) \frac{\partial(\Phi_1, \Phi_2)}{\partial(u_1, u_2)}]\, du_1 du_2$.

8-2.P3. $\int_\Phi \omega =$

$$\int_{[0,1]^2} (v_1 w_2 - v_2 w_1) \frac{\partial(x_1, x_2)}{\partial(x, y)}\, dx\, dy + (v_1 w_3 - v_3 w_1) \frac{\partial(x_1, x_3)}{\partial(x, y)}\, dx\, dy + (v_2 w_3 - v_3 w_2) \frac{\partial(x_2, x_3)}{\partial(x, y)}\, dx\, dy$$

$$= (v_1 w_2 - v_2 w_1)^2 + (v_1 w_3 - v_3 w_1)^2 + (v_2 w_3 - v_3 w_2)^2 \text{ because } \frac{\partial(x_i, x_j)}{\partial(x, y)} = v_i w_j - v_j w_i.$$

8-2.P4. $\Phi_1(r,\theta) = \dfrac{2r \cos 2\pi\theta}{1 + r^2}$, $\Phi_2(r,\theta) = \dfrac{2r \sin 2\pi\theta}{1 + r^2}$, $\Phi_3(r,\theta) = \dfrac{1 - r^2}{1 + r^2}$. So,

$$\frac{\partial(\Phi_2,\Phi_3)}{\partial(r,\theta)} = \frac{16\pi r^2 \cos 2\pi\theta}{(1+r^2)^3}, \qquad \frac{\partial(\Phi_3,\Phi_1)}{\partial(r,\theta)} = \frac{16\pi r^2 \sin 2\pi\theta}{(1+r^2)^3} \qquad \text{and}$$

$$\frac{\partial(\Phi_1,\Phi_2)}{\partial(r,\theta)} = \frac{8\pi r(1-r^2)}{(1+r^2)^3}.$$

Therefore, $\Phi_1(r,\theta)\dfrac{\partial(\Phi_2,\Phi_3)}{\partial(r,\theta)} = \dfrac{32\pi r^3 \cos^2 2\pi\theta}{(1+r^2)^4}$,

$\Phi_2(r,\theta)\dfrac{\partial(\Phi_3,\Phi_1)}{\partial(r,\theta)} = \dfrac{32\pi r^3 \sin^2 2\pi\theta}{(1+r^2)^4}$ and $\Phi_3(r,\theta)\dfrac{\partial(\Phi_1,\Phi_2)}{\partial(r,\theta)} = \dfrac{8\pi r(1-r^2)^2}{(1+r^2)^4}$.

The sum of these three is $8\pi\dfrac{4r^3 + r(1-r^2)^2}{(1+r^2)^4} = 8\pi\dfrac{r}{(1+r^2)^2}$. Therefore

$$\int_\Phi \omega = 8\pi \int_{[0,1]^2} \frac{r}{(1+r^2)^2}\,dr\,d\theta = 8\pi \int_{[0,1]} \frac{r}{(1+r^2)^2}\,dr = 2\pi.$$

8-2.P5. Let $(a_1,\dots,a_n)$ be any point in $\mathbb{R}^n$ and let $\Phi:[0,1]\to\mathbb{R}^n$ be the constant map with value $(a_1,\dots,a_n)$ everywhere on $[0,1]$. Then each component function of Φ is also constant and therefore has derivative (Jacobian) 0 everywhere. Consider any 1-form $\omega = \sum_{j=1}^n f_j dx_j$. Since $\int_{[0,1]} f_j(\Phi(u))\Phi_j'(u)\,du = 0$ for every j, we have $\int_\Phi \omega = 0$.

8-2.P6. Let $\Phi:[0,1]^3\to\mathbb{R}^2$ be given by $\Phi(x_1,x_2,x_3) = (x_1,x_2)$.

8-2.P7. To extend φ to the left of 0, take $\varphi(u) = \varphi(0) + u\varphi'(0)$ and to extend to the right of 1, take $\varphi(u) = \varphi(1) + (u-1)\varphi'(1)$. Since φ is the restriction of a C^1 map on $\mathbb{R}$, it follows easily that Φ is the restriction of a C^1 map on $\mathbb{R}^2$. Thus it is a 2-surface.

Its range is $\{(x,y)\in\mathbb{R}^2 : 0\le x\le 1 \quad \text{and} \quad 0\le y\le\varphi(x) \text{ or } 0\ge y\ge\varphi(x)\}$. The range of Ψ is the graph of φ, namely, $\{(x,y)\in\mathbb{R}^2 : 0\le x\le 1, y=\varphi(x)\}$.

Problem Set 8-3

8-3.P1. Let $\alpha = \sum_{i=1}^n f_i dx_i$ and $\beta = \sum_{j=1}^n g_j dx_j$. Then $\alpha\wedge\beta = (\sum_{i=1}^n f_i dx_i)\wedge(\sum_{j=1}^n g_j dx_j) = \sum_{i\ne j}^n f_i g_j dx_i\wedge dx_j = -\sum_{i\ne j}^n f_i g_j dx_j\wedge dx_i = (\sum_{j=1}^n g_j dx_j)\wedge(\sum_{i=1}^n f_i dx_i) = -\beta\wedge\alpha$.

8-3.P2. Put $\alpha = dx_1\wedge dx_2 + dx_3\wedge dx_4$. Then $\alpha\wedge\alpha = 2\,dx_1\wedge dx_2\wedge dx_3\wedge dx_4$.

Problem Set 8-4

8-4.P1. Since $D_i f = 2x_i$ for $1 \le i \le n$, then $df = 2\sum_{i=1}^{n}(x_i dx_i)$. Now, $D_j x_i = 1$ if $j = i$ and 0 otherwise. Therefore

$$d(df) = 2\sum_{i=1}^{n}((\sum_{j=1}^{n}(D_j x_i) dx_j) \wedge dx_i)$$

$$= 2\sum_{i=1}^{n}((D_i x_i) dx_i \wedge dx_i) \quad \text{because } D_j x_i = 0 \text{ if } j \ne i$$

$$= 0 \text{ because } dx_i \wedge dx_i = 0.$$

8-4.P2. We have

$$d\eta = d\left(\frac{x}{x^2 + y^2} dy - \frac{y}{x^2 + y^2} dx \right)$$

$$= \left(\frac{y^2 - x^2}{x^2 + y^2} \right) dx \wedge dy - \left(\frac{y^2 - x^2}{x^2 + y^2} \right) dx \wedge dy = 0.$$

The other terms vanish since $dx \wedge dx = dy \wedge dy = 0$. Thus η is closed.

Next. suppose η is exact; then there exists a C^2 function f such that

$$df = \frac{\partial f}{\partial x} dx + \frac{\partial f}{\partial y} dy = \frac{x}{x^2 + y^2} dy - \frac{y}{x^2 + y^2} dx .$$

Consider the 1-surface $\gamma:[0,1] \to U$ given by $\gamma(t) = (\cos 2\pi t, \sin 2\pi t)$. That is $\gamma_1(t) = \cos 2\pi t$, $\gamma_2(t) = \sin 2\pi t$. On the one hand, we have

$$\int_\gamma df = \int_{[0,1]} (\frac{\partial f}{\partial x}(\gamma(t)) \frac{d}{dt}\gamma_1(t) dt + \frac{\partial f}{\partial y}(\gamma(t)) \frac{d}{dt}\gamma_2(t)) dt = \int_{[0,1]} \frac{d}{dt}(f(\gamma(t)) dt$$

$$= f(\gamma(1)) - f(\gamma(0)) = 0 \quad \text{because } \gamma(0) = \gamma(1).$$

But on the other hand,

$$\int_\gamma df = \int_\gamma \frac{\partial f}{\partial x} dx + \frac{\partial f}{\partial y} dy = \int_\gamma \frac{x}{x^2 + y^2} dy - \frac{y}{x^2 + y^2} dx$$

$$= 2\pi \int_{[0,1]} (\cos 2\pi t) \frac{d}{dt}(\sin 2\pi t) - (\sin 2\pi t) \frac{d}{dt}(\cos 2\pi t) dt = 4\pi^2 \ne 0.$$

Problem Set 8-5

8-5.P1. As in Def. 8-5.1, denote the two component functions of T by t_1 and t_2. Then $t_1(x) = x^2$ and $t_2(x) = x^3$ and $T^*\omega = T^*(y_1 dy_2) = t_1 d(t_2) = x^2(3x^2) dx = 3x^4 dx$.

8-5.P2. In subscript notation, we have $T(x_1, x_2) = x_1 - x_2$ and we have to find $T^*(dy_1)$. Now, $T^*(dy_1) = d(x_1 - x_2) = dx_1 - dx_2 = dx - dy$ in xy-notation.

8-5.P3. $T^*\omega = T^*(dy_1 \wedge dy_2) = d(ax_1 + bx_2) \wedge d(cx_1 + ex_2)$
$$= (a\,dx_1 + b\,dx_2) \wedge (c\,dx_1 + e\,dx_2) = (ae - bc)dx_1 \wedge dx_2.$$

8-5.P4. $df_1 \wedge \cdots \wedge df_n = (\sum_{j=1}^{n} \frac{\partial f_1}{\partial x_j} dx_j) \wedge \cdots \wedge (\sum_{j=1}^{n} \frac{\partial f_n}{\partial x_j} dx_j)$. In view of Proposition 8-

3.5 and Proposition 8-3.7, this is a sum of various wedge products having n factors each. Those summands with repeated dx_j are zero and can be neglected. and we are left with a sum of wedge products

$$\frac{\partial t_{i_1}}{\partial x_{i'_1}} \cdots \frac{\partial t_{i_n}}{\partial x_{i'_n}} dx_{i'_1} \wedge \cdots \wedge dx_{i'_n},$$

where $\langle i'_1, i'_2, \ldots, i'_n \rangle$ is an n-index of distinct integers in $\{1, \ldots, n\}$, and all such n-indices occur. If σ is the permutation that rearranges $\langle i'_1, i'_2, \ldots, i'_n \rangle$ in ascending order as the k-index $\langle 1, \ldots, n \rangle$, then $dx_{i'_1} \wedge \cdots \wedge dx_{i'_k} = (\text{sign } \sigma) dx_1 \wedge \ldots \wedge dx_n$. Therefore the wedge product displayed above is equal to

$$(\text{sign } \sigma) \frac{\partial t_{i_1}}{\partial x_1} \cdots \frac{\partial t_{i_n}}{\partial x_n} dx_1 \wedge \cdots \wedge dx_n.$$

It follows from this that the sum of wedge products that $df_1 \wedge \ldots \wedge df_n$ equals (recall that all permutations of $\langle 1, \ldots, n \rangle$ must occur) is precisely

$$\frac{\partial(f_1, \ldots, f_n)}{\partial(x_1, \ldots, x_n)} dx_1 \wedge \cdots \wedge dx_n.$$

8-5.P5. If $k > n$, then it is obvious that $T^*\omega = 0$. So assume $k \le n$. By Def. 8-5.1 and Remark 8-5.2(e), $T^*\omega = (b_I \circ T) dt_{i_1} \wedge \cdots \wedge dt_{i_k}$. We can drop the factor $b_I \circ T$ in the rest of our computation. Now,

$$dt_{i_1} \wedge \cdots \wedge dt_{i_k} = (\sum_{j=1}^{n} \frac{\partial t_{i_1}}{\partial x_j} dx_j) \wedge \cdots \wedge (\sum_{j=1}^{n} \frac{\partial t_{i_k}}{\partial x_j} dx_j).$$

In view of Proposition 8-3.5 and Proposition 8-3.7, this is a sum of various wedge products having k factors each. Those summands with repeated dx_j are zero and can be neglected. and we are left with a sum of wedge products

$$\frac{\partial t_{i_1}}{\partial x_{i'_1}} \cdots \frac{\partial t_{i_k}}{\partial x_{i'_k}} dx_{i'_1} \wedge \cdots \wedge dx_{i'_k},$$

where $\langle i'_1, i'_2, \ldots, i'_k \rangle$ is a k-index of distinct integers in $\{1, \ldots, n\}$, and all such k-indices occur. If σ is the permutation that rearranges $\langle i'_1, i'_2, \ldots, i'_k \rangle$ in ascending order as the k-index $J = \langle j_1, \ldots, j_k \rangle$, then $dx_{i'_1} \wedge \ldots \wedge dx_{i'_k} = (\text{sign } \sigma) dx_J$. Therefore the wedge product displayed above is equal to

$$(\text{sign } \sigma) \frac{\partial t_{i_1}}{\partial x_{j_1}} \cdots \frac{\partial t_{i_k}}{\partial x_{j_k}} \, dx_J .$$

It follows from this that, in the sum of wedge products that $dt_{i_1} \wedge \cdots \wedge dt_{i_k}$ equals, those terms for which $\langle i'_1, i'_2, \ldots, i'_k \rangle$ is a permutation of any one particular ascending $J = \langle j_1, \ldots, j_k \rangle$ (recall that all its permutations must occur) add up to

$$\frac{\partial(t_{i_1}, \ldots, t_{i_k})}{\partial(x_{j_1}, \ldots, x_{j_k})} \, dx_J .$$

Besides, all ascending k-indices J are covered; therefore $dt_{i_1} \wedge \cdots \wedge dt_{i_k}$ is precisely equal to the summation in the statement of the problem.

8-5.P6. It is enough to argue the case when $\omega = b_I dy_I$, a simple k-form. Let $I = \langle i_1, \ldots, i_k \rangle$, an ascending k-index in $\{1, \ldots, m\}$ and denote the component functions of Φ by $\Phi_1, \ldots, \Phi_m$. We apply 8-5.P5 with $n = k$ and $T = \Phi$ to get

$$\Phi^* \omega = (b_I \circ \Phi) \frac{\partial(\Phi_{i_1}, \ldots, \Phi_{i_k})}{\partial(x_1, \ldots, x_k)} \, dx_1 \wedge \cdots \wedge dx_k .$$

Since the Jacobian of the inclusion map is 1, it follows from the above equality and the definition of integral of a differential form over a surface that $\int_\Phi \omega = \int_{1_k} \Phi^* \omega$.

8-5.P7. $d\omega = \left(\dfrac{\partial f_2}{\partial x} - \dfrac{\partial f_1}{\partial y} \right) dx \wedge dy + \left(\dfrac{\partial f_3}{\partial y} - \dfrac{\partial f_2}{\partial z} \right) dy \wedge dz + \left(\dfrac{\partial f_1}{\partial z} - \dfrac{\partial f_3}{\partial x} \right) dz \wedge dx .$

$$T^*(d\omega) = \left(\left(\frac{\partial f_2}{\partial x} - \frac{\partial f_1}{\partial y} \right) \circ T \right) T^*(dx) \wedge T^*(dy)$$

$$+ \left(\left(\frac{\partial f_3}{\partial y} - \frac{\partial f_2}{\partial z} \right) \circ T \right) T^*(dy) \wedge T^*(dz) + \left(\left(\frac{\partial f_1}{\partial z} - \frac{\partial f_3}{\partial x} \right) \circ T \right) T^*(dz) \wedge T^*(dx)$$

Now, $T^*(dx) = dT_1 = \dfrac{\partial T_1}{\partial u} du + \dfrac{\partial T_1}{\partial v} dv$, $T^*(dy) = \dfrac{\partial T_2}{\partial u} du + \dfrac{\partial T_2}{\partial v} dv$ and $T^*(dz) = \dfrac{\partial T_3}{\partial u} du + \dfrac{\partial T_3}{\partial v} dv$. Therefore

$$T^*(d\omega) = \left(\left(\frac{\partial f_2}{\partial x} - \frac{\partial f_1}{\partial y} \right) \circ T \right) \frac{\partial(T_1, T_2)}{\partial(u, v)} du \wedge dv$$

$$+\left(\left(\frac{\partial f_3}{\partial y}-\frac{\partial f_2}{\partial z}\right)\circ T\right)\frac{\partial(T_2,T_3)}{\partial(u,v)}\,du\wedge dv+\left(\left(\frac{\partial f_1}{\partial z}-\frac{\partial f_3}{\partial x}\right)\circ T\right)\frac{\partial(T_3,T_1)}{\partial(u,v)}\,du\wedge dv.$$

Thus $T^*(d\omega)$ is the required inner product.

Problem Set 8-6

8-6.P1. By Proposition 8-5.4, $\Gamma_{i0}^*(dy_j) = d(\Gamma_{i0}^*y_j) = d((\Gamma_{i0})_j)$ and similarly for Γ_{i1}. From the description of $(\Gamma_{i0})_j$ in Remark 8-6.3, we find that $d((\Gamma_{i0})_j)$ and $d((\Gamma_{i1})_j)$ are as required.

8-6.P2. $\partial\Phi = \sum_{i=1}^{k}(-1)^i(\Phi\circ\Gamma_{i0}-\Phi\circ\Gamma_{i1})$ and hence

$$\partial(\partial\Phi) = \sum_{i=1}^{k}(-1)^i\sum_{j=1}^{k-1}(-1)^j((\Phi\circ\Gamma_{i0}-\Phi\circ\Gamma_{i1})\circ\Gamma_{j0}-(\Phi\circ\Gamma_{i0}-\Phi\circ\Gamma_{i1})\circ\Gamma_{j1})$$

$$= \sum_{i=1}^{k}(-1)^i\sum_{j=1}^{k-1}(-1)^j(\Phi\circ\Gamma_{i0}\circ\Gamma_{j0}-\Phi\circ\Gamma_{i1}\circ\Gamma_{j0}-\Phi\circ\Gamma_{i0}\circ\Gamma_{j1}+\Phi\circ\Gamma_{i1}\circ\Gamma_{j1}).$$

8-6.P3. (i) $\Gamma_{j0}(x_1,\dots,x_{k-1}) = (x_1,\dots,x_{j-1},0,x_j,\dots,x_{k-1})$ and therefore

$$\Gamma_{i0}\circ\Gamma_{j0}(x_1,\dots,x_{k-1}) = \Gamma_{i0}(x_1,\dots,x_{j-1},0,x_j,\dots,x_{k-1})$$

$$= \begin{cases} (x_1,\dots,x_{j-1},0,0,x_j,\dots,x_{k-1}) & \text{if } i=j \\ (x_1,\dots,x_{i-1},0,x_i,\dots,x_{j-1},0,x_j,\dots,x_{k-1}) & \text{if } i<j. \end{cases}$$

$\Gamma_{i0}(x_1,\dots,x_{k-1}) = (x_1,\dots,x_{i-1},0,x_i,\dots,x_{k-1})$ and therefore

$$\Gamma_{j+1,0}\circ\Gamma_{i0}(x_1,\dots,x_{k-1}) = \Gamma_{j+1,0}(x_1,\dots,x_{i-1},0,x_i,\dots,x_{k-1})$$

$$= \begin{cases} (x_1,\dots,x_{i-1},0,0,x_i,\dots,x_{k-1}) & \text{if } i=j \\ (x_1,\dots,x_{i-1},0,x_i,\dots,x_{j-1},0,x_j,\dots,x_{k-1}) & \text{if } i<j. \end{cases}$$

Similarly the other relations.

8-6.P4. (i) The result is true for $k = 2$, because

$$\sum_{i=1}^{2}\sum_{j=1}^{1}(-1)^{i+j}\Phi\circ\Gamma_{i0}\circ\Gamma_{j0} = \Phi\circ\Gamma_{10}\circ\Gamma_{10} - \Phi\circ\Gamma_{20}\circ\Gamma_{10} = 0$$

by part (i) of 8-6.P3 (with $i=j=1$).

Assume that the result is true for $k-1$, that is,

$$\sum_{i=1}^{k-1}\sum_{j=1}^{k-2}(-1)^{i+j}\Phi\circ\Gamma_{i0}\circ\Gamma_{j0} = 0.$$

Now,

$$\sum_{i=1}^{k}\sum_{j=1}^{k-1}(-1)^{i+j}\Phi\circ\Gamma_{i0}\circ\Gamma_{j0}$$

$$=\sum_{i=1}^{k-1}\sum_{j=1}^{k-2}(-1)^{i+j}\Phi\circ\Gamma_{i0}\circ\Gamma_{j0}+\sum_{j=1}^{k-1}(-1)^{k+j}\Phi\circ\Gamma_{k0}\circ\Gamma_{j0}+\sum_{i=1}^{k-1}(-1)^{i+k-1}\Phi\circ\Gamma_{i0}\circ\Gamma_{k-1,0}$$

$$=\sum_{j=1}^{k-1}(-1)^{k+j}\Phi\circ\Gamma_{k0}\circ\Gamma_{j0}+\sum_{i=1}^{k-1}(-1)^{i+k-1}\Phi\circ\Gamma_{i0}\circ\Gamma_{k-1,0}$$

by the induction hypothesis.

But

$$\sum_{j=1}^{k-1}(-1)^{k+j}\Phi\circ\Gamma_{k0}\circ\Gamma_{j0}+\sum_{i=1}^{k-1}(-1)^{i+k-1}\Phi\circ\Gamma_{i0}\circ\Gamma_{k-1,0}$$

$$=\sum_{i=1}^{k-1}(-1)^{k+i}\Phi\circ\Gamma_{k0}\circ\Gamma_{i0}+\sum_{i=1}^{k-1}(-1)^{i+k-1}\Phi\circ\Gamma_{i0}\circ\Gamma_{k-1,0}=0$$

by part (i) of 8-6.P3.

Similarly the others.

8-6.P5. Follows from the results of 8-6.P4.

8-6.P6. Suppose f is a function of class C^1 on an open subset U of $\mathbb{R}^n$. Let ω be a $(k-1)$-form of class C^1 and Φ a k-surface in U. Under these conditions, the formula holds and the proof is as follows.

It is a consequence of the general Stokes theorem that $\int_{\Phi}d(f\omega)=\int_{\partial\Phi}(f\omega)$. Now, $d(f\omega)=(df)\wedge\omega+f\,d\omega$ and hence $\int_{\Phi}d(f\omega)=\int_{\Phi}(df)\wedge\omega+\int_{\Phi}(f\,d\omega)$. Substituting the previous equality into the aforementioned consequence of Stokes theorem, one gets $\int_{\partial\Phi}(f\omega)=\int_{\Phi}(df)\wedge\omega+\int_{\Phi}(f\,d\omega)$, which implies the required equality.

Now suppose $n=1$, $k=0$ and $U\subseteq\mathbb{R}$ is an open set containing the closed interval $[a,b]$. Then the k-form ω is a 0-form, which is a C^1 function g. Consider a k-surface $\Phi:[0,1]\to U$ such that $\Phi(0)=a$ and $\Phi(1)=b$. Then $\partial\Phi$ is the chain $\{\Phi(1)\}-\{\Phi(0)\}=\{b\}-\{a\}$. This means $\int_{\partial\Phi}(f\omega)=f(b)g(b)-f(a)g(a)$. Also, $\int_{\Phi}(f\,d\omega)=\int_{\Phi}(f\,dg)=\int_{[0,1]}fg'=\int_0^1 f(\xi)g'(\xi)\,d\xi$ and $\int_{\Phi}(df)\wedge\omega=\int_{\Phi}(g\,df)=\int_0^1 f'(\xi)g(\xi)\,d\xi$. Thus the equality proved reduces to that of the formula of integration by parts, but under stronger differentiability hypotheses.

8-6.P7. $\Gamma_{10}(t)=(0,t)$, $\Gamma_{11}(t)=(1,t)$, $\Gamma_{20}(t)=(t,0)$, $\Gamma_{21}(t)=(t,1)$. Therefore, for Φ as in Example 8-2.2(e), we have

$\Phi\circ\Gamma_{10}(t)=(0,0)$; range: just the origin;

$\Phi\circ\Gamma_{11}(t)=(\cos 2\pi t,\sin 2\pi t)$; range: circle of radius 1 about the origin;

$\Phi\circ\Gamma_{20}(t)=(t,0)$; range: segment between the origin and $(1,0)$;

$\Phi\circ\Gamma_{21}$ is the same as $\Phi\circ\Gamma_{20}$.

Only $\Phi\circ\Gamma_{11}$ has range contained in the boundary of the range of Φ.

For Φ as in Example 8-2.2(h), we have

$\Phi \circ \Gamma_{10}(t) = (-\frac{1}{2}\cos 2\pi t, -\frac{1}{2}\sin 2\pi t)$; range: the circle of radius $\frac{1}{2}$ about the origin;

$\Phi \circ \Gamma_{11}(t) = (\cos 2\pi t, \sin 2\pi t)$; range: the circle of radius 1 about the origin;

$\Phi \circ \Gamma_{20}(t) = (\frac{1}{2}(3t-1), 0)$; range: segment between $(-\frac{1}{2}, 0)$ and $(1,0)$;

$\Phi \circ \Gamma_{21}$ is the same as $\Phi \circ \Gamma_{20}$.

Only $\Phi \circ \Gamma_{11}$ has range contained in the boundary of the range of Φ.

8-6.P8. By the general Stokes theorem, $\int_{\partial c} \omega = \int_c d\omega$. However, $d\omega$ is an $(n+1)$-form and is therefore 0.

Problem Set 8-7

8-7.P1. By the divergence Theorem, (1) is the same as

$$\int_\Phi \left(\frac{\partial F_1}{\partial x} + \frac{\partial F_2}{\partial y} + \frac{\partial F_3}{\partial z} \right) dx \wedge dy \wedge dz$$

$$= \int_\Phi [-6(x+y)(x^2+y^2)^2 + 2x^2 z]\, dx \wedge dy \wedge dz$$

$$= \int_{[0,1]\times[0,2\pi]\times[0,1]} [-6r^5(\cos\theta+\sin\theta) + 2r^2(\cos^2\theta)z]\, r\, dr\, d\theta\, dz$$

$$= \tfrac{1}{4}\pi.$$

We next verify the answer by actually evaluating (1).

Now,

$$\partial\Phi = -(\Phi\circ\Gamma_{10} - \Phi\circ\Gamma_{11}) + (\Phi\circ\Gamma_{20} - \Phi\circ\Gamma_{22\pi}) - (\Phi\circ\Gamma_{30} - \Phi\circ\Gamma_{31}),$$

where

$$\Gamma_{10}(\theta,z) = (0,\theta,z), \qquad \Gamma_{20}(r,z) = (r,0,z), \qquad \Gamma_{30}(r,\theta) = (r,\theta,0)$$
$$\Gamma_{11}(\theta,z) = (1,\theta,z), \qquad \Gamma_{22\pi}(r,z) = (r,2\pi,z), \qquad \Gamma_{31}(r,\theta) = (r,\theta,1).$$

Therefore,

$$\int_{\Phi\circ\Gamma_{10}} F_1\, dy \wedge dz = \int_{[0,\,2\pi]\times[0,1]} (F_1\circ\Phi)(0,\theta,z)\frac{\partial(0\cdot\sin\theta, z)}{\partial(\theta, z)}\, d\theta\, dz = 0;$$

$$\int_{\Phi\circ\Gamma_{11}} F_1\, dy \wedge dz = \int_{[0,2\pi]\times[0,1]} (F_1\circ\Phi)(1,\theta,z)\frac{\partial(1\cdot\sin\theta, z)}{\partial(\theta, z)}\, d\theta\, dz = 0,$$

$$\text{because } (F_1\circ\Phi)(1,\theta,z) = 0;$$

$$\int_{\Phi\circ\Gamma_{20}} F_1\, dy \wedge dz = \int_{[0,\,1]\times[0,1]} (F_1\circ\Phi)(r,0,z)\frac{\partial(r\cdot\sin 0, z)}{\partial(r, z)}\, dr\, dz = 0;$$

$$\int_{\Phi\circ\Gamma_{22\pi}} F_1\, dy\wedge dz = \int_{[0,\,1]\times[0,1]} (F_1\circ\Phi)(r,2\pi,z)\frac{\partial(r\cdot\sin 2\pi,\,z)}{\partial(r,z)}\, dr\, dz = 0;$$

$$\int_{\Phi\circ\Gamma_{30}} F_1\, dy\wedge dz = \int_{[0,\,1]\times[0,\,2\pi]} (F_1\circ\Phi)(r,\theta,0)\frac{\partial(r\cdot\sin\theta,\,0)}{\partial(r,\theta)}\, dr\, d\theta = 0;$$

$$\int_{\Phi\circ\Gamma_{31}} F_1\, dy\wedge dz = \int_{[0,\,1]\times[0,\,2\pi]} (F_1\circ\Phi)(r,\theta,1)\frac{\partial(r\cdot\sin\theta,\,1)}{\partial(r,\theta)}\, dr\, d\theta = 0.$$

Thus $\int_{\partial\Phi} F_1\, dy\wedge dz = 0$. Of the six corresponding integrals occurring in the sum for $\int_{\partial\Phi} F_2\, dz\wedge dx$, the first two and last two turn out to be 0, but the middle two both turn out to be $-\frac{6}{7}$ and cancel out. So, $\int_{\partial\Phi} F_2\, dz\wedge dx$. We proceed to find $\int_{\partial\Phi} F_3\, dx\wedge dy$.

$$\int_{\Phi\circ\Gamma_{10}} F_3\, dx\wedge dy = \int_{[0,\,2\pi]\times[0,1]} (F_3\circ\Phi)(0,\theta,z)\frac{\partial(0\cdot\cos\theta,\,0\cdot\sin\theta)}{\partial(\theta,z)}\, d\theta\, dz = 0;$$

$$\int_{\Phi\circ\Gamma_{11}} F_3\, dx\wedge dy = \int_{[0,\,2\pi]\times[0,\,1]} (F_3\circ\Phi)(1,\theta,z)\frac{\partial(1\cdot\cos\theta,\,1\cdot\sin\theta)}{\partial(\theta,z)}\, d\theta\, dz = 0;$$

$$\int_{\Phi\circ\Gamma_{20}} F_3\, dx\wedge dy = \int_{[0,\,2\pi]\times[0,\,1]} (F_3\circ\Phi)(r,0,z)\frac{\partial(r\cdot\cos 0,\,r\cdot\sin 0)}{\partial(r,z)}\, dr\, dz = 0;$$

$$\int_{\Phi\circ\Gamma_{22\pi}} F_3\, dx\wedge dy = \int_{[0,\,2\pi]\times[0,\,1]} (F_3\circ\Phi)(r,2\pi,z)\frac{\partial(r\cdot\cos 2\pi,\,r\cdot\sin 2\pi)}{\partial(r,z)}\, dr\, dz = 0;$$

$$\int_{\Phi\circ\Gamma_{30}} F_3\, dx\wedge dy = \int_{[0,\,1]\times[0,\,2\pi]} (F_3\circ\Phi)(r,\theta,0)\frac{\partial(r\cdot\cos\theta,\,r\cdot\sin\theta)}{\partial(r,\theta)}\, dr\, d\theta = 0,$$

because $(F_3\circ\Phi)(r,\theta,0) = 0$;

$$\int_{\Phi\circ\Gamma_{31}} F_3\, dx\wedge dy = \int_{[0,\,1]\times[0,\,2\pi]} (F_3\circ\Phi)(r,\theta,1)\frac{\partial(r\cdot\cos\theta,\,r\cdot\sin\theta)}{\partial(r,\theta)}\, dr\, d\theta$$

$$= \int_{[0,\,1]\times[0,\,2\pi]} (r^2\cos^2\theta)r\, dr\, d\theta = \tfrac{1}{4}\pi.$$

Therefore, $\int_{\partial\Phi} F_3\, dx\wedge dy = \tfrac{1}{4}\pi$. It follows that the required value of the integral is $0 + 0 + \tfrac{1}{4}\pi = \tfrac{1}{4}\pi$.

8-7.P2. Using Green's theorem,

$$\int_{\sum_{i=1}^{2}(-1)^i(\Gamma_{i0}-\Gamma_{i1})} (5-xy-y^2)dx+(2xy-x^2)dy$$

$$= \int_{[0,1]^2} \left(\frac{\partial Q}{\partial x}-\frac{\partial P}{\partial y}\right)dx \wedge dy$$

$$= \int_{[0,1]^2} ((2y-2x)-(-x-2y))dx \wedge dy$$

$$= \int_{[0,1]^2} (4y-x)dx \wedge dy$$

$$= \int_{[0,1]} \int_{[0,1]} (4y-x)dxdy = 2-\frac{1}{2}=\frac{3}{2}.$$

$$y\to(0,y) \qquad y\to(1,y) \qquad x\to(x,0) \qquad x\to(x,1)$$

$$\int_{\sum_{i=1}^{2}(-1)^i(\Gamma_{i0}-\Gamma_{i1})} (5-xy-y^2)dx = -\int_{\Gamma_{10}} (5-xy-y^2)dx + \int_{\Gamma_{11}} (5-xy-y^2)dx$$

$$+\int_{\Gamma_{20}} (5-xy-y^2)dx - \int_{\Gamma_{21}} (5-xy-y^2)dx$$

$$= \int_{[0,1]} 5\,dx - \int_{[0,1]} (4-x)dx = 5-4+\frac{1}{2}=\frac{3}{2}.$$

$$\int_{\sum_{i=1}^{2}(-1)^i(\Gamma_{i0}-\Gamma_{i1})} (2xy-x^2)dy = -\int_{\Gamma_{10}} (2xy-x^2)dy + \int_{\Gamma_{11}} (2xy-x^2)dy$$

$$+\int_{\Gamma_{20}} (2xy-x^2)dy - \int_{\Gamma_{21}} (2xy-x^2)dy$$

$$= \int_{[0,1]} (2y-1)\,dy = 1-1 = 0.$$

The answer is verified too.

8-7.P3. $F_1 = y \qquad \dfrac{\partial F_1}{\partial x}=0 \qquad \dfrac{\partial F_1}{\partial y}=1 \qquad \dfrac{\partial F_1}{\partial z}=0$

$F_2 = z \qquad \dfrac{\partial F_2}{\partial x}=0 \qquad \dfrac{\partial F_2}{\partial y}=0 \qquad \dfrac{\partial F_2}{\partial z}=1$

$F_3 = x \qquad \dfrac{\partial F_3}{\partial x}=1 \qquad \dfrac{\partial F_3}{\partial y}=0 \qquad \dfrac{\partial F_3}{\partial z}=0$

By Stokes theorem the integral (1) equals

$$-\int_\Phi dy \wedge dz + dz \wedge dx + dx \wedge dy. \tag{2}$$

Observe that

$$\frac{\partial(y,z)}{\partial(r,\theta)} = \begin{vmatrix} \dfrac{\partial y}{\partial r} & \dfrac{\partial y}{\partial \theta} \\ \dfrac{\partial z}{\partial r} & \dfrac{\partial z}{\partial \theta} \end{vmatrix} = \begin{vmatrix} \sin\theta & r\cos\theta \\ \dfrac{r\sin 2\theta}{b} & \dfrac{r^2\cos 2\theta}{b} \end{vmatrix} = \frac{r^2}{b}(\sin\theta\cos 2\theta - \cos\theta\sin 2\theta)$$

$$= -\frac{r^2}{b}\sin\theta,$$

$$\frac{\partial(z,x)}{\partial(r,\theta)} = -\frac{r^2}{b}\cos\theta,$$

$$\frac{\partial(x,y)}{\partial(r,\theta)} = r.$$

The last integral (2) equals

$$\int_{[0,a]}\int_{[0,2\pi]}\left[\frac{r^2}{b}\sin\theta\,drd\theta + \frac{r^2}{b}\cos\theta\,drd\theta - rdrd\theta\right] = -\pi a^2$$

$$r \to (r,0) \quad r \to (r,2\pi)$$
$$\theta \to (0,\theta) \quad \theta \to (a,\theta)$$

$$\int_{\partial\Phi} y\,dx = \sum_{i=1}^{2}(-1)^i \int_{\Phi\circ\Gamma_{i0}-\Phi\circ\Gamma_{i1}} y\,dx = -\int_{\Phi\circ\Gamma_{10}} y\,dx + \int_{\Phi\circ\Gamma_{1a}} y\,d$$

$$+ \int_{\Phi\circ\Gamma_{20}} y\,dx - \int_{\Phi\circ\Gamma_{22\pi}} y\,dx$$

$$= \int_{\Phi\circ\Gamma_{1a}} y\,dx$$

$$= \int_{[0,2\pi]}(a\sin\theta)(-a\sin\theta)\,d\theta$$

$$= \int_{[0,2\pi]} -\frac{a^2(1-\cos 2\theta)}{2}\,d\theta = -\pi a^2\,;$$

$$\int_{\partial\Phi} z\,dy = -\int_{\Phi\circ\Gamma_{10}} z\,dy + \int_{\Phi\circ\Gamma_{1a}} z\,dy + \int_{\Phi\circ\Gamma_{20}} z\,dy - \int_{\Phi\circ\Gamma_{22\pi}} z\,dy$$

$$= \int_{\Phi\circ\Gamma_{1a}} z\,dy = \int_{[0,2\pi]} \frac{a^2\sin 2\theta}{2b}a\cos\theta\,d\theta$$

$$= \frac{a^2}{4b}\int_{[0,2\pi]}\left[\sin(2\theta+\theta)+\sin(2\theta-\theta)\right]d\theta = 0\,.$$

Similarly

$$\int_{\partial\Phi} x\,dz = 0.$$

The answer is thus verified.

References

1. Apostol, T. *Mathematical Analysis*. Narosa Publishing House, New Delhi, 1974.

2. Artin, M. *Algebra*. Prentice-Hall, Englewood Cliffs, NJ, 1994.

3. Berberian, SK. *A First Course in Real Analysis*. Springer Verlag, New York, 1974.

4. Brown, AL, Page, A. *Elements of Functional Analysis*. Van Nostrand Reinhold, London, 1970.

5. Burkill, JC, Burkill, H. *A Second Course in Mathematical Analysis*. Cambridge University Press, Cambridge, 1970.

6. Cheng, S. Internet article: Tests for local extrema in Lagrange multiplier method.
http://planetmath.org/?op=getobj&from=objects&id=7336

7. Choudhary, B, Nanda, S. *Functional Analysis with Applications*. Wiley Eastern, New Delhi, 1989.

8. Corwin, LJ, Szczarba, RH, *Multivariable Calculus*. Marcel Dekker Inc, New York and Basel, 1982.

9. Drager, LD, Foote, RL. The contraction mapping lemma and the inverse function theorem in advanced calculus. *American Mathematical Monthly*, 93(1):53–54, 1986.

10. Edwards, CH, Jr. *Advanced Calculus of Several Variables*. Academic Press, New York, 1973.

11. Gilsdorf, M. Internet article: A comparison of rational and classical trigonometry.
http://web.maths.unsw.edu.au/%7Enorman/papers/TrigComparison.pdf

12. Gopalkrishnan, NS. *University Algebra*, 2nd ed. Wiley Eastern, New Delhi, 1986.

13. Graves, LM. *The Theory of Functions of Real Variables*. McGraw-Hill Book Company, New York, 1956.

14. Hoffman, K, Kunze, R. *Linear Algebra*, 2nd ed. Prentice-Hall, Englewood Cliifs, NJ, 1971.

15. Kantorovich, LV, Akilov, GP. *Functional Analysis in Normed Spaces*, Pergamon Press, Macmillan, New York, 1964.

16. Kumaresan S. Implicit function theorem. *Bona Mathematica*, 9(2–3):52-78, 1998.

S. Shirali, H.L. Vasudeva, *Multivariable Analysis*,
DOI 10.1007/978-0-85729-192-9, © Springer-Verlag London Limited 2011

17. Lang, S. *Real Analysis*. Addison-Wesley Publishing Company, Reading, MA, 1983.

18. Lang, S. *Undergraduate Analysis*, 2nd ed. Springer Verlag, New York, 1997.

19. Loomis, LH, Sternberg, S. *Advanced Calculus*. Addison-Wesley, Reading, MA, 1968.

20. Protter, MH, Morrey, CB. *A First Course in Real Analysis*, 2nd ed. Springer Verlag, New York, 1991.

21. Pugh, CC. *Real Mathematical Analysis*. Springer Verlag, New York, 2002.

22. Rudin, W. *Principles of Mathematical Analysis*, 3rd ed. McGraw-Hill International Book Company, New York, 1976.

23. Shirali, S, Vasudeva, HL. *Mathematical Analysis*. Narosa Publishing House, New Delhi, 2006.

24. Singh, S. *Linear Algebra*. Vikas Publishing House Pvt Ltd, 1997.

25. Sohrab, HH. *Basic Real Analysis*. Birkhäuser, Boston, 2005.

26. Spivak, M. *Calculus on Manifolds*. W.A. Benjamin, New York, 1965.

27. Thomas, GB, Finney, RL. *Calculus and Analytic Geometry*, 9th ed. Addison-Wesley, Reading, MA, 1996.

28. Wildberger, NJ. *Divine Proportions: Rational Trigonometry to Universal Geometry*. Wild Egg Books, Sydney, 2005.

Index